Library of
Davidson College

STARS, MINDS AND FATE

STARS, MINDS AND FATE

ESSAYS IN ANCIENT AND MEDIEVAL COSMOLOGY

J.D. NORTH

THE HAMBLEDON PRESS
LONDON AND RONCEVERTE

Published by The Hambledon Press, 1989

102 Gloucester Avenue, London NW1 8HX (U.K.)

309 Greenbrier Avenue, Ronceverte, WV 24970 (U.S.A.)

ISBN 0 907628 94 X

© John D. North 1989

British Library Cataloguing in Publication Data

North, J.D.
   Stars, minds and fate: essays in ancient and
   medieval cosmology.
   1. Cosmology – History
  I. Title
   113'.09   BD494

Library of Congress Cataloging-in-Publication Data

North, John David.
   Stars, minds, and fate: essays in ancient and medieval
   cosmology/J.D. North.
   Includes index.
   1. Cosmology. 2. Astronomy, Ancient.
   3. Astronomy, Medieval. 4. Astrology.
   5. Science – History. I. Title.
   QB981.N773   1989   523.1 – dc19
   88-35243 CIP

Printed and bound by W.B.C. Ltd., Bristol and Maesteg

CONTENTS

Acknowledgements		vii
Preface		ix
List of Illustrations		xi
1	The Attractions of Past Science	1
2	Moon and Megaliths	11
3	By Direction from Above	21
4	Neolithic Newtons	29
5	Venus, By Jupiter!	37
6	On the Trail of the Comet	45
7	The Culmination of Ptolemy	53
8	Astrology and the Fortunes of Churches	59
9	Chronology and the Age of the World	91
10	Between Experience and Experiment	119
11	*Opus quarundam rotarum mirabilium*	135
12	Monasticism and the First Mechanical Clocks	171
13	Hierarchy, Creation, and *Il Veltro*: Three Footnotes to Dante's *Inferno*	187
14	The Astrolabe	211
15	Astrolabes and the Hour-Line Ritual	221
16	*Summa ratione confectum*: An Astrolabe Drawn by Computer *(with Ole Østerby and Kurt Møller Pedersen)*	223
17	Eternity and Infinity in Late Medieval Thought	233
18	Celestial Influence — the Major Premiss of Astrology	243
19	Intimations of Cosmic Unity? Fourteenth-Century Views on Celestial and Sub-Lunar Motion	301

20	Kinematics – More Ethereal than Elementary	313
21	The Alfonsine Tables in England	327
22	1348 and All That: Oxford Science and the Black Death	361
23	Nicolaus Kratzer – The King's Astronomer	373
24	The Medieval Background to Copernicus	401
Index		415

ACKNOWLEDGEMENTS

The articles in this volume first appeared in the following places. They are reprinted by the kind permission of the original publishers.

1 *Times Literary Supplement* (1972), pp. 1421-2.
2 *Times Literary Supplement* (1971), pp. 633-5.
3 *Times Literary Supplement* (1980), pp. 563-4.
4 *Times Literary Supplement* (1975), pp. 921-2.
5 *Times Literary Supplement* (1976), pp. 770-1.
6 *Times Literary Supplement* (1982), pp. 1407-8.
7 *Times Literary Supplement* (1977), pp. 467-8.
8 *Centaurus*, 24 (1980), pp. 181-211.
9 *Cosmology, History and Theology*, ed. by W. Yourgrau and A.D. Breck (New York, Plenum Press, 1977), pp. 307-33.
10 *Times Literary Supplement* (1983), pp. 1163-5.
11 *Physis*, 8 (1966), pp. 337-72.
12 *The Study of Time*, ed. by J.T. Fraser and N. Lawrence (New York, Springer Verlag, 1975), pp. 381-93.
13 *Annali dell' Istituto e Museo di Storia della Scienza di Firenze*, 7 (1981), pp. 5-28.
14 *Scientific American* (January, 1974), pp. 96-106.
15 *Journal for the History of Arabic Science*, 5 (1981), pp. 113-14.
16 *Archives Internationales d'Histoire des Sciences*, 25 (1975), pp. 73-81.
17 *Infinity in Science*, ed. G. Toraldo di Francia (Rome, Istituto della Enciclopedia Italiana, 1987), pp. 245-56.
18 *Astrologi hallucinati*, ed. by P. Zambelli (Berlin, New York, Walter de Gruyter, 1986), pp. 45-100.

19 *Cahiers du Séminaire d'Histoire des Sciences de l'Université de Nice* (Journées Oresmes, Juin 1983), ed. by P. Souffrin, p. 30-42.

20 *Machaut's World. Science and Art in the Fourteenth Century*, ed. by M.P. Cosman and B. Chandler (New York, Academy of Sciences, 1978), pp. 89-102.

21 *Prismata. Naturwissenschaftsgeschichtliche Studien.* Festschrift für Willy Hartner, ed. by Y. Maeyama and W. Saltzer (Wiesbaden, Franz Steiner Verlag, 1977), pp. 269-301.

22 *From Ancient Omens to Statistical Mechanics: Essays on the Exact Sciences Presented to Asger Aaboe*, ed. by J.L. Berggren and B.R. Goldstein (Copenhagen, University Library, 1987), pp. 155-65. *Acta Historica Scientiarum Naturalium et Medicinalium*, 39.

23 *Science and History: Studies in Honor of Edward Rosen* (Ossolineum, Orbis, 1978), pp. 205-34. *Studia Copernica* (1978).

24 *Copernicus Yesterday and Today*, ed. by A. Beer and K.A. Strand (Oxford, Pergamon Press, 1975), pp. 3-16. *Vistas in Astronomy*, 17.

PREFACE

The historical essays collected together in these pages are all in some way related to a lasting human preoccupation with cosmological matters.[1] Far from having been written for each other's company, they were published over a period of twenty years. If this does not make them radically inconsistent, the span is long enough to have exposed changes in my own attitudes to a university discipline which, as Marvell might have said, fate so enviously debars. A third of my chapters first saw the light of day as reviews in the *Times Literary Supplement*: they were usually written in haste, and were probably the better for it. It must be said that their titles were not of my invention, but they are kept, for reasons that should be obvious. While I cannot pretend to the rank of journalist, I take some pleasure in the thought that, judged by irate correspondence alone, those pieces were exceptions to the rule that the average readership of a scholarly article is 1.4 (and presumably falling).

The contents of the book must speak for themselves. They have been changed in only the most trifling respects, although comments have been appended to several of them, and some of the original illustrations have been replaced by more suitable ones. The early chapters tend to be of a more general kind than the later, and for the most part offer a survey of current scholarship rather than any original contribution to it They might often seem to be putting across ideas that are commonplace, but I do not think that this was so when they first appeared. It is not every reviewer who has had his work wrapped round the Sunday edition of a weighty and well known American newspaper and hurled at its editor. It is hard now to credit the hostility that greeted Alexander Thom's writings on what – for better or worse – became known as archaeo-astronomy. It is even harder to believe that one remembers correctly the strong reluctance shown even by many professional scientists to reject the pronouncements of Immanuel Velikovsky. These facts should be borne in mind when the relevant chapters are read. There are of course still those who scorn Thom's work in private, but they are almost invariably wrong, and I shall shortly publish evidence of a new sort, supplementing Thom's findings in a way that will make this more obvious than ever.

---

[1] A second collection, covering a later historical period, has been published as *The Universal Frame*, also by the Hambledon Press (1989).

Modern responses to such ideas as Thom's reflect a cultural divide between the humanities and the natural sciences that would have surprised those medieval scholars whose thoughts I discuss in the main part of the book. The eclecticism of the typical medieval scholar might now seem astonishing, regrettable, amusing, or derisory, according to one's view of how rigid intellectual barriers should be. We shall seriously misunderstand ancient and medieval thought, however, if we are not prepared to share a willingness to look across such frontiers as those dividing astrology from ecclesiastical history, biblical chronology from astronomy, and angelic hierarchies from the planetary spheres, theology from the theory of the continuum, celestial laws from terrestrial, or the work of the clockmaker from the work of God himself, namely the universe. These are among the themes of the middle chapters of the collection. Historians of medieval natural philosophy and science are themselves not without fault in this matter of boundaries. So successful have recent studies of medieval kinematics and the logic of the continuum been – in particular those relating to Merton College and the Parisians in the fourteenth century – that students may be forgiven for thinking that there was nothing else scientific of importance. I include two chapters (19 and 20) as a reminder of related astronomical subtleties.

Four or five of the essays in this book are concerned with rather more recondite astronomical material, but their contents are likely to be only marginally less accessible to the ordinary reader than the contents of the short general survey with which the book ends. This survey was written for the quincentenary of Copernicus' birth, and to that extent might seem to betray my view of the middle ages as a period of intellectual history with a vitality of its own, and not merely as a repository of an ancient wisdom that was awaiting a revolutionary reinterpretation. Not that the freshness of the Copernican outlook can be denied: it is, for instance, in marked contrast to the essentially medieval values of the craftsman and courtier, Nicolaus Kratzer, the subject of the penultimate chapter. There is an obvious bias in the subjects treated in the book. The Oxford perspective from which most of it was written leaves one, rightly or wrongly, with the feeling that the momentum of fourteenth-century science was largely spent by Kratzer's time. Some of the reasons will be found in chapter 22, on the effects of the Black Death. As in my *Richard of Wallingford*, so in chapter 21, evidence is presented showing how certain Oxonian ideas gained currency on the continent of Europe, only to be forgotten in their place of origin. Oxford wanted from Kratzer only the latest thing in sundials and had all but forgotten the likes of John Killingworth. It cannot be said that the value presently placed there on its fourteenth-century past has altered much, and I should be content if this collection were to promote, by however so little, the historical priorities of such men as Digby, Ashmole, Wood, and Tanner.

LIST OF ILLUSTRATIONS

Between Pages 213 and 214

1 Persian astrolabe, signed Muhammad Muqim, and dated A.H. 1067 (= A.D. 1647/8)

2 The back of the astrolabe in Plate 1

3 English astrolabe of the late 14th or early 15th century

4 The back of the astrolabe in Plate 3

Page 325

5 Memorial brass to John Killingworth from the transept of Merton College, Oxford

Between Pages 375 and 376

6 Autograph letter of Nicolaus Kratzer

1

THE ATTRACTIONS OF PAST SCIENCE

> Even *Mathematics, Natural Philosophy, and Natural Religion*, are in some measure dependent on the science of MAN; since they lie under the cognizance of men, and are judged of it by their powers and faculties.

Hume was writing in an age of cut and dried philosophy, but he knew when to throw in a cautious "in some measure". Scientific ideas do not occur *ex nihilo* or *ex nemine*, whether or not they are more interesting than the men who have them. "The autobiography of a man whose business is thinking," began the late R. G. Collingwood's autobiography, "should be the story of his thought." But how much detachment did he think possible? Write about what any man once did, and you are likely to be saying what he had a mind to do. You are on the way to becoming a historian of ideas, and adopting a style without realizing it, like Molière's Monsieur Jourdain. Whether or not you will be accepted as such seems to depend on how big the ideas are, and on certain other considerations which historians of ideas seem not prepared to articulate. Few historians take territorial claims very seriously, but since such claims are made from time to time – "the history of science is a segment of the history of ideas"; for example, or "the history of science is foreign to history proper" – I may as well begin by saying why I dislike these claims and these titles.

I dislike the titles "historian of science" and "historian of ideas", not because they are indefinable, but because they are constrictive. After having been applied for some time in a professional context, they harden, until the less imaginative member of the profession believes that his title tells him what he should and should not discuss, if he is to be true to his clique. Academic barriers, like national frontiers, have their uses, but it is a moot point whether these are compensation enough for the shelter they offer to mavericks, or for the obstacles they put in the way of so many laudable pursuits. The sciences are, and always have been, linked with other intellectual and cultural pursuits, the arts, religion, and philosophy, so that it is often impossible to say where one ended and the other began. Even between the natural sciences there are frontiers as fickle as the national sort, and if a modern historian of chemistry or physics, for instance, refuses to look beyond their present place on his map, he fails as a historian.

Some historical subjects lend themselves to a conceptual treatment more readily than others. Competent histories of specific philosophical ideas have been written from at least the early years of the eighteenth century (J. J. Bruckner, Charles Batteux, and so on), and by expecting less of a history one could push the date back 2,000 years. Although good histories of mathematics have never been numerous, there are here, as in philosophy, criteria of proof and acceptability which make adulation of the trivial more apparent, if not more difficult, than in less formal subjects. In a sense, the problem is there reversed; it is more congenial to become an expert in one system than a dabbler in many. Likewise in the empirical sciences, the more highly formalized a science is, the more intellectually satisfying it tends to be in isolation, with the result that a type of historian appears who seems to think that his only mission is to exhibit his hero's work, emend it where necessary, and commend it where commendation is due. He is a sort of chairman of examiners and a glorified proof-reader combined. Occasionally he will stick at editing texts(". . . no less than twenty emendations upon the German text of Schnupfenius and Schnapsius"). Because the work is so onerous this creates the illusion that it is a historical end in itself. For what the observation is worth, I note that those who call themselves historians of ideas do not, as a rule, tangle with the editing of texts. In view of the parlous financial state of our university presses, they are very wise.

The intrinsic attractions of past science are a significant snare in more ways than one, for they place the historian in one of Hegel's many dilemmas. (How, Hegel asked himself, could he, as a philosopher, participate in the bloody events of the Napoleonic wars, which were the only meaningful events of his world?) Should he too be disturbed at the thought of his non-combatant role? Is he nothing but a lapsed scientist? The thought might drive him deeper into mere technicalities, or it might lead him to look for increasingly abstruse historical determinants. The divergence leads to much dissension. It is easy enough to see how a history of the Darwins, Malthus, Haeckel, and Huxley for example, may require, and at times demand, a social, economic, and political treatment, and it is only marginally more difficult to concede that there could be something in Boris Hessen's approach to Newton's *Principia*; but there are in the history of mathematical and scientific ideas places where the intellectual structure is so rich, the social element so tenuous, so illusory, or so irrelevant, and the techniques for its extraction so crude, that only a bore or a fanatic would insist on applying them. It is not enough to argue vaguely that mathematics has repercussions in application, for the essence of pure mathematics is that it has none as a defining constituent.

Again, it is all very well to say that human thought is neither called forth, nor changed, without external stimulus, but psycho-history is not yet an exact science, and I imagine that few historians would like it if it were. Note Collingwood's reference to the "fashionable scientific fraud of the age", and more recently the mild hostility with which Frank Manuel's psychological studies of Newton have been received by Newtonians.

But perhaps the most pathetic gesture of all towards sociology is when the social milieu is equated with a few influential clubs and scientific societies. Even the British Association represents a pretty small section of the thinking and unthinking world. It is as though one were to write on the relations of parliament and the people, and mention only the Athenaeum.

Like Voltaire's Emilie, who had a penchant for translating Newton but an aversion from history ("which overwhelms the mind without illuminating it"), you probably find yourself without a taste for both subjects. On the other hand, the tastes of historians themselves are many. At one extreme there is the Cartesian dream, that everything within the province of human cognition is connected with every other thing, by chains of reasons of the sort to be found in geometry; and that philosophical rationalism should be somehow extended to historical events. There is the Humean method of toning down this extreme view: reason – in Hume's language – is and ought only to be the slave of the passions. And then, continuing to the other end of the spectrum, we find at last the dull, plodding chronicler of the particular, who proves with anecdotal detail – by Dickensian example, as we might say, rather than Kingsleyan precept. The more attention he gives to detail, the more readily can he refute other men's generalizations.

Now it is often supposed that the historian who meddles with the sciences simply must be of the Cartesian disposition, but if we except the sub-species with sociological tendencies, and a few others, this is untrue. Historians who write of science are no sinister phalanx with a characteristically scientific weaponry, but a group with widely divergent interests and facing in many directions. They are drawn from many different backgrounds – far more than is the case with the "historians of ideas" group – and they lack the *esprit de corps* of the generally well-disciplined members of our university schools of modern history, in which conformity may be largely illusory, but where there is at least a common denominator of training. There is, admittedly, a small historicist fringe, for the most part made up of scientists who have little to do with history, but who believe in the possibility of prediction from a logic of historical events. They write books about life in the year 2000, and such things, and, so far as I know, they are not taken very seriously in academic quarters, where even historism, the notion that the history of a thing is an explanation of it, is often regarded as dangerous and deviationist.

The act of generalization is, of course, an essential part of writing history, and those who try to write history entirely in the form of singular statements, perhaps for safety's sake, are generally singularly dull. The historian lives in a permanent state of apprehension lest historical evidence be selected simply to ensure the truth of some preconceived generalization. We all know the tranquillizer to dispel the fear. Settle for dull certainty, exclude nothing, leave the dangerous work to others, and steer clear of "ideas". The man who trafficks with the history-of-ideas crowd is sure to be tempted to categorize and classify. And that, too, is the temptation of a scientific subject-matter, which shows least historical complexity when human beings are least in evidence. Temptations are a matter of taste, but these are temptations into which all historians with taste will pray to be led.

This much having been said, I am not sure that I can discern any clear message from "historians of ideas" which could be related to the specific needs of scientific history. When Arthur Lovejoy wrote the introduction to the first number of his *Journal for the History of Ideas*, in 1940, it was evidently his chief aim to promote collaboration between scholars in different provinces of intellectual history. He mentioned so many different categories of what I suppose he meant by idea, that it might have been easier to limit his interests by saying what they excluded. Scientific discoveries and theories, philosophical ideas in literature and the arts, preconceptions, categories, metaphors, pervasive and widely ramifying doctrines, and even classical and modern thought as a whole, all fell into his scheme. It is difficult to escape the feeling that Lovejoy was brought up by Hegel, whose writings are permeated with a symbolism of spirit, thought, conception, and idea – and indeed, of an Idea so powerful that, through its realization in history, God rules the world.

For Lovejoy's followers, "idea" can mean anything from a narrowly defined concept to some indefinable "spirit of the age". It can be a passing mood, or one of a thousand "-isms", a tentative thought, or a generalization of Toynbeean or Kuhnian proportions. It can relate to absolutely anything. Ponder for a moment too long on whether a toothpick is a scientific instrument, accidentally come across a decorative specimen, preferably signed, and before you know what is happening you are writing a cultural history of the toothpick. Discover a chance medieval reference to the possibility of the Earth's movement, and it is only a matter of time before, by applying purely mechanical techniques known to most users of libraries, you have written a massive compilation purporting to show beyond all doubt the existence of a significant pre-Copernican preoccupation.

Of course these things can be done well, and of course the result need not be trivial. Lovejoy's *The Great Chain of Being*, for example, was well written and far from trivial, and he did history a service in drawing together scholars from adjacent disciplines on the evolution front, with excellent results to be seen, for instance, in the volume *Forerunners of Darwin* (ed Bentley Glass, Oswei Temkin, and W. L. Straus). But there is evidence enough elsewhere that the abstractive approach to history can misfire badly: it is often undisciplined, it can be obsessive (as collecting stamps depicting railway trains might be), and above all it lends itself to distortion, whether by introducing anachronistic conceptions or merely by inflating the importance of the trivial.

Jackdaws of the past
As against this approach, there is that of a number of men who were active especially between the two world wars and who resembled somewhat the great seventeenth-century English antiquaries. They noted everything down, as long as it was old, and preferably quaint. In some cases their virtues included a willingness to collect artefacts (to which historians of ideas seem generally oblivious), so reminding us that history is not necessarily founded wholly on literary sources. But unlike so many philosophers, who have been unburdened with the desire to know more than what the past *might* have been like, they dispensed with their critical judgment and acted like jackdaws. They forgot the words of the great philosopher Brillat-Savarin, to the effect that man lives not on what he eats, but on what he digests.

Brickbats enough have been aimed at George Sarton and Lynn Thorndike on this score, some of them well merited, although I have no wish to add to them, for both have saved many of us from untold drudgery. Their eclectic approach showed that invaluable insights into the history of past civilizations are to be had from a study of astrology, alchemy, Hermeticism, phrenology, witchcraft, and most of what is conventionally passed off as scientifically disreputable and worthless. *Ecrasez l'infâme!* (Actually, Sarton had some curious prejudices, and his dismissal of Galen's physiological writings as inferior is notorious.) Such systems of thought come into being to serve the needs of their creators, and were perpetuated by others of like mind. Deliberately to leave them out of the historical reckoning when writing of men who believed in them is to be more than selective; it is to be anti-historical. This does not mean that it is mistaken to be censorious, in the manner of Gibbon, for example. If you reproach past science, however, you do so as a scientist, and not as an historian.

The line of division is tenuous, and many have crossed it inadvertently. It seems that for many people history is pointless without some living purpose. We might mention Aristotle, whom many historians of ideas have no doubt put into their family tree, but who wished to reconstruct the past merely with a view to learning from it. To take a few other examples from the past: the great seventeenth-century Jesuit historian of astronomy, J.-B. Riccioli, was looking for the true system of the world, although he never seemed to be in much of a hurry to find it. His contemporary, Ralph Cudworth, used scientific, philosophical, and theological history to refute the materialism of Hobbes. Such would-be-scientific historians as Bolingbroke, Hume, and the French writers of *l'histoire raisonnée*, seemed able to turn descriptive historical propositions, by devious sleight-of-hand, into propositions with an evaluative or prescriptive (and usually ethical) form, and there are today several historians of the sciences who seem to be bent on committing the same fallacy.

Then there is the Machian fallacy, with which Collingwood seems to have concurred, that to understand a science it is first necessary to become acquainted with its history. Even Hegel could see that, however much philosophers used the history of philosophy for their mental development, it was none the less external to philosophy. Mach's style of "historical" presentation is harmless only if the historical reader refuses to be persuaded to make the past look so much like the present and only if the scientist refuses to allow outmoded and inconsistent doctrines to obscure what can be reasonably supposed relevant. As for the precept that one should put oneself, as far as possible, in the minds of the men of whom one is writing, this surely applies to every historian, and not only historians of ideas. It does not entail any one way of *writing* history, and certainy does not entail the "I was there" style, which has its own dangers; for the Galens, the Galileos, and the Gausses of the past faced scientific problems, whereas our problems are textual, palaeographic, linguistic, and historiographic. That much abused species, the professional scientist turned historian, is perhaps most abused when he fails to see the difference.

And yet even more problematical is the philosopher. The liaison between the philosopher and the historian of science has traditionally been strong: natural philosophy, after all, was once the staple of philosophy as a whole. It is not unknown in England for posts ostensibly in the "history and philosophy of science" to be filled without a single philosopher on the appointing board. Such attempts to break with philosophy are objectionable for the simple reason that there are many perfectly legitimate questions which the historian may ask which require a partly philosophical treatment. He may ask about the uses, the status, and the criteria for acceptance, of scientific laws, principles, and procedures in some historical situation. Leaving aside the debate whether a philosopher writing on scientific methodology need ever go outside make-believe examples, there is no doubt that many who call themselves philosophers do pose problems in connection with current science corresponding to the historical problems I mentioned. The solutions will be different in many ways, but the methods of reaching them will have much in common, and it will be a sad day when the existing barrier between history and philosophy, which many purblind and irresponsible academics are erecting as fast as their impaired vision permits, finally becomes impassable.

It is my impression that when historians who now write about science tie themselves to a philosophical school, it is usually unwittingly, which was certainly less true in the time of Ernst Mach and Pierre Duhem. It is difficult to assess what effect such a tie has on style. How much of Mach's positivism is revealed in his stolid "new text books for old" style is hard to say. It is often suggested that positivism, as Mommsen's guiding principle of historiography, fostered an excellence in the control of minutiae which helped to make him the foremost classical historian of his time, even though he was exhausted by the minutiae before the time had come for the grand synthesis he had in mind. Duhem, too, wrote at great length (especially on physical and cosmological thought from antiquity to the Renaissance), but his bias was too marked, and his syntheses were made without enough attention to the details, so that his works fill the reader with excitement of the sort to be had from standing inside an elegant building on the point of collapse. (And having mentioned Mommsen and Duhem together, it is hard to avoid recalling a touch of irrationality in both: Mommsen loathed the Gauls, and Duhem looked down on everyone else.)

Expertise in handling primary materials is obviously important to all historians, whether of ideas or of anything else, but it cannot alone ensure the requisite sympathy with the period. What can happen when a writer lacks both qualities has been revealed by Abraham Sachs, to take a single example, in connection with Velikovsky's imaginative and highly popular excursions into cuneiform texts. ("Wrong century, wrong country, wrong king, wrong sea.") When only the expertise is there, the distortion is harder to prove. It seems to me that the "philosophical" historian tends to be the man most lacking in sympathy with those in whose disembodied ideas he deals. This is often because he speaks of ideas in the same way as idealist philosophers spoke of Reason, namely as something with an existence and a history independent of the individual thinker. There were several notable writers in this vein between the two world wars, of whom I may name A. N. Whitehead, E. A. Burtt, Hélène Metzger, Emile Meyerson, Alexandre Koyré, and R. G. Collingwood.

Inasmuch as historians of ideas tend to jump from century to century, they are likewise liable to take a similarly unsympathetic view of the past, but they are unlikely to match the philosophical historian's determination to cram the past into neat categories. Collingwood's histories are blatant examples of this. The categories are very much those which he had successively learnt and taught in the Oxford Lit. Hum. School. Whitehead went one better, and offered a few of his own. All, however, shared one decided virtue: they actively used their own powers of reasoning in ways apparently unknown to so many of those who engage only in making rebarbative compilations of "positive" scientific knowledge.

The ideologies of idea historians
Philosophical historians are not the only propagandists. In the past you could write history to the glory of God, and no one questioned your motives, for your readers were on your side. It is not easy to detect any common creed, philosophical, sociological, or otherwise, within the history of ideas. Perhaps, as in the history of science, we should be looking for less idealistic springs – a man, a city, a century, or the subject in which the writer, proud of his intellectual pedigree, was himself nurtured – to find the true object of modern hosannas. Or perhaps the vision is too grand and too impressionistic for it to be easily seen, as in Burckhardt's *Civilisation of the Renaissance in Italy*, in which the Quattrocento meant not only a break with the Middle Ages, but nothing less than the birth of modern Europe. He almost entirely overlooked the nascent science, but no matter. The Enlightenment idea of progress, which spread from an eighteenth-century search for a Newtonian system covering all human activity, clearly infected Burckhardt, and well illustrates how a historian may be guided by ideological currents of the very sort he is studying. The idea of progress, incidentally, is notorious among historians of science as an infallible stimulus for feelings of guilt. The object is to have regular witch-hunts, looking for "Whig historians". Perhaps it is all a Tory plot, for under certain circumstances it can be of exactly the same character as the plot which so offends. "Newton is not the intellectual ancestor of your man, but of mine", and so on.

One of the largest of the ideologically coherent groups of historians of ideas, namely those with a Marxist philosophy, appears reluctant to lay claim to the title, no doubt because "the ideal is nothing but the material world reflected by the human mind and translated into forms of thought". (Nothing but!) Even so, there have been several notably balanced, sensitive, and consistent Marxist interpretations of science. Benjamin Farrington's *Greek Science* was one which reached an unusually wide public, while in a very different style Eugenio Garin's writings on humanism and science are among the most impressive products of contemporary Italian scholarship. The balance is somewhat upset by those who would like to base the whole of past human existence on a minimum of crude axioms, but Marxism is not the only ideology in which you are not free to choose the company you keep.

One Marxist historian, J. G. Crowther, has made an interesting comment on the history of science, which he sees as a self-conscious pillar of Establishment that has since 1940 given progressively less attention to the social relations of science, thus strengthening the traditional conservative theory of dominant interests. His sociological explanation could just possibly be a piece of biography in disguise: the social relations of science are said to be controversial, and therefore to have been ignored by historians of science seeking a long period of undisturbed study. We have so far had no Spengler to tell us that we cannot escape the destiny to which society commits us.

I am conscious of the folly of trying to relate a "history of ideas", which I cannot clearly distinguish from the rest of history, to the achievements of a group of historians whose work encompasses, for example, Akkadian cosmology and the morphology of Puritan intellectual society, Erasistratan physiology and the mathematics of Hua Lo-Keng, prehistoric megalithic monuments and the structure of DNA. If what has been written hitherto as the history of science is to be subordinated to some more extensive commonwealth, it must be to history as a whole, and neither to the "history of ideas" nor to the natural sciences. This is not to argue either against or for academic autonomy. If the only historians ever to take the sciences into their reckoning had been those within university departments of modern history, we should be in a sorry state of ignorance, but this is not to say that our separation is inevitable, or that it should continue except as an administrative convenience. If it has anything to do with an incompatibility, it is not of subject-matter, but of persons and unbelievably inflexible institutions. What we most need is not a revival of the historiography of Humboldt, Boeckh, and Droysen but a visit from the Hauptmann von Köpenick.

2

MOON AND MEGALITHS

'I AM FORCED to the conclusion that nothing of any great moment has been established by the astronomical *nouveau vague* [sic] flowing over Stonehenge.' Thus wrote Jacquetta Hawkes in 1967, commenting upon certain theories of Stonehenge which had been proposed, in *Antiquity*, *Nature*, and elsewhere, by G. S. Hawkins and Fred Hoyle. She had begun with the observation that every age has the Stonehenge it deserves—or desires. Having found that cries of delight were provoked from a general audience when she contrasted the wildly jagged stones seen by the Romantics with the smooth regularity perceived by the eye of the classicist, she interpreted the Hawkins-Hoyle stir merely as a sign of our scientific times. Needless to say, she was out of sympathy if not with the times, at least with astronomers who could not even reach 'close agreement ... on the inevitability of their interpretations'.

Putting aside the rather charming supposition that astronomers live in a harmonious world of objective and inevitable truth, it must be pointed out that Professors Hawkins and Hoyle were, directly or by implication, setting themselves up not as astronomers but as archaeologists. They were applying simple astronomical techniques, but only to offer certain interpretations—largely incompatible interpretations—of the use, if not necessarily the principal use, of the Stonehenge remains.

In fact neither showed himself to be historically very well informed. Much of the time Professor Hoyle managed to dodge the issue by purporting to answer not the archaeologist's problem but the simpler problem of how *we* should design a Stonehenge-type monument. How should we use it to predict eclipses, for instance? Within the outer bank at Stonehenge is a circle of fifty-six holes, dug and

---

A. THOM: *Megalithic Lunar Observatories*. 127 pp. Clarendon Press: Oxford University Press. £3.

ALEXANDER MARSHACK: *Notation dans les gravures du paléolithique supérieur: Nouvelles méthodes d'analyse*. Translated by Mme. J.-M. Le Tensorer and others. 123 pp. Bordeaux: Delmas.

refilled soon afterwards, and now named after John Aubrey, who drew attention to them in the seventeenth century. In articles, and in *Stonehenge Decoded*, a book first published in 1965, and now available in paperback (Fontana, 40p), Professor Hawkins had proposed a most implausible theory of the Aubrey holes. They were seen simply as an aid for use in a moderately complicated counting procedure for eclipse prediction, which depended for its validity on the fact that there are certain natural periodicities of eclipse recurrence.

The method proposed was unfortunately of very dubious astronomical value. In spite of its supposed 'remarkable accuracy', at one point Professor Hawkins casually applied one of his rules to our own century, deriving lunar eclipses for 1926 and 1908, both at best imaginary. The trouble with eclipses is that there are infinitely many eclipse cycles (based on a combination of the *saros* and J. N. Stockwell's twenty-nine-year cycle), although most of them are of very little value. At all events, an appropriate eclipse rule could be found to fit almost any large number of holes. Miss Hawkes's fears were understandable: the Aubrey holes had begun to look like one of those hateful digital computers. But then, as every child knows, we might say the same of our fingers without supposing that they have no other use. Have archaeologists not always tended to take labels too seriously? The Beaker People must be turning in their barrows at the thought that posterity remembers them thus.

Eclipses, especially of the Sun, but also of the Moon, are difficult to predict without a tolerably extensive knowledge of astronomical theory, unless it is somehow appreciated that after certain periods of time the eclipse cycle tends to recur. And 'tends' is the right word. Both Professors Hoyle and Hawkins decided that the Aubrey holes told their user that there was a *strong likelihood* of an eclipse at certain times. Now neolithic man might conceivably have formulated general rules concerning eclipses on the basis of eclipses he had actually observed, but he would have been hard pressed to make generalizations from essentially undetected 'danger periods'. How, then, did he proceed? Neither author explained the problem from this end. Professor Hoyle suggested that the circle of Aubrey holes was a representation of the ecliptic (the apparent circular path of the Sun as it moves once a year round the sky). He suggested that there would have been a marker for the position of the Sun on the ecliptic, and others for three highly conceptualized points connected with the Moon.

We know something of the intellectual struggle of the Greeks when they arrived at similar notions more than a thousand years later, but this is itself no argument one way or the other. Even so it is a piece of history worth bearing in mind when we consider Professor Hoyle's claim to see significance in the fact that he was led to an understanding of his own technique from a study of Stonehenge. This, he said, would have been 'strange indeed if Stonehenge had no astronomical connotation'. But the problem was to decide on a historically plausible connotation. The well-documented Babylonian methods of 600 B.C. had seemed to him 'no more than obscure numerology'. He was also constrained to add at the end of a paper describing his own solution: 'a most remarkable point still remains. The method could not work.' No wonder that professional archaeologists were sceptical.

Despite all this, it was really Professor Hawkins's show, a fact of which we are reminded by quotations from the *Daily Express*, the *Daily Mirror*, and *The Observer*, on the covers of the re-issued *Stonehenge Decoded*. The real secret of Stonehenge, as every news editor knows, is that the whole world loves a Druid. (The Druids of course had nothing to do with the building of Stonehenge.) Professor Hawkins produced his book out of a handful of articles, slender, but of some originality. In them he had shown restraint, but the book, hastily assembled with the collaboration of one 'neither astronomer nor archaeologist', threw sobriety to the winds. Setting aside the melodramatic and autobiographical style, and even overlooking its central astro-archaeological theses, the book was filled with misunderstandings, half-truths, and irrelevancies. Its artless but condescending tone was well adapted to a popular audience, but surely its author cannot have remained so untouched by the purely archaeological criticisms made at the time as to think that a re-issue is a substitute for a new edition. The tragedy is that he has something worth saying, and that by refusing to refine it he is failing to communicate with those whose judgment is worth having.

It has long been recognized that the 'avenue' at Stonehenge is directed towards midsummer sunrise, and seventy years ago Sir Norman Lockyer even tried to calculate from this the date of construction. (His method was at best probabilistic, since we do not know whether sunrise was taken as the first glint of the Sun, or the appearance of the full Sun, or some intermediate state.) More than a century ago Edward Duke had noted other stones aligning with the sunrise at the time of the solstice. Greatly moved by a dawn visit to the monument in 1961 Professor Hawkins and his assistants calculated the directions of 120 'significant' lines drawn between the positions of pairs of stones as marked on relatively small-scale charts. A quite irrelevant fact, of which archaeologists were so suspicious, and of which he was inordinately proud, was that the directions were calculated with the help of a computer.

It was found that some directions occurred more commonly than others, and that these coincided with the computed directions of sunrise and sunset at the times of winter and summer solstice, and also with certain extreme directions of the rising and setting Moon. Which calculation was done first is a matter on which the book and the original article do not agree. Between writing the two, he heard that C. A. Newham had found an alignment with sunrise at the equinoxes. Returning to his charts he found more significant alignments which he had previously overlooked, and in the book was able to add eight equinoctial directions to his earlier lists.

Although the archaeologists were irritated by the overall crudity of the book, when their annoyance had subsided they began to nibble away at its astronomical core. They pointed out that Professor Hawkins had mixed different building periods in his alignment points; that the plans he was using were not reliable, and that many of the stones had been disturbed; that the horizon selected was a crucial factor in deciding the direction of a rising or setting object in the sky, and yet that the bank around the monument might well have been high enough to provide the horizon in the west. It was even suggested that some of the points through which Professor Hawkins drew his lines were not stone-holes at all, but had a natural origin, which implied that many of the supposed alignments were fortuitous. But no one can really be said to have demolished the general thesis that Stonehenge incorporated *some* astronomical alignments.

It is extremely unlikely that this will ever be rigorously disproved: and that we can say as much is almost entirely due to the labours of Alexander Thom, until ten years ago Professor of Engineering Science at Oxford. Professor Thom has always studiously avoided Stonehenge, with all its uncertainties, but over the years has collected together an immense amount of information on more than 600 megalithic sites throughout the British Isles, having personally visited more than three-quarters that number, and having surveyed more than half. In this way he has made possible a statistical analysis of his findings, which are remarkable not only in themselves but for the relative absence of irrelevant historical speculation.

Megalithic Sites in Britain appeared in 1967, in the wake of *Stonehenge Decoded*, and at first attracted unfavourable comment, largely by association. There is little doubt that most of Professor Thom's book will stand the test of time. In it he shows that the men responsible for the circles built nearly 4,000 years ago were capable of working within very narrow limits. Professor Thom underscores the urgency of accurate surveys. He shows that the Avebury site, for instance, was set out with an accuracy approaching one part in 1,000; and yet even scales of yards and metres on some modern plans might differ by more than one part in 100. Thanks to the accuracy with which the rings of megaliths were set out, Professor Thom was able to show that they were designed on at least six non-circular geometrical plans—as ellipses, for instance, or flattened 'circles', with component circular arcs drawn from different centres.

These centres were themselves at the vertices of right-angled triangles of at least four or five different sorts, but always having sides made up of simple multiples of a fundamental unit—what Professor Thom called the 'megalithic yard' (2.72 modern feet). This is not to suggest that the theorem of Pythagoras was known, for it was evidently believed, for example, that a triangle with sides 41, 71, and 82 was right angled, which it is not. Here for comparison it is worth remembering the Old Babylonian 'Pythagorean triangle' tablet, Plimpton 322, which is of roughly comparable date, or a little older.

Professor Thom showed how the architects of the megaliths were obsessed with lengths of integral numbers of megalithic yards, even to the point of distorting the circles—with due attention to a neat geometrical construction—to give an integral periphery. He showed conclusively that multiples of the unit ($2\frac{1}{2}$ and 10 yards) and sub-multiples (down to 1/40, in the cup and ring markings) had been used.

If Professor Thom's conclusions as to neolithic metrology and geometry were astonishing, what he discovered of the astronomy of the time was even more so. He showed for site after site a preoccupation with the Sun at the solstices. He proved that a calendar had been in use in which the year was divided into eight, sixteen, or even thirty-two parts. But he also left no doubt in the mind of an impartial reader that some of the alignments were lunar, while some pointed to the rising or setting of one or other bright star. Dates are deducible from the directions of the latter, dates which tend to be within a couple of centuries of the nineteenth century B.C. The lunar orientations were touched on in *Megalithic Sites*, but only briefly, and *Megalithic Lunar Observatories* now supplements the earlier volume. The author's results are, if anything, more interesting than before.

To appreciate the achievement of those who observed from the megaliths, while keeping their achievement in proper perspective, we should at least keep half an eye on the probable evolution of their methods. Once it was noticed that the Sun rises and sets over a different point of the horizon on successive days, that at midsummer it reaches its greatest displacement along the horizon in one direction (north), and that in midwinter it reaches the other extreme, it would have been natural to fix the two turning points with permanent direction markers. These might have been a pair of stones, or an avenue, or a stake or stone taken in conjunction with some distant natural object, such as a mountain top, for a foresight. (This last method, found in many of Professor Thom's examples, is obviously potentially very accurate, a minute of arc being easily achieved. One foresight, Mount Leinster in Ireland, is ninety-one miles from Parc-y-Meirw. It is not absolutely clear, incidentally, that Thom is right to ignore refraction in azimuth when sighting over a coastline.) It seems likely that some calendrical or religious motive—or both—might explain the wide acceptance of this first type of observation.

Perhaps comparable lunar observations were made almost simultaneously, but it is unlikely that the nature of the Moon's fluctuating habits of rising and setting were fully appreciated at all quickly. Although in any month the Moon's rising and setting points fluctuate between certain limits, as do the Sun's over a year, yet the limits themselves now vary as between successive months. The full cycle of the second fluctuation takes more than eighteen years. And then again, there is another very slight oscillation of the limits, due to an effect which the sixteenth-century Danish astronomer Tycho Brahe is generally supposed to have been the first to discover.

Professor Thom shows that several megalithic observatories (a name which is certainly not out of order) give conclusive evidence that the first two effects were allowed for, and very strong evidence that some alignments can only be explained in terms of the effect discovered by Tycho. Since there is a temptation to see in this some sort of priority of discovery, it is worth adding that Tycho described the effect, and separated it from cognate effects, in the conceptual terms of the geometry of the sphere. The megalith builders, with a tenacity, perseverance, and minuteness of which Professor Thom's own work is reminiscent, at first merely staked out lines pointing to extreme positions of the Moon as it crossed the horizon. The two sorts of activity were not at this stage really comparable.

But this is not to say that genius was wanting in the ancient world, for the precise positions of the theoretical extremes were not in all strictness immediately observable. Professor Thom holds that in Caithness certain fan-like arrangements of stones are devices for calculating the true, but generally unobservable, lunation extremes. The problem is a highly complex one, and his views are not likely to go unchallenged. He offers two modern solutions, and finds that one may be fitted to the archaeological remains; but he can hardly be said to have explained by what route the originators of the Caithness rows arrived at the supposed solution. This detracts from the completeness of his argument, but certainly does not refute it. There were, as he shows beyond all doubt, giants in the earth in those days.

Professor Thom is usually almost too cautious in his claims, but on one or two occasions he allows himself to speculate freely. He clearly knows at first hand the hazards of sailing in western Scottish tidal waters; but is it legitimate to assume, as he does, that the tides must have been *correlated* with the Moon's motion 4,000 years ago? Not improbable, perhaps; but would this explain the motive for building so many observatories? If so, what price the thesis of Mediterranean influence? And, if so, was there any appreciation of the far subtler correlation between *eclipses* and the different periodicities which may in principle be guessed at rather than deduced from the lunar observatories? It is extremely difficult to know where to stop. It is all too easy to accept without a second thought even such a sentence as 'The object then was to predict the dates of the maxima of the perturbation cycle.' The megalith builders marked out the maxima, certainly, but the claim that they dated them, that is, that they recorded long time intervals, is one which demands at least an argument about plausibility. And then, as argued before, is not the ascription to these people of the very concept of a 'danger period' for eclipses lacking in plausibility?

Professor Thom's brilliant work leaves us with many unanswered questions. At a simpler level, are there road or barrow alignments to be found? (There is a suspiciously straight footpath running from a tumulus in North Tidworth in the direction of the avenue at Stonehenge.) Why so many observatories? Is Silbury Hill—giving it a structure on top, perhaps—an articificial astronomical sight? (Just think of all that recent digging when a theodolite and a step-ladder might have revealed all. How natural that there should be nothing *inside* it!) Was the intellectual coherence evidenced by neolithic design mirrored by a deep social conformity? Was there perhaps an academic rivalry between different groups, or did some sort of central neolithic National Trust supervise their activity? Where did astronomy fit into the religious life of the community? And, above all, where did it all start?

As Professor W. Hartner has explained, an early awareness of the movements of the constellations is apparent in an Elamite seal impression of the fourth millennium B.C.; but this is far from being the earliest artifact of prehistoric astronomy. In his large quarto volume, beautifully illustrated with photographs, and—like Professor Thom's books—with excellent line drawings, Mr. Alexander Marshack presents material evidence for systematically recorded lunar observation extending in an unbroken line from Aurignacian and Perigordian cultures to the end of the Magdalenian, that is, the upper (i.e. later) palaeolithic period, from about 36,000 to 10,000 B.C. This all strongly suggests a source for the calendars of the later agricultural peoples.

For ten years Mr. Marshack has been studying engraved bone and stone artifacts, on which are sequences of marks which have been seen by others as evidence for a decimal system of counting, as hunting tallies, or as a ritual, ceremonial, or sexual symbolism. Although he is apparently aware of thousands of examples, very many of which support his claims, his book deals with only six, all from the Musée des Antiquités Nationales, Saint-Germain-en-Laye. His commentary (originally written in English but first published in this French version) admirably explains, and in detail, his methods of microscopic analysis of the marks (suggesting, for instance, that they were put on a given object at different times). He also presents his own notation for summarizing the intentions of the engraver.

The reasons for thinking the marks to be a lunar notation are strong. The length of a month is approximately twenty-nine-and-a-half days, but the precise time of a new Moon (or of a full Moon, for that matter) is not easy to distinguish, and a casual observer unaware of the precise period will often introduce extra days of lunar invisibility. Reckoning to or from the first time he sees the new crescent, to or from the last crescent visible, and making a mark for each day, he will score perhaps between twenty-seven and thirty-one marks in a 'month'. The artifacts discussed do indeed have markings in multiples of this sort, while the marks seem also to be grouped fairly naturally into quarter months. Although this summary might seem intolerably vague, the book itself should go far to convince the sceptic. It should be added that the markings occasionally have the appearance of the appropriate lunar phase.

There is obviously much to be learnt from Mr. Marshack's meticulous approach. When he finally publishes his evidence for the ubiquity of the markings, and for their association with animal and anthropomorphic drawing and symbolism, it should be possible to add substantially to our knowledge of the cosmic and spiritual attitudes of palaeolithic hunting communities. And rather than scatter his pearls in divers obscure articles, may we not hope that he will write a sequel for the same publisher?

Our knowledge of prehistoric life, and pre-eminently of its astronomical aspect, has come a long way during the past decade, and for this we should be grateful to the enlightened few who have suppressed a human tendency to read one's own sermons in stones, and unhistorical truth in everything. The time has come for professional archaeologists to throw off the feeling that they are being ever persecuted by occultist pyramid-measurers.

Addendum

Since writing this and the two reviews that follow, I have discovered some important principles according to which Stonehenge and many other monuments of the late neolithic and early bronze ages were very accurately aligned with astronomical events. I am now convinced that the Aubrey holes at Stonehenge were positioned for alignments, and not for eclipse prediction – although by the nature of things that possibility can never be ruled out absolutely. There is what seems to me conclusive evidence for the use there of Thom's so-called 'megalithic yard'. These and other related problems will be the subject of a forthcoming book of mine on Stonehenge.

3

BY DIRECTION FROM ABOVE

"In vain do we extend our view into the heavens . . . in vain do we consult the writings of learned men, and trace the footsteps of antiquity." George Berkeley's complaint was that a curtain of words gets in the way. Had he lived today he might have complained of talk of "astronomy" whenever a crescent moon is found scratched on a rock, or of buildings and towns described as "observatories" whenever they are aligned with some special direction of the Sun. This particular curtain of words was already being thrown over Stonehenge in Berkeley's day, by William Stukeley, with a religious warp and an astronomical weft. Druidism was then all the rage: no sooner had Stukeley finished laying out a private druid's temple than Archbishop Wake recommended that he take orders.

Now it is astronomy's turn, and there are more efficient forms of publicity than the pulpit. Of course a more important point is that the Druid hypothesis was wrong, while the best astronomical explanations currently being offered for the orientations and certain structural elements of many prehistoric monuments are undoubtedly right – give or take a few words much too casually used. On the other hand, no one can overlook the veritable flock of black sheep hiding the ewe lambs from view. How is the ordinary reader to sort out the sense from the nonsense?

E. C. Krupp's book is as honest and safe an astronomical guide to the monuments of ancient Britain and Brittany, Egypt, and North and Mesoamerica as any yet written at an introductory level. Of its seven chapters, four are by the editor himself (he is Director of the Griffith Observatory in Los Angeles), while the others are by authorities on particular regions. The book contains virtually no information that is new, but gives a critical evaluation of most recent work – and of some not so recent – as well as a very temperate critique of the lunatic fringe. Despite its rather vague and at times even ramshackle chronological structure, this is a thoroughly enjoyable survey written in a clear and easy style.

KRUPP, E. C. [1979], *In Search of Ancient Astronomies*, London: Chatto and Windus, pp. 277.

It is weakest at its point of contact with the history of that ancient astronomy for which documentary evidence survives. Whatever Berkeley thought, words are not without their value; and with a little more about Babylonian astronomy, for example, a few of the arguments might have been strengthened. Babylonian procedures had something in common with those of the builders of the textless monuments, and one does not have to be a pan-Babylonian to believe that a wider comparative study is called for. Its title notwithstanding, a book of this sort is not calculated to break such difficult new ground – and no one is likely to be seriously misled by the jacket, with the Callanish stones at Lewis on the front and the Avenue of the Dead at Teotihuacán on the back. What a welcome change from the hackneyed and irrelevant "mid-summer sunrise over the heel stone at Stonehenge"!

The work opens with a crash course in the basic astronomical ideas, and a sprinkling of historical remarks which are not always helpful, and at times incautious. The illustrations are usually extremely good – although there is an exception here, showing the sun and Sirius rising over the pyramids in a way suggesting heatstroke on the part of the artist. The book really gets into full swing with a summary by Alexander Thom, and his son A. S. Thom, of their extraordinary surveys and analyses of the stone rings, rows, and other "megalithic" (which is to be taken to include minilithic) monuments of northern and western Europe. There are more than 900 rings in Britain, and Alexander Thom, who has spent fifty years on the subject, has made accurate surveys of more than 300 of them – not to mention the sites in Brittany. Those who know his books (*Megalithic Sites in Britain* and *Megalithic Lunar Observatories* have now been joined by *Megalithic Remains in Britain and Brittany*) will still find this summary chapter refreshing. It gives the reasons for thinking that the circles were constructed to a ubiquitous unit of length ("the megalithic yard"), circumferences often having been made an integral number of "megalithic rods" (of two and a half megalithic yards, or 6.80 feet). There is an honest admission that statisticians disagree about the constancy of the unit of measure of the British circles as a whole, although there is substantial agreement over the Scottish circles.

Can it be that those uppish Rhineland Beaker People, trying to tell the rest of Europe how to do things properly, were confronted by native conservatism? Daydreams about future European standardization are prompted by the statistics: 67 per cent true circles; 17 per cent flattened circles of types A and B; and 5 per cent eggs of types I and II – in one case achieving a ratio of perimeter to diameter of three precisely (or at least so within one part in 5,000). Woodhenge is a nest of eggs; is that flat and polished stone a drawing board for it? Note the ellipse halves at Knock (Wigtownshire) paired (on a petroglyph) to form a spiral pattern spaced in intervals of one megalithic inch. Many rings are perfect ellipses, although perhaps seventy times as big. A megalithic yard is deemed to be 40 "inches". Was it perhaps originally ten hands?

The circles are impressive as architecture; but as such they are a northern European phenomenon, whereas in their astronomical properties they follow a pattern shared by structures in many parts of the world. In short, they are in some way directed towards places on the horizon at which significant celestial events take place.

Seen from a particular place, the sun and moon rise and set at different points on the horizon according to the season or time of month. These points fall within certain limits. The sun rises and sets at its most northerly points at the summer solstice, and at its most southerly at the winter solstice. Four fundamental directions are thus defined, and these often coincide with monumental orientations that are in some sense clearly "special" – as when there is an axis of symmetry. With the moon the problem is more complicated, by dint of complexities in its motion, but the principle is similar – there are merely more fundamental directions. With a star the rising point is fixed, to all intents and purposes. (It changes appreciably only with the centuries.) There are many simple star alignments for a handful of bright stars (Aldebaran, Capella, and Regulus, for example) corresponding to their positions in about 1800 or 2000 BC.

Solar and lunar directions are more numerous. Often a distant mountain peak will serve as a foresight against which the celestial object is seen, from a menhir backsight (or along a line of menhirs). Lest there be any doubt about the correctness of the basic principle, consider the menhir, cairn, and ring at Kintraw, which – taken together with a notch on the Jura horizon – point to the winter solstice sunset. The arrangement at first seemed as though it might be fortuitous, since a foreground ridge blocked the view of the notch. Looking back along the alignment, however, a ledge was found on the hillside which could have served as the viewing platform, and excavation by the archaeologist E. W. MacKie proved it to be artificial.

The time is long since past when such dramatic evidence was needed to bolster confidence in the general principles of monument orientation, but there is plenty of room for controversy over the extraordinarily rich detail of the Thom account. (There are of course many independent scholars in the field, and a new journal *Archaeoastronomy* has recently appeared as supplement to the *Journal of the History of Astronomy*, in which so many of the Thoms' articles have been published.) Little is said in the book of the real difficulties of the problem of eclipse prediction, or of the most daring of Thom's hypotheses, relating to the fanlike arrangements of stones in the several lunar sights in Caithness. Still more complex rows, no doubt similar in function, cover the landscape at Carnac in Brittany – with egg-shaped rings at each end of vast grids of stones which are all part of an enormous complex of megaliths surrounding Quiberon

Bay. Here the Grand Menhir Brisé, once more than sixty feet high and weighing over 340 tons, served as a universal foresight for several monuments.

When Alexander Thom's graphs of alignments show peaks at points suggesting that the year was divided into eight (or even sixteen or thirty-two parts), one gets the uncomfortable feeling that the prehistoric world was another place; but then you see that the points can be labelled with such familiar names as May Day, Lammas, Martinmas, and Candlemas, and a sense of familiarity and continuity – perhaps not entirely spurious – takes over. Look at an aerial photograph (or Thom's surveys) of the stones, and ask yourself whether they are not as immediate (and remarkable) in their way as a manuscript drawing by Pythagoras might be. And if you give the palm to Pythagoras, remember that they are half as old again.

A chapter by the editor himself fills in some of the background to Stonehenge, referring to chronicles ancient and modern, the nonsense quotient by no means diminishing monotonically with time. It is good to see a little posthumous credit going to that modest amateur, C. A. Newham. Archaeologists are fond of saying that there is no evidence that Stonehenge was ever a scene of human sacrifice, but then contradict themselves by their treatment of a succession of astronomer victims, attracted to Stonehenge as moths to a candle. The best survive; and for a rough idea of some of the sins of the worst, Krupp's account is a useful introduction. Do the Aubrey holes serve to predict eclipses, or tides, or nothing at all? Was the spherical temple of the Hyperboreans a vast dome? (There's a word, "spherical", that Krupp should have silently changed to "circular". The Loeb translations are not yet sacred books.)

If the archaeologists are occasionally forgiving, perhaps that is because their world too is insecure. Dr Krupp has an outline of recent British Museum Radiocarbon Laboratory evidence for the dates of the different phases of Stonehenge. Coach tour operators are going to have to revise their patter, push the dates back, and reconsider the significance of the "Mycenaean" daggers carved on one of the sarsen stones. Evidence too recent for inclusion in the book affects remarks about Peter's Mound, now known to be modern.

The Colarado astronomer J. A. Eddy writes about sites where North American Indians have left clear evidence of an interest in celestial alignments. We know that they showed some interest in the sun, moon, and stars – what cultures have ignored the sky entirely? – but it must be said that there is very little evidence of what one might call "theoretical depth" in their remains. There are pictographs of crescent moons (near which a star or other mark is sometimes shown) in rock art from the American South-west, and William Miller, of the Hale Observatories, suggested more than twenty years ago that some of these might possibly depict the conjunction of the crescent moon and the supernova of 1054

AD (the exploding star the remains of which are visible as the Crab nebula). This hypothesis has attracted a great deal of attention recently – out of all proportion, it seems to me, to its intrinsic probability.

There is something to be said for thoroughness, on the other hand, which was not the hallmark of much of the early archaeology of North America, and the conflicting announcements made at various times about relatively well-defined structures have been at times exasperating. One *cause célèbre* is the Sun Temple in Mesa Verde (Colorado) – which takes its name by the way, from a "sun symbol" etched on one of the stones, a pattern later found to be a fossil imprint of a leaf. There are many interesting alignments, even so. Earthen mounds in Newark, Ohio, have a lunar orientation. Posthole circles at the Cahokia mounds in southern Illinois have been interpreted as solar (for sunrise at solstices and equinoxes).

Dr Eddy's own studies have concentrated on the rock patterns on the Western Plains, especially the "medicine wheels", where solar and stellar alignments are apparently present. Is it at all remarkable that these relatively modern alignments (less than 1,000 years old, it seems) should apparently follow in a tradition carried on so long before, an ocean away? Did some early Briton cross the Atlantic in a gigantic beaker, perhaps? It seems far more plausible to classify "pointing" to the solstices (and so on) as a humanly obvious thing to do, at least when your night sky is not filled with the glare of street lights and your horizon not obscured by buildings or trees. The North American Indian failed to go further. When the Mesoamerican peoples went into greater theoretical depth, their procedures differed from the European – although they found cyclical relationships in the same way as astronomers in other literate cultures. Thus five cycles of Venus correspond to eight years of 365 days. Without the confidence that this sort of procedure is also fairly obvious, one could hardly begin to entertain the idea that ancient peoples could predict eclipses. Perhaps one should not.

Most of the Maya writings were burnt by their Spanish conquerors. Fragments of only four original codices survive, and all relate to the heavens. (They include, notably, lunar and solar almanacs and a century-long Venus ephemeris.) There is more complete information about the Aztecs, and an account by the Spanish historian Juan de Torquemada reports on the style of astronomical observations made from the roof of the palace of the king of Texcoco, a report corroborated by pictures from various codices. A. F. Aveni provides an excellent guide to the many puzzling aspects of the orientations of the remarkable buildings left by these peoples. There is even a literary reference to Montezuma's wish to pull a building down and set it right because it did not align with the sun at the equinox. Venus and the Pleiades provide good explanations for some of the alignments – and are supported by folklore and legend. With Sirius and Capella the ethnohistorical sources do not as yet support the computer. It is good to see the distinction drawn.

One of the strangest findings in the area is that there are several families of sites with shared orientations, these orientations having no immediately obvious significance. There is the family (which includes Teotihuacán) of sites directed seventeen degrees east of north – preserving, perhaps, directions important only at other sites and long before. There are no natural foresights in the flat terrain, and purely architectural codes of practice might well have been handed down to people who did not properly understand their justification.

The editor himself takes on the task of trying to restore a little of the old lustre to Egyptian astronomy, which has suffered a number of indignities during our own century by unfavourable comparisons at a mathematical and theoretical level with the work of the Babylonians. There is plenty of Egyptian celestial mythology, of course, and the occasional celestial ceiling. There are the "star-clocks", tabulations of constellations inside coffin lids (2100 to 1800 BC) from which we are thought to inherit the twenty-four divisions of our day. An equally important (but fortuitously so) Egyptian contribution to astronomy proper is the year of fixed length, namely of 365 days, with its apparent disdain for such irrelevancies as agriculture, religion, and political change.

But what about the alignment of buildings? There is the ceremony called the "stretching of the chord", illustrated in a relief of perhaps 2500 BC. It shows the Pharoah and the goddess Seshat grasping mallets and stakes, around which a cord is looped as they lay a foundation stone. There are other, similar, illustrations of the ceremony from later times. We are told that plumb-lines (the *merkhet*) and stars were also involved, but not exactly how the orientation of the temple axis was precisely found. Or was it, perhaps, not found very precisely? Holidays in Egypt in search of temple alignments with compass and theodolite have been popular, off and on, for over a century, and numerous solstitial alignments have been claimed, not to mention many stellar alignments. Norman Lockyer produced a few plausible results among a morass of implausible ones.

Gerald Hawkins has recently confirmed, revised, and expanded earlier lists, and claims to have found some lunar directions (in the Great Temple of Amen-Ra at Karnak, rebuilt 1480 BC). I must confess that the Egyptian material rarely leaves me with a sense of ineluctable astro-archaeo-logic. Quite irrelevant, I know, but I cannot help thinking of that text that tells of star transits taken with reference to the head, ears, and shoulders of a sitting man. Not much precision in that, although there is no doubt that the Egyptians could align a building accurately when they tried – as the East-West orientation of the Great Pyramid shows. (It is true to two or three minutes of arc.)

The history of astro-pyramidology, mostly disreputable, is worthy of a book in itself. Krupp avoids most of the sillier episodes neatly enough, and reserves a final chapter for what he rightly judges to be, or to border on, fantasy. There are the complex geometrical networks of trackways, supposedly consciously planned (if we are to believe Alfred Watkins and others) in prehistoric times, linking up monuments and natural characters of the horizon. Later writers have argued for connections with geomancy, including Chinese versions, and with circuits of subtle telluric energy that tends to accumulate in such spooky places as prehistoric passage graves. Like all ineffable things, this sort of energy is not easily controlled, and it would probably be unwise to try to harness it, bearing in mind that even one so skilled as Bladud the Druid crashed during a flight on his levitated stone.

As he looked down, did Bladud ever see the figures of the zodiac found by Katherine Maltwood in 1929 in the Glastonbury landscape, or any of the rival zodiacs found since by patriots of Wales, Surrey, and Cambridgeshire? Did he ever try to use the "landing strips" on the plains of Nazca (Peru) which Erich von Däniken – in books which by some sort of magic have sold millions of copies, and have inspired at least a couple of films – has claimed for his astronaut gods? His argument, broadly speaking, was that the Maya were too stupid to have done their remarkable work without extraterrestrial help. This story surely all began as a monumental leg-pull, which only later assumed its megalithic proportions. One thing is certain: von Däniken himself did not have astronaut gods to help him with his astronomy. But who, if anyone, helped the Dogon (tribesmen from what is now Mali) to their valid beliefs that Sirius has an invisible companion with a fifty-year period, that Jupiter has satellites, and that Saturn has rings? According to an American orientalist, R. K. G. Temple, all was revealed to the Dogon by extraterrestrial amphibians 5,000 years ago or more. The Dogon beliefs were first communicated by two French anthropologists in the 1940s, who alleged only that they were more than 800 years old. What is the real explanation? A passing missionary with an interest in astronomy and his enthusiasm aroused by a newspaper report of the discovery of the invisible companion to Sirius?

Whether or not this is right, surely the problem is trifling by comparison with that of explaining precisely why such a trifling problem – like the miscellaneous phenomena which provided Immanuel Velikovsky with his starting point – should call forth explanations which are from beginning to end extravagant and fantastic. It is now thirty years since Velikovsky's *Worlds in Collision* first appeared and yet still it has a loyal following of those who simply *want* it to be right. If his "comet Venus" really did alter the geographical location of the earth's poles or the direction of the earth's axis at the times Velikovsky claims, then Thom's work must be discarded. Here is one of a hundred reasons for putting an end to the whole foolish debate.

4

NEOLITHIC NEWTONS

Historians who hazard writing about five consecutive centuries are not exactly numerous, but those who are prepared to unite fifty centuries by a single theme are running academic risks quite beyond the call of duty. Most of the hostility shown towards them, when it does not stem from some vague philosophical bogey about the propriety of a history of ideas, seems to be grounded on the fear that there is some sort of plot afoot to fashion a "straightjacket for humanity" (to steal George Meredith's phrase). "We desire to find admired qualities in the past", remarked Stuart Piggot, one of the British Academy's organizers of the 1972 symposium, held jointly with the Royal Society, and whose proceedings are now published as *The Place of Astronomy in the Ancient World*; Professor Piggot went on to add that "mathematical and scientific qualities are admired today". Were these qualities really there in the past? He alluded to the long and patient work of Alexander Thom on which was based "a thesis which if accepted demands the recognition of considerable mathematical skills among the non-literate societies of northwestern Europe from the fourth to the second millennia BC; mute inglorious Newtons . . .".

The trouble with words like "mathematical" and "astronomical" is that they are as heavily laden with meaning as the word "historical". Not long ago an Oxford economist was to be heard maintaining that "forty per cent of forty per cent is forty per cent – that's just simple mathematics". No doubt in a sense he was right: and presumably he was able to do a lot of good and rational economics, advise the government and so forth, without understanding the multiplication of fractions. The purist may object to a rope and pebble view of mathematics, and yet peoples ancient and modern alike have done a great deal which most would accept as astronomy, with a rational, and in a sense mathematical frame under an empirical exterior. There is no need to see a contradiction (as does R. J. C. Atkinson) "between the positive evidence for prehistoric mathematics and astronomy on the one hand, and the negative evidence

KENDALL, D. G., and others (editors) [1974], *The Place of Astronomy in the Ancient World*, London: Oxford University Press, for the British Academy, pp. 276.
PEDERSEN, O. [1974], *A Survey of the Almagest*, Odense: Odense University Press, pp. 454.

for recorded numeracy on the other", although there is a problem, certainly, in explaining how so much tacit mathematics can be present without real "numeracy". But Professor Atkinson is not out to raise arid questions of definition, which, with their overtones of amazement at the idea that a lifter of stones should have something in common with a Nobel laureate, are pale and uninteresting in comparison with the rich and varied evidence brought together in the proceedings of the 1972 symposium. The literate societies of the Babylonians, Egyptians, Chinese, and Maya there provide a hard documentary core of mathematical astronomy on which the sceptic can cut his teeth, while the prehistoric evidence from Western Europe will allow him to match his imagination against mathematical analyses of a sort offered by Professors Thom and D. G. Kendall.

The "unwritten evidence" is not all ancient. The conceptual frame of the world of Polynesian and Micronesian navigator–priest–astronomers is still, in a manner of speaking, there to be examined in use. D. H. Lewis explains some of the elements of their secret atronomy, more or less sufficient for their navigational purposes, and serving as a basis for myth and religion. It seems that – notwithstanding stories of the Hawaiian magic calabash – Oceania has known nothing of navigational artefacts for use at sea; but a new account of a carefully sited land-based stone canoe, for training purposes in the Gilberts, adds an unusual dimension to the book. (Is it now only a matter of time before some Mystic Press gives us a book on the stone coracle of Salisbury Plain?) The difficulties of the anthropologist in Oceania include the secrecy of the navigators, our unfamiliarity with their cosmology, and its contamination with modern Western ideas. The evidence is strong that voyages directed by astro-navigation were longer and more frequent in earier times, but Professor Lewis is at pains to avoid the fallacy of supposing that he has opened "some kind of ethnographic window into the neolithic past of continental cultures, earlier by millennia and up to half the earth's circumference away".

The window in question has been pushed open rather more by Professor Thom than by any other person. His great strength is the large number of megalithic sites he has patiently surveyed. His contribution to the symposium was to reinforce conclusions set out earlier, especially in his two books, and to add the results obtained by groups working with him in Brittany and Orkney. His most striking results relate to Le Grand Menhir Brisé (Er Grah), seen as a universal lunar foresight, and his most controversial include his ingenious explanation of the alignments of the rows of stones at Le Ménec. He argues, very plausibly, that these could have been used to find the extrapolation distance of any of the four backsights to the west of the Sea of Morbihan, and that the rows replaced the simple sectors at Petit Ménec and St Pierre.

E. W. MacKie takes up the problem set by Professor Thom's work, and looks for tests through excavation. Professor Thom had predicted an observing platform of some kind at Kintraw, and all the evidence now marshalled suggests that there was in fact an artificial boulder notch and platform of the sort required. Dr MacKie ends with some speculation about the character of the society which gave birth to, and preserved, the sophisticated astronomical techniques under review. He draws an analogy between them and the procedure which enabled a Tamil calendar maker to make a mental prediction, accurate to four minutes, of an eclipse in 1825, without really appreciating the initial theory. The analogy is one which should not be pressed very far, if we are to keep a proper sense of perspective over the different orders of theoretical structure underlying the ancient and the not-so-ancient routines.

Dr MacKie draws a useful distinction between orientation and alignment. The difference is one of intention: a building may, for magical, religious, or traditional reasons, be given some important astronomical direction without its having been used as an astronomical instrument. "It seems very doubtful", he explains, "whether any useful, accurate observations of the Sun could be made simply by looking along the straight side of a masonry building." G. S. Hawkins, in confirming P. Barguet's belief that the Temple of Amon-Re at Karnak was astronomically aligned, is speaking, in Dr MacKie's terminology, of an orientation. Of course, there are countless overtones to the notion of a "useful, accurate observation", and many of them would have been worth considering further.

The evidence of climate and its influence on ancient civilizations is considered in a masterly way by H. H. Lamb. Like what is known of the changing nature of the landscape (alluded to by Professor Atkinson), it poses several imponderable questions, but tallies with the evidence that the sites in Britain and Brittany claimed as astronomical date only from the end of the neolithic period or from the Early Bronze Age. So much for the necessary prerequisites of an apparently astronomically orientated society. With all the prior conditions satisfied, that society seems to have taken its work seriously – for quite apart from the extraordinary accuracy of the alignments found by Professor Thom and others, there were engineering feats such as that which resulted in the erection, for example, of a menhir (Locmariaquer) nearly as tall as Cleopatra's Needle, and more than half as heavy again.

Professor Atkinson's summary of neolithic technology goes beyond the demands of astronomy, but contains an apt comparison between past and present. Silbury Hill, he suggests, probably required the energies of 500 men for fifteen years. "In view of the small size of the neolithic population, this represents a fraction of the 'gross natural product' at least as great as that currently devoted by the United States of America to the whole of its space programme." The greatest thing before Jesus Christ, as President Nixon might have said.

In the last of the papers read to the symposium on the unwritten evidence for ancient astronomy, Professor Kendall, a statistician, goes hunting quanta, and especially the quantum of length (the "megalithic yard") which Professor Thom has claimed is to be found in the stone circles. This paper is a superb example of the art of simplifying material which is far from simple. Professor Kendall, while pleading for costly aerial surveys of as many monuments as possible, seems reasonably content with a quantum of 5.445 ft (standard deviation 0.0181 ft). Even in this paper there is a sidelong glance at ordinary human kind, or at least at the Brigade of Guards, whose members are said to be capable of maintaining a pace of thirty inches on the parade ground, plus or minus half an inch.

In the discussion of this part of the symposium, L. E. Maistrov, from Kalingrad, mentioned the inhabitants of North Ossetia, some of whom still remember the way of fixing holidays and the seasons by solar sightings from established positions. The alignments of the Dorset cursus were also introduced into the discussion. Since the 1972 meeting, stone cairns in Wyoming, perhaps no more than two centuries old, have been claimed as the first observatory attributed to nomadic people (the Plains Indians) rather than to a settled agricultural community. Denmark has now found traces of a demonstrably astronomical monument comparable in area with Stonehenge. This is a potentially enormous branch of archeology, and yet, in Professor Thom's words, "there is so much to be done, and there are so few people with the ability to make the necessary measurements".

The situation is hardly better vis-à-vis the written evidence for ancient astronomy, and any student who wishes to rattle off a quick thesis is recommended to avoid this part of the symposium. A. Sachs writes on the cuneiform texts of Assyria and Babylonia, of the Venus observations from the reign of a king of the First Dynasty of Babylon, and of the later so-called "astronomical diaries" (from the eighth century BC to the first). The enormous care and industry on which his contribution is based are as well concealed as the messages he deciphers. Some of the documents would seem familiar on Fleet Street, with their mixture of astronomy, meteorology and commodity prices. A. Aaboe's sketch of the character and content of the known Babylonian mathematical astronomy of the last five or six centuries BC not only conveys a very clear idea of what is involved in the analysis of the documents in question, but sets the Babylonian achievement in excellent perspective against that of Greek and Islamic astronomy.

R. A. Parker explodes once again the widely held belief that Egypt was the repository of all ancient astronomical wisdom. Not until influenced by Hellenistic science do we find anything approaching a theoretical astronomical treatise. But when we find that there was a decent calendar in use in the third millennium BC, that star clocks for telling the time by night were in use before the twenty-fourth century BC, and that the division of the day into twenty-four (equal) hours has its origin in Egypt, then history and civilization seem to shrink a little.

The same is true when Chinese and Mayan ideas are brought in to the picture, by Joseph Needham and J. E. S. Thompson respectively. The Chinese records ("broadly speaking... either printed or lost") stretch to the fourteenth century BC and are comprehensive from the third. Astronomy in China was equatorial rather than ecliptic, and was a product of state bureaucracy rather than of priest or individual scholar. Among its most impressive achievements are those relating to instrumentation, whether for observation or for simulation of the movements of the heavens – as on the hydro-mechanical clock tower of the eleventh century AD. As one suspects was the case in other cultures, secrets were for keeping. "From now onwards", runs a ninth-century Imperial edict, "the astronomical officials are on no account to mix with civil servants and common people in general."

As in all of the more impenetrable branches of history, when secrets and mysteries are wanting there will be those who will supply them. J. E. S. Thompson's account of the Maya calendars is a model of lucidity, not least valuable for his warning against playing number games with the Dresden codex in hand, or, for that matter, against finding alignments among the buildings of the Maya, in which right-angles seem to have occurred more by accident than by design. The recent astronomical claims of C. H. Smiley are heavily criticized.

The Maya emphasis (unlike the European – *pace* those who are looking for cultural diffusion) is apparently centred on heliacal rising after inferior conjunction, and this for astrological reasons. Neither the names nor the glyphs of other planets than Venus are known. The Venus tables were in fact solved by John Teeple, an American chemical engineer who whiled away long train journeys with the problem, in 1930. Teeple also offered an ingenious explanation of Maya eclipse techniques, which Professor Thompson accepts.

R. R. Newton opens the book with a simple astronomical introduction for those less at home in the subject than he, and contributes another paper in which history is turned to astronomical use, as it were. He investigates the non-gravitational acceleration of the Earth and Moon, and this through ancient eclipse records. He throws much cold water on the records in question, and their use of modern ends, and here his conclusions are strengthened rather than weakened by G. J. Toomer's recent demonstration that he should have placed the so-called "eclipse of Hipparchus" in 190 BC rather than 129 BC.

In the narrowest sense, what Professor Newton is doing is irrelevant to a historian's history. But to read Einstein is to recognize a debt to Newton, and to read Newton is to be so often at only one or two removes from Ptolemy. And from Ptolemy to Hipparchus and the

Babylonians is a step so short that no one could possibly dismiss out of hand Professor Aaboe's claim that "Babylonian mathematical astronomy was the origin of all subsequent serious endeavour in the exact sciences . . ." And this is a historical assertion. No considered synthesis along these lines could be offered at a two-day symposium covering such a span as did this. But after reading its proceedings, only the uncomprehending are likely to be left without a strong feeling of real continuity with the past, and one based on material every bit as enduring as potsherds.

For all the brilliance of the Babylonian achievements, it is generally agreed that the culmination of scientific astronomy in antiquity coincided with the synthesis offered by the Alexandrian astronomer Ptolemy, of the second century AD. Olaf Pedersen's *A Survey of the Almagest* offers a detailed analysis of Ptolemy's greatest work. Heiberg's edition of the Greek text of *Almagest* has been available for more than seventy years, and there are translations to be had in French, German, and English – the best of them by Karl Manitius into German. There are, however, many parts of the work which remain opaque, even in translation, and some such comprehensive guide as Professor Pedersen's to its theoretical structure has long been sorely needed.

His *Survey* is far more than a mere synopsis of the contents of Ptolemy's work, for it includes most of the biographical information now available, and gives a very good idea of the ways in which *Almagest* passed into the scientific culture of Islam and Christendom. He reminds us of the reliance Copernicus placed on it; and "there is no question that it was a greater scientific achievement than the *De revolutionibus* which has obliterated it from fame." Most important of all, Professor Pedersen helps to reinforce a thesis, the truth of which is becoming increasingly clear as time goes by, namely, that the mathematical methods of modern science are in a direct line of inheritance from the geometrical and kinematic methods of Ptolemy.

Almagest contains neither astrology nor physical cosmology; for the first one must turn to the *Tetrabiblos*, and for the second to the *Planetary Hypotheses*. *Almagest*, as its Greek title tells us, is a mathematical summary. It begins with what amounts to a treatise on trigonometry, and in it Ptolemy painstakingly rehearses complex geometrical techniques for the prediction of the positions of Sun, Moon, stars and planets.

Although his style is intensely mathematical, and although at times he seems to be disenchanted with the evidence of the senses, a list of ninety-four dated observations abstracted from *Almagest* by Professor Pedersen leaves us in no doubt about his priorities. More than half of these observations were made before his own lifetime and the earliest, from 721 BC, takes us straight back to the Babylonian astronomy of the London meeting. To the Babylonian tradition Ptolemy added the characteristically Greek techniques of eccentric and epicycle, as developed by Apollonius and Hipparchus.

The figure of central importance in the transmission from Babylon seems to have been Hipparchus, whose solar and lunar theories were taken over and modified by Ptolemy, and who had (more than two centuries earlier) created an amalgam of his own systematic observations with the Babylonian eclipse records. Ptolemy also borrowed much of his mathematics from Hipparchus, and some knowledge of star positions, but this is not to say that he was simply a plagiarist, as J-B. J. Delambre suggested. Professor Pedersen is very successful at showing him to have been, on the contrary, a man of genius and originality, and he quite convinces us that posterity was very discerning when it made *Almagest* canonical. In much the same way, his own exegesis is likely to become canonical in the hands of those looking for an easy but detailed introduction to Ptolemy's genius.

5

VENUS, BY JUPITER!

The story so far: in 1950 Immanuel Velikovsky published *Worlds in Collision*, a book which has since been reprinted at least seventeen times. The book did not, as its supporters would have us believe, "shake the scientific establishment to its very foundations", but it did anger a number of establishment scientists, whose uncharitable reactions, suitably dramatized, have since been used to enhance the attractions of the Velikovsky thesis. His claims were, on the face of it, all too easy to comprehend. At some time before 1500 BC a brilliant fiery object was expelled from the planet Jupiter, to enter into a long elliptical orbit around the Sun. This object became the planet we know as Venus. Terrifying those who saw it crossing the heavens, it came close to the Earth in about 1450 BC, and as they passed through its cometary tail our forefathers experienced such catastrophes as are reported by the author of Exodus – namely the plagues of Eygpt, and all that.

As the comet's head came nearer, the Earth was caught in its gravitational and electromagnetic grip, so that the very axis of the Earth changed direction, while cities were laid waste and all manner of devastation was caused. The rotation of the Earth was so disturbed that men the world over left records of darkness persisting for an unnaturally long time, or of the Sun standing still – all according to geographical location. The night sky shone as the comet's head and its writhing serpentine tail exchanged colossal electrical bolts. And with an eye to the main chance, when most of the Egyptians were no doubt pulling their heads under the bed-clothes, the Israelites slipped through a gap in the Red Sea, a gap created by the pull of the comet. The Egyptian army, not realizing that the comet was working for the other side, was not so fortunate.

A whole series of close approaches followed, and the Earth was wreathed in a nauseous haze, mankind finding its salvation in the manna (ambrosia) which fell from Venus. (This food was essentially a sweet carbohydrate formed by bacterial action in the hydrocarbons –

PENSEE, The editors of [1976], *Velikovsky Reconsidered*, London: Sidgwick and Jackson, pp. 260.

pitch, petrol, and so forth – of Venus's atmosphere.) Some fifty years after the exodus of the Israelites there was another approach. Again the Earth's axis was tilted, its surface was riven, cities burnt and fell, and Joshua the while got the Canaanites on the run. (Nasty things were happening elsewhere in the world, but space is at a premium, and the Old Testament story will serve to illustrate the drift of the Velikovskian thesis.)

Venus's next violent assignation was with Mars, who was pulled out of his orbit by her in the middle of the eighth century BC. Mars then drew close to the Earth, and another series of cataclysms followed. Rome was founded, in 747 BC, with Mars as its god. Again the Earth altered course. In 721 BC a Martian approach shifted the Earth's axis yet again, and the year was lengthened somewhat. Mars made a last fateful approach in 687 BC, when a great thunderbolt passed between him and the army of Sennacherib, an army which was up to no good outside the walls of Jerusalem. (Things were happening the world over, but nowhere so *providentially* as in the Middle East.) Again a tilt of the Earth's axis, and a disturbance of its rotation. "So the Sun returned ten degrees, by which degrees it was gone down", says Isaiah 38, viii. Or, if you prefer the *New English Bible*, the shadow of the Sun went up ten steps on the stairway of Ahaz. If we are to feed Isaiah to our computers, we had better first decide which translation is right.

Again Mars and Venus did battle in the sky. It is all to be found in the *Iliad* (which, *ergo*, was written after 747 BC), no less than in records from the Far East and America. Mars was thrown out of the ring, and Venus became the lovable planet we all know, with a near-circular orbit between Mercury's and ours. And that was that, and the world has been a relatively quiet place ever since.

Dr Velikovsky's *Worlds in Collision* was followed by *Ages in Chaos* (Volume 1, 1952, seems to be all that has yet appeared), *Earth in Upheaval* (1955), and *Oedipus and Akhnaton* (1960). The first of the sequels proposed a radical revision of ancient history, as indeed did *Thesis for the Reconstruction of Ancient History* (1945), a work by Dr Velikovsky now rarely mentioned by his followers. The argument is that some six hundred years are to be eliminated from the standard historian's chronology of Egyptian history. The other books bring stone, bone, and psyche into the reckoning; but all are essentially props for the story told in *Worlds in Collision*, and as far as that is concerned Dr Velikovsky stands by all but a few minor details.

Pensée was the magazine of the Student Academic Freedom Forum, published from Portland, Oregon, and its editors "sought fair play" for Dr Velikovsky between 1972 and 1975, when publication ceased. *Velikovsky Reconsidered* is for the most part a collection of twenty-nine articles drawn from issues of *Pensée* between 1972 and 1974. This is of course only one of many journals to have given space to the Velikovsky phenomenon, and a book with the title *The Velikovsky Affair* (1966) was spawned by an issue of the *American Behavioral Scientist* (September 1963).

Although perhaps less numerous than adherents to Kung Fu, there is no doubt that the Velikovskyites are a powerful force in the undergrowth of academe. They support two main causes, and it is difficult to decide whether that of proving Dr Velikovsky right is more or less important to them than that of proving the existence of a "scientific mafia" (to take a phrase from the first paper of the new book, by David Stove) intent on doing the Master down. There is even a British "Society for Interdisciplinary Studies", founded in 1974 "with the aim of encouraging a rational approach" to Velikovsky's theories. However well justified the endless charges of undercover censorship by the scientific community – and many of the charges do suggest unwise intemperance and hasty *ad hominem* judgment – it can hardly be said that the Velikovskian cause has languished for want of publicity. Why, then, does it still look like a lost cause? Part of the answer is that neither Dr Velikovsky nor his followers have properly understood the priorities of the argument offered in his writings. In a nutshell, the argument cannot properly enter a scientific phase until the history is sorted out. And this is never likely to happen.

Part 1 of *Velikovsky Reconsidered* is a piece of *ad hominem* sociological argument, proving in a mildly interesting way that heretics tend to get burnt. Did the astronomer Harlow Shapley really – through the old-boy network – force the sacking of the senior editor of Macmillan who was responsible for accepting the Velikovsky manuscript? Whatever the answer, it has little to do with the plagues of Egypt or the eruption of Jupiter. Who, in an academic environment, is not surrounded by a measure of dishonesty, insecurity, and intrigue? As Einstein, shortly before his death in 1955, wrote to Velikovsky: "Ich möchte glücklich sein wenn auch Sie die Ganze Episode von der drolligen Seite geniessen könnten." In the same letter he conceded: "Ich bewundere Ihr dramatisches Talent." But I take it that there is more at stake than drama and applause, and that the remainder of the new volume is more important than the polemical opening.

Part 2 of the collection is a motley of quasi-historical material, mostly by Velikovsky himself, showing that, for example, the pyramids give evidence that any change in the geographical positions of the Earth's poles during the pyramids' history must have been small and temporary. Readers of a sceptical bent will not find this hard to accept. (Dr Velikovsky's earlier works ascribe lasting change only to the *direction* of the axis in space, with the consequence that the heavenly bodies, including the Sun, sweep out different paths in their daily motions.)

He goes on to pay homage to Stonehenge, and writings by G. S.

Hawkins. He must perforce argue that Stonehenge was repeatedly remodelled to cater for the celestial rearrangements postulated in his history. How he would cope with the more accurately and closely argued work of Alexander Thom, done over a much wider area, I cannot easily imagine, although Dr Velikovsky is no stranger to *ad hoc* hypotheses, and no doubt some of the faithful are at this very moment rewriting Thom. There is a slight piece by Dr Velikovsky on the historical (i.e. documentary) evidence for an Earth without a Moon, and there is an utterly irrelevant make-weight by A. M. Paterson on Bruno's view of that subject. There is an essay by L. E. Rose suggesting that the Venus tables of Ammisaduqa have nothing to do with that ruler, or with his time (c 1570 BC ?), but are from the eighth century. The usual analyses are based, we are told, on the astronomers' dogma, "the uniformitarian attitude that the solar system has for untold years been just as it is now".

This injured cry, which characterizes a true Velikovskian as surely as it might once have done a Darwinian, is not likely to take in anyone who can see both sides of the argument. For, putting aside the palpable inadequacies of Professor Rose's study of the Venus "observations", what is at stake is again *historical* evidence for the relatively recent discontinuities posited for the solar system. Laplace, the arch-uniformitarian, with his proof of the inherent stability of the solar system, has nothing to do with the case. It goes without saying that, if he had had reason for supposing that Jupiter was capable of exploding, he would have added a caveat, or even what his critics seem to be incapable of giving in print, namely the mathematics of the ensuing behaviour. The fact is that the whole of the ramshackle edifice of nonsense to be found scattered throughout the Velikovskian corpus is purported to have a historical (including archaeological and geological) foundation, but that it has none. Parts 3–5 of *Velikovsky Reconsidered* are of a more or less scientific character. They are there to confirm a historical theory, to confirm, if you like, the boundary conditions of a motley of scientific theories. If suitably orchestrated they would confirm almost anything.

This is just as well, for the most conspicuous trait of Dr Velikovsky's fantastic story is its imprecision. But before anyone sets to work to decide what it would be reasonable to demand of its author by way of epochs, velocities, directions, charges, and so forth, he should firmly reject the argument that this is worth doing because "on the strength of Velikovsky's theory" such and such a phenomenon was predicted and was subsequently observed. Dr Velikovsky is undoubtedly an astute scientific eclectic with a flair for predicting (in a rather general way) the unexpected. He has to his credit the correct predictions that the Moon's rocks would be magnetic (and didn't the Nasa officials kick themselves?), that Jupiter would be found to be sending out radio signals, and that Venus has a retrograde rotation. He also has to his name many mistaken predictions, and he usually manages to brush aside the counter-evidence as inconclusive or unimportant. What he does not give is any very profound nexus between his theory and his predictions. There is much glib talk of electrodynamic forces, orbital perturbation, Van Allen belts, and so on, and a reader as deficient in a sense of scientific history as most of Dr Velikovsky's followers might be forgiven for thinking that he invented these things. This is not his fault, any more than the manifold weaknesses of his supporters are his fault, but the science – if we except some of the cosmic geology – is glib, none the less.

Part 3 of *Velikovsky Reconsidered* is a series of articles aimed at showing that Velikovsky's story is, as far as the planetary orbits are concerned, astronomically possible. Laplace might have been puzzled by the absence of equations, but I think he would have been broadminded enough to admit, even without equations, that in a system with infinitely variable parameters, infinitely many things are possible, and that the Velikovskian ballet of the planets might well be among the possibilities. Part 4 concerns the atmosphere of Venus, and all those hydrocarbons. For the moment things are not going too well for Dr Velikovsky, but time will tell. Part 5 deals with the Moon, and magnetic remanence. And that, in brief, is how Velikovsky is here "reconsidered", by a group of writers who are, with very few exceptions, committed believers. It is they, and not the scientific mafia, who have been "shaken to their foundations" by Dr Velikovsky's bizarre rendering of his ancient texts.

How do such people become so deeply involved in spinning new scientific myths out of such tangled wool? They can hardly be put in the category of those legions who leaf through the pages of *Worlds in Collision* in search of excitement on a wet Sunday afternoon, and who may be excused a little incredulity. Anyone taking the game seriously must face up to a book in which a constant stream of quotations from most of the world's ancient cultures has had imposed upon it an arbitrary and extravagant interpretation. Alternatives are rarely considered, and – for what the observation is worth – the interpretation tends to come before the evidence. Perhaps that is the hypothetico-deductive method at work, so we must not grumble. At all events none of us is equipped to weigh judiciously all of the textual evidence, although anyone with half an eye for the truth should realize that when a drama of Velikovskian dimensions is made to hang on the translation of only a word or a phrase in most of the sources quoted, it is not enough to use antiquated editions, or translations made by scholars ignorant of the strains to which their works would one day be subjected.

One can scarcely excuse Dr Velikovsky's ignorance of individual cultures on the grounds that he is acquainted with so many. He appears to be almost totally ignorant of studies of the interdependence of the world's mythologies – although of course to reveal the connections would not help the thesis that the events described in the myths of the world are independent descriptions associated with universal catastrophes. Even granted that he has a penchant for nineteenth-century texts, could he not have considered Max Müller? He seems to be completely unaware that when one author repeats uncritically a story from another, the number of authorities is not thereby increased. And can he really expect us to admit Jonathan Swift's allusion to the then unseen satellites of Mars as relevant to a prehistoric state of affairs? Out of the same window should be thrown references to Hevelius, Rockenbach, Bochart, Pomponius Mela, and a score of other irrelevant accretions. The book would become shorter and more transparent; but we should lose Hesiod's *Theogony*, which would be a pity, since this provides some of the finer patches of colour for *Worlds in Collision*, and thereby makes the book more bearable. As for the oral traditions of sixteenth-century South America, who is to say that they are not the product of some earlier Velikovsky?

The answer to the last question is obvious. But why should the historian who is expert on this aspect of South American culture waste his time on the Velikovskians? He might as well try to explain the Maya calendar to a South American football crowd. For no matter how cogent the complaint about persecution in the course of their search for the truth, Dr Velikovsky and his supporters have shown no inclination to heed the balanced criticism of the expert. Eleven years ago, Abraham Sachs of Brown University, Rhode Island, spoke for a quarter of an hour on the shortcomings of Dr Velikovsky's use of Mesopotamian material, which is of course drawn from cuneiform tablets (c 3000 BC to the first century AD). With a minimum of effort he exposed utterly the absurdities not only of numerous details but of the entire "new" chronology of *Ages in Chaos*. He showed that Dr Velikovsky, following a book written in 1915, was right to suppose that Venus is missing from a certain text; but that this is no argument for a solar system without Venus, since the other planets are also missing. On such slender evidence had *Worlds in Collision* been based. Professor Sachs gave short shrift to the work of Hommel ("senile by 1890 . . . his condition had certainly not improved perceptibly by 1920"), from whom Dr Velikovsky had taken some of his ideas about the Venus tablets of Ammisaduqa. Professor Sachs's general conclusions were that "in Dr Velikovsky's works, one finds a wasteland strewn with uncritically accepted evidence that turns to dust at the slightest probe", and that it is advisable to be a cuneiformist if you are going to write about cuneiform texts. Dr Velikovsky, who was present at this talk, has to the best of my knowledge never answered the criticisms made in it. They can hardly be dismissed as the work of a cuneiformist mafia.

If *Worlds in Collision* leads many of its readers astray, this is no doubt because they are too generous. There are scientists who accept it as good history, and historians who assume that it must be good science. It is of course fiction, implausible, incoherent, and untrue to human nature itself. Can anyone believe that the Egyptian army, or any army of men, was once so brave as to charge between two walls of water in the circumstances described by Dr Velikovsky, even granted that the Israelites had gone through first? Faced with the problem of making sense out of Exodus 13-14 without recourse to miracles has for most critics been an exercise in restraint. For Dr Velikovsky it is an exercise in melodrama. Speaking quite generally, are not earthquake, fire, tidal wave, hurricane, and electric storm ingredients enough to inspire most of the stories of natural violence recorded in ancient literature? The sound of one of the 1883 Krakatoa eruptions was so loud that it was to be heard 3,000 miles away. The sea waves generated travelled even further, and clouds of debris were scattered over much of the Earth's surface, bringing with them sunsets the very colour of the jacket of *Velikovsky Reconsidered*. I would not wish for one moment to foist such *dramatis personae* on Dr Velikovsky (who has in any case made some use of them in the sub-plots), but a very substantial part of *Worlds in Collision* might have been as easily cast in terrestial as in celestial dress.

And as for the celestial scenes, when they are not vague or positively wrong, they are wildly speculative. They also show a marked tendency to suppress the unpalatable. Two or three examples must suffice. "And there was a thick darkness in all the land of Egypt three days" (Exodus 10, xxii); but was there not light where the Israelites lived? Augustine wrote: "Minerva [Athene] is reported to have appeared . . . in the times of Ogyges." So much for the birth of Venus in historic times; but look to the source and you find that Augustine makes Mercury much later than Minerva – besides which he is talking about real human beings who were later deified, not about planets, by any stretch of my imagination. "Augustine also synchronized Joshua with the time of Minerva's activities." Again this is no model of honest quotation, for all that Augustine says is that between the exodus of the Israelites and the death of Joshua, ceremonies were instituted by the Greek kings in honour of false gods. But who carries Augustine on a railway journey, let alone his Chinese, Indian, Mexican, or Babylonian counterparts?

And so I could go on. Literary permutations on gods, planets, archangels (Gabriel represents Mars, by the way), and men, are fuel for any burning imagination, and Dr Velikovsky knows better than most of us how to feed the flames. Eighteen printings of the key text! What university press would nowadays turn away such genius? But then, Dr Velikovsky is "a psychiatrist by vocation and a historian by avocation". To him it is natural to ask to what extent "the terrifying experiences of world catastrophes have become part of the human soul and how much, if any, of it can be traced in our beliefs, emotions and behaviour as directed from the unconscious or subconscious strata of the mind". Will this be the last refuge when the scientific mafia has done its worst? In the last essay of *Velikovsky Reconsidered* William Mullen points out: "If biological experimentation offers concrete proof that instincts acquired under catastrophic circumstances might be transmitted genetically, then the whole psychology implicit in [Dr Velikovsky's statement] is objectively grounded." So perhaps we shall live to see university chairs in archaeo-psychogenetics. For my own part the only worlds I can see in collision are those of reason and unreason.

6

ON THE TRAIL OF THE COMET

"The Cosmic Serpent" is a name given to a hypothetical giant comet, a fiery dragon of a thing that hurled thunderbolts and generally caused mayhem in prehistoric times. If you find the hypothesis acceptable, you can say that three thousand years have all but erased the terror of those events from human memory; but that Victor Clube and Bill Napier, two professional astronomers, have caught the comet by its hypothetical tail, before it slipped from human awareness completely. They have written an exciting book about it, a book which, with its carefully calculated exterior, contains many new and challenging arguments. Patrick Moore declares on the jacket that it is one of the most extraordinary books he has ever read. The astronomical part is perhaps more sober than the title suggests – a wolf in serpent's clothing, in fact. The clothing is at times plainly uncomfortable, but was presumably donned to attract an excitement-loving public that would have been bored by the sub-title: "A Catastrophic View of Earth History". From an astronomical point of view, this is in a very different class from most popular catastrophist writing. Velikovskians will feel more at home if they read the work backwards, for it is at the end that the more sensational rewriting of history comes, not to mention a few kind words about Velikovsky's historical excursions. Faint-hearted readers be warned.

The Cosmic Serpent is a splendid source of object lessons in how and how not to write for a wide public. Should the university world ever go into another phase of reckless expansion, some research institute for scientifico-rhetorical orthopraxis might well put it under the microscope, to study its peculiar blend of fine-grained argument and fantasy. Structurally speaking, it fits together beautifully. There are touches of condescension in the fine detail in the form of phrases like "0.1 per cent of a millimetre", but they are offset by passages such as that taking for granted the notion of "hydrogen burning" in stars. Generally speaking, it is a model of plain exposition. It is rather a bore when it moralizes about the Establishment, about bandwagons falling over cliffs,

CLUBE, V., and B. NAPIER [1982], *The Cosmic Serpent: A Catastrophist View of Earth History*, London: Faber, pp. 299.

the unwisdom of a prescribed wisdom, and so forth; but it has a trendy scientific touch in its excessive use of the word "scenario", which is sprinkled over its pages like tektites over the globe. Historians will have to bear in mind that Copernicus and Tycho are lunar craters. How difficult to remember every reader, even of a review; and yet if the thesis of the work is acceptable, there is a reason why every reader should be concerned, for the history of the world is at stake.

The book opens with a lengthy sermon on scientific scepticism. It is pointed out that astronomers disagree about many things – for example, about quasars, spectral displacement in the light from distant star systems, and the evolution of them, the galaxies. Our Sun is of course a member of a star system ("the Galaxy") with a characteristically spiral form. It is generally accepted that the universe of galaxies is expanding. Dr Clube has argued elsewhere against this view, and in favour of the view that *our* Galaxy is in a state of rapid expansion. He believes – and this idea is at the very heart of the book – that there are sporadic bursts of activity within the nuclei of individual galaxies, at intervals of, say, 100 million years, resulting in the ejection of material with enormous velocities.

Briefly, the spiral arms of our Galaxy are young, and our Sun is old. The Sun moves steadily through the spiral arms, crossing them every 50 million years – and having crossed what is known as Gould's belt about ten million years ago. Now the Galaxy might have seen three explosions in its nucleus in the last 100 million years, the last perhaps 30 million years ago. As for the consequences of crossing the spiral arms, some astronomers have thought that proximity to a supernova, an exploding star, might have upset the history of life on Earth; others that interstellar gas clouds might have provided us with our ice ages; but here it is argued – and at a qualitative level very convincingly – that our solar system acts as a large gravitational scoop, as it were, for millions of large solid bodies. These "planetesimals", or conglomerates of ice, dust, and rock, include our comets. Recent telescopic evidence is put forward for the existence of gigantic interstellar comets, and a sketch is included of the way they might grow in interstellar space. To a catastrophist Earth historian, though, the question of paramount interest concerns their potential bombardment of the Earth.

Captured by the Sun's gravitational field, and put into orbit round the Sun, their acquisition by our solar system seems to be erratic, but their loss from the system fairly steady. Some (unconventional) statistics for survival imply that the last batch to have been captured was acquired a few million years ago. It must be said that much of the book's scientific colour comes from its use of sparkling new observations and its authors' willingness to mention alternative interpretations. Evidence for the rates of acquisition of comets is drawn from meteoroids found on the Moon's surface. But do these come from the asteroid belt, are they an interstellar dust, or of cometary origin? Spacecraft have already provided clues allowing some of the alternatives to be ruled out. Some comets get into the asteroid belt. Perhaps asteroids are inactive comets. There are fifty or so satellites, ring systems, and Chiron sized bodies in the solar system. (Chiron is in orbit between Saturn and Uranus, and is of the size of a large asteroid – say as big as a good-sized mountain.) Five or ten of the fifty are in unstable orbits, and are likely to be lost to the system, so the picture of a steady-state solar system seems to be wrong. This is a key point in the argument, for episodic capture of such objects by the solar system implies occasional *bombardment of the Earth*.

How often, and how catastrophic, are these encounters likely to have been? The suggestion is that we are likely to have been hit by objects the size of the asteroid Apollo a handful of times in the last few hundred million years. (Apollo was discovered in 1932, and was the first asteroid known to be in an Earth-crossing orbit, that is, to be a potential collision hazard.) More specifically, we are told that in the last 600 million years the Earth has probably been struck about ten times by missiles with energies of the order of ten or twenty million hydrogen bombs, and once or twice by objects with twenty times as much energy. Tennyson's "Nature, red in tooth and claw" suddenly becomes an almost cosy image.

How plausible is all this? I can only say that, irritating as I find their occasional side-swipes at mainstream opinion, their vague talk of a "new physics", and their tendency to introduce quite irrelevant bits of scientific history whenever things are getting exciting, I am willingly carried along by the general drift of the two writers' astronomical arguments. Craters seem to be ubiquitous in the solar system. Studies of the lunar surface suggest that the cratering rate was once higher than it is now. Hudson's Bay and the Gulf of Mexico are perhaps impact structures, and there is reason for thinking that there are many large impact structures on the Earth's surface of the order of a thousand kilometres across.

Catastrophes on the scale suggested would of course lead to the sudden extinction of species, and provide an explanation of the apparently erratic extinction rates revealed by palaeontologists, with "brief episodes of mass extinction of organisms followed by invasions of new forms into vacated ecological zones" (N. D. Newall). Does the fact that major extinctions seem to be associated with the principal geological boundaries (Permian to Tirassic, and so on) mean that they share a common cause? In the case of one boundary there is a sudden jump in the concentration of iridium in clay (at the Cretaceous-Tertiary boundary), iridium perhaps of interstellar origin.

But there are possible effects other than at the Earth's surface. Quite modest (and hence more frequent) impacts could well re-align the circulation currents in the core which generate the magnetic field. The magnetization of rocks has long been known to give evidence for a reversal in the magnetic field rather more often than once in a million years, and the coincidence of falls of tektites with the reversals might be thought to support the idea that associated impacts are their cause. Then there is the coincidence of dinosaur extinction with the greatest period of vulcanism in the Earth's history; and the occurrence of vulcanism generally in episodic bursts, something an impact theory seems to explain more effectively than the theory of plate tectonics. (The theory of continental drift, powered by slow currents within the mantle, is not itself at issue.) It all hangs together very plausibly: huge impacts in the remote past, and much more frequent encounters with smaller bodies in Earth-crossing orbits, even in relatively recent times. Objects in this second class would devastate areas a few hundred miles across.

This is where the short-period comets come into the story. There are over a thousand members of the family of Apollo-type objects with a diameter of more than a kilometre, and these are explained as long-period comets, pulled into their short-period orbits chiefly by Jupiter, and boiled dry, so to speak, relatively quickly. It is hard to explain why there are so many short-period comets. The suggestion made in the book is that they result from the fragmentation of a single large comet; but no matter. The important historical point is that one or two of them could well have come close to the Earth, with spectacular effects, both visually, and from impacting debris. It is calculated that within the last five thousand years there must have been about fifty impacts in the energy range 1-100 megatons of TNT, about five in the range 100-1,000, and an even chance of an impact in the range 1,000-10,000. Three quarters of them will land, like American astronauts, in the oceans. Appropriately, the best-recorded in recent times landed in Siberia on June 30, 1908, near the Tunguska river, flattening a forest to a distance of seventy kilometres from the centre, and causing such a commotion that a driver on the Trans-Siberian Railway 600 km away was forced to halt his train. This was perhaps one of the fifty smaller events, say a comet fragment with impact energy between 40 and 100 megatons. Such missiles are too small to be seen in space, but seismic stations on the Moon bear witness to trails of boulders in the wakes of comets, and of course meteor streams are evidence of old comet trails. Such trails last for a thousand years and more, and from them (and early historical records of them are not uncommon) we can say that the Earth has run across the orbits of something like fifty of them during the past two thousand years. The comets Halley and Encke were in all probability sources of real terror on the Earth more than once in the past five thousand years.

This is where the confident astronomical trail ends, and the somewhat impressionistic historical and mythological trails begin. In striking contrast with the earlier sections of the book, there is rarely a clear display, or mention, of alternative interpretations. Not that there is anything lacking in range. Take natural philosophy, for example: the rise of materialism in classical times is associated with the passing away of important prehistoric gods which were comets in the sky. Lucretius prompts the hypothesis; but before reaching for a translation, you must bear in mind that it will have been done by someone for whom comets are not what they were in the mind of man. (Had it been otherwise, the somewhat dark comment goes, "scholarship would already have foundered".) The drift of the argument is to show that although comets were once among our chief deities, later moves against naturalistic religions (by such as the Greeks, Amos, Zoroaster, and the Buddha) might have come because some comet got thrown out of the solar system, or because the fuel for its tail was used up. This unlikely tale hangs by a very slender thread, and Lucretius is the sceptic who gets most of the limelight in its telling. He comes along, refusing to find *mind* in lands and Sun, sky and sea, stars and Moon, and telling us that it wouldn't be right to impose a punishment "as on the rebellious Titans, on all those who by their reasoning . . . seek to darken . . . the Sun". The comment offered is that the argument "can make reasonable sense only if comets were among the gods and the Titans were a special group of comets which could somehow, on occasion, dim the Sun".

The excessively enthusiastic sleuthing doesn't stop there. For it to be thought possible that a comet may darken the Sun, "the knowledge that some very large comet indeed had at some time appeared must have been available". It doesn't seem to worry the two writers that such knowledge is not reported in a less oblique manner in other surviving texts. And almost as a justification for the tendentious argument that has just gone, they add quite gratuitously: "In like manner, one cannot casually reject the claim by Diodorus of Sicily that the Chaldeans for example knew about the regular return of periodic comets". Not casually, perhaps. (Lucretius is presumably meant to have known the principle since when the Sun caught Phaeton in his fall, and set him and his steeds on their proper course, we are invited to read the description of Phaeton, "the everlasting torch of the firmament", as that of a regularly recurring comet.)

The Cosmic Serpent is filled with this sort of wishful thinking:

> It is remarkable indeed how few are the recognizable references to comets as such in Babylonian and Egyptian records. This cannot be because they did not exist, so it must be because *they were generally described as something else.*

Once we recognize this fact, "we are obliged to see them as being among the most important and fundamental elements of the ancient sky", and "the Hellenic philosophers were thus responsible for a really quite major revolution in human thought: they were the first to describe comets in particular much as they appear to us, the first to make rational attempts to explain their origin in terms that we recognize as scientific". Such honesty in the declaration of one's strategy in exegesis is rare. The strategy is certainly powerful. The cosmic god Ptah spews forth the deities Nenet and Nun, and this suggests that either Jupiter spews forth comets as the result of a close encounter, or simply that a large comet split into two. Leto gives birth to Apollo and Artemis. Not quite Phaeton coming from Apollo who was fathered by Zeus, "but the underlying themes are not dissimilar when once they are recognized as derived from different accounts of a comet breaking up". (Zeus is of a much greater antiquity than Homer, by the way, since Greek, Latin, and Sanscrit are descended from a common Indo-European language.) Hesiod's battle of the Titans begs to be interpreted as a shower of comets doing battle with pre-existing ones; the course of evolution of Chinese dragons (taking 3000 years) suggests short-period comets; the World Tree of Norse cosmology is seen as a giant comet; and so the list goes back and forth through historical times, with its hints of knowledge lost but terror preserved. Was Apollo's bow the crescent head of a huge comet, and was Typhon originally a huge fall of meteorites? Comets are like snakes in a psychoanalyst's consulting-room: everywhere for the asking.

It is hard to avoid scepticism, and ironically, the harder the two writers try to convince, the harder it is to accept their story. There is surely "something there", behind so many early myths, but the question is how many layers of telling and re-telling conceal the events to which they ultimately bear witness? Occasionally one has the feeling that the events are very near to the surface. It seems very plausible that the pillars of cloud and fire in *Exodus* (13:21-2, and 14:18-19) were cometary, and somehow connected with comet Encke. But to quote H. S. Bellamy, who quotes a tenth-century Italian, who quotes "mythological sources now lost", to the effect that the Sun, Moon, planets, and signs of the zodiac are connected together by a fiery dragon, ("like a weaver with his shuttle") as the only "literal interpretation of the dragons as comets" is doubly odd. First, it makes no mention of comets; and second, the dragon is the conventional name for the lunar nodes, which do connect together, as it were, the ecliptic and the Moon's orbit. One should beware, even more, what are elsewhere in the book called "hypothetical mediaeval monks". The real ones are bad enough.

The denouement of the work comes with the arguments for two important historical episodes of cometary bombardment, one circa 2500 BC, and one in 1369 BC, the first responsible for the Flood, world-wide, the second confined to the eastern Mediterranean. Their thesis obliges the two authors to revise standard Egyptian chronology, by dropping about four centuries, much as Velikovsky did before them. Certainly the arguments here are much more carefully drafted. Radio-carbon dating, bristlecone pine, volcanic acidity in Greenland ice cores, traces of vulcanism in the Mediterranean – it is all made to hang together in something like the style of the early chapters, even when it doesn't quite convince. (I don't see that the Karnak clock marks can tell us very much, and all the evidence for the abandonment of the wandering year in Egypt is very ambiguous; but I must leave these matters to experts.) There is even a most beautiful spanner poised over the works, with the thought that large fireballs might upset Carbon-14 dating through modification of the atmospheric C-14 content. All the sadder, therefore, that the book is packed with so much modern mythology.

Perhaps it is inevitable that the megaliths be brought into the picture. They are old, they have to do with the heavens, and they're not even associated with an ambiguous literary tradition. Their common orientation on solar and lunar horizon phenomena is beyond doubt. Why not, then, orientations on phenomena associated with Olympus, *alias* World Tree, *alias* Cosmic Serpent? The lack of obviousness in the regularity of such phenomena, and their infrequency, combined with the slow erection of the monuments, make this a pointless trail; but what of the stones as seen from above? Did not Stukeley say they were serpent temples, and was he not "closer in time to local tradition which had by then not been silenced by the arrival of a new rationalism"? The conjecture is that the earthworks and avenue at Stonehenge might be a portrayal of a comet with a stubby tail, rather like one described by D. Gill in 1882. A problem that offers itself at once is that of the date of most of the European stone circles – namely well before 1369. "It is tempting", comes the answer, "to see the 1369 BC catastrophe as the culmination of a period during which the need to appease the sky god and/or predict its imminent arrival became ever more pressing".

Everyone to his own temptations. A more attractive temptation, by my taste, and one put in my way by the book, is to see at least some of the Neolithic "cup and ring" markings as comets. Similarities with some of the Han tomb paintings of comets are quite striking. Less plausible is the claim that abrupt changes in the orientations of Stonehenge and the pyramids represent an attempt to track one particular comet. The hypothesis is described as "interesting and mathematically testable", but it might have been added that the second is true only given an excessive number of *ad hoc* assumptions. We have no really reliable information with the right degree of precision about prehistoric comets, and the mathematical test in question would itself run the risk of circularity.

One of the book's recurrent themes is that of the prevalence of ideas post-1369 BC as to the sky-gods, "supported by knowledge of what happened in and before 1369". This "eventually formed the basis of what we now recognize as astrology". Here is a fanciful reading of history for which there is no evidence, unless the word "eventually" be allowed to swallow up a thousand years, or thereabouts. This illustrates an important point incidentally, and that is the ease with which writers in this genre tend to pass backwards and forwards through time and space finding whatever it is they set out to find. The *tzolkin* calendar used by Aztecs and Toltecs, for instance, had a 260-day period – which happens to be half the period between oppositions of comet Encke, more or less. (It also happens to be precisely twice the sum of the squares of seven and nine, numbers found mysteriously in many of the world's calendars; but let me not substitute one heresy for another.) You can fit together the periodic times of the planet Mars and Halley's comet – so this opens up a way of bringing at least one planet into the cometary picture, and so extending its scope. More important, it allows you to bring comets into texts mentioning Mars. (Whether the association "may eventually be verifiable from Babylonian astrolabes" is something I simply cannot understand.) An eclipse cycle of the Babylonians is seen as nine cycles of Halley's comet. (Could 684 years not be thirty-six times the calendar cycle of nineteen years?) Comets are found to be everywhere, whether in the Beltane fires of the Celts, the disaster at Mycenae, the fire-god of the Maya, or the Feast of the Repelling of the Troglodytes in the reign of Amenhotep I. No doubt more careful work, perhaps even prompted by this, might turn up more convincing proofs, but they will certainly not be proofs if they are only in the (infectious) style of "Could it not be the case that . . .?", a style that pervades *The Cosmic Serpent*. The division of the stars into constellations *might* have followed from the fact that comets turned astronomers' attentions to the sky. There *might* indeed have been a link between a story in Plato and inhabitants of the Atlantic seaboard driven thither by a Flood in 2500 BC. It *might* have been the case that records allowed "astrologers" to link events in 2100 BC, 1369 BC, and 689 BC, and to predict "great happenings around 0 BC". But is any of these things likely? To support the Messianic conjecture, for instance, it is not enough to say that "Some 2000 years closer in time to these events than us, it is reasonable to suppose memories of what went on at these epochs still survived". That is just starry-eyed nonsense. If wild guesses like this are to be justified, as they might prove to be, it will be as a result of hard historical work.

The pity of it all is that here is a truly important subject, important for anyone with a concern for the history of civilization, at least. It is a subject in which a number of excellent scholars have learned the difficult art of walking slowly. A racy book like this, exciting though it will plainly be for many a casual reader, can only spoil the going for the more committed.

7

THE CULMINATION OF PTOLEMY

A History of Ancient Mathematical Astronomy is without question the most important study of its subject yet written. It is not an easy work – more like a mathematical Pauly-Wissowa than like Gibbon, in fact. I suppose it could be said that most historical writing passes (to use phrases from George Meredith) somewhere between ascetic rocks on the one hand and sensual whirlpools on the other. As all who know his work will agree, it is hard to find scholarship more ascetic than Otto Neugebauer's. By his own confession, he refuses to hide his ignorance behind "a smoke-screen of sociological, biographical and bibliographical irrelevancies"; but when it is plainly *not* a question of ignorance, he still prefers to avoid these particular historical dimensions, content to reveal to those who are prepared to penetrate the technical detail "a kaleidoscopic picture . . . of the history of the first and oldest natural science".

His is a history of intellectual achievement, rather than of such important biographical matters as Archimedes' bath or Tycho Brahe's silver nose. It is not a compilation from existing encyclopedias, such as is Sarton's *Introduction* (here qualified as overburdened with irrelevant detail, not to speak of its "absurdly rigid chronological arrangement and the historical introductions which are reminiscent of the mentality of Isidore of Seville"). It is not a web spun from the history of Greek philosophical speculation. (And here Laplace is quoted, to the effect that "au milieu des rêves philosophiques des Grecs, on voit percer sur l'astronomie des idées saines".) It is no mere enumeration of arguments concerning the movement of the earth or the centrality of the sun. Such sensual whirlpools are as necessary to Professor Neugebauer as are neon lights to the Acropolis.

So much for style. The overall plan of *A History of Ancient Mathematical Astronomy* is well conceived, but not immediately obvious. Ptolemy's *Almagest* (second century AD) comes first, "since it is fully preserved and constitutes the keystone to the understanding of all ancient and medieval astronomy". There follows an investigation of earlier work, in particular

NEUGEBAUER, O. [1975], *A History of Ancient Mathematical Astronomy (3 vols)*, Berlin and New York: Springer, pp. 1457.

Babylonian astronomy, which is then related as closely as the evidence allows to early Greek astronomy. The last broad historical division covers Hellenistic astronomy, as it is known from papyri, Ptolemy's minor works, and the so-called *Handy Tables* of Ptolemy. (Professor Neugebauer incidentally discredits the arguments of those who would have Alexandria a crucial centre of astronomical research and influence during the five centuries before Ptolemy.) There is much on the theory of instruments, especially the gnomon and astrolabe, despite a disclaimer at the outset, but the promise to avoid Chinese astronomy is kept. The influence of Chinese on Islamic (and hence Western) astronomy was in any case probably not visible before the creation of Mongol states in Western Asia. Within its terms of reference, *HAMA* (as the work was designated by the *cognoscenti* even before it was published) is comprehensive to an extraordinary degree. Without wishing to destroy any of its kaleidoscopic quality, I shall put the pieces back into a roughly chronological order.

One of the main obstacles to a historical understanding of Babylonian astronomy is that we (and the Greeks were in the same situation) know nothing of the motives or of the arguments, mathematical or astronomical, which led to the procedures so painstakingly decoded by Professor Neugebauer and fellow historians. As he explains, the influence of the available mathematical tools on the shaping of scientific theories is all-important. Babylonian astronomy operates without any model of a spherical universe and without circular motions – both of which have seemed necessary *a priori* to those who only knew the astronomy of the Greeks. From *HAMA* it is possible to see very clearly how the arithmetical astronomical systems of the Babylonians functioned, and at the same time to marvel at the tenacity of such pioneers as the three Jesuit fathers, Strassmaier, Epping, and Kugler, who first brought the Babylonian methods into the light. Others who should be mentioned for their devotion to the task of publishing cuneiform texts are Pinches, Sachs, Thureau-Dangin, and of course Neugebauer himself. The sheer physical problem of matching up disordered fragments of broken clay is enormous. One of the most notable texts, a lunar ephemeris, now almost completely restored from nine fragments, took some seventy-five years in the rebuilding.

The theory of the moon was the most highly developed part of Babylonian astronomy. The lunar ephemeris mentioned above runs to eighteen columns, while most planetary ephemerides need only four or five. None of these ephemerides, nor any of the procedure texts, contains any hint of astrological application such as we find in the Hellenistic and Roman period. The Babylonians predicted phenomena (e.g. consecutive stationary points for a planet) rather than numerical coordinates (e.g. longitude on the ecliptic). All planetary theory must begin with the experience of periodicity in observed phenomena, and the weight of evidence amassed, sifted, and evaluated by Professor Neugebauer shows clearly that the Babylonians were masters of the analysis of observed periodicities. But speaking generally, did they observe as much as good astronomers are supposed to do? Professor Neugebauer gives ample evidence (as, for example, from what is known of the nineteen-year intercalation cycle) to shatter the belief, "inherited from late antiquity", that they observed extensively, and over great periods of time. Mathematical schematization became, it seems, something of a straitjacket, limiting the empirical freedom of the Babylonian astronomer, but making the historian's task marginally easier.

Whatever the limitations of Babylonian mathematical astronomy, Egyptian equivalents – notwithstanding numerous ancient and modern myths – were virtually non-existent. Egypt is given a section in *HAMA* only "in order to draw the reader's attention to its insignificance". Among the vast collection of inscriptions and papyri now extant, there is not one record of a fundamental astronomical observation; and most of what now passes as "Egyptian astronomy" is a crude astronomical arrangement for the division of the night into "hours". Why all those ancient references to Egyptian astronomy? As Professor Neugebauer points out, all inhabitants of Egypt, including the Alexandrian Greeks, were "Egyptians". As for the modern historian who wishes to postulate secret knowledge, "nothing is less secret than 'secret' literature". There is a cautionary tale here which concerns an absurd interpretation by Duncan Macnaughton (1932) of hieroglyphic signs supposingly recording an Egyptian horoscope to be dated July 14, 2034 BC. In AD 1962 Macnaughton's interpretation was used (at the California Institute of Technology) as the most distant checking point for some highly accurate planetary tables.

The early Greeks got better marks, although Professor Neugebauer has no time for Greek philosophy as an early stage in the development of science. He has an inbred dislike, for example, of the "philosophical palaver" of Theon, and the "pythagorean" speculations which so often took precedence over accurate observation. Greek science inherited from the Babylonians sexagesimal computation and other arithmetical techniques, as well as large numbers of parameters for lunar and planetary motions. The time of closest contact seems to have been the second century BC, and the course of Hellenistic astronomy was clearly influenced for three centuries thereafter. (The first evidence for the Babylonian "system B" outside Mesopotamia is of great interest. It is in the astronomical poem of Manilius, written at the beginning of our era.) The Greeks were inspired above all by their geometrical view of a spherical earth and sky, and the compounding of circular motions. The extraordinary achievements of Eudoxus, with his model of a universe comprising homocentric spheres, is hardly undervalued by Professor Neugebauer, who nevertheless points out that "purely geometrical ingenuity leads nowhere", adding that historians' fascination with the geometrical aspects of planetary theory has obscured the decisive role played by computational procedures.

Aristarchus's treatise on the sizes and distances of the sun and moon (well known from the book by Thomas Heath, but here reduced to seven pages as a proof of Professor Neugebauer's synoptic powers) is described as a "purely mathematical exercise" having little to do with practical astronomy, Hipparchus is at the meeting of the ways. Apollonius had mastered the kinematics of eccentric and epicycle; Archimedes had investigated the length of the year; and much Babylonian material was at last to hand.

A History of Ancient Mathematical Astronomy is a valuable antidote to many popular historical myths. Remarks made on Heraclides ("precursor of Copernicus", as the myth goes) are likely to stir up controversies not yet extinct. A determined attempt is made to erode the idea that the astrology of our era is a Babylonian thing. "Before the fifth century BC celestial omina probably did not include predictions for individuals, based on planetary positions in the signs of the zodiac and on their mutual configurations." A crude astrology applied to Mesopotamian society as a whole was transformed in Greek hands into a universal system applicable to individuals. This "truly Greek creation" was "in many respects parallel to the development of Christian theology a few centuries later".

Much of the mythology attaching to the name of Hipparchus is sifted for any value it might have. His importance is seen to lie in his new methodology of exactness, observable numerical data providing a decisive criterion for the correctness of theories which are now thought to have been not of his own devising. There is a breath of fresh air (admittedly inspired by Vogt's work) blowing through those pages in which the star catalogue is discussed, and in which it is concluded that the first catalogue of stars (but one not yet based on orthogonal ecliptic coordinates) is the Hipparchian, while the second is Ptolemy's of the *Almagest*, and that this could not have been derived from the first. There is criticism of the way in which the history of Hellenistic mathematical geography has been obscured by the ascription to Hipparchus of all concepts found in Ptolemy. Limits are hard to set to the undoubted genius of Hipparchus, somewhat overplayed by Delambre, who was nevertheless only following Pliny's extravagant lead. According to Pliny (whose modern apologists are dismissed by Neugebauer), Hipparchus predicted the eclipses of the sun and moon for six centuries. This prediction is not plausible, nor would it have made any sense. Another crown of laurels to be removed from Hipparchus's brow is that given to him by historians for his "invention of spherical trigonometry" – although it is allowed that he might have used the so-called analemma methods which dominate later Indian trigonometry.

The Culmination of Ptolemy

As for the theory of the precession of the equinoxes, there is of course no question of removing this, one of the greatest discoveries of ancient science, from the list of Hipparchus's achievements, and indeed Professor Neugebauer gives reasons for proposing him as the inventor of the so-called theory of trepidation (an oscillatory precession).

Between Hipparchus and Ptolemy, it seems that the sort of mathematical ingenuity typified by Eudoxan astronomy was dead, but that the spread of astrological practice created a need for tables and other adjuncts of calculations. Ptolemy quotes not a single observation from the two centuries between Hipparchus and Menelaus. The miracle is not that the old techniques were preserved, but that men capable of transforming the grey historical uniformities should so suddenly appear, and almost as suddenly disappear, so that Bīrūnī, for example, can be described by Professor Neugebauer as possibly Ptolemy's only peer before Kepler and Newton. To appreciate the role of genius in the evolution of scientific thought – which some among us seem anxious to brush aside – one might consider the essentially barren nature of that extraordinary ritualization of calculation by the Tamil eclipse calculators, whose methods were observed by Le Gentil in 1765 and also by John Warren in 1825. Amazingly accurate results were to be had from computation with cowrie shells on the basis of memorized tables and with only such theory as was embodied in the procedural rules. Again, Babylon is at the beginning of the trail, but at the end there was little beyond the ritual. Needless to say, there is in *HAMA* a good account of P. Ryl. 27 and that group of Greek and demotic papyri which provide an understanding of Babylonian influences on Greek astronomy and the transmission of Babylonian methods to India – which country they had reached by the second century AD, that is, in pre-Ptolemaic form.

The high-water mark of any work on classical astronomy must inevitably be set by the *Almagest* of Ptolemy. With his customary thoroughness, Professor Neugebauer explores the work at length and relates it to its precursors no less than to Ptolemy's own *Analemma, Planisphaerium, Tetrabiblos*, and the rest. There is a painstaking analysis of the letter of Synesius relating to the anaphoric clock, and there are many salutary remarks over the pathetic scholarly treatment thus far accorded to the *Geography*. Ptolemy's great achievements in the theory of terrestrial mapping were scarcely appreciated before the sixteenth century, and the virtues of his writings on optics not properly before our own time. As for the so-called Ptolemaic system of the universe, a crypto-Aristotelian physical cosmology superimposed on Ptolemy's abstract mathematical frame, the evidence that Ptolemy was its originator is scarcely a decade old. Summarized here, it is neatly related to Proclus. Perhaps the comments made on the philosophical fragments are too dismissive:

> It is not surprising that Ptolemy followed peripatetic or stoic doctrines but

with a certain eclectic attitude. Fortunately these philosophical theories are without importance for his actual astronomical work.

One has the feeling that where some people have nightmares about elephants, Professor Neugebauer has them about philosophers. But lest he be thought insensitive to the overall tastes and attitudes of Ptolemy, I note the judgment that the three books of the *Harmonics* (relating musical and planetary harmonies) very probably "record thoughts for us which he must have felt to be central to his whole life's work". The analogy with Kepler is inescapable.

After Ptolemy there is a feeling of anticlimax, heightened by the neglect of Byzantine materials – and especially of Islamic influences on them – by modern scholars. Why were Ptolemy's observations not repeated and refined, and his mathematical theories further developed? Professor Neugebauer notes briefly the unfavourable cultural climate of the Later Roman Empire, some unfortunate didactic trends, and "admiration for the philosophical tradition" which began to create an "increasing opposition to the method and the spirit of scientific astronomy". The astrological market helped texts to survive, and Ptolemy's *Handy Tables* (themselves to this day awaiting a good critical edition) carried a disproportionate burden of responsibility for the transmission of Ptolemy's ideas. The transmission to India, to Islam, and to Christendom, is not Professor Neugebauer's subject, but these matters are touched upon lightly and entertainingly in the first dozen pages of *HAMA*, which the historian of a later period will ignore with impunity.

HAMA is a work of rare erudition, and no historian of the ancient world should be ignorant of its existence and its general character. It can be read at many levels, and even those whose concern is with ancient societies at large will be advised to try to square their ideas with Neugebauer's. (At a more trivial level, the man who wants guidance over the calculation of ancient celestial phenomena should search out note 16 on page 98.) The neutrality of Professor Neugebauer's mathematical style makes his strong human likes and dislikes seem the stronger by contrast. He castigates, for example, the

> ridiculous provincialism of classical scholarship which is willing to compile tomes listing every appearance in Homer of the names Achilles or Agamemnon but has not yet found the time to produce, e.g. a reliable edition of Ptolemy's "Geography", a work whose influence on the civilization of the Middle Ages cannot be overestimated.

He is disdainful of philosophy, the "history of ideas" (at least in one connotation of the phrase), the astronomy of ancient Egypt, Hegel, and charlatans generally. But his work is a history of ideas in the very best tradition: it is in fact the *Almagest* of ancient history, and will not soon be equalled.

8

ASTROLOGY AND THE FORTUNES OF CHURCHES

When Jerome Cardan asserted that the breach of Henry VIII with Rome was caused by the conjunction of Jupiter, Mars, and Mercury in the constellation of Aries in 1533,[1] he was doing his utmost to offer an interpretation of the historical event in terms of an astrological doctrine of conjunctions which by then had become commonplace. Imported into Christian Europe with the writings of decidedly unchristian astrologers – notably Abū Maʿshar (Albumasar) – the doctrine offered simple procedures for deciding the fortunes of churches and religions on the strength of conjunctions of the superior planets, in particular of Saturn and Jupiter. This was an idea which many who were unschooled in the higher reaches of astrology felt they could grasp, and one which lent itself to the support of much religious prophecy, sometimes restrained, often wild, but usually comprehensible. I shall indicate its chief sources, and some of the ways in which it was developed to serve religious ends, especially in the fourteenth and fifteenth centuries. It was still taken seriously in certain university circles, I should add, at the end of the seventeenth century. I do not know who was the last notable theologian to give the idea credence, but I shall end with an account of Pico della Mirandola's critique of it. When a religious event is correlated with a particular planetary conjunction only after the whole thing has happened, there is a sense of anticlimax, if not of plain deceit. The writings to which I shall refer can at best serve as sources of ecclesiastical history of a rather low grade.

When they are prognostications made before the event, they are not in the strictest sense sources of ecclesiastical history at all; but they are in themselves, after all, a part of the history of religious thought which it would be dishonest to brush aside.

I cannot begin to do justice here to the commitment of the medieval church to astrology as a whole, but I will begin with some remarks of a general sort, to indicate the size of the problem. Many a theologian whose writings seem to be untainted with astrology may be found slipping unawares into a doctrine of astrologically propitious times, whether the relatively mild one that prayers are more effective if offered at one time of day rather than another, or the belief that the stars pour out their influence in astrologically determinate ways to whose rhythm man, if he is wise, will learn to relate all his religious actions. 'Join then', said William Laud to king Charles's Third Parliament in 1628, 'and keep the unity of the Spirit, and I will fear no danger though Mars were lord of the ascendant, in the very first instance of this Session of Parliament, and in the second house, or joined, or in aspect, with the lord of the second, which yet Ptolemy thought brought much hurt to commonwealths'.[2] Another man in an influential theological position – although no doubt lacking Laud's fine library of works by Ptolemy and his legions – was the philosopher Pietro Pomponazzi of the previous century, who firmly believed that the cross and the name of Jesus had power because, for a certain time, the stars looked with favour on the new religion, and gave power to the symbols to produce miracles.[3] On a lower plane, the ordinary learned clerk of the centuries between the thirteenth and the seventeenth might have strongly disapproved of the practice, but he is likely to have known of at least one or two of the many doctrines of propitious times. He might have tried his hand at the art of deciding on a suitable day and hour for a journey, for example, basing his calculation on the position of the Moon; or he might simply have been guilty of the avoidance of 'dismal' (or 'Egyptian') days, such as were listed in calendars and almanacs. If we are indeed to distinguish, it is an uncomfortably fine distinction we must make between, on the one hand, an Archbishop who offers up a prayer to avert the influences of the stars, and, on the other, a man who makes an incantation on an occasion which is chosen on similar astrological principles, but who now hopes to draw stellar influences to some avowedly magical end.

The flirtation of Mother Church with astrology in all its guises was an affair of the heart – as in the exquisite pagan decoration of ecclesiastical buildings and books – and an affair of the head – as when abstract notions of astrological determinism were introduced into the predestination debate. If we are to be generous, we can grant that a good Christian must have hoped that astrology might tell him something of the ways in which the Divine providence acts – on the assumption, that is to say, that God acts through natural intermediaries. Even a Protestant could believe this source of knowledge open to him, without his having to deny the thesis that the divine grace was beyond human powers of direction. For those who believed in it, astrology was, at the very least, a source of knowledge. It did not have to be used as a means of manipulating personal fortune. As for Fortune, Chance, and the Fates, no matter what the best of Christian thinkers might have argued, astrological influences were as often as not linked in the mind with them, rather than with the dictates of God or the derivative certainties of nature. And when the post-Reformation theologian inveighed against Fortune and comparable notions, inviting his congregation to substitute the irresistible hand of God for these creations of the pagan mind, not only was he prone to do scant philosophical justice to a subject which had been under scrutiny from the time of the Fathers, but he had as little influence on his hearers as Augustine had had before him.

It is not surprising, therefore, that the most important discussions of the compromise between the Christian faith and astrology should have centred on philosophical questions of fatalism and determinism. There were always, of course, the clever mathematicians like Nicole Oresme, who could point out the illogicalities of supposing that the celestial motions repeat themselves; but Oresme's influence on astrology was hardly greater than the influence of Zeno on men who shot arrows. Fatalism was another matter entirely, and one which was serious enough without such added complications as whether or not Christ's character was related to the constellations at the time of his nativity, or whether even the birth of the Virgin might have been free from celestial influence. After all, as Pierre d'Ailly asked, why should she and her son be exempt from natural laws?[4] Did the Sun in the sky not warm her as it warms others? And Pierre d'Ailly was a man safe enough to have presided at the Council of Constance at which in fact he gave the opening sermon (2 December 1414).

Once the Church was persuaded that the Christian subjected to astrological law was in a situation not different from that of the Christian subjected to other natural laws, such as those of Aristotelian natural philosophy (which Richard Lemay has convincingly shown to have been introduced in some measure into the West through the astrological work of Abū Ma^{c}shar),[5] astrologers were able to extract themselves from the predicament in which the Fathers of the early Church had left them. They were able, that is to say, to dissociate themselves if they so wished from the stigma of a diabolic art. A man like Cecco d'Ascoli might have been guilty of the heretical teaching that malign spirits could be conjured from the air, and if he were unlucky he might suffer the same sort of death at the stake (1327), but by steering clear of all that savoured of necromancy the astrologer usually contrived to die in his bed, and often with honour.

The distinction between natural astrological causation and what could be effected through incantation was a spiritually, if not scientifically, important one, and is well brought out in a commentary on the *De sphaera* of Sacrobosco, written by Robertus Anglicus in 1271. If unfortunate stars are in conjunction with the Moon at the time of an eclipse, he said, the effects of the eclipse may be disastrous, especially when one is sowing seed, as he had learned in England, where barley seed had been sown in the most fertile soil at the time of an eclipse. 'Yet the seed died as if none had been sown, and the people of the country believed that there had been an incantation, until I showed them that such a happening was natural.'[6] In dealing with astrology one must never lose sight of his concept of natural causation.

It must seem strange to find that the doctrine of planetary conjunctions was disliked by some who were perfectly happy with the idea that individual nativities could be significant. One might have imagined that some sort of statistical argument would have been called into play, to compound the destinies of men who were of a mind to form a religious association. This was not in fact the *raison d'être* of the doctrine. More to the point was the explanation implied by Regiomontanus when he wrote to Jacob von Speier of the great conjunction signifying Christ's advent – that is, when he made the *founder* of the religion subject to the celestial configuration.[7] The weakness of this argument will become evident in due course. That Regiomontanus knew the literature and the rationale of the doctrine perfectly well is shown by his having recommended to Speier the basic treatises by Albumasar, Massahalla, John of Ashenden, Pierre d'Ailly, and Anthony of Mount Ulm. The last-mentioned, which Regiomontanus seems to have esteemed most highly, is an extremely sinister blend of demonology and astrology. Like Jakob von Speier, we should begin with Albumasar (to use the name I shall adopt in a western context).

It is doubtful whether any astrologer had greater influence on the medieval west than Abu Ma^{c}shar, whose *Introduction to Astronomy*, written in Baghdad in 848 A.D., was twice translated into Latin in the twelfth century (in 1133 by John of Seville and in 1140 by Hermann of Carinthia), whose *Shorter Introduction* had been translated by Adelard of Bath even earlier (c. 1120), and whose many other writings include the *Great Conjunctions*, also translated by John of Seville.[8] The Latin version, *De magnis coniunctionibus*, enjoyed considerable popularity, and was twice printed, first by E. Ratdolt (Augsburg, 1489) and later for Melchior Sessa (by Jacobus Pentius de Leucho, Venice, 1515).

Abū Ma^{c}shar's work on conjunctions seems to have been much influenced by his teacher al-Kindī, the first great philosopher to have written in the Greek tradition in Arabic. Al-Kindī firmly believed that knowledge through revelation and prophecy is superior to the truths of unaided reason. He wrote on astrology, and, in a letter which is still extant, attempted to predict the duration of the empire of the Arabs on the basis of planetary conjunctions. Broadly speaking, the rarer the conjunction affecting political and religious events, the more potent its force. This is the doctrine central to my survey, and al-Kindī's editor, O. Loth, thought it began with him.[9]

In order of the slowness with which they traverse the zodiac, Saturn comes first (29.5 years), followed by Jupiter (11.9 years) and then Mars (1.88 years). Roughly speaking, Saturn and Jupiter meet every twenty years, while Mars meets with Jupiter about every two years and with Saturn marginally more often.[10] Although it is impossible to be very precise without being inordinately tedious, we can say that the conjunctions of Saturn and Jupiter which are separated by a triple period, that is, about sixty years, will occur rather less than ten degrees apart. They can therefore occur within a single sign (30°) of the zodiac, if the first of the trio is close enough to the beginning of the sign. The character of the sign in which the conjunction occurred was thought to be important, but for the moment we need not go into details. Since the three signs in which successive conjunctions occur are likely to be equally spaced, comprising what is called a triplicity, three suitably chosen conjunctions in a series will be associated with one triplicity — which like the signs themselves was thought to have characteristic properties (fiery, earthy, aery, or watery, and so on). The successive conjunctions, occurring every 20 years, are simply the 'great conjunctions' of the title of Albumasar's book. There was also defined a *coniunctio maior,* which took place every 240 years, and a *coniunctio maxima,* which occurred every 960 years. The definitions of these were frequently misunderstood. Bacon, for example, gives a very hazy account in his *Opus maius,* and leaves us with the impression that he was paraphrasing Albumasar very carelessly. I will explain the periods of 240 and 960 years briefly, and those whose sympathies are with Bacon may omit the next paragraph.

Albumasar opens his *Great Conjunctions* with a detailed and rather precise account of the mean movements of Jupiter and Saturn, which we can summarize by saying that the conjunctions marked '0' and '3' in the series on the accompanying diagram will be thrice times $2°25'17''10'''6^{iv}$ apart, or about $7\frac{1}{4}°$.[11] We begin as near as possible to the beginning of Aries, in the fiery triplicity. Conjunction number 13, however, will fall into a new triplicity. If we reckon 20 years between great conjunctions, then the 'greater conjunction', when we pass from one triplicity to the next, should happen after 260 years; or, as Albumasar has it, the conjunctions may stay within a triplicity for about 240 years. They will then, by a loose extension of the argument, be in other triplicities for three similar periods, making 960 years in all, after which they will return to the original triplicity. *Coniunctio maxima* is that conjunction which marks the return to Aries. By virtue of the retrogradations of the planets in practice, and the fact that the figures quoted for the angles are not convenient sub-multiples of 360°, the argument is not precise and not worth discussing in any greater detail.[12] What matters is that henceforth those who wished to find patterns in history had three convenient historical periods to conjure with, namely periods of 20, 240, and 960 years.

This is all reminiscent of Neoplatonic writings on the Great Year, of which the immediate source was *Timaeus* 39^D, but which was in reality a much older idea. The Great Year was the time taken for the planets to return to the same relative positions, and it must therefore be a

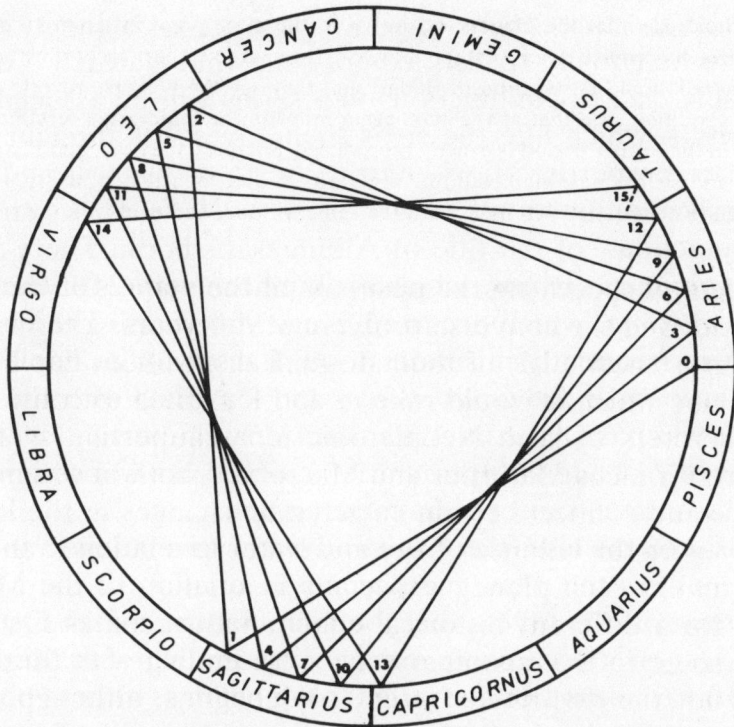

Fig. 1

multiple of the time between great conjunctions. Plato gives no estimate of its length, although some of his modern interpreters have abstracted a figure of 36,000 years encouraged by the fact that several early astronomers took this as a truly Platonic value. While they are almost certainly wrong,[13] and while the figure of 36,000 years seems always to come from the Hipparchian values for the slow movement of the eighth sphere[14], it is of interest to see that this precessional movement was often linked with the doctrine of the *Timaeus*. The figure occurs in Albumasar's *De magnis coniunctionibus*.[15]

Of far greater interest are the historical associations of the two distinct traditions, the one of the eighth sphere and the other of cycles of planetary conjunctions. The problem is one of great complexity, and I must restrict myself to a single quotation from a classical quasi-astronomical source which relates to historical catastrophe and renewal. According to the late fourth century Neoplatonist Nemesius, in his *De natura hominis*:[16]

> The Stoics say that the planets, returning to the same point of longitude and latitude which each occupied when first the universe arose, at fixed periods of time bring about a conflagration and destruction of things; and they say the universe reverts anew to the same condition, and that as the stars again move in the same way everything that took place in the former period is exactly reproduced. Socrates, they say, and Plato, will again exist, and every single man, with the same friends and countrymen; the same things will happen to them, they will meet with the same fortune, and deal with the same things.

The problem of specifying the positions of the planets at the creation of the world, and the converse problem of determining the age of the world from a knowledge of their original disposition, has a long history and one which it would take us too far afield to consider here. Two interpreters of such Neoplatonic ideas important in medieval Europe are Firmicus Maternus and Macrobius, both of whom claimed that they could connect certain cataclysmic changes in the history of civilization with the balance of fire and water in relation to the stars.[17] Firmicus makes each planet, especially in relation to the Moon, responsible for a different historical epoch. Saturn comes first, Jupiter next, and so on in the customary sequence ending with Mercury.

This is not the doctrine of great conjunctions, although it has so many points of resemblance that it should not be impossible to trace a connection. In Māshallāh it was a change from aery to watery triplicity which marked the Flood, and the same shift nearly four thousand years later which heralded the rise of Islam. E. S. Kennedy has written an extremely valuable survey of the concept of the World Year in Islamic writings, especially as they relate to Babylonian, Indian, and Persian sources.[18] 'One often feels', wrote Kennedy, 'as Birūnī would say, that the only reason for studying the subject is to be able to warn the reasonable man away from it'. He nevertheless introduces order into a mass of material (in a summary which it would be impossible to condense further), and it is worth noting that the great conjunction singled out by Māshallāh as signifying the Flood in 3102 B.C. is the beginning of the *kaliyuga* of Indian chronology. Kennedy finds no trace of the doctrine of conjunctions in Hindu or Hellenistic astrology, however, and with David Pingree argues for its invention in Sasanian Iran. To Abū Maʿshar's *Book of Thousands* he conjecturally assigns ultimate responsibility for the world-period tables found in so many *zījes* (collections of astronomical tables).

Albumasar it was who gave the doctrine of great conjunctions to the

West, both directly in the work already discussed, and through Māshallāh, whose work *De ratione circuli* (as it was known when translated into Latin by John of Seville (?) in the twelfth century), while differing in many ways from Albumasar's was clearly derivative in its central theme. The later work, printed under different titles (Venice, 1493; Basle, 1533; Nuremberg, 1549), is short, written in an abbreviated style, and much easier to follow than Albumasar's. The four grades of conjunction there distinguished, for example, are simply associations of Saturn, Jupiter, and Mars in a single term or face with the Sun in aspect (*maximae*), or conjunctions of Saturn and Jupiter (*maiores*), of Saturn with Mars (*mediae*), and of Jupiter with Mars (*minores*).[19] According to Kennedy,[20] the original work on conjunctions is not extant, although a fragmentary astrology of Ibn Hibintā is apparently drawn from it. Kennedy's table summarizing the 'conjunction horoscopes'[21] computed by Māshallāh makes it clear that the work circulating in the West was a pale shadow of Māshallāh's thought, which was apparently closer to Albumasar's than the Latin book suggests. The table, from which I have already cited conjunctions foretelling the Flood and the rise of Islam, also includes conjunctions indicating the birth of Christ, and several political events of Abbasid times. It is all very reminiscent of the Albumasar treatise, in which very precise dates[22] are given for the same events, as well as general rules for the advent of prophets and religions. These are scattered and arbitrary, but they play an important part in medieval chronology, and it will be necessary to mention some of the ways in which they were taken up by others in the West. To one such writer, namely Roger Bacon, I now turn.

References to Albumasar's theory of great conjunctions are to be found scattered through Bacon's writings,[23] but there are several coherent passages in the *Opus maius* which show how seriously he took the doctrine. He was particularly enamoured of the idea of applying the mathematical sciences to Church government, and to the conversion of infidels, and believed that since astrology could show the superiority of Christianity to other religions, its use could only strengthen the Church's faith. The conjunction of Jupiter with each of the other planets signified, he said, a different religion. Since there are six other planets, this meant that he must find six principal religions. Jupiter's conjunction with Saturn signified, he said, the divine books and Judaism – not because the Jews were saturnine, but because, like Saturn, they come first in the sense of being most remote. Jupiter with Mars signified the law of the Chaldees; with the Sun, the law of the Aegyptians; with Venus the law of the Saracens (*quae est tota voluptuosa et venerea*); and with Mercury Christianity.[24] 'And they say that the law of Mercury is more difficult to believe than others, and presents many difficulties to the human intellect', wherefore it rightly represents the Christian faith, with its hidden truths and profundities.[25] The reference is to the complexity of the Ptolemaic theory of the motion of the planet Mercury, which is much more difficult to understand than the theory of any other planet, except possibly the Moon. As for the 'law of the Moon', this is the law of corruption and foul things which violates all other laws and suspends them. This, the law of Antichrist, of lying, of necromancy, and magic, will not last long, a fact we may judge from the Moon's rapid motion.[26]

This is nothing but a much embroidered and christianized version of Albumasar, to whose work on conjunctions (especially books 1 and 2) Bacon refers, before he goes on to explain the three types of Jupiter-Saturn conjunction which supposedly fixed precisely the historical epochs of the six faiths.[27] The great conjunction of the year 6 B.C. (actually December of 7 B.C.) signified the birth of Christ. The conjunction of 26 B.C. had been in the fiery triplicity. That of 7 B.C. was in the watery triplicity. The next, in A.D. 14, would again be in a fiery triplicity, thus making nonsense of the theory; but no matter. Precision was not in Bacon's best interests. As he tells us, Albumasar forecast that Islam would last for 693 years, a figure which he suggested agreed with the Number of the Beast in *Apocalypse* xii, namely 663 (*sic*), 'less than the aforesaid by 30 years'. 'But', as he explains, 'scripture in many places subtracts something from the whole number, for that is the way of scripture, as Bede says'.[28] This is hardly the scientific spirit for which Bacon's name is – or at least used to be – remembered; but always he had the problem of reconciling traditions which were barely compatible. As he himself said in the same place:

> Nolo hic ponere os meum in coelum, sed scio quod si ecclesia vellet revolvere textum sacrum et prophetias sacras, atque prophetias Sibyllae, et Merlini et Aquilae, et Sestonis, Joachim et multorum aliorum, insuper historias et libros philosophorum, atque juberet considerari vias astronomiae, inveniretur sufficiens suspicio vel magis certitudo de tempore Antichristi.[29]

Bacon was not deterred from a sketchy astrological analysis of Christ's birth,[30] using the argument that this was a natural event and therefore subject to the stars. He was, of course, writing for the Pope himself, and it was with circumspection that he offered an apology for astrology. Theologians feared, he said, that in studying the subject they might be called magicians, but the Church should not dispense with weapons merely because they may be used for ill, any more than a man should dispense with the law merely because some lawyers are dishonest. The Church, he believed, needed all the power it could muster, including the miraculous power of words uttered at astronomically propitious times, to overcome the power of Antichrist. Bacon went on to say that the Tartar conquests had already been helped by the confidence engendered by their astrologers, and that the danger was great that the Tartars would shortly join forces with the Saracens. The Church must patronize astrology if the Infidel was to be resisted.[31] Using similar arguments we now send rockets out into space.

Bacon was not the only Christian scholar of note in his century to accept the doctrine of great conjunctions. Albertus Magnus might have protested that neither fate, nor providence, nor the stars, can deprive the free human will of its liberty of action, but he certainly believed in the potential efficacy of stellar influence.[32] One must simply resist sin, and all will be well. Although he seems to have disliked the Great Year motif,[33] yet in his *De causis et proprietatibus* he made the conjunctions of the planets effect great changes in the world, especially in regard to flood, showing that he had imbibed the standard meteorological doctrine.[34] It would have been quite inappropriate in such a work, of course, to have raised the matter of religious change, and I am not aware of any other occasion on which he brought up the subject in his writings.

That there was something less than satisfactory about an argument based only on the authority of a pagan astrologer does not seem to have struck either Bacon or Albertus Magnus very forcibly. Others too were content merely to repeat what they had read in Albumasar or Massahalla. I might mention from the same century Petro d'Abano, who in his *Conciliator* managed to link the Great Year doctrine to that of great conjunctions, which he thought to control the careers of great prophets. Petro's bones were burned for heresy some forty years after his death.³⁵ Cecco d'Ascoli – who, as I have remarked already, died (1327) in the flames – did at least try to offer an explanation of these same ideas. In no less a place than a commentary on a relatively sober text-book of astronomy (as opposed to astrology), namely the *De sphaera* of Sacrobosco, Cecco introduced a wholly alien argument when discussing the meaning of the word 'colure'. (Colures are great circles on the sphere, passing through the north and south poles.) He quoted a pseudo-Hipparchian book *De hierarchiis spirituum* to the effect that in the colures there is an *incubus* and *succubus* by whose virtue at the time of a great conjunction men as of the Godhead are born.³⁶ Such was Merlin, and such will be Antichrist. (There is no suggestion that an incubus visited Mary.) At the close of the commentary Cecco again rehearses the same idea, but then wanders off into a different and rather garbled doctrine of the Ages of the World, this time ascribed to Zoroaster.³⁷ The notion that the angelic spirits populate the heavens, to be brought down by incantation at the time of great conjunctions, was a not uncommon ingredient of that large class of treatises 'on the constellations'.

There is little point in listing the names of men who repeated the doctrine of great conjunctions, but there are two names in particular which should not go unnoticed – John Ashenden and Nicole Oresme, Ashenden because he wrote an important *summa* of astrology which was authoritative enough on this point for it to be listed by Regiomontanus as one of three western sources, and Oresme because he is so often represented simply as an opponent of astrology. Oresme has all the *ad hominem* arguments against astrology – that it fosters deceit and fraud, the illicit use of sorcery, and so forth – as well as the 'rational' arguments already alluded to. He was sceptical of the possibility of rational prediction and of the true recurrence of phenomena, granted the premise that motions are incommensurable, and yet when he wrote his book on divination, he does seem to have believed in the possibility of forecasting events from conjunctions, events of all the usual kinds – flood, pestilence, war, and political and religious change of the sort Albumasar had in mind.³⁸

John Ashenden must be placed in a rather different category. Far from being sceptical of the claims of astrologers, his *Summa iudicialis de accidentibus mundi* shows him to have been a compiler of numerous sources of a wide range of qualities, sources which he mixes uncritically, except that he has especial reverence for Albumasar.[39] The *Summa* was completed at Oxford in 1347 (book one) and 1348 (book two), and, as, its title suggests, gives an account of only a single branch of judicial astrology, namely that which deals with the universal accidents of the world, as judged by revolutions and conjunctions. Despite its limited theme, the *Summa* runs to approximately 375,000 words. Much of it concerns astrological meteorology and such natural disasters as flood, famine, war, and inflation, and indeed the second book was completed while the plague was sweeping through Europe, a plague its author claims to have predicted in connection with the great conjunction of 1345.[40] The mortality was produced, said Ashenden, 'by God in the first instance, namely through an eclipse of the Moon [cf. Robertus Anglicus' remarks] and through the aforesaid great conjunctions as by natural instruments'.[41]

The first book of the *Summa* is appreciably shorter than the second,[42] and is a sort of *theorica* underlying the whole. It is in the first book that we find a pedantic blend of Albumasar on great conjunctions with the utterances of other writers, but for his more revealing utterances of a prophetic nature we must turn to his other writings. The *Summa* opens on a defensive note, drawing attention to the backbiting of envious contemporaries – unfortunately they are not named – and recounting a number of stories in support of the idea that astrology is a powerful and lucrative pursuit.[43] It goes on to decide, by reference to many different authorities,[44] at what time of year the world was created, and where the planets were at the time.[45] In the second chapter of the first *distinctio,* John Ashenden considers different opinions about the age of the world. Here his authorities include Orosius, Methodius, Eusebius, Gerlandus, Jerome, and *Policronicon,* in addition to many of those used before.

Inevitably, biblical chronology – especially of the sort based on generations – predominates over more properly astronomical material. The six ages of the world are introduced from Vincent's *Speculum historiale* and from *Policronicon* without astronomical support. There is here a morass of chronological information which I think one would have to be a Registrar of Births to appreciate. Only towards the end of the chapter is a respectable astronomer introduced, and then, as we should have expected, we meet not only Albumasar, with his book of great conjunctions, but also Alfonso X of Castile, and the chronological tables which are to be found at the beginning of any complete set of Alfonsine planetary tables. There is no point in my reproducing Ashenden's digest of ancient chronology, but I would like to comment on the uncritical way in which he mixes his sources. He will, for example, take Albumasar's interval between creation and the flood (2226^y 1^m 23^d 4^h) and the Alfonsine interval between the birth of Christ and the flood (3101^y 318^d), and add them (giving 5328^y 16^d 8^h 30^m!),[46] overlooking the fact that the two figures were arrived at from substantially different theories.[47] He has more faith, he admits, in Alfonso's figure, which is 'true and precise', than in the other, over which there is much doubt, many writers[48] preferring 2242^y to 2226^y. I think most of us would now agree, however, that sixteen years is not a serious error when the age of the universe is at stake. In the third chapter of the first *distinctio* (I.i.3), Ashenden makes heavy going of ten errors which he claims to find in discussions of the age of the world. Again Albumasar's doctrine is cited in support of the chronology, and for most of the time Ashenden is merely playing with numbers, and fitting conjunction periods (occasionally modified) into historical intervals. An example of his own, for 1345, is a rare personal intrusion (I.i.4) into a jungle of other men's ideas.

In Ashenden's digest of incompatible constituents he tries, not surprisingly, to get Ptolemy's support, quoting *Quadripartitum* (I.ii.2), but not convincingly. He drifts into the Great Year theme, mentioning Macrobius, Hermes, Trismegistus, and Firmicus Maternus. His explanation of the way in which great conjunctions change sign and how sects are decided (I.ii.3) adds nothing to Albumasar, and has none of Bacon's embroidery. Consequences for the reigns of secular rulers (I.ii.4) and alternative definitions of the different classes of great conjunction (I.ii.5)[49] again tell us nothing of Ashenden or his times. The remainder of the first tract deals with other basic astrological theories, while the longer second book for the most part avoids such human affairs, in favour of meteorological and related questions – under which heading we may include the plague.

John Ashenden's *Summa* was neither his only work on conjunctions, nor was it his first, for he wrote a number of actual prognostications, not only for the great conjunction of 1345, but in due course also for 1349 and 1357 (lesser conjunctions), and 1365. The astronomical predictions are of some intrinsic interest, but are irrelevant for our present purposes.⁵⁰ The strictly astronomical predictions for 1345 and 1365 (and possibly all) were done for Ashenden by William Reed, another fellow of Merton College.⁵¹

The predictions based on the 1345 and 1349 conjunctions were for the most part concerned with war and weather – with the pestilential aspect greatly magnified retrospectively. That of 1345 had said that the English king would have victory as long as the effects of the conjunction lasted,⁵² and when writing his 1357 prognostications – after Crecy and Poitiers – Ashenden was obviously doubly pleased that the campaign in France was going so well, although he warned against fraudulence and deceit on the part of the French in future offers of peace. Perhaps he had in mind a letter from Jean de Murs sent to Clement VI (d. 1352) in which the ruin of France was said to be a probable outcome of the 1365 conjunction, if peace was not made between her and England.⁵³ It is with the eclipse for 1365 that Ashenden's thoughts turn to religion, for as he and many of his contemporaries recognized, it would occur (with Mars in the same sign) in a new triplicity (watery, rather than aery), and therefore, according to the rules of Albumasar's game, this meant at the very least a change in the fortunes of some religious sect, and might even signify the advent of a new prophet. As it happens, the conjunctions for 1385 and 1405 were back in the aery triplicity, but it is doubtful whether Ashenden realized this complicating fact.

Ashenden thought that the least likely outcome was the advent of a prophet, since the conjunction was in the sign of Scorpio, house of Mars, but that if a new prophet – and hence a new sect – were indeed to arise, both would be characterized by cruelty, malevolence, and deceit. There is no suggestion of Antichrist, and indeed Ashenden thought that the scriptures forbid us to predict the coming of Antichrist. As a good astrologer Ashenden put his money on each of the two chief faiths contending – at least within his experience – for religious power. He was optimistic about the destruction of the Saracens, since – as I have already pointed out – Albumasar had said that Islam would last for 693 years, and the Number of the Beast was 666. These two numbers are close, he said, and this must be significant.⁵⁴

One wonders when Ashenden thought the religion of the Saracens to have properly begun. In 1365 half a century had passed since the

693rd anniversary of the Prophet's flight from Mecca, but the *conoscenti* would have known that the conjunction announcing the rise of Islam occurred in A.D. 571, no less than 794 years previous to 1365. It seems that Ashenden simply could not subtract. But — apart from the 'correction' which Ashenden made to Bacon's text — this has all the appearances of a passage from the *Opus maius* which I have already quoted,[55] and which made some sense a century earlier. Islam might suffer destruction, Ashenden argued, but the Church of Rome would not escape oppression by secular rulers.[56]

This oppression was passed off, however, as something of little more moment than all the other disasters attendant on a great conjunction — flood, shipwreck, battle, famine, and the like. Given more prominence was the downfall of the kingdom of Scotland, which would suffer a decline in population.[57]

In all that Ashenden was saying it is easy to imagine parallels with Joachimist oracles, and yet he quite explicitly rejects the association. Joachimist predictions, as he points out, did not have a basis in astrological calculation. The Joachimist work he quoted relating to the years 1357–65 was not genuine, although it is known to have been written before 1250,[58] and to have been updated in the following century. In its original form its predictions were for the years 1254, 1256, 1260, and 1265, during which period the Greeks would recover Constantinople (1254), there would be two popes, a just one at Lyons, the other, an unjust and iniquitous pope, at Rome (1257). The Church and clergy would be trodden down (1260) more than at any time since the reign of Constantine, but in the end all Greece would return to the obedience of the Roman Church (1265). Then would come the preachers of Antichrist.

Ashenden regarded such prophecy as based on 'illicit sciences' and the pure imaginings of the supposed prophets. Although, as we have seen, his own arithmetic was not flawless, he no doubt felt that that of Joachimist prophecy, with its patterns of twos and threes, was both trivial and spiritually dangerous. The periodicities of Saturn and Jupiter were not, after all, likely to be confused with the doctrine of the Trinity. The essential simplicity of Joachimism was one of the secrets behind its large following. If the astrologer lived in a more rarefied academic milieu, it was one which offered more potential shelter. The astrologer was not, for example, obliged to tangle with the concept of a Messianic Age, a Millennium, a Sabbath Age, or any other of those ends to history which seem to have been an almost inevitable ingredient of the simplistic numerical division of past and future into a small and finite number of periods.[59] The motions of the planets might be combined with history of this sort, but there was nothing about Ptolemaic astronomy which obliged one to combine the two subjects. Ashenden was perhaps disturbed by the low professional quality of those who preached of the Second Coming, or the end of the world – charlatans and deceivers, as he calls them.[60] He was irritated by a report that an Oxford lecture had recently given 7500 years as the interval from Noah's flood to the conflagration by which the world would end. Even on the most conservative fourteenth century estimates, this would have given three clear millennia of future history. It was not the proximity of the end of the world, but the temerity of those who predicted it, which roused Ashenden's indignation. In other words, a dyed-in-the-wool astrologer, a man whose whole art was directed at prediction of the future, could at the same time be theologically conservative. Perhaps he lived with the memory of his fellow Mertonian Thomas Bradwardine, who had accused Joachim of Arianism, and even of Pelagianism.[61]

Whether or not it has anything to do with Ashenden's strictures it is hard to say, but some manuscripts of his prognostications omit them in part or in whole. A manuscript now in Paris[62] leaves out the attack on Joachim, while two Ashmole manuscripts[63] break off before the criticism of the Oxford lecture.

Ashenden's final prediction was a half-hearted affair,[64] covering weather conditions for 1368 to 1374, and having nothing to do with the doctrine of great conjunctions in which he was so obviously counted an authority. Other adherents to the doctrine there were in plenty, however, and even a cursory survey of Lynn Thorndike's book[65] yields at least a score of prognostications on this basis for the fourteenth and fifteenth century. Jean de Murs, to whose prognostication for 1365 I alluded earlier,[66] used his letter to Clement VI as an occasion for recommending a vigorous crusade against followers of the perfidious Mahomet, and an end to strife within Christendom – by which I take it he means France. Geoffroi de Meaux, Firminus de Bellavalle, Levi ben Gerson, Jacobus Angelus (astrologer to Leopold of Austria, himself an astrologer of sorts), Jehan de Bruges, and Matteo Moreti of Brescia, are among those who treated of this sort of prediction in a rather predictable way. Vague promises about war, flood, and religion, and occasionally an allusion to inflation: this was all fast becoming as much a cliché as Fortune's Wheel.

Not all prognostications are so stereotyped. The Minorite Jean de Roquetaillade (or Rupescissa), who was for long imprisoned, first in Franciscan convents and later in the papal prison at Avignon, prophesied the destruction of the religious orders, and even of the main part of the Order of St. Francis. Only a few true Minorites would survive, seed to be sown to replenish the whole world.[67] His writings were inspired by a variety of sources. He often used the concept of Antichrist, and that of a millennium, and was somewhat influenced by Joachimism. His best known work was his *Vade mecum in tribulacione,* covering prophecies made for the period 1356–69. For between 1360 and 1365, for example, he foresaw a cruel conflict in the world of nature, great snakes consuming beasts of prey, while birds, even songbirds, would attack one another; and so his forecast went on. The populace will attack the nobility, and an Antichrist will appear in Jerusalem, attracting the Jews to his teaching. During the first part of the period the princes of the Church will prepare to flee Avignon. Muslim will fight Christian, but a Spanish king will be victorious against the Moors. A western Antichrist – a heretical Emperor, a new Nero – will appear at some time after 1362, soon to be followed by the Angelic Pope, *reparator* of the world, and marking the beginning of the third *status* of the world in the Joachimist sense. Together with a specifically French Holy Roman Emperor, he will first take temporal form before 1365, both of them as abject Minorites. They will destroy the power of Mahomet, free the Greeks from the Turks, and henceforth only appoint cardinals from the Greek Church.

Astrology cannot be held responsible for the greater part of this wild apocalyptic extravaganza, but numerical calculation, at least, was held by Jean to be a more reliable guide than 'miracles and signs', and in the second tract of his *Liber ostensor* he shows that he was not averse to ensuring that his prophecies were compatible with astrology.[68] In this last mentioned work he cites Albumasar's book of conjunctions. I have not seen the sole surviving manuscript of the *Liber ostensor,* but from the description by Mme Bignami-Odier it seems very likely, bearing in mind its general subject matter and the fact that Albumasar is quoted (at ff. 21r–22r), that Jean will prove to have been influenced, albeit slightly, by the doctrine of great conjunctions.

Jean de Roquetaillade's prophecies might have been sooner forgotten had they not been used by Telesphorus, described by Miss Reeves as 'a Francophile Joachimite, inspired by the *Oraculum Cyrilli* and Roquetaillade's interpretation of it'.[69] Of more relevance to my present theme, however, are the views of a distinguished fourteenth century scholar who happened to attack not only Telephorus, but also Joachim, Cyril, many other 'recent' prophets, and also, in a separate treatise, the doctrine of conjunctions. This man was Heinrich von Langenstein, successively professor of theology at Paris and Vice-Chancellor of the University of Vienna. Although he attacked the Joachimites in 1392,[70] Henry of Hesse, as he is now more often called, had as recently as 1390 shown a certain admiration for the methods of Joachim, although by then he was an old campaigner against astrological prediction, while yet being adept at astronomy and the mathematical sciences. His earliest dated astronomical work was in fact on the comet of 1368, and in his *Quaestio de cometa* he argued that prognostications based on comets were of no value.[71] His later *Tractatus contra astrologos coniunctionas de eventibus futurorum* was called forth by predictions based on a Saturn-Mars conjunction of 1373. Astrologers, he said, had proved themselves, by their example, to be incapable of accurate prediction (pars II, passim); and the phenomena which conjunctions and eclipses were supposed to foretell could be better foretold by the kind of physical astrology to which he adhered, and astrology which placed no reliance on the traditional division of the zodiac into houses, faces, etc. – on which system he pours scorn (I.ii, II.iv).[72] The first part of his treatise was, ironically enough, an excellent text-book for the theory he was setting out to demolish,[73] but the arguments he gave must, nevertheless, have made the best of contemporary natural philosophers stop and think. Why should the rarity of a phenomenon be related to its power? Why, because Saturn and Jupiter may be joined three times in a year in the same sign of the zodiac, should the devil be bound?[74] Why speak as though a planet is powerful at its apogee? When most distant, its action should be feeblest.[75] Here speaks the writer of a treatise on optics, who knew a law of the weakening of light with distance. There are several comparable objections of a 'rational' sort, but in the end he left his readers high and dry, astrologers without rules by which to work. This, rather than any weakness in his arguments, no doubt explains why his protestations, so different in character from those of his ally Nicole Oresme, created very few echoes during the two centuries following.

This is not to say that the doctrine of great conjunctions went uncriticized. Pico della Mirandola's polemic I shall discuss later, for he was in many ways the most cutting critic of all. A scarcely worthy partner to these two, however, was Pierre d'Ailly, whose weakness was perhaps that he gave too much thought to the *concordantia astronomie cum theologia* – to take the title of one of his works.[76] Miss Reeves has said of him that, writing as he was on the eve of the Council of Constance, 'Even while operating on a practical level to end the Schism, d'Ailly was alive to the possibility that the Church might be in the grip of cosmic forces to which the only certain guides were the prophets'.[77] Pierre left many writings which hover on the borders of astrology and astronomy, and among other things he pressed for calendar reform.[78] This was not the only respect in which he showed himself a disciple of Roger Bacon, although it was as a critic of Bacon that he wrote his critique of the doctrine of great conjunctions.

The work, under the title *De legibus et sectis contra superstitiosos astronomos,* was written in 1410, four years before his *Concordantia.*[79] It is essentially a defence of true astrology against those who, in practising the art, contravene Christian teaching, and he seems to be particularly concerned to prevent the Baconian version of the doctrine of great conjunctions from taking root. He grants that religions – both 'laws' and 'sects' – are influenced by the heavens. His *Concordantia,* indeed, is a piece of standard chronology of world history, based on this idea, and his *Elucidarium* (printed with the former work by Ratdolt, and directed at clarifying it) carries through his quite unambiguous purpose. But Pierre d'Ailly now shows that he thinks some religions to be more resistant to influence than others. The Jewish and Christian faiths are not subject to such influence, but the Muslim and all idolatrous faiths are so affected. (As for the natural effects of great conjunctions – floods, and so forth – he firmly believes in them.) This is all hard to reconcile with his belief that Christ's own nativity was subject to celestial influence, and is indeed flatly inconsistent with the belief he expressed in his *Concordantia* and elsewhere, that great conjunctions entail important changes in the world, among which were the rise of the kingdom of Israel, the law of Moses, the dominion of Charlemagne and the Franks, the rise of the Franciscan and Dominican orders, and the Tartar empire.[80] He listed past conjunctions and corresponding events, and in more than one place extended his study into the future, to the coming of Antichrist, the destruction of Islam, and the end of the world.[81] It has been observed that he twice predicted great changes for the year 1789.[82] If we are to offer an apology for the Cardinal's apparently inconsistent 'harmony' of astrology and theology, it must be that he distinguished between the natural and the supernatural elements of Christianity, and made only the former subject to astrological explanation and prediction. But as to where he drew the line, it is doubtful whether even he could have explained. The usual implications of the idea that stellar influences acted on Christ were that the passion and crucifixion were not undertaken voluntarily. The Inquisitor who sent Cecco d'Ascoli to his death was in no doubt about this.

Prognostication from great conjunctions went on unchecked throughout the fifteenth century. There is, for instance, an interesting anonymous prediction from the Library of the Rhenish Palatinate for a great conjunction, and two other conjunctions involving Mars, of 1425, all three in Scorpio. Sinister events are foreseen, and are linked with the Hussites. Another anonymous source of the same period announces that certain pseudo-astrologers had connected a Venus-Jupiter conjunction before Lent in 1430 with the Hussites.[83] Later in the century Jerome Torrella, who prophesied that the Jews would suffer much from the conjunction of 1464, was telling his patron, Ferdinand of Aragon, precisely what he wanted to hear.[84] Another example of a work embodying this type of prediction, and one which was destined to achieve enormous popularity (judging by the numerous printed versions of it), was Johannes Lichtenberger's *Pronosticatio*. This work, by the astrologer to Frederick III, has been extensively studied, and there is nothing I can add to a proper understanding of its mainly political purpose,[85] but I will observe that even though Lichtenberger's eclecticism introduced inconsistencies into his astrology, no-one at the time can have cared very much. He draws on almost every important European medieval source of prophecy, in addressing himself to Pope, Emperor, and laity, and yet at the core of his ill-organized central argument for the imminent collapse of the Church there is the belief that the conjunction of 1484 will give rise to new prophets, one of whom, born about 1496, would stir up revolts and bring about a revision of the law. There would later be the monk in white, inclining to the Chaldean faith, burning men under the pretext of religion, and finally dying a shameful death. Melanchthon, Cardan, and Gauricus are all said to have believed that Luther's birth was connected with the 1484 conjunction. Luther knew of the prediction, 'und sie auf sich und Ereignisse seiner Zeit gedeutet, er hat sich aber auch kritisch mit ihr auseinandergesetzt'.[86] There is in Lichtenberger's work some play on the number 19, reminiscent of Jean de Murs and even Albumasar, and certainly related to astronomy, but in a totally unrestrained way, if we are to judge by the more rigorous rules of Albumasar's making. It is well known that Paul of Middelburg accused Lichtenberger of plagiarism,[87] and of disregard for true astrology. The accusation was not answered, but the popularity of the *Pronosticatio* was an answer of sorts, and shows clearly enough that astrological refinement was not what the public was seeking. As for pope Innocent III, to be on the safe side Lichtenberger made him subject only to God, and not to the stars. (This was an interesting departure from a not uncommon fourteenth and fifteenth century practice, whereby the pope was identified with Jupiter.)

These examples could be easily multiplied, although eventually there is a tendency to monotony of a kind less common in other forms of prophecy from the same period. As astrology was more exact than, for example, the typical Joachimist imagination, so it was more readily refuted. When Henry of Harclay, Chancellor of Oxford University, took the side of the scholastics against Arnold of Villanova, and wrote a *quaestio* (on the coming of Antichrist) in which Joachim's own writings were also examined, it is significant that the more precisely calculated a prediction, the more easily did he undermine it.[88] On the other hand, there was no unqualified opposition to astrology as a whole, from anyone familiar with the subject under attack, and this for the simple reason that no really effective theory of natural causation could account for all that men wished to explain. The last polemic I shall discuss, one book of which (out of twelve) was directed against the doctrine of great conjunctions, illustrates well the pressures to which a scientifically inspired prophet was subjected.

Pico della Mirandola has something of a reputation for what Lynn Thorndike has called 'the most outstanding theoretical and literary attack upon the art [of astrology] since the treatises of Oresme and Henry of Hesse'.[89] The title of the work in question, *Disputationes adversus astrologiam divinatricem,* shows that it is only specifically directed against *astrological* divination.[90] Pico accepts stellar influence on the terrestrial world, although he thinks the standard methods of analysing this influence are foolish and mistaken. D. P. Walker has observed that Pico uses what almost amounts to Ficino's theory of a type of astral heat, an influence carried down to us by a celestial spirit, permeating the universe,[91] and elsewhere he quotes the Dominican theologian at Florence, Tommaso Buoninsegni, who was never able to convince himself, he said, that 'Pico, Savonarola and other excellent men wished to condemn true and legitimate astrology'.[92] Buoninsegni begs the reader to excuse those who, carried away by their righteous anger against bad astrology, went too far and attacked what was good. Pico's biography is important to an understanding of his book, which shows all the enthusiasm, not to say detailed knowledge, of an author guilty of apostasy. Having, as a younger contemporary of Ficino, added a cabbalist magic to the latter's *magia naturalis,* and having, in 1486, gone to Rome with 900 theses which he offered to prove and defend, he was accused of heresy. Innocent VIII appointed a commission to investigate the orthodoxy of the young nobleman, and in due course he was unwise enough to publish an *Apologia* for those of his theses which had been categorically condemned. He fled to France, was imprisoned there, and returned to Florence where he had the support of Lorenzo de' Medici. With the accession of Alexander VI as pope – a man steeped in occultism – he obtained absolution from charges of heresy, and a letter saying as much was to be printed with all his later works. This was dated 18 June 1493. Before the end of the following year Pico had died, leaving his rather disorganized *Disputationes* unpublished.

In a sense, Pico's attack marks the beginning of the end of the doctrine of great conjunctions. It is not that scholars thereafter relinquished the doctrine. On the contrary, it was perhaps used in the sixteenth century more than ever before, and, indeed, only a few months before Pico died the Siennese Tizio tells of a widespread anxiety caused by conjunctions, eclipses, and other portents.[93] Works like Pico's could not alone bring an end to astrology – for that one must look for pressures from without, for new sciences which would render astrology superfluous. But Pico's book must have weakened the resolve of many intelligent individuals who might otherwise have followed the old astrology blindly, and without any thought of subjecting it to criticism by its own criteria. Albumasar, said Pico, was the greatest astrologer since Ptolemy, and yet even he was guilty of error.[94] The most respected of later astrologers have likewise made serious mistakes.[95] Thus Arnold of Villanova predicted the coming of Antichrist in 1345; Abraham Judaeus foretold their Messiah's coming in 1464;[96] Pierre d'Ailly wrote that in the year of the Council of Constance astrologers predicted that the Church would no more be at peace, that dissension and the fall of religion would be brought about by the retrogradation of Jupiter in the first house in the figure of the heavens for the year, and that schism would herald Antichrist.[97] All these predictions were inaccurate, as we know from experience. The nativity which astrology would have given to Christ would not have made him a great prophet, nor have brought him to a violent end. What Albumasar said about Mahomet we can likewise dismiss in the light of experience. And if we take as significant the interval of approximately 300 years,[98] why was there no great religion between Christ and Mahomet? The doctrine fails on empirical grounds alone.

More significant, however, are the many telling theoretical objections. If the conjunction signifying Islam is that of fifty years before it, what are we to say of the intervening great conjunctions? They occur, after all, every twenty years. Pico thought that the doctrine derived from a misunderstanding of Ptolemy's *Quadripartitum* II,4, and had difficulties in reconciling his veneration for Ptolemy with two much quoted aphorisms from *Centilogium*[99] – which was then thought to have been written by Ptolemy. More important, however, was Pico's clear demonstration of the intolerable vagueness of the later doctrine. Why should the effects of a conjunction take many years to become manifest – and an indefinite number of years into the bargain? How can we trust astrologers who derive what they claim to be significant results – for example, the horoscope of the world – which later turn out to have been based on erroneous mean motions for Jupiter and Saturn? Judged against Bede and Rabanus, Pico maintained that Pierre d'Ailly – his *bête noire* – made several chronological errors. It was folly on d'Ailly's part to accept Bacon's equation of the six planets with six religions, for this, said Pico, showed a lack of understanding of the diversity of religions and sects of idolaters. He shows here his close familiarity with the subject as it was discussed in the Ficino circle.

Had Pico lived longer, it is not improbable that he would have expanded his remarks in the eighth chapter of his fifth book, where he reveals a better historical judgement than either Nicole Oresme or Henry of Hesse before him. Here he shows that the times of past events – such as had been controlled by reference to the conjunctions of Saturn and Jupiter – are to be found only from historical, and never from astrological sources, and that the astrologer covertly acknowledged the fact. This is now perhaps self-evident, but in Pico's time it was not so. He held too many personal beliefs which we should now dismiss as foolish, but there was an admirable objectivity in his scrutiny of medieval astrology. One should never forget that his criticism was directed more often at great authorities than at small, and that he did not excuse a weak argument merely because it favoured a Christian cause. Looking back over the ways in which Albumasar's relatively sober ideas were subsequently livened by the Christian imagination, Roger Bacon's additions à propos of Mercury and Christianity are as unrestrained as any, and Pico scorned them accordingly.[100] But there were others, said Pico, in his final remarks to Book V, who believed that Bacon was right, since Mercury was the librarian of the gods, and the Christian religion had an abundance of books. 'Haec est admirabilis sapientia astrologorum; obliti scilicet sunt gentium bibliothecas, Arabum et Hebraeorum libros fere nullos adhuc viderunt'. Pico, for all his faults, overlooked none of these books willingly in his critique of a doctrine which was always in a sense aimed at nothing less than the reduction of history to natural philosophy. Had short titles been in vogue, he might have called his work *The Poverty of Historicism*.

NOTES

1. D. C. Allen, *Doubt's Boundless Sea*, Baltimore 1964, pp. 50–1.
2. The opening of parliament was 17 March 1628. See *The Works of William Laud*, ed. W. Scott and J. Bliss, vol. I, Oxford 1847, p. 169. I have altered the editors' punctuation.
3. I here follow D. P. Walker, *Spiritual and Demonic Magic from Ficino to Campanella*, London 1958, p. 110.
4. *Opus maius*, ed. J. H. Bridges, vol. I, Oxford 1897, p. 267. On Pierre D'Ailly, see Louis

Salembier, *Petrus de Alliaco,* Insulis 1886, p. 182; and also, for a general study, the same author's Pierre D'Ailly, Tourcoing 1932.
5. *Abū Maᶜshar and Latin Aristotelianism in the Twelfth Century,* Beirut 1962.
6. L. Thorndike, *The* Sphere *of Sacrobosco and its Commentators,* Chicago 1949, pp. 244, 196 (text): '... manifestavi quod taliter accidere debeat per naturam'.
7. Maximilian Curtze, 'Der Briefwechsel Regiomontans mit Giovanni Bianchini, Jacob von Speier, und Christian Roder', *Abhandlungen zur Geschichte der mathematischen Wissenschaften,* XII (1902), 306 (letter to Speier).
8. For further details see Richard Lemay, *op. cit.,* Introduction.
9. 'Al-Kindī als Astrolog', *Morgenländische Forschungen, Festschrift für H. L. Fleischer,* Leipzig 1875, pp. 261–309. Loth shows that Abū Maᶜshar incorporated his master's letter, unacknowledged, into his own book on great conjunctions. Most of al-Kindī's work deals, however, not with great conjunctions, but with conjunctions of the maleficent planets Saturn and Mars in the sign of Cancer.
10. If the planets moved at uniform speeds, which of course they do not, the intervals would be approximately 19·9 years, 2·2 years, and 2·0 years respectively.
11. There are inconsistencies in the Ratdolt edition which I have used, and the first quoted figure becomes $2°25'1''10'''6^{IV}$ in one passage. A similar diagram to ours is to be found in the Preface to Kepler's *Mysterium Cosmographicum,* (ed. Ch. Frisch, vol. I, Frankfurt 1858, p. 108; ed. F. Hammer, vol. VIII, Munich 1963, p. 26) of 1621. Kepler tells us that he was lecturing with this diagram on 9 or 19 July 1595 when he was led by it to a scheme of cosmic harmony. The diagram is not to be found in Albumasar, and is included here for purely explanatory purposes. Kepler's parameters differed somewhat from Albumasar's. Although I have not yet introduced Kepler's terminology, for which see n.19 below, this is a suitable place at which to point out that he half-heartedly believed in a correlation of 'maximum conjunctions' and historical periods. The epochs which are apparently implied there are 3975 B.C. (or Creation), 3182 B.C. (Enoch), 2388 B.C. (Noah), 1514 B.C. (Moses), 794 B.C. (Isaiah), 6 B.C. (Christ's coming), 789 A.D. (Charlemagne), 1583 A.D. (stella nova); for which see the Epitome of Copernican astronomy ed. Max Caspar, *Ges. Werke,* vol. VII, Munich 1953, pp. 477–8.
12. The general pattern breaks down rather seldom, but several writers of the middle ages seem to have been aware of the fact that it did so.
13. See, for a summary of the evidence, Thomas Heath, *Aristarchus of Samos,* Oxford 1913, pp. 171–3. Heath is wrong to accept the supposed Sacrobosco quotation as coming from the *De spera.*
14. According to Hipparchus the stars change their celestial longitudes at a rate of 'at least' one degree per century. Ptolemy, in *Almagest,* repeated the argument but later took the figure to be exact. The stars should accordingly make a circuit of the sky in 360×100 years. This has nothing to do with the planetary motions with which we are concerned.
15. Ratdolt edition, first two pages of sig. a.4.
16. Ed. Matthaeus, 38, p. 309; quoted by F. E. Robbins in the Loeb edition of Ptolemy, *Tetrabiblos,* p. 15.
17. Macrobius, *Comment.* II. 10 Opera, ed. F. Eyssenhardt; Leipzig 1893; Firmicus Maternus, *Matheseos,* ed. C. Sittl; Leipzig 1894, III. 1 ('Thema mundi').
18. 'Ramifications of the World-year Concept in Islamic astrology', *Proceedings of the Tenth International Congress of the History of Science, Ithaca 1962,* Paris 1962, pp. 23–43.

Astrology and the Fortunes of Churches 85

19. There are other terminologies, e.g. in a commentary on *Centiloquium* (pseudo-Ptolemaic) by Haly, *coniunctio maior* is a conjunction of Saturn and Jupiter in Aries, *media* means a change of triplicity, and *minor* is a conjunction in any other sign than Aries, and with no change. Note also Cardan's three types of Saturn-Jupiter conjunction, at intervals of 794^y 214^d (*magnae*), 198^y 236^d (*mediae*), and 19^y 315^d 19^h (*minores*). Rheinhold gave periods close to Cardan's. Kepler (see n. 11 above) rounded off to 794 years for the *coniunctio maxima*, made no mention of *con. media*, and loosely defined *coniunctio magna* as a conjunction of Saturn and Jupiter with Mars nearby.
20. *Op. cit.,* p. 42, n. 33, and pp. 33–4.
21. *Ibid.,* p. 34. These horoscopes are cast, not for the moment of a conjunction, but for the moment of the Sun's entry into Aries in the year in which the conjunction took place. They were to be criticized by Pico della Mirandola.
22. Intervals of time may be specified to the hour, for example.
23. Apart from those mentioned below, see especially the passage in the *Metaphysica*, published by R. Steele, *Opera hactenus inedita Rogeri Baconis,* I, Oxford 1920, pp. 43–6.
24. Ed. Bridges, I, p. 255–7. (Cf. II, pp. 371–2).
25. *Op. cit.* I, p. 257.
26. *Ibid.,* pp. 261–2.
27. Dicamus quoque quia cum Juppiter per naturam significet fidem et diversitatem legum in temporibus et vicibus atque sectis ex complexionibus Saturni et ex complectionibus ceterorum planetarum cum eo scilicet Jove ... Si fuerit complexus Saturnus, significabit quod fides civium eiusdem sit Judaisma, quod congruit planete Saturni eo quod omnes planete iunguntur ei, et ipse nemici illorum iungitur. Et similiter iudaica fides; omnes cives ceterarum confitentur ei; et ipsa nulli confitetur ... Et si complexus ei fuerit Mars significat culturam igneam et fidem paganam. Et si complexa ei fuerit Venus significat fidem unitatis et mundam, ut fidem Saracenorum et ei similem. Et si complexus ei fuerit Mercurius significat fidem Christianam ... Et si complexa fuerit ei Luna significat dubitationem ac volutionem et mutationem ac expoliationem a fide ... (Albumasar, *De magnis coniunctionibus,* II. 8; quoted from Garin, ed. cit., n. 90 below).
28. *Ibid.,* p. 266. Cf. Ashenden's version of the idea, p. 34 below.
29. *Ibid.,* pp. 268–9.
30. *Ibid.,* p. 267.
31. *Ibid.* 394–403.
32. *De fato,* ed. Borgnet, vol. XXXI, pp. 694–714. He excepted Christ from stellar influence.
33. *Ibid.,* p. 708.
34. Ed. Borgnet, vol. IX, pp. 618–21. Albertus knew Albumasar's work on great conjunctions. See the work quoted in n. 32, at p. 713.
35. He died circa 1315.
36. L. Thorndike, *The* Sphera *of Sacrobosco and its Commentators,* Chicago 1949, pp. 387–8: 'Incubus et succubus coluros tenent et quandoque in maiori coniunctione eorum virtute velut divinitatis homines oriuntur'.
37. *Ibid.,* pp. 408–9. The pseudo-Zoroastrian work is entitled *Liber de dominio quartarum octave spere.* A quadrant of the eighth sphere is said to dominate every historical period of 12,000 years and the turning points of which men of divine attributes are born of *incubi* and *succubi,* etc. etc. This has more the appearance of an Indo-Persian doctrine than of anything based on a theory of the precession of the eighth sphere.

38. This is a part of astrology, 'qui est des grans aventures du monde, puet estre et est assez souffisament sceue en general tant seulement'. See the edition of the *Livre de divinations* in G. W. Coopland, *Nicole Oresme & the Astrologers,* Liverpool 1952, pp. 54–5.
39. The *Summa* was printed at Venice in 1489. The Bodleian copy (Ashmole 576) is much corrected at the outset from a Merton manuscript, probably now MS Oriel 23. The author's name is rendered 'Eschuid' in the printed edition. The copy in the Kongelige Bibliotek, Copenhagen, is sparsely annotated, and then chiefly in the sections on great conjunctions.
40. A general idea of the contents of the *Summa* can be had from L. Thorndike, *A History of Magic and Experimental Science,* vol. III, New York 1934, ch. 21 ('John of Eschenden'), where reference is also made to the manuscripts (cf. Appendix 20 and pp. 329–34).
41. Ms. Oriel 23, f. 222v.; quoted by Thorndike, *op. cit.,* p. 332, n. 12.
42. But note the erroneous pagination of the printed edition near the end. The books are in the ratio of 3:5, approximately, and not 1:3, as Thorndike suggests.
43. I have quoted passages from this introduction (in Middle English translation) in my *Richard of Wallingford,* vol. II, Oxford 1976, pp. 86–89.
44. For example, Macrobius, Augustine, Josephus, Firmicus, 'Hermes', Isidore, Severinus, Rabanus, Comestor, Vincent of Beuvais, Grosseteste, Sacrobosco, Robert of Leicester, Bacon, Walter of Odington, and an anonymous computist.
45. I earlier alluded to this well-worn tradition, in connection with the quotation from Nemesius.
46. There are a number of inconsistencies in the manuscripts and printed edition, and I am not sure that the figure represents the author's intentions; but he makes so many mistakes of a kind which are clearly his, that there is no point in aiming at consistency here.
47. Both were of course trying to reconcile their results, as best they could, with Old Testament chronology. Albumasar is supposedly quoting 'Belenus Bentenniz', probably Apollonius of Tyana (1st cent. A.D.).
48. Septuagint, Isidore, Eusebius, Bede, and Augustine, for example.
49. Cf. n. 19, above.
50. They are summarized in the following table, in which S, J, and M denote Saturn, Jupiter, and Mars, respectively:

Ms.	Planets	quoted time of conjunc.	actual day	quoted longitude	actual longitude	sign of zod.	other considerations
A	(eclipse)	1345 Mar.20,19^{h}46^m	Mar.19	202°	186°.5	Libra	lunar
A	S J	1345 Mar.23,4^{h}46^m	Mar.24	319°	319°.0	Aqu.	
B	S M	1349 Mar.23,10^{h}16^m	Mar.23	4°	3°.2	Aries	
B	(eclipse)	1349 Jul.1,11^{h}43^m	Jul. 1	287°	286°.6	Cap.	total lunar ed.
B	J M	1349 Aug.7,21^{h}49^m	Aug. 6	101°	99°.8	Can.	
C	S M	1357 Jun.8,10^{h}30^m	Jun. 5	111°	109°.8	Can.	
C	S J	1365 Oct.30,2^{h}22^m	Oct.25	218°	217°.0	Sco.	
C	(eclipse)	1357 Jul.31,18^{h}46^m	Jul.31	317°	316°.2	Aqu.	lunar

Key to fols, all of MS Digby 176., 14th cent.: A-9r. to 16r; B-30r. to 33r; C-34r.-49v. Other MSS in which some of the material is reproduced are Ashmole 393[i], ff. 81v–86r;

Ashmole 192, pp. 1 to 106; B.M. Royal 12 F XVII, ff. 172r–180r; B.M. Sloane 1713, ff. 1–14; BN 7443.

Notes: Times given are from midnight. Ashenden's times under C are in astronomical reckoning, i.e. from the noon of the previous day, and he usually gives *completed* days, not dates. In the first line I follow Thorndike's emendation of 9^h to 19^h (vol. III, pp. 326–8). Invariably I modify to give ordinary civil dates. The 46^m on line two follows only because $2^d 9^h$ are added to the figure on the first line. Thorndike is in error by a day here.

51. The best MS., namely Digby 176, belonged to Reed, who assembled it piecemeal in an interesting way, explained in a note on f. 1 v; and in a table of contents on the same page the two statements are made about the authorship of the calculations. There is, even so, room for doubt. I have recalculated all the conjunctions in the table (n. 50), and found that (I) the first for 1349 is very close indeed to a prediction on the basis of the Alfonsine tables; but five days out by the Toledan tables; (II) the second for 1349 would have been predicted for (the correct) 6 August from both tables; (III) the other two were badly calculated, but in both cases the results quoted come appreciably nearer to Alfonsine predictions; (IV) for 1345, the Alfonsine prediction would have been for 17 March, the Toledan for more than 3 weeks earlier, the actual conjunction March 24, and the quoted conjunction March 23, which therefore has an *ex post facto* look about it. The agreement may simply be a fortunate chance. These figures may also have been covertly borrowed (see n. 53 below). The precise prediction of a conjunction along traditional lines is a tedious business, with ample room for mistakes.
52. Thorndike, vol. III, p. 340.
53. A French translation of a large part of this letter is given by Pierre Duhem, *Le Système du Monde*, vol. IV Paris, repr. 1954, pp. 35–7, although as Thorndike indicates (*op. cit.*, vol. III, p. 319, n. 81), Duhem's interpretation is vitiated by the fact that he misread the symbols for Jupiter and Saturn, and interchanged meanings. (From a French standpoint, the worst mistake was to have made France subject to Saturn and England to Jupiter. Duhem, who was intensely nationalistic, probably never made a more sinister error.) Jean predicted the conjunction for 30 October 1365 in 8° Scorpio, and the later Ashenden/Reed figures were in suspicious agreement with these. Both figures are in error (25 October, 7° Scorpio), but presumably because based on the same (Alfonsine) tables. Note that the *Chronica Johannis de Reading*, ed. J. Tait (Manchester, 1914), p. 166, records the conjunction as *having happened* on 30 October in 8° Scorpio. One must obviously treat chronicles with caution, when looking for firm astronomical data.
54. MS. Digby 176, f. 36v.
55. P. 68 above.
56. MS. Digby 176, ff. 36v–37.
57. *Loc. cit.*
58. Majorie Reeves, *The Influence of Prophecy in the Later Middle Ages*, Oxford 1969, p. 84. The text in question was *De concordantiis*, and an updated copy is to be found following an excerpt from Ashenden's *Summa* in MS Ashmole 393, at f. 80 v. For other manuscript references see Reeves, *ibid.* pp. 50–1, 84.
59. The system laid down by the Dominican St. Vincent Ferrer, in which the twelve *status* of the Church are likened to the twelve signs of the zodiac, was a not untypical *jeu d'esprit*, but was not a part of the (relatively) hard core of astrology. For references to the system, see Reeves, *op. cit.*, p. 171.

60. MS. Digby 176, ff. 39v–40r.
61. Reeves, *op. cit.*, p. 84, n. 3.
62. B.N. 7443, ff. 221r–227v.
63. Ashmole 393, ff. 81v–86r and Ashmole 192 (I) pp. 1–106 (a copy of the former).
64. For a synopsis, see Thorndike, *op. cit.*, vol. III, pp. 343–5.
65. *History of Magic and Experimental Sciences,* vols. III and IV, *passim.*
66. See n. 53, above.
67. See especially J. Bignami-Odier, *Études sur Jean de Roquetaillade,* Paris, 1952; M. Reeves, *op. cit.*, especially pp. 225–8 and 321–5; and E. F. Jacobs, *Essays in Later Medieval History,* Manchester 1968, ch. IX (reprinted from an article of 1956–7).
68. For a summary of the *Vade mecum* see Bignami-Odier, ch. 5 (pp. 157–73). For the *Liber ostensor* and its sources see pp. 142–56.
69. *Op. cit.*, p. 325.
70. *Ibid.*, p. 426.
71. The text on comets and that on conjunctions (mentioned below), together with several other texts (e.g. by Jean de Murs & Nicole Oresme) and a useful commentary, are in Hubert Pruckner, *Studien zu den astrologischen Schriften des Heinrich von Langenstein,* (Studien der Bibl. Warburg, XIV; Berlin 1933). Henry accepted the Aristotelian meteorological explanation of comets as exhalations of vapour from the earth to the region of the upper elements.
72. The strongly Aristotelian second part is an interesting piece of natural philosophy, discussing as it does properly physical influences (like magnetic influence) as possibly radiating from the stars. It ends with the admission: 'Nolo tamen negare, quin sint in superioribus multe habitudines causales et inclinative inferiorum ad diversas dispositiones et effectus, ex quibus habitudinibus, si constarent simul, et ex dispositione inferiorum possent aliqui effectus futuri propinqui pronosticari et rationabiliteter coniecturari' (II. VIII; *ed. cit.*, p. 193).
73. Strangely enough, although he quotes Albumasar's *Liber de coniunctionibus* often enough in other connections, his authority for the rise and fall of religions (I. IV) is Alkabitius (ed. cit., p. 144).
74. I. VII; *ed. cit.,* p. 147.
75. I. VIII; *ed. cit.,* pp. 149–52.
76. First published c. 1480 by Joh. de Paderborna, Louvain, but a better edition is that by Erh. Ratdolt, Venice 1490; written 1414 (together with the work that is often treated as a separate item, but which is a part of the first, *Concordantia astronomie cum hystorica narratione*).
77. *Op. cit.,* p. 423.
78. Pierre Duhem, *op. cit.*, vol. IV, pp. 175–83 reproduces much of Pierre d'Ailly's thought on this subject.
79. Printed in the Louvain edition of c. 1480; see n. 76 above.
80. For some of the many possible references, see Thorndike, *op. cit.*, vol. III, p. 107.
81. N. Valois, 'Un ouvrage inédit de Pierre d'Ailly, le *De persecutionibus ecclesiae*', *Bibliothèque de l'École des Chartes,* LXV (1904), pp. 557–74.
82. See Charles Guignebert, *De imagine mundi ceterisque Petri de Alliaco geographicis opusculis* Paris 1902, p. 20, where he quoted the *Concordantia,* cap. 60, See also Salembier (*op. cit.*, n. 4 above, 1932), pp. 357–9. Canon Salembier almost appears to believe that the prediction was meritorious.

83. For both sources see Thorndike, op. cit., vol. IV, p. 93.
84. Hieronymus Torrella, *Opus praeclarum de imaginibus astrologicis* Valencia, Alfonsus de Orta, 1496, sig. d, ff. 5–6.
85. See, for a recent view, the very detailed work by D. Kurze, *Johannes Lichtenberger. Historische Studien,* Heft 379; Lübeck and Hamburg 1960, and Majorie Reeves, *op. cit.,* especially pp. 347–74.
86. See Kurze, ibid., pp. 58–9.
87. See Kurze, *op. cit.,* pp. 16, 34–5. The year was 1492.
88. Reeves, *op. cit.,* pp. 315–7, referring to F. Pelster, 'Die Quaestio Heinrichs von Harclay ueber die Zweite Ankunft Christi...', *Archiv. italiano per la storia della pietà,* I Rome 1951, pp. 33–5.
89. *Op. cit.,* vol. IV, p. 529.
90. The best modern edition, with a useful introduction (and translation into Italian), is by Eugenio Garin, Florence 1946, in 2 vols.
91. *Spiritual and Demonic Magic from Ficino to Campanella,* London 1958, pp. 55–6; the reference is to Garin's ed., pp. 194 seq.
92. Ibid., p. 58. The occasion was the publication (1581) of his Latin version of Savonarola's Italian treatise against divinatory astrology.
93. H.-François Delaborde, *L'Expédition de Charles VIII en Italie,* Paris 1888, pp. 317–8 reports Tizio's memoirs as recounting strange happenings – eclipses, conjunctions of stars, dissension among the cardinals, and rumours of French invasion.
94. Ed. Garin, p. 72.
95. This, and all other material to which reference is made, is from Book V, ed. Garin, pp. 520–622.
96. Ed. Garin, p. 522; but cf. p. 592, where the year is given as 1444 A.D.
97. *Concordantia, ed. cit.,* cap. 59.
98. Cf. Albumasar: cum complete sunt 10 revolutiones ex revolutione sue essentie et permutationes multe ex apparitione prophetie et permutatione vicis et sectarum et consuetudinem secundum quod narrabimus, ut in exemplo in temporibus futuris quod est quia complete sunt 10 revolutiones Saturni in diebus daribindar fuit apparitio Alessandri filii Philippi nobilis et remotio vicis Persarum; et quia complete sunt ei 10 revolutiones alie ex revolutione sua apparuit Jesus Filius Marie super quem fiunt orationes cum permutatione secte; et quia complete sunt 10, alie revolutiones ex revolutione sua apparuit Meni et venit cum lege que est inter Paganos et Nazarenos; et quia complete sunt 10 alie revolutiones ei ex revolutione sua venit propheta cum lege... (*De magnis coniunctionibus,* II, 8; quoted by Garin, *ed. cit.,* p. 54 n.).
99. Ashenden mentions nos. 58 and 64 in his *Summa* on more than one occasion.
100. Ed. Garin, pp. 132, 622.

Addendum

Much literature of relevance to this chapter has appeared in recent years. A work that chronicles seventeenth-century echoes of the principles here discussed is E. Labrousse. *L'Entrée de Saturne au Lion,* The Hague, 1974.

9

CHRONOLOGY AND THE AGE OF THE WORLD

> The *Europeans* had no Chronology before the times of the *Persian* Empire: and whatsoever Chronology they have now of ancienter times, hath been framed since, by reasoning and conjecture.[1]

Isaac Newton's two categories of chronological argument, by reasoning and by conjecture, are the two poles around which belief about the age of the world and the chronology of its history have always turned. But one man's conjecture is another man's reason, and Newton's contemporaries often seem to have had as much difficulty in deciding whether a particular chronological or chiliastical argument was conjectural, literal, cabalistical, revelational or rational, as they had in first formulating it. Borrowing from these categories, however, I might say that I am to consider arguments about the age of the world which were 'rational' in the sense that Newton's chronology was thought to be so.[1] In other words, I shall try to bring together a group of specifically astronomical arguments developed out of premises which were for the most part conjectural, if not positively nonsensical.

Writers today who wish to pour scorn on chronology in this 'rational' tradition tend to select as their target James Ussher (1581-1656), Archbishop of Armagh—a man, I might add, of an erudition seldom matched by that of his critics. Ussher's supposed folly is to have set a date to the creation of the world. This event he placed at the beginning of the night preceding 23 October 4004 B.C.[2] But this date, assigned to the Creation by 'acotholicorum doctissimus' (as he was called by his worthy Jesuit opponent Henry Fitzsimon), whether or not it may be properly seen as a symptom of aberrant intellectualism, like millenarianism among Anglicans, was neither new with Ussher nor was it even the

---

[1] Newton was more concerned about the end of the world than its beginning, and in his published works does not specify its age; but it was a simple matter for his followers to extrapolate to the Creation, using the *Short Chronicle* which prefaces his *Chronology*.[1] He usually follows the time intervals of the Jesuit Petavius (and not, as one might have expected, Ussher's), but his dates are four years higher than Petavius', suggesting a Creation date of 3988 B.C.

[2] The year 710 of the Julian Period, for which see below. My edition is that of 1722.[2] The preface is rather pedantically dated 1650 in vulgar reckoning, and 1654 from the true nativity (Ussher's reckoning).

product of characteristically seventeenth-century thought.³ Chronology had been the key to universal history from the time of antiquity, and Old Testament genealogies stretching from the time of Adam himself, and reinforced by other Hebrew and patristic sources, could always be cited in refutation of Epicurus, Lucretius, and all who considered the world eternal, or even excessively old. The roots of Christianity, however, were to be clearly established as older than the philosophy of the Greeks; and yet it was thought essential that the Christian historian harmonize with Holy Writ the chronicles of the different nations—the Persians, 'Chaldeans,' Egyptians, Greeks, Romans, and the rest⁴—in order to strengthen Faith in a literal reading of scripture. That this judgement was right may be seen from the fact that atheist and theist alike, especially in the eighteenth and nineteenth centuries, chose chronology as a point at which to attack the authority of the Bible.⁵ These were the tactics of Thomas Paine in *The Age of Reason* (1795-6): 'that witling Paine' was a description offered of him by a chronologist, William Hales.

The fundamental method of determining the epoch of the Creation from the Bible was, of course, to reckon by generations and reigns. Even where the ages of named individuals are given, there usually remains the problem of assigning an interval between the birth of members of successive generations, and early historians tend to accept three generations to the century, as an average. Reckoning by reigns gives remarkably consistent results over different cultures and historical periods, taking an interval of about 22 years as the mean.⁶ A more serious obstacle to accuracy is the large number of discrepancies which exist among the different versions of scripture—the Septuagint, the Samaritan version, and the Massoretic text—not to mention such historians as Philo Judaeus, Josephus, and Theophilus, bishop of Antioch.⁷ The literature is truly vast, but the method adopted from the first in all comparative chronologies was to seek one acceptable datum (or more) which enters into the chronologies of two or more nations. Thus the birth of Cyrus, which allows of a correlation with Greek

---

³Ussher's chronology is widely known by virtue of the fact that it was added (by unknown authority) to the margins of the King James Bible, which first appeared in 1611. It is still therefore readily accessible.

⁴The Jews did not establish an era in the precise sense of the word. The chief of the ancient eras were the era of Nabonassar, 747 B.C.; the era beginning with Coroebos' victory in the Olympic Games, 776 B.C.; and the founding of Rome, for which Varro gave the date 753 B.C. The Olympiads were possibly the last system of the three to be actually used, and the era of Nabonassar was probably used from his own time. It was adopted by Ptolemy for use in the *Almagest*.

⁵For this later history of the subject see the excellent *The Age of the World: Moses to Darwin* by Haber.⁽³⁾

⁶Newton was certainly conjecturing when he took 19 years.

⁷Speaking generally, the Septuagint tends to assign longer lives to those between the Flood and Abraham, and adds other individuals to the genealogies. The discrepancy between the derived total span from the Creation to Abraham is that between 1948 years and 3334 years, or thereabouts. From the Creation to Christ's birth those relying on the Septuagint arrive at about 5500 years, while the majority give an interval of the order of 4000 years. For further data see p. 318 below.

chronology, offers a way into the chronologies of the Persians, Medes, and Assyrians, and in the last of these the destruction of Solomon's Temple by Nebuchadnezzar links with Hebrew sources. Thence the chronology may be extended to the foundation of the Temple, to Exodus, Abraham's birth, and—for the intrepid believer—to the reign of Nimrod, the Deluge, and the Creation of the world.

The synthetic procedure thus outlined was not of a sort likely to recommend itself to men of a rationalist temperament, men to whom the symmetries of the world are more appealing than its historical oddities. The simplest way of introducing *a priori* symmetries into a temporal analysis of world history is of course to make use of temporal cycles, and there are no more notorious cycles connected with world history than those which go under the generic name of 'Great Year.'

The expression 'Great Year' has meant many things to many people. Especially associated with Plato's name,[8] the idea of recurrences in history came in for trenchant criticism by St Augustine, who saw that it would not allow any historical event a unique importance.[9] He saw that the uniqueness of Christ's redemptive power thus ran the risk of being undermined by neoplatonic thought, which was in any case incompatible with the very first verse of the Old Testament. The idea of a Great Year is nevertheless closely related to others which were in due course seen as guides to the age of the created world, and its ancestry is important. It has a Pythagorean look about it, for since the Great Year is the interval between periodic recurrences of grand conjunctions of the Sun, Moon, and all the planets, it must be the least common multiple of the periodicities of those bodies; and the Pythagoreans—so the common treatment of the subject runs—had a weakness for numbers, wherefore . . . and so on. At all events, the third-century (A.D.) grammarian Censorinus tells us that the Pythagorean Philolaus had made a Great Year consist of 59 years, including 21 intercalary months, and an ordinary year of $364\frac{1}{2}$ days.[10] Censorinus gives other figures for the period, quoting various authorities, but all possibly taking the period as an integral multiple of solar and lunar periods only.[11] Censorinus also mentions that Oinopides of Chios, a younger contemporary of Anaxagoras, made the length of a year $365\frac{22}{59}$, while Aelian and Aëtius ascribe to Oinopides a Great Year of 59 years. These two items of information fit together rather well: If

---

[8] The principal source is *Timaeus*, 39D. Cf. Cicero, *De natura deorum*, II.51-2.
[9] See especially *De civitate Dei*, XII.18, but also XII.14, where Augustine has a little trouble with Origen (*De principiis*, 3-5, 3) and a passage from *Ecclesiastes* (1, 9-11): "The thing that hath been, it is that which shall be. . . ."
[10] *De die natali*, 18,8. I owe this reference, and the remainder of the paragraph, to Dicks.[(4)]
[11] Dicks,[(4)] p. 76, points out that Schiaparelli, dividing the Great Year of Philolaus ($21,505\frac{1}{2}$ days) by suitably chosen integers, derives data for the planetary periods in very close correspondence with modern values. This would suggest—and Thomas Heath follows Schiaparelli—that his 'Great Year' was such in the Platonic sense. Dicks disagrees, but it is difficult to see how one should adjudicate in the debate, since it is at least arguable that Censorinus' information is likely to have been incomplete on a matter involving lengthy numerical data. In 59 years, Saturn completes approximately 2 revolutions, Jupiter 5, Mars 31, the Sun, Mercury, and Venus 59, and the Moon 729.

the common figure of $29\frac{1}{2}$ days is taken as the length of a lunation, the number of days in a Great Year (of the Sun and Moon only) must be a multiple of 59, and it must also be a multiple of a number close to 365, the approximate length of a year. A period of 59 years totalling 730 months is an obvious one to take; but how Oinopides knew that this contained 21,557 days is not at all clear.[12]

The connection of these ideas with calendrical problems, and especially with that of intercalation, is never far to seek. Recall the intercalation cycles associated with the names of the fifth-century (B.C.) Athenian astronomers Meton and Euctemon.[13] To correlate the lunar months and the solar year, 235 months (= 228 + 7 intercalary)[14] were equated with 19 years and 6940 days. Better agreement with the truth was attained by Callippus, a century later, when four Metonic cycles of years, that is, 76 years, were made equal to one day fewer than the Metonic total (27,759 as opposed to 27,760). Another advantage of the Callippic cycle is that the number of leap years in it is constant, and this is obviously not true of the Metonic cycle, to its detriment.

The Metonic cycle was used in early Jewish and Greek calendars—but sporadically, not regularly even in Athens—as well as in the ecclesiastical calendars of Christian Europe. Long after its invention the cycle was taken to have been begun at 13 July 432 B.C., which meant, for example, that on 13 July 1 B.C. the 14th year of a cycle began, and that the first new cycle in our era began on 13 July A.D. 6.[15] Tedious and unmemorable though these facts may be, I must point out that other source-dates were accepted for the 19-year cycle throughout the greater part of the Christian era—and these as the result of writings ascribed to Bishop Cyril of Alexandria and Dionysius Exiguus. The cycle assigned to Cyril was supposedly issued in A.D. 437, expiring in 532.[16] It is now thought to be a seventh-century Spanish fabrication.[17] Some later writers took A.D. 532, the year with which Dionysius started his 19-year cycle, as a basis for the continuation of a 532-year cycle. For a cycle beginning with A.D. 532, the year 1 B.C., the nominal year of Christ's birth, is very fittingly the first year of a 19-year cycle, A.D. 1 being the second year, and so on.

The Metonic cycle had by this time lost its prime calendrical purpose, and was no longer associated with rules of intercalation. It had become merely a device for ordering the years with a view to determining the date of Easter, and into the tangled web of Easter computation I certainly have no wish to enter. There is involved in it the cyclical period of 532, however, which I should not pass by without further comment. The most famous of the rules for Easter were the Roman and the Alexandrian, and the practice of the early Church was at first fairly evenly divided between them. The Roman rule took an 84-year cycle (of

---

[12] Dicks,[4] p. 89, leaves this matter very much in the air.

[13] The cycle was probably borrowed from Babylon, although in extant sources it first appears there c. 383 B.C. See Parker and Dubberstein.[5]

[14] Of these, 110 were 'hollow' (29 days) and 125 were 'full' (30 days).[6]

[15] The first Callippic Cycle ran from 330 B.C.

[16] In short, it covered 5 × 19 = 95 years.

[17] The exposure is due to Bruno Krusch. See Jones[7] for further details.

3 × 28 years), and the Alexandrian rule the 19 years of the Metonic. I will first explain the 28 year cycle, and then say how the cycles thus far mentioned were to be compounded into yet longer cycles.

The 28-year cycle is to modern eyes a simple consequence of the adoption of certain calendrical conventions—those of the Julian calendar, combined with the selection of a period of seven days for our week. Even the Christian who believes that the week was divinely ordained is bound to admit that the Julian calendar had more than a trace of paganism in its institution. There is therefore nothing either empirical or sacred about a rule ascertaining periodicity of recurrence of the phenomenon of a particular day of the month coinciding with a particular day of the week. The cycle is one of 28 years.[18] I have emphasized its conventional character, because even in the nineteenth century we find chronologists marveling at the fact that a creation of the world in 4004 B.C. implies that in the year 1 A.D. the days of the week repeated their order at the Creation. In the words of Edward Greswell, whose life at Oxford (according to the *Dictionary of National Biography*) was spent "in the systematic prosecution of his studies," and who was writing only a few years before Darwin's *On the Origin of Species:*

> this is another of the singular and hitherto overlooked coincidences, which a true scheme of chronology brings to light; and one among other proofs, thereby supplied, that in the Divine mind, and in the providential constitution and adjustment of time and of its several relations from the first, the Julian reckoning of time itself... must have been contemplated from the first. It is no accidental coincidence that B.C. 4004 was thus the first year of the cycle of 28... [8]

A sceptic might be excused for wondering whether 4004, a figure at least as early as Basil (A.D. 330–379), was first chosen (in preference to numbers close to it, which could have been fitted to the Bible equally well) for the very reason that it was a multiple of the 'solar cycle' of 28 years.

Combining the 28-year and 19-year cycles gives the cycle of 532 years, a cycle which is often—but mistakenly—said to have been introduced into the West by Victorius of Aquitaine.[19] One of its advantages was that it went some way toward reconciling the computus of the Alexandrian church with that of the early Roman church. According to Greswell, Georgius Syncellus used the 532-year period for the measurement of time from the Creation downward.[20]

There were many other calendar-cycles than those I have mentioned. Few were of greater potential historical use than the Julian Period—as it was named by Joseph Justus Scaliger, who recovered it from an earlier work by Roger of

---

[18] A common year exceeds 52 weeks by one day. Month-day and week-day would coincide every 7 years, were it not for the leap-year's extra day. The concept of the 'concurrent' arises in this connection. The concurrent is the excess of the annual Julian cycle over the hebdomadal in the 28-year period. Its numerical values (with multiples of 7 removed), in successive years, are 0,1,2,3,5,6,7,1,3, ...

[19] Jones,[7] p. 64. The old version is that Victorius (Victorinus, according to Scaliger) derived the cycle from Arianus of Alexandria, a contemporary of Theophilus, and introduced it in A.D. 457. It is clearly unknown, however, either to Victorius or to Dionysius.

[20] Greswell,[8] p. 191, note. Syncellus, who was alive in 810, compiled his general account of Creation from *Jubilees*. He will be referred to on a number of occasions below.

Hereford.[21] This period of 7980 years was a product of the solar cycle (28 years), the Metonic cycle (19 years), and the Roman Indiction cycle (15 years).[22] Every year of the Julian cycle is characterized by a different trio of numbers belonging to the three cycles, beginning (perhaps only theoretically) in 4713 B.C. and ending—or recurring, if the world survives—in A.D. 3266. The Julian Period has never been regarded as a Great Year in the Platonic sense of a period after which were reproduced the original positions of two, or more, and preferably all, of the planets, and perhaps their nodes and apogees as well. The calendar cycles to which I have referred were less comprehensive, being merely lunae-solar, but they did permit the Christian chronologist in principle to rectify the dating of the first Easter week, that is, to make it compatible with the chosen date of Creation.[23] I shall now consider very briefly some of the principal early references to Great Years in the Platonic sense, including eastern examples, before turning to medieval Europe, and ways in which Christian thought became reconciled to one or other *numerical value* for the periodicity, without accepting the underlying rationale of a universe with an ever-repeating history.

Plato gave no value for the length of the Great Year, although his editors have fostered on him such figures as 36,000 years, or 12,960,000 daily rotations.[(9)] There is no justification for any such ascription. The references to Philolaus and Oinopides by Censorinus, as already explained, yield a figure of 59 years, and a reference to Democritus by the same author yields a rather puzzling 82 years.[24] Censorinus again is the source for a figure of 2484 years, drawn from Aristarchus; but Tannery has argued that this is a mistake for 2434 years, 45 times as great as a triple ἐξελιγμός, which is the period defined by Geminus as the shortest time containing a whole number of days, a whole number of lunations, and a whole number of anomalistic months.[25] Whatever the truth of Tannery's suggested correction, there seems to be no ground for thinking that Aristarchus' cycle is anything but lunae-solar.

This rather disappointing beginning to the career of so grand a conception was changed with the infusion of a rather ludicrous historical tradition which began, it seems, with the historian Berosus (fl. *c*. 290 B.C.). Berosus, priest of Bel, was one of those principally responsible for transmitting Babylonian history

---

[21] See Ussher's preface (*Lectori*), the third page, in the *Annales*.[(2)] Ussher has 'Robertus' rather than the correct 'Rogerus.'

[22] The last of these was a taxation period, introduced in A.D. 312 by the Emperor Constantine the Great.

[23] The precise date (within the year) of the world's first week must for this purpose be known. Opinions on this matter are discussed at pp. 328-329 below. The doctrine of the relation of the date of Creation and that of the Easter-moon first appeared in the writings of the African authors pseudo-Cyprian and Hilarianus. See Jones,[(7)] p. 39, and cf. p. 64 on Victorius.

[24] Paul Tannery suggested that this was a mistake for 77 years. See Heath,[(9)] p. 129.

[25] See Heath,[(9)] pp. 314-6. On the error in calling the basic period by the name of 'Saros,' see Neugebauer.[(10)] It has long been widely—but wrongly—assumed that period of 223 synodic months (= 242 anomalistic or draconitic months) was the basis of the prediction of eclipses by the Babylonians and early Greeks. (It is, of course, a genuine eclipse period.)

and astronomy to the Greek world, and his work began with the origins of the world and the Flood. He spoke of Babylonian astronomical observations carried out over a period of 490,000 years.[26] Epigenes of Byzantium increased this figure to 720,000 years, while Simplicius made it 1,440,000 years. All were treated with proper scepticism by Cicero, Diodorus, and later by Georgius Syncellus; but with the intrusion of Stoic ideas, what had originally been perhaps no more than an excessively long historical interval became transmuted into a value for the Great Year of astronomy. The Babylonians, moreover, were now imagined to have made astronomical observations since the commencement of a Great Year, and Palchos, in the fifth century, believed that they had observed in every climate and almost every day during that time! (Bouché-Leclerq,[11] p. 575).

Writers mentioning the concept 'Great Year' are almost invariably reluctant to show how an actual value for its length is arrived at, and this for obvious reasons. By the time of late antiquity, very precise parameters for the planetary motions were known, but this only made the discovery of a least common multiple of the periodicities more difficult. Macrobius, for example, in his commentary on Cicero's *Somnium Scipionis*, would obviously have liked to set forth the arguments, but in the end he leaves it that 'the philosophers' reckon the Great Year as 15,000 years.[27] This is no easier to justify than most of the figures quoted above. The 490,000 for example, involving a square of 7, a 'week of weeks', or ten thousand jubilees, must appear numerologically inspired, to those who know the symptoms. The 473,000 might just possibly be an approximation to the product of the square of the *exeligmos* 223 lunations, about 18 years, and the Sothis period of 1461 Egyptian years, but this is very unlikely.[28] One certainly cannot rule out the use of the Sothis period in connection with Great Years, for the figure 1461 is the length of the Sun's 'maximum year' in the Great Introduction (*Kitāb al-ulūf*) of Abū Ma'shar.[29] Yet another explanation which

---

[26] Variant quotations from different sources give in addition figures of 470,000, 473,000, 468,000, and 432,000. The last two are from Syncellus, and the last is discussed again at p. 315 below. See, for further information, Bouché-Leclercq,[11] pp. 38-9, where figures of 300,000 years (Firmicius Maternus, 4th century A.D.) and 17,503,200 (Nicephorus of Constantinople, c. A.D. 750-829) are mentioned. Nicephorus wrote a work on chronology, translated into Latin by Anastasius the Librarian, and included in his *Chronologia tripartita*. Notes by Syncellus were added to it, and together they appear in Jacob Goar's edition (Paris, 1652), and its successor, in the series *Corpus Scriptorum Historiae Byzantinae*, 2 vols. (Weber, Bonn, 1829) under the title ΕΚΛΟΓΗ ΧΡΟΝΟΓΡΑΦΙΑΣ (Chronographiae). I quote Goar's Latin translation from the latter.

[27] For an English translation see Stahl,[12] pp. 220-2. Macrobius is arguing that Cicero's statement to the effect that a man's reputation cannot endure for a single year refers to the Great Year of the *Timaeus*. Cicero was obviously more realistic than Macrobius.

[28] The idea is to be found in Hales,[13] Vol. 1, pp. 40-41. The Sothis period is the time taken for the seasons (e.g., the summer solstice) to return to the same civil date in the Egyptian calendar. The shift of date is due to the discrepancy of about a quarter of a day between the actual year and the 365-day Egyptian year. (We have $1460 \times 365\frac{1}{4} = 1461 \times 365$).

[29] The table in which this figure occurs was copied out by Ashenden, as I shall mention later. For the original context of this material, see Pingree,[14] p. 64.

has been offered for one of the figures quoted involves the 18-year cycle. According to William Hales, writing in the early nineteenth century, the period of 432,000 years is 'Chaldean', and is made up of cycles of 18 and 24,000 years.[30] Hales makes it clear that he takes the longer period to be a 'Hindu' parameter for the precession of the equinoxes—or, as it should be called in this historical context, the 'movement of the eighth sphere.' An explanation with better historical credentials will be given later (see the paragraph after the next).

This brings us, however, to an unavoidable clouding of the very conception of a Great Year. Even a modern editor of Plato's *Republic*, J. Adam, has suggested that when Hipparchus found a figure of one degree per century for the movement of the eighth sphere, thus implying that the stars circuit the sky in 36,000 years, he was influenced by Plato's perfect number of the *Republic*, which he interpreted as the square of 3600.[31] Quite apart from the fact that Hipparchus merely set one degree per century as a lower limit to the phenomenon he had discovered, the empirical character of his result is not in doubt. It would also be surprising if Hipparchus had understood the *Republic* passage in question, since it seems to have been beyond the resources of most of Plato's editors to do so.

Later writers did introduce the movement of the eighth sphere into the periodization of world history, as I shall have occasion to show, and there is nothing intrinsically improbable about a medieval 24,000-year cycle explained along these lines, since this corresponds closely to the data accepted by several Islamic astronomers. Even so, the explanation of the 432,000 offered in the twelfth book of the Sanskrit *Mahābhārata* and in the first book of the laws of Manu (both stemming from a source earlier than the second century A.D.) is no doubt the original one, and is very different (Pingree,[14] p. 28; van der Waerden[16]): There, a 'year of the gods' is equal to 360 ordinary years, while 12,000 of these, that is, 432,000 years, make up a *yuga* of the gods—called by later astronomers in India the *Mahayūga*, or Great Year. This in turn was divided into ten parts, in the ratio 4:3:2:1, and the last part, the historical period in which we are now, on this system, living, was thus to be of duration 432,000 years—the period we find in the early ninth-century Western writer Georgius Syncellus. It might well be the case that different explanations were found at an early date for one and the same number. Priority certainly seems to belong to the system as it was found in India, where a *Kalpa* or 'day of Brahman' (containing a thousand *yugas*) is mentioned, according to Pingree, more than two centuries before Christ.[17] That it is a Babylonian system in origin seems to follow from the statement reported by Syncellus from Berosus, that the total length of the reigns of Babylonian kings before the Flood was 120 saroi, where the (true) saros is

---

[30] Hales,[13] Vol. 1, pp. 41–42. cf. n. 67 below.
[31] See Heath,[9] pp. 171-2. A better commentary on the whole subject is to be found in Cornford.[15]

3600 years.³² This is the earliest implied explanation of the figure of 432,000 years, although it is worth observing that, sexagesimally written, the number becomes 2,0,0,0, which in itself seems almost to guarantee a Babylonian origin. The interval's occurrence in Berosus as a historical period commencing with the Creation of the world makes it proper—if that is the right word—that it should have passed into Islamic, Byzantine, and Western chronological reckoning.

As an example of the way in which the bare bones of this now Indian and Iranian doctrine crept surreptitiously westward across half the world, we might first consider Pingree's masterly study of the *Kitāb al-ulūf* (Book of the Thousands)³³ by Abū Ma'shar (A.D. 787-886). The sources for this book were held to have gone back to antediluvian times, and it was supposedly written on the basis of written records of 'cycles of the Persians' discovered in a hole in Isfahan, where they had been preserved from natural disaster from the time of the ancient kings of Persia (Pingree,⁽⁴¹⁾ pp. 2-4). This is all reminiscent of Plato's reference in *Timaeus* to the antediluvian wisdom of the Egyptians.³⁴ In a *zīj* (a set of astronomical tables) by the same author, which again purported to transmit to posterity the astronomy of the prophetic age, some of the methods are of Indian origin, and introduce the *yuga*. Certain planetary parameters (in the *zīj*) were received indirectly from Persian sources, while the planetary models were Ptolemaic. On the score of antiquity, the most one can say of the writings supposedly recovered from the Isfahan hole is that they were certainly older than Piltdown man.

At the root of Abū Ma'shar's astronomical system, that is, of a system which was essentially Indian, was an assumption that the planets are in (mean) conjunction at the first point of Aries at regular intervals of time. Of four Indian methods he could have followed,³⁵ he takes one which assumes grand conjunctions of the mean planets at Aries 0° in 183,102 B.C.; at midnight of Thursday/Friday, 17/18 February 3102 B.C., the epoch of the Flood; and in A.D. 176,899. The *yuga* is now clearly one of 180,000 years. Most of the planetary parameters

---

³²In Goar's Latin version (*ed. cit.*, footnote 26 above), Syncellus writes: Sed et Berosus Saris, Neris et Sossis annorum numerum composuerit, quorum quidem Sarus ter millium et sexcentorum, Nerus sexcentorum, Sossus annorum sexaginta spatium complectitur. Saros autem centum et viginti, decem regum aetate, hoc est myriadum quadraginta trium et duorum millium annorum collegit summam Annos porro illos Historicorum nonnulli dies coniiciunt, ex quo Eusebium Pamphili causati, qui Sarrorum annos dies esse non animadvertit. Nullo tamen consilio inscitiae accuusant. Qui namque quod non erat, vir alioquin eruditus, assereret? Vir ille, dico, qui Graecorum sententiam, plure saecula, annorumque portentosas myriadas ex fabulosa Zodiaci per partes adversas in idem signum conversione, ab arietis, inquam, termino ad eandem metam revolutione, iam a mundi natalibus praeteriisse asserentem probe dignosceret? Qua vero necessitate pressi mendacium veritati coacervare excogitaverunt?

³³Pingree.⁽¹⁴⁾ The work is lost, but summaries by other writers allowed Pingree to reconstruct most of the argument of the work, which is conjectured to belong to the period A.D. 840-860.

³⁴See the long tale about ancient cultures by Critias near the beginning.

³⁵Pingree,⁽¹⁴⁾ pp. 28-9. The other three methods were known to the Arabs as those of the *Sindhind*, *al-Arkand*, and *al-Arjabhar*.

come from the *Sindhind*.

Of the three other Indian systems, two share with that followed by Abū Ma'shar a Flood date of 17 or 18 February 3102 B.C.[36] For the date of the Creation, al-Hāshimī tells us that Abū Ma'shar took this (in effect) as 4693 B.C., in his Book of the Thousands. In the *Kitāb al-qirānāt* (The Book of Conjunctions), however, Abū Ma'shar sets 2226 years, 1 month, 23 days, and 4 hours between the birth of Adam and the Flood, yielding a Creation date of 5328 B.C.[37] Dates coinciding with these, or near to them, appear in western writings, either borrowed or from a common source. As Pingree explains of the *Kitāb al-ulūf*, "there is an overwhelming mass of religious references which clearly belong to an Islamic milieu, not a Zoroastrian one," and that "though some of his astrological material may go back to a Sasanian source through the unknown scholar of the time of Hārun al-Rashīd, he has largely revised it to conform to the conditions of his own age" (Pingree, [14] p. 58). It is not surprising, therefore, that at least some Jewish and Christian authors should have come close to sharing epochs with Abū Ma'shar. From the Septuagint, Eusebius, Augustine, Isidore, and Bede we find from the Creation to the Flood 2242 years, only sixteen years or so removed from Abū Ma'shar's figure of the *Kitāb al-qirānat*. The discrepancy is easily explained in terms of the different approaches to the problem, the scriptural approach (by generations), and the astronomical (by conjunctions), which we must suppose was aiming to replicate it. As for the inconsistent Creation-to-Flood interval of 1591 years, from 4693 B.C. to 3102 B.C., this fits no well-known source very well. If a comparison with Josephus—of the early writers—seems profitable, this is probably for the wrong reason (see footnote 37). When we consider the Indian Flood-date of 3102 B.C., however, we find it frequently mentioned in Western astronomical works, which it apparently enters not only through the writings of Abū Ma'shar himself, but through the astronomical tables of al-Khwārizmī (*c.* A.D. 820) in the version of the Spanish astronomer Maslama al-Majrītī (*c.* A.D. 1000). Adelard of Bath put this into Latin early in the twelfth century.[38] Essentially the same table, slightly expanded but much more convenient to use, occurs in the astronomical tables produced for Alfonso X of Castile (*c.* 1272). There the intervals between eleven different eras are given in days (decimally as well as sexagesimally expressed).[39] The era of the flood is

---

[36] There are problems over the precise day of the Flood. Most writers took the Thursday, and Abū Ma'shar is said to have been alone in taking Friday. See Pingree,[14] p. 38.

[37] Pingree,[14] p. 38. It seems that 4693 B.C. is the beginning of the second century of the millennium of the Sun, according to Māshā'allāh's chronology. It is close to a figure for the Creation (4698 B.C.) abstracted by some commentators from Josephus' history of the Jews, and to a figure (4697 B.C.) in Magnus Aurelius Cassiodorus. But no one should ever pin an argument on Josephus, whose text is very corrupt and inconsistent.

[38] The tables were edited by Suter.[18] See especially p. 109. Neugebauer[19] has since translated them into English and supplemented them. See especially pp. 82–4.

[39] Judging by the copy in the Bodleian Library,[20] the table is completely accurate, and (although in a different form) consistent with the Khwārizmī table. The only additional material is an era 'last of the Persian Kings,' 27 August 274, and of course the Alfonsine era, 1 June 1252.

17 February 3102 B.C. Time and again one finds this particular date quoted from the Alfonsine work, and combined with one of the several scriptural Creation-Flood intervals, thus resulting in a rather foolish blend of incompatible elements, but one unlikely to offend readers ignorant of its origins.

One author of such a blend was John Ashenden, the fourteenth-century astrologer of Merton College, Oxford. In an exceedingly long *summa* of judicial astrology, running to 375,000 words or so, Ashenden devotes perhaps 20,000 words to the age of the world and the time of year of its creation.[40] He is especially indebted to Abū Ma'shar's work on great conjunctions (but not *The Thousands*), and he is even conscious of the infusion of ideas from India. He quotes a work (no longer identifiable) by Walter of Odington (a monk of Evesham of the previous generation) on the age of the world, to the effect that philosophers, and Indians in particular, maintain that all planets are at the head of Aries at their creation.[41] Over the discrepancies between his principal authorities for the age of the world he is far from complacent, but rather than offer a critical account of his own he is content to produce a muddled digest of the works of others—among which may be mentioned the Septuagint and Hebrew Bible, Josephus, Orosius, Methodius, Eusebius, Jerome, Augustine, Julius Firmicus Maternus, Martinaus Capella, Isidore, Bede, Rabanus Maurus, Helperic, Gerlandus, and such 'moderns' as Grosseteste, Bacon, and Robert of Leicester.

Ashenden's chronology is inevitably cut down to essentials. There is no question of his listing the generations very systematically. He speaks in rather general terms when he divides history up into the 'seven ages of the world,' the first from the Creation to the Flood, the second from thence to the birth of Abraham, and so on. Touching lightly on the discrepancies among his many sources, he quotes Walter of Odington, who had said that there is nothing reprehensible about accusing an evangelist of lying.[42] This remark is a necessary preliminary to the act of collecting epochs from as many sources as possible, with a view to reconciling them. The Evesham monk's lost work clearly supplied Ashenden with the greater part of his material, and it is perhaps significant that Walter of Odington was, like him, an astronomer.[24] The work must have been a morass of genealogical comparisons and it is not difficult to find even among the products of that favorite nineteenth-century pastime of compiling from different authorities lists of biblical dates (of the Creation, the Flood, and so forth) signs

---

[40] The first part was written in 1347, as is mentioned in the first chapter of the first *distinctio*. The whole was later printed under the name of Johannes Eschuid (1489).[21] All references given below to this work are to the better manuscript copy in Oriel College.[22] A general idea of the contents of the *Summa* can be had from Thorndike,[23] Vol. iii, Ch. 21, where references are also made to the MSS (Appendix 20 and pp. 329-34).

[41] MS Oriel 23, f. 3v. Ashenden goes on to report that the Jews and Christians differ in their interpretation of what comprises 'first moon.' Adam could not have seen its first conjunction, . . . , and so on.

[42] MS Oriel 23, f. 6v (I.i.1): Sed dicit Odynton quos fas non est dicere vel existimare quemquam evangelistum fuisse mentitum et allegat hoc ab Augustino.

of Walter of Odington's hand, doubtless known through Ashenden.[43] Some scores of dates are recorded, which would indeed have been more convenient if tabulated. Scriptural dates given for the Creation according to the authorities quoted vary from 3752 B.C. to 5530 B.C. When he comes to the astronomers, he complains that Albumazar is the only one whose book (on great conjunctions) he has seen which calculates the beginning of the world.[44] The others go no further back than the Flood. This is not surprising since events in their world histories were usually heralded by 'great conjunctions' of Saturn and Jupiter, conjunctions which could hardly take place before the Creation of the heavens. From Abū Ma'shar, Ashenden quotes the figure of 2226 years, 1 month, 23 days, 4 hours already mentioned for the Creation-Flood interval.[45] To this he adds the Indo-Alfonsine 3101 years, 318 days, and obtains for the period from the Creation to Christ's birth, 5328 years, 16 days, 8 hours, 30 minutes![46] This, he wistfully complains, differs from all authorities; and yet he sets great store by the precision and truth of the Alfonsine interval, while he is reasonably satisfied that the other interval is not far removed from the 2242 of Eusebius, Bede, and so many others.

If it is difficult for us now to feel any sympathy with works of synthesis like John Ashenden's, this is perhaps because the precise timing *per se* of a historical event does not seem to us to be a particularly significant characteristic of it. Accuracy is a minimum requirement of history. It may have prevented error in the ordering of events, but it did not provide an understanding of the pattern which some commentators, at least, were seeking. It was all very well for Robert Grosseteste to protest that men should stop querying *Genesis* on the subject of Creation.[47] (A bishop, Synesius, had argued for the eternity of the world and the preexistence of the soul, but that was in the fifth century.[48] Nearer home

---

[43] In Hales,(13) the name is simply rendered 'Odeaton astrologus.' In the printed edition of Ashenden ('Eschuid') the name is almost unrecognizable as 'eduiconien'! The printed edition is sometimes described as accurate, but there are scores of mistakes on every page, many of them numerical.

[44] See the discussion of this work at p. 64 above.

[45] MS Oriel, 23 f. 9r and 9v (I.i.2), has both 22 days and 33 days, and it is possible that the discrepancy stems from the author's own carelessness.

[46] The month has 30 days. Carrying over $365\frac{1}{4}$ days as a year (as he seems to do) should surely leave 5 days, 22 hours. There is discussion of a calculation involving 6 days (rather than 16) in I.i.3, but this is scarcely worth discussing here. (See MS Oriel 23, f. 13r.) The precision to which Ashenden pretends is even more ludicrous in the light of the ten errors he warns against in I.i.3. They include the confusion of different calendars and years, as well as the acceptance of spurious precision in chronological statements.

[47] Et sic desinant admirari cur mundus non sit antiquior, quam dicit scriptura, et cur non prius incepit quam dicit scriptura, quia non potest intelligi incepisse prius, quam incepit, ab intellectu comprehendente totum tempus praeteritum terminatum, sicut non potest intelligi mundum alibi esse, quam sit, ab intellectu, qui comprehendit extra mundum non esse spatium, cum tamen necessarium sit, ipsum posse esse alibi, quam sit, apud imaginationem ponentem spatium extra mundum.(25)

[48] There were writers who were not above suggesting that Aristotle himself had denied the eternity of the world.

was Daniel of Morley, denying the eternity of the world, and yet in the next breath saying that God created everything simultaneously.) Grosseteste was no doubt right to protest. But in the protesting, as he also did, at doubts about the age of the world, he underestimated the obvious pleasure that was to be had in using mathematics in sacred chronology. It was the sort of pleasure some people now get from electronic calculators; and Roger Bacon had no doubts about the spiritual value of such mathematical activity—witness many passages of the *Opus maius*. This activity went wildly astray, however, once the groundplan of sacred history was established—not uniquely, but as one of a handful of possible systems, reconciled as far as possible in the Ashenden manner. It went astray because it fell into the hands of scholars with a taste for Great Years, Great Conjunctions, and a number of similar astronomical designs for the introduction of structure into history.

Perhaps the most pervasive of these designs was that which concerned conjunctions of the two most distant planets then known, Jupiter and Saturn. I have described at length elsewhere how the history of religions and sects was supposedly governed by so-called 'great conjunctions' of these planets, and how past religious change was thus interpreted, and how prophecies were derived.[26] I have touched upon some of the consequences of this doctrine already. It owed much to the philosopher al-Kindī, in the work of whose pupil Abū Ma'shar it was chiefly introduced to Christian Europe. Three signs of the zodiac symmetrically placed (i.e., each separated from the next by three signs) are called a *triplicity*. By chance the mean periods of Jupiter and Saturn are such that twelve successive conjunctions of those planets (conjunctions which occur at intervals of about 20 years) may all occur within a single triplicity. They will be in other triplicities for similar periods of time, about 240 years in each, and after approximately 960 years conjunctions will return to somewhere near the place of the first. There are certain differences in terminology between Abū Ma'shar and the second most important source of the doctrine, Māshallāh, but these are unimportant in comparison with the fact that chronologists were provided with three time intervals to conjure with: 20 years, 240 years, and 960 years.

A change of triplicity at a great conjunction might signify the end of the world, but, as I have already pointed out, there could be no such event to herald its creation. According to Māshallāh, the conjunction of 3321 B.C. (with change of triplicity) had indicated the Flood, and another conjunction of the same planets had taken place in 3301 B.C., the year in which he believed the Flood had occurred.[49] (And lest this date seem less important than 1776, I will remind you that in 1776 America was separated only from England, whereas at the time of the Flood, according to the Premonstratensian Francois Plaçet, America was separated from the rest of the world.[28]) Extrapolating back to the

---

[49] Kennedy.[27] This is an important contribution to a very extensive subject. Kennedy quotes several other suggested spacings between great conjunction and Flood (pp. 25-6). Ashenden (I.i.3; MS Oriel 23, f. 12v) discusses several different opinions on the timing of Flood and conjunction.

Creation from the Flood by astronomical means was done in various ways, all of them somewhat reminiscent of the Great Year scheme. The Indo-Persian versions, with their enormous time scales, did not find a ready acceptance in the West. An interval of 180,000 years from Creation to the Flood was, after all, utterly irreconcilable with the Old Testament.[50] From an era 180,000 years before the Flood-Kaliyuga date (17 February 3102 B.C.)[51] a variety of world-periods had been reckoned: the *tasyīrāt*, *inticha'āt*, and *fardārāt*. Kennedy lists eleven different sources, all Persian or having a strong Persian connection.[52] The periods of the maximum, great, medium, and small *tasyīrāt* are (for one revolution) 360,000, 36,000, 3600, and 360 years, respectively. The four grades of *intihā'āt* move through a zodiacal sign in 12,000, 1200, 120, and 12 years, respectively. There seem to be several variant systems of *fardārāt*. One *fardār* listed by Kennedy, for example, is of 75 years duration, and assigns a different number of years to each planet (including the lunar nodes). This system was known in the west through Abū Ma'shar's *Great Introduction*, and a table was copied by John Ashenden[53] from this work listing five sorts of period, all made up of lesser periods allocated to the planets, under the names *anni fardarie* (total 75 years), *anni maximi* (4588 years), *anni maiores* (588 years), *anni medii* ($298\frac{1}{2}$ years), and *anni minores* (129 years). The *anni maximi* listed correspond precisely with what Pingree calls the 'mighty years' of the *Kitāb al-ulūf*.[54]

There is not much consistency between the Ashenden–Albumasar table and the reports by Kennedy and Pingree, and there is little to be gained by pursuing the matter here. Western scholars were dimly aware of the Indo-Persian eras, but there are few signs, if any, that the underlying theory of planetary longitude was understood. The West had its own thousands, not overtly astronomical, but capable of being made so. The world had been created in six days, and this the apocalyptic writers took as an indication that the world would endure for six ages, each of a thousand years, before at last came the Day of Judgement, to be followed by a millennium of rest.[55] A thousand years is one day "in the testimony of the heavens" (*Jubilees*, 4:30); and one day "is with the Lord as a thousand years" (II *Peter*, 3:8). As a measure of the influence of this mode of interpreting, perhaps it is not without relevance that Creation dates tend to be brought forward so as to place the Day of Judgement in the future while preserving the assumption that the world's duration will be 6000 years. In other words, the later an author, the later he is inclined to date the Creation. (Most Jewish

---

[50] According to Kennedy,[27] p. 24, this was ascribed to the people of Pārs in Persia, in a Persian zīj.

[51] Misprinted 1302 in Kennedy,[27] p. 27.

[52] See Kennedy,[27] pp. 26–30 for the source of this and most of the following information.

[53] MS Oriel 23, f. 7v (Ashenden's *Summa*, I.ii.3 and not I.ii.2, as in the printed edition). The table as included in the Latin Albumazar is accompanied by an utterly unintelligible explanation, and in consequence John Ashenden's version can make little sense to an uninformed reader. See Ref. 29. The table in the printed edition does not agree with the MSS I have seen, but I have not pursued the matter.

[54] Pingree,[14] p. 64; see footnote 29 above.

[55] For an elaboration of this point, see Robbins,[30] p. 27.

chronologists have been exceptions. They have always left an excessive margin of safety.)

Bouché-Leclerq[11] (p. 499) writes of a system, claimed as Tuscan but found again in Mazdean cosmogony, in which each sign of the zodiac was supposed to rule the world for a thousand years. Gayomart (the first man, in ancient Persian mythology) was born under Taurus, Adam and Eve under Cancer (horoscope of the world), and so on. Kennedy[27] (pp. 37-38) summarizes the evidence for the same system, which in Iran goes back to Sasanian times, and possibly the fifth century B.C. The 12,000 years is apparently never presented as a *cyclical* period. The 'mighty *intihā*' takes a thousand years to move through a sign, but there is considerable uncertainty about the point in time from which the Persian thousands were to be reckoned, and different chronologists took it in different ways. It is possible that several versions of this scheme, together with that rare thing, a comprehensible explanation, found their way into the European tradition. On the whole, the works of Abū Maʿshar, as they appear in Latin translation, are too obscure on the points at issue to have been very influential. The difficulty of recognizing the doctrine is that thousands, and high powers generally, are popular in all cultures as an expression of immensity in space and time alike. (Need I do more than recall the fascination of the number 10^{39}?) From Bardesanes, last of the great Gnostics, who put the duration of the world at 6000 years,[56] to the astrologer who either did or did not have to do with the dark lady of the Sonnets, namely Simon Forman,[57] a scheme with thousands was always at a premium. And of course it remains so in some religious circles to our own day.

If Christendom was not anxious to accept the Great Years of the Greek and Indo-Persian worlds, there were plenty of reasons. Many a writer touched on the 36,000-year period of the eighth sphere—and regarded it as a Great Year—with impunity. Alexander Neckham, William of Auvergne, Sacrobosco, and Bartholomaeus Anglicus all did so, for example, but not to make any doctrinal use of it. Others saw the same spiritual danger of repetition in history as Augustine had written against, and the objection was codified in 1277, when Étienne Tempier, bishop of Paris, made as the sixth of the famous 219 condemned opinions:

> That when all the celestial bodies return to the same point, which happens every 36,000 years, the same effects will recur as now.[58]

There were subsequently many voices raised against the idea, with, for instance, Nicole Oresme—in the footsteps of Haly—suggesting its illogicality, and Pico della Mirandola that it was irreligious. This did not prevent the movement of the eighth sphere from being used as a guide to the age of the world. John Ashenden referred to its use. With his customary facility for introducing something of

---

[56] He supposed that in 60 years the planets made an integral number of revolutions. Thorndike,[23] vol. i, p. 376.

[57] There is nothing very new or remarkable about Forman's system, to which I shall refer again below.

[58] For this and others of the articles with a bearing on astrology, see Thorndike,[23] vol. ii, pp. 710-12.

everything, he quoted a period of 300,000 years (from 'Hermes Book 2' and from 'Julius Firmicus Book 3') and also a period of 535 years, as a Great Year in the original Platonic sense, ascribing it to 'Johannes Rocensis' (John of Rupescissa, or Roquetaillade?)[59] This 535 is obviously an error for 532, the product of 19 and 28.

Two centuries later the game was still being played according to similar rules. Pierre Turrel was an influential French provincial astrologer of the early sixteenth century, who in a printed work of 1525, or thereabouts, considered the duration and end of the world, as reckoned in four different ways.[60] The first was by the motion known as the trepidation of the eighth sphere, a somewhat complex movement according to which the motion of the sphere of stars with respect to the quinoxes was doubly periodic. Without going into details, it should be enough to say that in the Alfonsine theory, which Turrel was clearly using, a secular movement—equivalent to a revolution of the sky in 49,000 years—was superimposed on a periodic (trepidational) movement with a period of 7000 years. Turrel maintained that there were four stations, one at the end of each quarter (i.e., at 1750-year intervals), marked by the Flood, Exodus, destruction of Jerusalem, and the end of the world. Although I have not seen the work, it seems that Turrel exchanged the Alfonsine radix of A.D. 15 for the date of the destruction of Jerusalem by Titus, namely A.D. 70. The date of Creation thus astronomically derived was presumably 5181 B.C., and that of the Flood 3431 B.C.[61] The first date was reasonably close to figures given by scholars of the early Church, while the Septuagint puts the Flood in or around 3246 B.C., so Turrel's sleight of hand was quite as good as that of his twentieth-century successors in cosmology.

According to Thorndike, Turrel's method had previously been advocated by Jean de Bruges, in a work written in 1444 under the title *De veritate astronomiae* (Thorndike,[23] Vol. V, p. 311, n. 11). There are reasons for thinking that this fifteenth-century work might stem in part from John Ashenden's *Summa*, and it is no surprise to find that Ashenden refers to the theory of trepidation, with a period which in his case was of 640 years, a period which certain unnamed individuals are said to have used as a Great Year. This period was supposedly derived from the theory of trepidation of Thābit ibn Qurra, an earlier and marginally simpler theory than the Alfonsine. The details are clearly ill understood by Ashenden, although interesting in themselves.[62] No one could possibly

---

[59] The printed edition has 'Cretensis' and 525 years. The manuscript reads: Et mag. Johannes Rocensis dicit quod post 535 annos omnes stelle et omnes planete ad eundem punctum redeunt, et per easdem lineas ut prius vadunt (*ibid*. f. 15r).

[60] The work is of great rarity, and I follow Thorndike,[23] vol. v, pp. 310–11, who was himself describing the work at second hand. It is said to have been from a translation by Turrel of a Latin work in the monastery des trois Valées. On the history of its printing, see Thorndike,[23] vol. v, p. 310, n. 10.

[61] These figures do not tally with Thorndike's remark that Turrel (*c*. 1525) believed the world to have 270 years before its dissolution. This would put Creation at c. 5206 B.C.

[62] MS. Oriel 23, f. 15r and v.

have used this passage of the *Summa astrologie* to learn how to apply Thābit's theory of trepidation. Ashenden, furthermore, fails to show how the anonymous 'others' to whom he refers actually arrived at a Creation date on the basis of the theory; but the Turrel method was obviously intended. Yet again we find, when looking for the source of the idea that trepidation could be applied to world history, that Albumasar is in the West almost always at the end of the trail. The references given in Ashenden are to his *De magnis conjunctionibus*, II.8 and to the work *De vetula*, lib. III, which Ashenden—like many, but not all, of his contemporaries—thought to have been written by Ovid. In fact, this work also alludes to the doctrine of great conjunctions in a form which we know stems from Abū Ma'shar.

When we compare Albumasar's work with Ashenden's we find that the relevant passage (about 700 words in all) has been copied out almost verbatim—if that is the right description of a copy in which almost every sentence is misconstrued, and in which the numerical data are almost totally at variance.[63] Two interesting points are worth noticing: Ashenden interpolates the name of Thābit, which does not appear in those copies of Albumasar I have seen. (The absence of the name is not surprising, since the parameters quoted do not appear to be Thābit's.) And second, the era named is not the Hejira, but that of Yazdigerd, about ten years later. The Latin work by Thābit on the movement of the eighth sphere is based on the former era, and if Ashenden had read more carefully one of his principal sources, Walter of Odington, he would have known as much.[(24)]

Pierre Turrel's borrowing from Albumasar, direct or indirect, did not stop at the theory of trepidation (access and recess). Not only does he reproduce the doctrine of great conjunctions (with a period of 240 years within a given triplicity), but he makes use of Albumasar's division of history into 300-year periods, one such interval being the approximate time for Saturn to circuit the sky ten times.[64] As a typical example of Albumasar's chronological use of this there is the interval between Alexander the Great and the 'son of Mary.' (The conventional era of Alexander was in 312 B.C.)

The idea of a Great Year, manifesting itself as a motion of the eighth sphere,[65] had in astrological quarters by the seventeenth century become a well-established dogma. Nicole Oresme, Pico della Mirandola, and Francesco Piccolomini were three powerful opponents of the idea, and yet none of the three can be considered a complete sceptic as to all astrological influence. A powerful spokesman on the other side had been Petro d'Abano, from whom, in all probability, Pierre Turrel took a fourth period—one of 354 years and four lunar

---

[63] The printed edition I have used for the *De magnis conjunctionibus* (1515)[(31)] is none too perfect. See sig. C viii (*r* and *v*).

[64] *De magnis conjunctionibus* II.8,[(31)] f. C.8r. There seems to have been no attempt to make the different world-periods commensurate.

[65] Only after Copernicus was this accepted as a precessional movement of the equinoxes, in the modern sense.

months, for which time each of the planets was in turn to govern the world.[66] Petro d'Abano proposed a number of different astronomico-chronological schemes, and was particularly fond of a (steady) movement of the eighth sphere, to the periods of which he half-heartedly attempted to reconcile the classical chronologists.[67] Through his writings, some readers at least must have become dimly aware of the vast chronological cycles of the 'Indians' (*Indi*), for in one place he mentions "the beginning of the movement on Sunday, at sunrise, 1,974,346,290 Persian years having passed up to the present."[(32)] As an instance of Petro d'Abano's influence, we find the Elizabethan astrologer Simon Forman writing a short piece which comes to us among a collection of apocryphal tracts and genealogies, much concerning the book of *Genesis*. The item in question places the Creation at 3948 B.C.,[68] and the duration of the worlds as 6000 years, divided into twelve periods. During the course of world history, man's expectation of life was held to fall steadily, in a way which—however ludicrous—appears less arbitrary when located in Petro d'Abano's works *Lucidator* and *Conciliator*.[69] There the notion is linked with that of a golden age which gave way by degrees to a degenerate modern world, the diminishing life span of man playing its part as a symptom of the decline. There had always been those who, in comparing macrocosm and microcosm, wished to maintain that the ages of the world can parallel the ages of man, and who thought that when the senility of the world was evident in all secular affairs, then the last hour of the last age must be at hand. One did not have to be a seer or an astronomer to grasp the broad principle that the world had spent most of its force. In the words of C.W. Jones[(7)] (p. 133):

> the traditional chronology had combined in the popular mind with Augustine's doctrine of the Six Ages of the World, ably spread by Isidore, to create a millennial dogma which became a pseudo-scholarly fetish. To be sure, both Augustine and Isidore plainly stated that there was no predictable millennium. 'The end of the Sixth Age is known to God alone,' they repeated. But the pseudo-scientific saw that if five ages had lasted approximately five thousand years, the Sixth Age would last a thousand.

In the end, the habit died away as people tired of unfulfilled prophecies of doom, much as people tired of listening to the boy who cried wolf.

An argument concerning the age of the world highly reminiscent of that

---

[66] A year of twelve lunar months, each 29 19/36 days, is equal to $354\frac{1}{3}$ days. The quoted period is, as it were, a lunar year of years.

[67] Bede, Septuagint, Abraham Judaeus, and Josephus, for example. As the duration of a complete cycle of the stars he cites al-Battānī (23,760 years) and 'Azolphi' (25,200 years).

[68] His great near-contemporary, Joseph Justus Scaliger, gave 3950 B.C.

[69] Bodleian MS Ashmole 802, f.86, beginning "The wordle [*sic*] is divided into 3 partes. The firste from the Creation unto Noah his flod and after then untille 2000 years . . ." The age to which a man might live was supposed to reduce in steps from 1200 years ("and that begane in the head of [Sagittarius]") to a mere 75. The top of the page has been torn away, and this might have given some clue as to what 'it' was which was supposed to move steadily round the zodiac, starting at the head of Sagittarius at the Creation, occupying each sign for 500 years, and finishing at the end of Scorpio at the end of the world. Anno Mundi 6000.

from the eighth sphere was put forward by a number of writers, the most notable of whom was Johannes Kepler. It was proposed that the solar apogee, at the time of the Creation, was at the head of Aries. Its rate of movement was known to Kepler, as well as its current position. From these two data he concluded that the Creation had taken place in 3993 B.C., at the summer solstice (the Sun then being at the head of Cancer).[70] By the same argument Longomontanus had five years previously found a Creation date of 3964 B.C.[71] Lesser intellects were grateful for yet another easy passage into the secrets of the divine strategy. Henry Power, for example, Yorkshire physician and early fellow of the Royal Society, was to write "An Essay, to prove the World's Duration, from the slow motion of the Sun's *Apogaeum*, or the Earth's *Aphelion*"[72]: but his proof was in reality that of Longomontanus, and occupied scarcely a dozen lines. He confessed that "We take it for granted, from the Scripture Account, that the world is about 5000 years old," and that the Sun's apogee at the Creation was at the head of Aries. Since these two assumptions were compatible with its present position and motion, that is, since the astronomical calculation "draws nigh to the Scripture Account," it was assumed that all were acceptable. Power now went further, and suggested that "in all likelihood, he that made this great Automaton of the world will not destroy it, till the slowest Motion therein has made one Revolution." This, he thought, would take about another 15,000 years. The word 'prove' in the title of the essay was obviously capable of many shades of meaning in Henry Power's time.

I have no idea where the Keplerian idea began, but there is little doubt that he and Longomontanus were writing in a tradition which took in Georg Joachim Rheticus, protégé and spokesman for Copernicus. A thoroughgoing astrologer, Rheticus had managed to work one or two astrological passages into the *Narratio Prima*. The Copernican system has a changing eccentricity and direction of solar apogee, and Rheticus related the rise and fall of kingdoms and faiths to the changes. "We look forward to the coming of our Lord Jesus Christ," he wrote, "when the center of the eccentric reaches the other boundary of mean value, for it was in that position at the creation of the world."[73] Rheticus went on to say that his calculation was in close agreement with Elijah's divinely inspired prophesy that the world would endure only 6000 years. He should perhaps have said,

---

[70] Keplerus,[34] p. 42 (second pagination, that of the tables themselves). See the foot of the table 'Epochae seu radices': Ante Christum Anno 3993. die 24. Iulii, H.0.33′.26″. Medius [Solis] 0.0′.0″ [Canc.] Apog. 0.0′.0″. [Arietis]. The clipped style (of which this is a sample) was later copied by Hevelius in his *Prodromus astronomiae*,[35] to the same effect. Kepler put the annual motion of the line of apsides as 1′2″.

[71] Longomontanus,[36] put the annual motion of the line of apsides at 1′1″ 54′″ 5′ᵛ, a figure which would ensure that Christ's Passion was at 4000 A.M. and that Tycho Brahe's 'accuratissimi observationi' of the apsides for A.D. 1588 (95° 30′) were correct.

[72] Power[37] placed the apogee at "about the sixth degree of *Cancer*," and the motion per century as 1° 42′ 33″. There is a manuscript version among the Sloane MSS at the British Museum.

[73] Quoted from the Translation of the *Narratio Prima* (first printed in Danzig, 1540) by Rosen.[38]

more precisely, that the implied 'Copernican' duration of the world was 6868 Egyptian years (of 365 days), assuming two full revolutions of the mean Sun on its small central circle.[74]

Bouché-Leclercq has shown that after the invention of eccentric and epicycle, in the ancient world, astrology adapted itself quickly to the concept of apogee at which a planet reached its peak of influence.[75] Nor for many centuries to come was it realized, however, that the apogees of the planets moved (apart from movement with the eighth sphere of stars as a whole), and when we encounter any doctrine of what might be called a variable 'physical' influence in medieval astrology, it usually involves the position of the planet on the epicycle, or the position of Mercury's deferent center on its small auxiliary circle. Perhaps this was the source of Rheticus' inspiration. At all events, his idea caught the imagination of sixteenth- and seventeenth-century astrologers, and it would not be difficult to name half a dozen contemporaries of Newton who subscribed to it. The last of any note, so far as I am aware, was Henri de Boulainvilliers, Comte de Saint-Saire, whose *Histoire du mouvement de l'apogée du soleil* was written in 1711.[(39)] Of Boulainvilliers, Feller's *Biographie Universelle* holds that, "il n'en voyait les évènements qu'a travers le prisme de son imagination." While not quarreling with this judgement, I would like to point out that his imagination did not stretch to altering much the words Rheticus had written on the subject, 171 years earlier.

Throughout the long history of 'rational' attempts to solve the problem of the time of Creation, few writers appear to have had a very keen awareness of the difference between an argument based—even in part—on empirical premises and one based wholly on convention. The matter might have been clearer, had human and divine conventions been distinguished. The seventeenth-century Wittenberg theologian Aegidius Strauchius gives us an idea of what part a typical chronologist of the time thought God was playing in the arrangement of the cycles of the calendar. At the beginning of the fourth book of his *Breviarium chronologicum*,[(40),77] after paying his respects to Holy Writ and profane histories alike, he wrote that it seemed probable that God had deliberately so arranged matters that the first year of the world was the first also of the Sabbatic and Jubilean cycles (*anni Sabbatici et Jobelaei*); and that on the first day, the Sun was either at an equinoctial or solsticial point. (Strauchius himself favored the Autumn.) Another hypothesis proposed by the best chronologists, namely that the Moon was placed so as to show one of her principal phases immediately after the Creation, ought—said Strauchius—to be considered carefully; and the same attention should be given to the common belief that the hebdomadal cycle

---

[74] Rosen suggests the *Babylonian Talmud* as the source of the Elijah reference (Rosen,[(38)] n. 56). The equivalent period (two revolutions) in Julian reckoning is approximately 6863.3 years. Philip Lansberg, writing at the same time as Kepler, made the double period of the eccentric anomaly *exactly* 6000 Julian years, to conform with conventional chronology!

[75] Bouché-Leclercq,[(11)] p. 194, quotes Cleomedes (first century A.D.) and Theon of Smyrna (second century).

[76] My list omits some of the points Strauchius raises.

has continued since the first week of the world.

So much for God's design. But as for such arguments as those of Longomontanus, these were clearly now supposed figments of the minds of men. The apogee, Strauchius says, is a mere imaginary point, invented in order to save the celestial appearances, and the best astronomers are of the opinion that we know too little of it to make it the basis of a firm opinion (Strauchius,[40] p. 296). (This was a trifle disingenuous, since his contemporaries were quoting the movement, and by implication claiming an accuracy of one part in a quarter of a million.) Alternatively, as he might well have said, "let us stick to the week, the bare essentials of the astronomy of the sphere, and those lunae-solar cycles which allow us to correlate the phases of the Moon with the solar calendar." Of the voluminous effusions from the pens of biblical chronologists down the ages, by far the greater part are uneasy with methods falling outside the terms of this description. For ourselves, I hope we are all of a mind in dismissing all inference to the Creation on the basis of 'cyclical' relationships, whether of apogee or Easter, as equally delusive.

For an early example of a way in which the precise date of the Creation became a matter for scholarly debate, we need only consider the problem of Easter, a problem which has obsessed the Christian Church almost from its foundation. Hippolytus, the third-century bishop and martyr, had mistakenly supposed that after 16 years the moons recurred in order in the Julian calendar.[77] In the year 243, under the name of Cyprian, a computistical doctrine appeared according to which the earliest Paschal full moon fell on March 18. The argument for this date is intricate, but depends on the premisses that the first day of Creation was March 25, the supposed (Julian) spring equinox, and that the Moon was created at full, on March 28. The doctrine appears in every purely Roman computus, and in Bede's highly influential writings; and F. E. Robbins cites many medieval instances of its acceptance. In the words of C. W. Jones[7] (p. 13), "This constant recurrence of the notion shows how the [computists of the Roman church] struggled to support with confirmed Biblical doctrine the calculations which they could not easily support with astronomical knowledge." Inconsistent applications of these principles abound, but this fact seems to have been considered less important than that contemporary Hebrew usage be avoided, unless it be vouchsafed by divine sanction or revelation. (The main issue was "Could Nisan precede the equinox?")

It is difficult to comprehend the fervor with which scholars addressed themselves to the problem as to the season of Creation. They would cite the examples of the ancients, of the Chaldeans, Babylonians, Medes, Persians, Armenians, and Syrians, who were all said to have begun the year in spring; or, if they believed in an autumn Creation, they might refer to the Romans (before Numa's correction), the Egyptians, and those Jews who followed the Egyptian civil year. From Eusebius and Ambrose to Melanchton, Scaliger, and Kepler, the list of authori-

---

[77] For further information on the subject raised in this paragraph, see Jones,[7] pp. 12-13, 63.

ties in favor of spring was weighty; but so, too, was that stretching from Jerome and Josephus to Scaliger (who changed his mind for the second edition of his great work) and Ussher, all of whom favored autumn. There were those who favored one of the solstices. Vergil was quoted endlessly in favor of spring, and Solinus and Macrobius for summer. Arguments were culled from evidence of the slenderest sort. A discussion beginning with the testimony of *Genesis*, 8:11, that the dove returned to the ark with an olive leaf would end—after much horticultural discussion—with a precise day and month, and possibly even hour and minute for the Creation of the world.

No short summary can do justice to this enduring pseudoscholarly activity. John Ashenden, in the mid-fourteenth-century *Summa astrologie* to which I have already referred, thought fit to devote a large chapter (I.i.1) to its history up to his time. Most of his authorities I have listed in another connection. The arguments are, as one would expect, repetitive. Spring is a time of generation, said some, a time from which calendars begin, a time when the Sun is in the lamb, which came to take away the sins of the world.[78] The empirical, the conventional, and the religious. But what was the precise phase of the Moon at the Creation? And when were the days Creation reckoned to begin? On the answers to these questions hang the workings of the Easter *computus*, and certain more taxing arguments for an autumn Creation. There were biblical texts, for example, *Exodus* 23:16 and 34:22, concerning the ingathering of the fruits of labor, with which Roger Bacon, among others, thought to reinforce the choice.[79] And Adam, no doubt, needed food from the beginning. John Ashenden could not himself decide by reason, although he inclined to the vernal equinox, since Grosseteste had favored it, since the Church had asserted it, and since most astronomers were of the same opinion.

Throughout more than sixteen centuries of theological debate, this was the level of discourse. The most curious aspect of the entire situation was that almost invariably it was taken as axiomatic that Creation was with the Sun at one of the cardinal points of the ecliptic. Strauchius was often cited in the eighteenth and nineteenth centuries as one who could offer ten arguments in favor of autumn; and yet in not one does he explain why the *precise* moment of the equinox should have been chosen. John Ashenden brings Abū Mu'shar to the rescue, with an argument compounded from astrology and Aristotelian physics.[42] The argument was not calculated to find much favor with Christian computists.

Few medieval writers seem to have been conscious of the gross imperfections of the Julian calendar over four or five millennia, and of the shift in the equinoxes. Once the problem was widely recognized, however, there were no lengths to which a computist would not stretch his little astronomical knowledge in an

---

[78] The third argument is unusual. It is ascribed to Walter of Odington. According to Robbins,[30] p. 59, the notion that the Creation took place in the spring (in fact on 25 March) can be traced back to Annianus, while Ambrose was the first to introduce it into the Hexaëmeral tradition.

[79] Richard Holdsworth (1590–1649) reproduced much the same arguments as Bacon's.[41]

Chronology and the Age of the World

endeavor to give a precise minute to the Creation. Examples based on the solar apogee we have already seen. For parallels to Ussher's famous "beginning of the night preceding 23 October 4004 B.C." we need only consult the books of Joseph Justus Scaliger, Seth Calvisius, Johannes Rodolphus Faber, Franciscus Allaeus,[80] Dionysius Petavius, or a hundred lesser writers from an age when the whole subject had run riot. What conceivable spiritual or historical enlightment could Petavius, for example, suppose that he was granting when he calculated that the mean full Moon at the Creation of the world was on 27 October at 9 hours, 5 minutes, 42 seconds after midnight, and that on the fourth day, when the Moon was eventually created, it was somewhat decreasing?

In Ashenden's *Summa* there are one or two statements which he might have reported less casually had he been adept at calculating precise planetary positions in the distant past. As already explained, he quotes Walter of Odington as saying that the Indians thought all the planets to be at the head of Aries at the Creation (Ref. 22, f 3*v*). He quotes Julius Firmicus and Macrobius as saying that at the beginning of the world the planets were in such and such specific positions—and each of the planet's was in fact said to be at the fifteenth degree of its domicile.[81] To extract a precise date from such information as this was to all intents and purposes beyond the mathematical and astronomical resources of the middle ages. Had things been different, God's universe would have suffered at the hands of chronologists in much the same way as the Pharoah's pyramids have suffered at the hands of astroarchaeology.

It is because the correlation of planetary longitudes with times past is so difficult that the method of Great Years was so well liked. A "Great Year," wrote Thomas Lydyat of New College, Oxford, in 1628, being "a period of the Sunne and Moone," is the "trew, right and onelie foundation of this business." The business here was that of calendar reform; but Lydyat did not refrain from fitting the chief points of history into integral multiples of 592 years, with occasional resort to half-periods.[(46),82] Two centuries later, Richard Greswell of the same university was to be found practicing a remarkably similar art. He observed that 129 mean solar years were just a day less than 129 Julian years. In (31 × 129 +

---

[80] Allaeus, a 'Christian Arab,' traced history astrologically up to the birth of Christ and then to Calvin. His book, *Astrologiae nova methodus*,[(43)] was burnt at Nantes, as heretical. Since, in the copy I have seen, the diagrams have not been inserted, and the block for the grandiose initial capital has been cut in mirror image, there might have been more than one reason for the burning.

[81] Firmicus in his *Matheseos*[(44)] simply states in which signs of the zodiac the planets have their domiciles. Macrobius in his *Commentarii in Somnium Scipionis*[(45)] says of the World Year that it begins when anyone wishes it to begin (as opposed to when the planets are at the head of Aries, or wherever). On the other hand, in I.21 he reported that the early Egyptians said that at the very hour of birth of the world Aries was in mid-heaven, the Moon was in Cancer, the Sun in Leo, and so on, and that each planet was considered lord of the sign in which it was then found.

[82] Lydyat set the Creation at 4004 B.C., and the end of the tenth period at A.D. 1916. The system is explained in his book *Solis et Lunae periodus eruditae Antiquitati appellatus Annus Magnus*, etc.[(47)] In part this is a defence against attacks by Scaliger.

1) years, that is, 4000 years, the Sun will come back to its original position with respect to the meridian at the time of equinox (Greswell,[8] vol, i, pp. 234-36; vol. ii, pp. 32-34). The year 4004 B.C. having been accepted *a priori*, B.C. 4 is distinguished by this remarkable coincidence. Was this not the year of Christ's actual birth? Of the two, Lydyat's efforts were perhaps the more excusable, caught up as he was in the vortices of a fresh and strong scientific current. He set forth his chronologist's testament in these words, dedicated to King Charles I (16 September 1626):

> But concerning those Reckonings and notes of mine in the end of mine Emendation of Times, I have not made them, nor these neither last set downe, as determining anything for certaine, or foresetting the time of any future event; which God hath put in hisowne power, and lockt up as it were in the closet of his Privie Councile: ordering his workes in so wonderfull proportion of time, rather that men should acknowledge and admire his Providence upon their so coming to pass, then upon any imaginarie proportion conceited by themselves, presume to foretell the time thereof, before the same come to passe. Although as by long and manyfold experience the learned and wise Physicians have found that there are indeed by Gods ordinance, certaine Criticall dayes in the diseases of mens bodies, and Criticall Climactericall years with distinct Characters of severall Ages, in the course of their lives; so experienced States men have noted the like in the bodies Politike of Kingdomes and Commonweals; and worthie Divines also in the state and ages of the Church; which rightlie to discerne, it behoves warilie and circumspectlie to consider the same, not in the mere speculative abstract, but alwayes with respect to the predominant or peccant humor of the body. As for Astrologicall or other like vaine predictions or abodes, I thank God, I was never addicted to them...

Lydyat's final disclaimer was largely superfluous. Chronology was too dull a subject for most practitioners of astrology. (Not many could bring themselves to enliven their tables by commencing with the Creation, and ending with such names as Ben Jonson, Shakespeare, Beaumont, and Fletcher—as did Philipp Kynder, a decade or two later.)[48] Chronology was fast becoming the province of dry mathematical astronomers with a firm religious purpose—the very men Roger Bacon had wanted for the government of the Republic of the Faithful. These men were generally too astute to rest their case on the flimsy foundations of other men's intuitions. Seth Calvisius, for example, was one of a new breed, who took extraordinary pains to regulate his chronology by computing more than 200 eclipses.[49],83

---

[83] On the use of eclipses for chronological purposes, Philipp Kynder has some apposite remarks which might be nailed to several famous doors[50]: "It is amazement to me to see how the most learned & wise should be carried away with tradition or receaved custome taking them upon trust without examination. As namely the Eclypses which generally they should to be the Bases, & muniments and infallible Characters of all cronologie, where you may see the simplicity of the Ancient, when with astonishment they take notice of an Eclypse in many yeers 10, 20, or 100. And tell us of praediction upon the tyme of Cyaxaris and think to astonish us, as Columbus did the ignorant Westindians at his discovery and conquest. [He continues, pointing out the high frequency of eclipses.] And how much, I pray, doe the most learned and acute computants differ? Josephus (lib. 17, 18) makes mention of an Eclypse A.M. 3949. Kepler, the greatest reformer, refers this to the yeere 3946, three yeeres praevention. A.D. 238 an Eclypse Bunting thinks this to be refer'd to the former yeere."

Chronology and the Age of the World

Isaac Newton struck out in a different direction. We may as well end with him as we began. Commencing with a loose description (by Hipparchus) of the way in which Eudoxus had made the colures of the solstices and equinoxes pass through the constellations, he derived what he considered a precise enough position for the equinoxes to allow the sphere of Eudoxus to be dated (at 939 B.C.) using an argument from precession. The sphere could not have been made by Eudoxus himself: It must be the first ever fashioned by the Greeks. Other evidence suggested to him that it was the time of the Argonautic expedition.[84] With these fanciful ideas he did a little to turn scholar's thoughts away from the Creation, even though some were deflected only as far as the book of *Job*.[85] "Canst thou bind the sweet influences of Pleiades, or loose the bonds of Orion?" A more difficult problem than *Genesis*, perhaps, but one which offered a refuge safe from free-thinking evolutionists and the new astrophysics in the centuries to come.

---

[84] See, for an extended account, Manuel.[51]

[85] William Hales,[13] tells that with the help of Dr. Brinkley, Professor of Astronomy at Dublin, he calculated that the vernal equinox was in the constellation of Taurus in 2337 B.C., the date he had assigned to the trial of Job. (See *Job* 9:9, and 38: 31–2.) He published his results in the *Orthodox Churchman's Magazine* (1802),[52] only to find that he had been anticipated by one Ducoutant in a Sorbonne Thesis of 1765!

1. Isaac Newton, *The Chronology of Ancient Kingdoms Amended*, new ed. (T. Cadell, London, 1770), p. 45.
2. James Ussher, *Annales veteris et novi testamenti a prima mundi origine deducti*, etc. (Gab. De Tournes, Geneva, 1722), p. 3.
3. Francis C. Haber, *The Age of the World: Moses to Darwin* (The Johns Hopkins Press, Baltimore, 1959).
4. D. R. Dicks, *Early Greek Astronomy to Aristotle* (Thames and Hudson, London, 1970), pp. 75-6, 88-9.
5. R. A. Parker and W. H. Dubberstein, *Babylonian Chronology, 626 B.C.-A.D. 75* (Brown Univ. Press, Providence, 1956), p. 2.
6. Geminus, *Isagoge*, cap. 8, ed. by K. Manitius (Teubner, 1898).
7. C. W. Jones, *Bedae, Opera de temporibus* (Publication 41 of the Medieval Academy of America, 1943), p. 38.
8. Edward Greswell, *Introduction to the Tables of the Fasti Catholici* (Oxford University Press, 1852), p. 149.
9. Thomas Heath, *Aristarchus of Samos* (Clarendon Press, Oxford, 1913), pp. 171-3.
10. Otto Neugebauer, *The Exact Sciences in Antiquity* (Brown Univ. Press, Providence, 1957), pp. 141-3.
11. A. Bouché-Leclercq, *L'Astrologie grecque* (Paris, 1899), pp. 38-9.
12. W. H. Stahl (transl.), *Macrobius: Commentary on The Dream of Scipio*, (Columbia Univ. Press, New York & London, 1952).
13. William Hales, *A New Analysis of Chronology and Geography* (London, 1830).
14. David Pingree, *The Thousands of Abu Ma'shar* (Warburg Institute, London, 1968).
15. F. M. Cornford, *Plato's Cosmology* (London, 1937).
16. B. L. van der Waerden, *Science Awakening* (Noordhoff, Leiden, and Oxford Univ. Press, New York, 1974), Vol. II, pp. 306-8.
17. David Pingree, Astronomy and Astrology in India and Iran, *Isis* lxiv, 238 (1963).
18. H. Suter, *Kgl. Danske Vidensk. Selsk. Skrifter*, Raekke, hist. og filos. Afd. iii, no. 1 (Copenhagen, 1914).
19. Otto Neugebauer, *Kgl. Danske Vidensk. Selsk. Skrifter, Raekke, hist. og filos. Afd.* iv, no. 2 (Copenhagen, 1962).
20. Bodleian Library MS. Canon. Misc. 499, f. 2r. the Khwārizmī table. The only additional material is an era "last of the Persian kings," 27 August 274, and of course the Alfonsine era, 1 June, 1252.
21. Johannes Eschuid, *Summa astrologiae judicialis de accidentibus mundi* (Joh. Lucili. Santritter, Venice, 1489).
22. Oriel College, Oxford, MS 23, complete.
23. L. Thorndike, *A History of Magic and Experimental Science* (Columbia Univ. Press, New York, 1934).
24. John D. North, *Richard of Wallingford* (Clarendon Press, Oxford, 1976), Vol. iii, Appendix 38.
25. Robert Grosseteste, *De ordine emanandi causatorum a Deo*, ed. by Ludwig Baur, *Beiträge zur Gesch. der Philosophie* (1912), vol. ix, p. 147-50.
26. J. D. North, Astrology and the Fortunes of Churches, in proceedings of the International Colloquium in Ecclesiastical History (CIHEC) held in Oxford, 1974 ; see, above, chapter 8, pp. 59-89.
27. E. S. Kennedy, Ramifications of the World Year concept in Islamic astrology, *Proceedings of the Tenth International Congress of the History of Science, Ithaca, 1962* (Hermann, Paris, 1962), p. 34.
28. François Placet, *La Corruption du grand et petit monde* (G. A. & G. Alliot, Paris, 1668).
29. *Introductorium in astronomiam Albumasaris Abalachi* (Erhard Ratdolt, 1489), sig. h.3.

30. F. E. Robbins, *The Hexaemeral Literature* (Chicago, 1912).
31. John Ashenden, *De magnis conjunctionibus* (Melchior Sessa, Venice, 1515).
32. Bodleian Library MS. Canon. misc. 190, f. 83v. The *explicit* announces that this work is *Tractatus de motu octave spere secundum Petrum Padubanensem*.
33. Bodleian Ms Ashmole 802.
34. Joh. Keplerus, *Tabulae Rudolphinae* (Jona Saurius, Ulm, 1627).
35. Hevelius, *Prodromus astronomiae* (Danzig, 1690), p. 88.
36. Christianus S. Longomontanus, *Astronomia Danica* (Caesius, Amsterdam, 1622), I.2, pp. 46-7.
37. Henry Power, *Experimental Philosophy* (T. Roycroft, London, 1664), pp. 188-93.
38. Georg Joachim Rheticus, *Narratio Prima* (first printed in Danzig, 1540), transl. by Edward Rosen, in *Three Copernican Treatises*, 2nd ed. (Dover, New York, 1959), p. 122.
39. Henri de Boulainvilliers, *Histoire de mouvement de l'apogée du soleil* (1711), ed. by Renée Simon (1949).
40. Aegidius Strauchius, *Breviarium chronologicum* (Wittenberg and Frankfurt am Main, 1686), pp. 264-5.
41. Bodleian Library Ms Sancroft 129, f. 1v.
42. John Ashenden, *Introductio maior*, II.5.
43. Franciscus Allaeus, *Astrologiae nova methodus* (Rennes, 1654).
44. Firmicus, *Matheseos*, ed. by W. Kroll and F. Skutsch (Leipzig, 1897), II.2.
45. Macrobius, *Commentarii in Somnium Scipionis* ed. by F. Eyssenhardt (Leipzig, 1893), II.11.
46. Bodleian Library, MS Bodley 662, f. 2. Manuscript of a book to be printed, the prefaced petition dated February 1628 (O.S.), "Five years before mine enlargement" (f. 1v).
47. Thom. Lydyat, *Solis et Lunae periodus eruditae Antiquitati appellatus Annus Magnus*, etc. (G. S., London, 1620).
48. Bodleian Library, MS Ashmole 788, f. 175r.
49. Seth Calvisius, *Opus chronologicum* (Frankfurt, 3rd ed. 1629; 4th ed. 1650).
50. Bodleian MS Ashmole 788, f. 175r.
51. F. E. Manuel, *Isaac Newton, Historian* (University Press, Cambridge, 1963), chapters IV and V.
52. William Hales, *Orthodox Churchman's Magazine* ii, 241 (1802).

10

BETWEEN EXPERIENCE AND EXPERIMENT

When Adam Marsh – a close associate of Robert Grosseteste – was asked by certain friars what, precisely, was the nature of the active intellect, he answered that it was "the crow of Elias". The story is Roger Bacon's. Adam, he said, meant that it was either God or an angel, but he did not wish to say so, since the question was a taunt. When we in turn are asked what, exactly, was thirteenth-century science, the safest reply might likewise be "The crow of Elias" – meaning, perhaps, theology and philosophy, but leaving room for further options. But do we give such a safe, dull, answer? The thirteenth century can show us precious little original empirical science, but there is no doubt at all that a truly scientific spirit was stirring, that Grosseteste, William of Auvergne, and Bacon were among those responsible for it, and that even if it is to be cloaked in crows' feathers there need be no shame attached to a forward-looking estimate of the direction in which it was flying. By their nature, these four new books – all of them important in their different ways – provide us only incidentally with a sense of direction in the history of science, but they are all to be welcomed, even so, for a different reason. They all re-examine, or allow us to re-examine, presuppositions that were central to earlier, strongly directional history of science, with results that are certainly not trivial.

William of Auvergne was a Paris master, born in Aurillac around 1180, and bishop of Paris from 1228 until his death in 1249. Robert Grosseteste was born perhaps a dozen years earlier and died in 1253. This great ecclesiastic hardly needs an introduction, and yet there are uncertainties enough in his early life. He was seemingly connected with Hereford, and the famous legal and medical school there. He seems to have had a Paris connection, and his letters show that he numbered William among his friends, but Steven Marrone advises us to play safe, and consider the ideas of the two men as though they were elaborated simultaneously, since neither refers explicitly to the ideas of the other. How trying is the task of the historian of medieval ideas may be judged

McEVOY, J. [1982], *The Philosophy of Robert Grosseteste*, Oxford: Clarendon Press, pp. 560.

MARRONE, S. P.[1983], *William of Auvergne and Robert Grosseteste: New Ideas of Truth in the Early Thirteenth Century*, Guildford: Princeton University Press, pp. 319.

GROSSETESTE, ROBERT [1982], *Hexaëmeron*, ed. by R. C. Dales and S. Gieben, London: Oxford University Press for the British Academy, pp. 418.

BACON, ROGER [1983], *Roger Bacon's Philosophy of Nature. A Critical Edition, with English Translation Introduction and Notes, of "De multiplicatione specierum" and "De speculis comburentibus"*, ed. by D. C. Lindberg, Oxford: Clarendon Press, pp. 420.

from the very context of Adam Marsh's retort. Bacon said he had twice heard William defend the idea that the agent intellect is not a power of the soul, and he claimed that Grosseteste agreed – which is fairly certainly a misrepresentation of the facts. It used to be thought that Bacon studied under Grosseteste, but he clearly did not know a large part of his writings, and probably never met the older man.

Grosseteste was Master of the Schools at Oxford and later Chancellor. He was elevated to the See of Lincoln, the largest English diocese, in 1235, having a few years earlier been made first Lector to the Oxford Franciscans. Bacon saw him as having by-passed Aristotle, and as having beaten his own scientific path, but this is again a mistaken judgment. Grosseteste studied astronomy, and although he contributed nothing original, knew this subject better than most of his contemporaries. James McEvoy is convincing when (in a footnote, no more) he gives his reason for thinking that Grosseteste's book *De sphaera* pre-dates, or at least is independent of, the famous text-book of the same name by John of Sacrobosco. "Grosseteste as he aged", we are told further, "became sceptical of the philosophical accounts of the heavenly spheres, their number and kind of motion, and, borrowing a phrase from St Ambrose, he once referred to them as 'no more substantial than spiders' webs'". Perhaps the allusion to the gossamer-like quintessence was somehow a pun on a name for a part of the astrolabe, if it was not to be taken literally, but in any case he was maybe less sceptical than resigned to getting nowhere on this difficult and most speculative science. He was visibly bowled over by Aristotle's *Meteorologica*, from which stemmed his tracts on the comets, tides, heat, colour, and the rainbow. With some notable exceptions, he preferred to develop Aristotelian themes in monographs rather than write minute commentaries after the style of the later scholastics.

In *De luce*, according to McEvoy, Grosseteste was at his speculative best. His last works were mainly geometrical in form – geometry offering, as he thought, a clue to all action and passion. Most modern philosophers would regard his commentary on Aristotle's *Posterior Analytics* as his greatest work. After Grosseteste, Oxford took the lead for a while in Aristotle commentary, but Grosseteste had set philological standards that could not be kept, what with his new-found interest in Greek when he turned sixty – used when he wrote a commentary on the *Physics*, but also to underwrite his concern for Christian origins, "a new instrument of self-reformation", through his translations from the Septuagint and Pseudo-Dionysius. A few years before his death he began Hebrew. One might say that he preferred scholarship to dialectic, and perhaps even agree with McEvoy that "his blend of lively, imaginative curiosity and broad human experience is at the opposite remove from the dialectical habit of mind", but it was not without reason that Matthew Paris called him a hammer and despiser of the Romans. A theological conservative, Grosseteste was also a critic of the papacy, especially on the issue of plentitude of power, and he so ruthlessly applied acknowledged moral and canonical principles that Pope Innocent must have felt in his presence a little of what General Jaruzelski has felt more recently.

A scholar who remains not only active but positively vigorous into his eighties has every excuse for changing his persona from time to time. One of the many points of disagreement as to how Grosseteste is to be interpreted concerns his allegiance to what some might consider to be a grandiose light metaphor and others a "metaphysics of light" (as his first modern editor L. Baur called it). Professor Marrone holds that this is not at all central to Grosseteste's philosophy. He seems to be reluctant to disagree with other historians except in footnotes, which means that some of the most interesting parts of his book are to be found there, and in one such note he criticizes – but all too briefly – Palma, McEvoy, and A. C. Crombie for supposing that Grosseteste intended his metaphysics of light to extend *literally* to his epistemology, and that the one is needed for an understanding of the other. Marrone admits, though, that when Grosseteste wrote his Commentary on the *Posterior Analytics* he was still thinking of God, or God's Truth, as the light whereby men knew the simple truth. He admits also that passages making use of the image of light do reveal ambiguities. Perhaps an investigation of Grosseteste's vision of philosophy as a whole is called for, and of the functioning of analogy within medieval philosophy generally.

McEvoy turns the metaphysics of light to nice advantage in the structuring of his book: first comes a section on the "angelic light", then one on the "light of nature", and finally one on "the light of intelligence". Angelic knowledge was a theme at the heart of Grosseteste's *Hexaëmeron*, written when he was around seventy, with yet lots of time ahead of him for revision in the light of his encounters with Pseudo-Dionysius. There is no obvious mention of angels in the creation narrative of Genesis, but if we are to accept Grosseteste's word for it, they were generated from the first light on the first day as pure intelligence. There is an obvious analogy between light and intelligence, but for God, Grosseteste tells us, the relation is one of identity.

It was Grosseteste's work of translating and commenting on the four principal tracts by Pseudo-Dionysius that led Beryl Smalley to designate him the Erasmus of the Middle Ages. He was not always as sympathetic to the original as Erasmus tended to be, however. He was not above importing Augustinian ideas to fill it out, especially on the central question of sacred order. For Pseudo-Dionysius, hierarchy had been a *cosmic* property and had to do with the structure of things, but to an Augustinian and scholastic the theological categories of grace and love were needed to explain it in the most fundamental way. Again, unlike the original author, Grosseteste could introduce an Aristotelian element – distorted, it is true – to explain the workings of the cosmos in the manner of the astronomers.

Perhaps the most characteristic of all Grosseteste's cosmological writings is the little tract *De Luce*, described by McEvoy as "one of the few scientific cosmologies, and perhaps the only scientific cosmogony, written between the *Timaeus* and early modern times". (It occupies only nine pages in Baur's somewhat faulty edition.) The central idea is that light is the first form of corporeity, a simple substance without dimensions until they are introduced by self-diffusion. Creation was said to amount to the infinite multiplication of light from a point, equally in all directions, so as to give rise to a (finite) sphere. McEvoy is not always intelligible on scientific matters – on space, for instance – and he seems to have an unduly great respect for Grosseteste's mathematics of infinity. It is hardly possible to make a mathematician out of Grosseteste, I regret to say.

The cosmology of the *De Luce* is neat, but a little too tidy. Light (*lumen*) from the outermost sphere streams to the centre of the universe, producing by successive condensation and rarefaction a series of spherical shells, almost like a series of standing waves. These, the planetary and lower spheres, get their motion from the mover of the first sphere. The entire picture is of an outdated, homocentric and not epicyclic, universe, the universe of Eudoxus and Aristotle rather than of Apollonius, Ptolemy, and their successors. Grosseteste's sources, direct and indirect, include the Bible, the Fathers, Avicenna, Algazel, Averroes, Alpetragius, and Thabit, but Aristotle is the only source acknowledged. Grosseteste even had a vague idea of the doctrine of access and recess of the eighth sphere, which provided what we should regard as a variable equinoctial precession; but his system could not possibly have explained this in terms of his optical model. His conscious "synthesis of the cosmogony of Genesis and the cosmology of the *De caelo*" (McEvoy) was none the less influential for another reason, for it helped to persuade at least some of its readers, Bacon in particular, that natural philosophy cannot be properly understood without the same sort of geometrical constructions that are used in optics. This attitude towards geometry, according to McEvoy, "was to prove fundamental to the metaphysics of early modern science". One only has to read those of Bacon's writings newly edited by David Lindberg to realize just how influential were the writings on geometrical optics by Ptolemy, Alhazen (Ibn al-Haitham), al-Kindî, Tideus, Euclid, Apollonius, and the rest that struck such a resonant chord in the thirteenth-century West; but the more fundamental point at issue is whether the conscious elaboration of a metaphysics that "makes reality geometrical" was instrumental in advancing natural philosophy, or whether it stood more or less on the side-lines.

McEvoy thinks the former, and that in Grosseteste's works "we can discover . . . an originality and an importance which in the long run did have a bearing on science, in so far as science came into being in dependence on certain metaphysical beliefs". This is not a particularly bold generalization, and it doesn't add much to note the geometrical simplification of such "physical realities" as the climates, heat, and the rainbow, with which Grosseteste made some progress, for there were many scholars, before and after him, who effected a far more powerful application of geometrical methods without a conscious element of metaphysical panegyric, indeed without making any overt reference to their beliefs as to the "nature of reality". Of course, historians of ideas will always be able to piece together a likely picture of those beliefs: the detection of metaphysics as a sort of collective subconscious of the scientific community is nothing new. In the case of Grosseteste, Bacon, and their like, the game is easier because they made such a song and dance out of their metaphysical propensities. The fact, however, that we can find a metaphysical interpretation for a science proves neither that the science was created nor that it was first justified with that in mind.

I happen to think that the received version of this slice of medieval history is misleading – at best only a small part of the story. Geometrical techniques were "applied" by every student of the quadrivium, that is to say in astronomy, and although this was not the same as natural philosophy, there were many points of contact. According to McEvoy, Grosseteste probably preferred the universe of homocentric spheres to the complex system of Ptolemaic eccentrics and epicycles because he believed that "reality is mathematical", but he seems not to know of Ptolemy's own Aristotelian prejudices, for he speaks of the "positivism [sic] of the sophisticated Hellenistic astronomers" who, while saving the appearance, renounced "the effort at physical understanding" as well as "faith in mathematics as the key to nature's real functioning". It is true that the medieval scholar with an interest in the heavens had to wear two hats, one when he was calculating with and developing further Ptolemy's geometrical algorithms, the other when he was professing the metaphysical complexities of what, geometrically speaking, was a childishly simple (and much debased) Aristotelian world. The question is, which proved to be the more valuable tradition, scientifically speaking? I have no doubt that it was the Ptolemaic, and this even, in the long term, as regards natural philosophy.

For all his loyalty to Aristotle, Grosseteste often strained the words of the Philosopher to breaking point – on infinity, for example, where he seems to have thought he could save the day by introducing the infinite mind of God into his account. I think Aristotle would have accused him of asking the conceptually impossible of God. It is of some interest to read of Grosseteste's way of qualifying Aristotle on the nature of the heavens. He acknowledged that the spheres carrying the planets are of a non-elemental material, the so-called quintessence; but he supposed the planets themselves to be elemental, thus calling in doubt, if not breaking down, the conceptual barrier Aristotle had imposed between the celestial and the sub-lunar realms. This he did quite consciously, but it is perhaps too much to speak of a presaging of the unification of physical theory aimed at in later centuries, merely because he connected the heavens and the earth in a number of simple ways – for instance, through the astrological influences in which he believed, and the influence of light on the lower elements. It is wrong to say that Grosseteste "opened a radical breach in the theory of Aristotle" (McEvoy), or that Buridan, Oresme, and Cusanus did essentially the same sort of thing later, let alone under his prompting.

It is now thirty years since Crombie proposed a rather strong thesis, to the effect that the experimental method given by Grosseteste to the Oxford school marked the beginning of the modern tradition of experimental science, and that the methodological principles evolved at this time were at once applied with some success to optical science, as a first step in the new direction. McEvoy, who couples Crombie with Duhem on the question of the continuity of scientific progress since the Middle Ages, discharges both barrels – the first at Duhem, whose conventionalist philosophy he seems to misunderstand, and then at Crombie, on the basis of a doctrine of his own, introduced so obliquely as to be difficult to follow. If I understand it correctly, it is that "in scientific progress it is advances in actual methods of inquiry rather than in abstract methodology that are of determinant value"; and that the *sterility* of science, after the date (say 1240) from which the requisite methodology was available, confounds the Crombie thesis if taken in conjunction with the principle that it is methodology that really counts.

On the whole it would have been wise to leave this thorny subject alone, for it is not one to be tied up in a neat logical package. The very premise of sterility requires a careful and extensive historical study. It is true that too much has been claimed for the use of an "experimental method" in the Middle Ages, and that the word "experimentum" and its cognates had then as much to do with common sense and experience as with controlled experiment, and Crombie himself would be the first to admit that one should not overstate one's claims; but by the criterion of "advances in actual methods of inquiry", late medieval Oxford did produce some remarkable results. Grosseteste was often enough quoted for us to think of him at least as a sort of figurehead of the Oxford scientific movement, even though, as I have already suggested, it would be wrong to suppose that it would not have happened without him. McEvoy's "demythologizing" will be for the good if it turns historians' attentions towards other measures of scientific progress than simple experimentalism. (Does he, by any chance, have an experimental basis for the proposition that "admirers of Grosseteste . . . have to struggle with the workings of an inferiority complex that affects all historians of medieval science *vis-à-vis* their counterparts in the ancient classical or early modern periods"? All?)

Grosseteste wrote more than a dozen works with a psychological or epistemological content – the dividing-line was as fine then as now. He began from an Augustinian position, with a soupçon of astrology and medicine thrown in, but later moved over to Aristotle's stance in the *De anima*, interpreted through Avicennian spectacles. Gilbert Ryle would have had a field-day with the metaphorical element in the resulting philosophical tradition: the soul as incorporeal, but united to body; the "separate forms" united to the stars extrinsically, but needing no contact; the "first form", exemplar, and cause of a thing, as something abstract, simple, and separate. Two generations later the Oxford Franciscans had a theory of plurality of forms, which D. E. Sharp thought to have been taken over from Grosseteste; but McEvoy is rightly sceptical. For Grosseteste, the body was taken to be constituted from the four elements plus light (itself the closest approximation to incorporeal nature, close to the wholly immaterial soul) so body was somehow the product and the image of the whole cosmos. As Bonaventure said later, our bodies are "somehow all things". Sensation was for Grosseteste, as for Augustine (but not Aristotle), an active process – compare the old emission theories of vision, with rays leaving the eyes. With reservations, though, he adopted Aristotle's doctrine of the active intellect. (Strictly speaking, there is more than one concept of intellect in his work, one from Aristotle, and one from pseudo-Augustine.) Below the intellective and intellectual virtues (the understanding) there comes scientific reason (*virtus scitiva*), capable of constructing demonstrative knowledge of the permanent aspects of the world, and presupposing an immutable object. It is rather reminiscent of Kant.

Grosseteste could accept Aristotle in many respects, at one level of truth, but there was always that uneasy contrast with Christian tradition – for instance on the dualism of soul and body – a contrast heightened by the extended light metaphors. In *De veritate*, for instance, he has it that just as the light of the Sun is needed for objects to become visible, so the Creator's light of supreme truth – available to all human beings – is needed if the being of objects is to be revealed to the knower. The truth of anything created consists in its conforming to its exemplar cause in the divine mind – a common scholastic doctrine in Grosseteste's time, although it would be misleading to say that there was ever a coherent standard version.

Here are some of the basic ingredients of Marrone's monograph, which is focused on a very much narrower band of philosophy than McEvoy's – namely, the process of "putting Augustine aside" and "opting for science", in the elaboration of a theory of truth and cognitive evidence, especially in the thirty years or so from around 1220. Marrone begins with William of Auvergne's *Magisterium divinale ac sapientale*, a treatise in seven section, of which two are late (say 1231 to 1236) and are considered in great detail. William took simple truth to be a question not of intellectual judgment but of being – whether it related to the objective existence of the thing (*res*) or the being of the thing (*ipsum esse rei*). William broke with the Aristotelian tradition of an active intellect (sometimes taken to be a tenth celestial intelligence) that puts the images of things in the mind. Instead, not unlike Augustine, he held that the senses influence the intellective power not by impressing forms on it, but by exciting it to generate the forms itself.

William of Auvergne's favoured definition of truth seems to have been from Avicenna: the truth by which statements are said to be true is a conformity of speech with reality (*adequatio orationis et rerum*). You don't have to go outside the created world for the conditions of truth, so God is out of the picture. How about external truths, if only God has eternal existence? William saw that one might make the "reality" side of the equation the totality of eternal truths God has asserted in the eternal past, but he was not happy with the solution, since the reality that was thus signified, while statable, would not exist in the truth of being. He decided that truth, a relation, was like love: it could exist without its object existing; it was a relation of privation, establishing a lack of something shared. A true assertion refrains from implying something about reality that is not the case, so one might hold that there are many things true from eternity, without conceding that there are real things ("truths") that have eternal existence.

However vague he might have been about one sort of eternal truths (universal truths; other sorts were negative assertions, and statements of future fact), it has to be admitted that his way of phrasing a theory of truth so as to by-pass traditional ontology and natural theology, and to make use only of logic and language and a dash of reality, has a modern look about it. But of course the same goes for other philosophers from Plato to Abelard, given a sympathetic interpreter. Without suggesting for a moment that Marrone is being over-kind to his subject, what many readers will undoubtedly find lacking in his book is a sketch of the background history. Even a passing reference to the various notions of analytic truth, and a historical sketch of that, would have been welcome.

Perhaps one shouldn't press the point about God's disappearance from the scene. God was taken to be a living book and mirror full of forms for the intellect – let us call his the upper world, by contrast with the lower world of sensibles. The mind, as it were between them, looks to both worlds, and takes intelligible signs for combination into complex evident truths. Looking up gives us rules of truth, right living, and revelation – not so alien to William's culture, after all. It is interesting to see that both ways of looking were taken to be necessary for the acquisition of natural science. We get to know the proper principles of efficient and final causation in natural phenomena by looking to the world of sensibles, but the form of scientific principles (ordering and relating terms) will not be easy to understand, it will not be immediately evident from the meaning of the words of which they are composed. Perhaps William spoke here from the heart, for it seems that he might have tried his hand at understanding natural science late in life, and I can see no evidence that he had much idea of what it was all about. His key to the formulation of scientific explanation was more or less Aristotle's. One was to discover a concept to serve as a middle term relating the two objects one knows. One was supposed to do this by simple cognition, by a disposition, a habit, the end product of a process of repetition of the same cognitive experience. (Note his analogy with frequent reflection in a mirror.) There is, to my mind, nothing particularly new or startling here.

With Grosseteste it is otherwise: he seems at once more conservative and more original. His theory of truth changed as between his early theological works and his commentary on Aristotle's *Posterior Analytics* (1228-30, according to Dales, McEvoy, and Marrone; Callus and Crombie had preferred pre-1209 for a first version). Both were influential in Paris. At first Grosseteste had followed Anselm: simple truth was *rectitudo*, being right, conforming to a rule of what ought to be, and so effectively conforming to God's word. Later came the idea of a light of divine truth illuminating created truth, which then shows the mind the true thing. Presumably Grosseteste thought his two accounts compatible. The advantage of the second was that it did not demand a direct knowledge of God as a condition for knowing the truth. By the time of the commentary, God had disappeared from the definition: simple truth was the substance itself, the *esse, illud quod est*. Definitions of things make them known, reveal their essences. Science requires more, namely a knowledge of universals, and what the mind grasps when faced with them depends on what sort of understanding you've got.

This is where Christian tradition crept in, and in ways that most modern philosophers would find bizarre. First there are pure intellects, free from the distraction of sensible images, and able to contemplate God directly, thus grasping universals at once. (If not of Dante's Beatrice, perhaps one is reminded of Kant's imaginary example of a mind for which intuitions and concepts are one and the same.) Less pure intellects may receive illumination from a lesser light, an intelligence. Lower intellects still might contemplate the stars, and so find causal reasons for what is on Earth. Knowledge of a yet lower order is that of earthly things; and finally there come the intellects so weak that they can grasp only the outward accidents, those acting as direct referents for universal concepts. While on Earth, men were supposed to have access only to the last two ways of knowing universals. Science was for low grade – but not lowest grade – intellects. Before the Fall it was otherwise, Grosseteste thought, even conceding that there might be especially pure souls around exempt from confinement to grades four and five.

Grosseteste's theory of universal predication that seems so "totally new" (Marrone) is, in its essentials, that universal predication differs from particular predication according as one regards the form, or the essence it stands for, as one single thing or as a general characteristic also possessed by other objects. The universal is neither one nor many, but one *and* many – it is like the light from the Sun says Grosseteste (in an analogy that Marrone is right to hold inessential to his thinking), in that it generates other light and is generated. Having seen the same nature several times in several different objects, the mind decides that the nature is characteristic of a class, and should thus be recognized as a simple universal – thus cutting short the potentially endless task of enumerating individuals. Complex knowledge was for him, as for William of Auvergne, an accommodation (*adequatio*) between language and reality. Simple cognition concerns the essences of its objects, while complex cognition concerns actual existence, truth being the "proper signification of being". The paradigm of scientific demonstration was still taken to be something that depended on finding a middle term expressing the formal nature of the object to be understood. Science in the highest degree yields necessary truth as conclusions from premises drawn from proximate formal causes, while second-class "rational science", empirical science as we should call it, has universal terms referring (as before) to real things in the world, deals in regularities in their behaviour and prediction of it through laws of nature, but is *not* capable of revealing the natures of simple essences.

So much of all this is Aristotle or near-Aristotle, and so much of it resembles views held by philosophers who were not especially inclined towards the sciences, that I can see no special reason for linking it with Grosseteste's confidence in a scientific programme. On the contrary, there is a sense in which confidence in such a programme might be expected to have diminished. The principles of natural science, "experimental complex universals", were supposedly found in the second of a two-stage process of analysis and induction, but even then they were only taken to be propositions whose universal truth is ready for testing through experiment. Marrone insists that the word "experiment" was used in much the same sense as it is used today, and Grosseteste's now somewhat notorious example of the relation between scammony and red bile, traced by Julius Weinberg to Avicenna's logic, rather suggest that this is reasonable; but are we to go further with Marrone when he represents the arrival of Grosseteste's doctrine as a "truly significant moment in medieval thought", and "the first attempt in the medieval West to formulate a theory of the induction of complex scientific laws and to suggest an experimental method for veryifying them"? Did it really "change the way men of science explained the processes of their thought, the way they defended what they held to be true, and the way they taught such truths to others"?

On the point of Grosseteste's priority, the claim is reasonable enough, although presumably others were able to read the works in translation that Grosseteste read. (This is another point where one would like more attention to indebtedness. Marrone drops a few hints in footnotes, but generally leaves the problem to the reader.) On the question of a change in the way science was conducted, where is the evidence? The problem is much the same as that raised earlier in connection with McEvoy's book: What influence does the philosopher of science have on scientific progress? Suppose there were a consensus of historians as to the most significant advances in science, or experimental science, or inductive experimental science, or what its practitioners had represented as such, in the two centuries following Grosseteste's commentary – an arbitrary figure for the period during which it might have been read. How much of it would seem to have been even remotely spawned by that commentary, or by its successors? I think very little. The exceptions might be the works on optics that were written by men with Grosseteste's own optical writings to hand. What of Grosseteste's optical works? Surely *they* ought to provide a test case, since they were in many cases written after the commentary on the *Posterior Analytics*, in which many optical examples are given.

Sure enough, there is evidence that Grosseteste presented his material so as to fit, as well as may be, with the ideals set out in his methodological writings. Not that there was a flood of material. In fact the analysis he offered of the rainbow is just about the only example likely to persuade a modern reader that we are not forcing the argument. There he offered an analysis in terms of subsidiary causal relationships – laws such as of refraction, and of coloration in refraction – which he could in principle have studied separately under controlled experimental conditions. There is precious little hard evidence that he ever did so, being content rather with remarks like "rays of light passing through a spherical flask of water behave in such a way, as so and so has said, and as anyone can confirm to be the case". What of his follower, Roger Bacon, however? Did he not once claim to have spent more than two thousand pounds on books, tables, instruments, and other things necessary to his studies? Does this fact alone not speak loudly for the arrival of a new attitude to the empirical sciences? So far as I know, Grosseteste's writings contain no similar remark, but of course the two scholars started and finished their careers in very different social circumstances, added to which there were marked differences of temperament.

In David Lindberg's new critical edition of the Bacon texts on the multiplication of species and on burning mirrors we find copious reference to experiment – but usually in the same sense of confirming what had already been found and confirmed by Ancient and Muslim scholars. But first a word about the new edition, which is a replacement for the older defective editions by J. H. Bridges and I. Combach. It is admirably done, and the accompanying translation is all that it should be, true to not only the letter but the spirit of the original. By "species" Bacon means, briefly "the first effect of any naturally acting thing". It may be thought of as a virtue or likeness of a natural power. Thus he spoke of light, *lux*, in the body of the Sun, but of *lumen*, the species of the solar *lux*, multiplied by the air, the medium through which it passes. The two tracts are essentially geometrical optics with an admixture of Aristotelian natural philosophy.

It was evidently in the 1240s that Bacon went to Paris, where he lectured in the faculty of arts – he was one of the first to lecture on Aristotle's natural philosophy after the bans of 1210, 1215 and 1231. Back in Oxford not very long before Grosseteste's death, he doubtless knew the great bishop only through his writings. In or around 1257, Bacon joined the Franciscan order with its vows of poverty, but none, it seems, restricting the discussion of past wealth. He seems to have been the subject of growing disapproval by his superiors. Transferred to Paris, perhaps for closer discipline, he began raising money with all the fervour, if not the success, of his scientific descendants. Professor Lindberg puts the two works he has here edited in the early 1260s or, perhaps, the late 1250s. Bacon's more famous writings, the *Opus maius* and others written for the Pope, to whom he had addressed overtures for funds whilst he was still a cardinal in Paris, and the rather obscure details of his later life (which ended around 1292), really add very little to what we can discover about his experimental practice from the two newly edited works. Although I cannot guarantee that I have caught all examples of words cognate with "experiment" and "experience", I counted seven places where Bacon was citing experiments that others (Ptolemy, Ibn al-Haitham, al-Kindî, Euclid) had performed, some to do with atmospheric refraction; four cases of common experience ("*experimentum*" rather than "*experientia*" was used in one instance), three of them sufficiently artificial to count as simple experiments, perhaps; and just one example where a simple experiment/experience was used deliberately and systematically to dismiss a hypothesis. This last example is indeed rather obscure and indecisive – it has to do with the mode of multiplication – but it *is* the real McCoy. It would be hard to scrape together half a dozen comparable late medieval examples, and even harder to show that their authors had Grosseteste's *Posterior Analytics* in mind as they put Nature on the rack. (Theodoric of Freiberg is a notable instance of one who did.)

What William of Auvergne and Grosseteste knew full well – because

Aristotle had taught them – was that (as they would have said) in the leap from sensation to causal principles more is needed than an experimental method. A quickness of mind, *sollertia*, is called for. They were not themselves lacking in *sollertia*, and yet in the last analysis they were not scientists through and through, either in a modern or in an Aristotelian mould. There was an element of scientific redundancy in their writings, and this had largely biblical, patristic, and neo-Platonic sources. (This is not, of course, to say that everything from those sources was redundant – far from it.) That was far from being the case with Bacon, who would have been at home in the Merton College of the mid-fourteenth century in a way Grosseteste would not. Grosseteste stood at a parting of the ways, even helped to create the parting, but he retained an affection for the old path, too strong an affection, perhaps. This is well illustrated by the microcosmism in a short fragment ascribed to him, and in other writings, as well as by his magnum opus, *Hexaëmeron*. The idea of the microcosm was that man is somehow analogous to the universe. The analogy might be a very weak one, and it was not always clear whether a bodily or a spiritual parallel was intended. The idea of the World-Soul had caught on in the twelfth century, and there are signs of it in Grosseteste's *De sphaera*, but McEvoy shows how Grosseteste pruned some of the more exaggerated symbolic outgrowths of earlier microcosmic speculation, and shifted his own position at least twice.

The nearest we can come to a synthesis of Grosseteste's thought as a whole is in the *Hexaëmeron* – "On the six days of creation" – a work now meticulously edited by Richard C. Dales and Servus Gieben. As they explain in their short introduction, it was written as a book rather than as lectures, most probably between 1232 and 1235. They used the seven surviving manuscripts for their edition, one of them invaluable because much corrected by Grosseteste himself. This they assume to have been a personal copy, so its orthography is followed. There is evidence that a set of chapter headings (listed but not used) was provided by Adam Marsh; but the division into eleven main sections was evidently Grosseteste's own. It would be difficult to convey the range of the 219 chapters of *Hexaëmeron* in a short space, but at its centre there is the minute commentary on the opening of the book of *Genesis* by one who can now throw in Greek etymologies, as well as explain why scripture does *not* say this and that (why it does not say "Vidit Deus" in place of "Dixit Deus", for instance). Grosseteste is a scholar's scholar, and his editors have not let him down. I have only one reservation, and that concerns the infuriating method of indexing the main text by part, chapter, and section – but not page number – presumably because that saved collating typescript with page proofs.

By the time he was writing *Hexaëmeron*, Grosseteste had decided that astrology was pernicious, and written at the dictation of the devil; he was naturally hostile to the Peripatetic doctrine of the world's eternity; and speaking generally he had swung away from his earlier allegiance to many Muslim authorities. Running through his book is the parallel between the six natural ages of man, the microcosm, and the six ages of the world, the macrocosm, both of them reflecting the six days of creation, and both to be fulfilled and perfected by a seventh term. The work is pervaded by number mysticism, seven being the number of the virtues, the gifts of the Holy Spirit, the planets, the metals they govern, the days of the week, the stages in the growth of the human foetus, types of physical change, and so on. In consequence of all this, Grosseteste could expatiate on theology, astronomy, ethics, biology, and physics – in fact on most of the scholastic curriculum, and more besides – drawing numerous analogies as he did so, and thus obtaining at rather low intellectual cost a unified vision (theory would be too strong a word) of God and his creation. As for creation, it included the invisible, angelic, order and the visible, material, world; but both of these were seen as hierarchical, ordered, and sacral, not to say governed by teleology.

This was not merely an ivory tower of the intellect, for it had important moral, ecclesiastical, and political repercussions. He thought that the ecclesiastical order, for example, was derived from that of the celestial choirs. The prelate should be as a fixed star, with no more ambition to rise in the world's eyes than has a star to rise in the empyrean. That wasn't a hard thing for a prelate to say. The faith that existing hierarchies had been arranged by God in the best possible way lent support to the urge to reform. As McEvoy says, following Grosseteste – who did not manage to impose his philosophically grounded reforms on the Church – "failure in action done out of idealism and the highest love is better than most success". Those who came after him were content to fence off the parts of the curriculum more thoroughly, to exclude man from any of them, to separate physics from ethics, and so forth. Grosseteste would have been downcast, had he foreseen this turn of events. To say this is to comment, I suppose, upon at least one aspect of the thesis that he was a key figure in the development of Western science; but as we have seen, there are many other sides to the question.

I have said nothing here of the practical, technological side, which Roger Bacon had very much in mind when he wrote on natural philosophy, but of which Grosseteste seems to have been largely oblivious. Not without a practical side to his character, his was more the practicality of the administrator or the lawyer, and is well illustrated by his letter emphasizing the importance of the doctrine of the soul for the reform of the law of bastardy. This is not to say that he philosophized like a lawyer – as did that other writer on scientific induction, Bacon's namesake Francis. In Roger Bacon's view he was one of the few supremely wise human beings to attain to a level of philosophy reached by Solomon, Aristotle, and Avicenna. The future of his particular sort of learning lay, in thirteenth-century Oxford, with the mendicant friars rather than with the secular clerks. If we stretch a point, and regard the Franciscans as Grosseteste's pupils, Bacon's remark well illustrates how important an ingredient of academic fame famous pupils are.

11

OPUS QUARUNDAM ROTARUM MIRABILIUM

SUMMARY — A text, of the first years of the Fourteenth Century, or earlier, provides evidence of the design of astronomical clocks of a highly complex sort, which were previously known only from vague general descriptions. The provenance of the three known manuscripts containing this text, together with internal evidence, suggest a North Italian origin, although the text may well amount to an Italian astronomer's description of an earlier instrument, perhaps from the Islamic world. After attempting a reconstruction of the device, some probable exemplars are briefly considered. Apart from the possibility that that instrument follows the tradition of the anaphoric clocks of antiquity, its most notable characteristic is the accuracy with which the mean motions of the planets are represented, bearing in mind the remarkably economical arrangement of wheels used for this purpose. Nothing is known of the way in which the instrument was driven, but it may in any case be considered as an important antecedent to those designed by Giacomo and Giovanni de' Dondi, in the first half of the Fourteenth Century.

INTRODUCTION [a]

There was a temptation to give this paper the title 'The first planetarium', but this would have been to claim at once both too much and too little. The text presented here, abbreviated and cryptic as it is, describes a geared planetarium in some detail. But are we to discount the half dozen or so classical references to a device invented by Archimedes, for representing the 'various and divergent movements [of the Sun, Moon and planets] with their different rates of motion'? [b] Even if we dismiss

---

[a] I have to thank Mr Francis Maddison, Curator of the Museum of the History of Science, Oxford, for ideas and for many of the references to works cited below.
[b] For the full quotation from Cicero, together with others, see pp. 89 sqq. of D. J. Price's *On the*

these obscure citations on the grounds that no gearing is mentioned — and it should not be forgotten that the remains of a geared instrument of a most intricate sort survive from the First Century B.C.[c] — history nevertheless provides us with others, from the Middle Ages, where it is hard to avoid the conclusion that geared planetaria were known.[d] Even so, no precise figures are known for the wheel ratios of any fully fledged geared planetarium older than that introduced here. Something is known of those in the Antikythera instrument, and in the geared lunar phase indicator of al-Bîrûnî,[e] but these are not really comparable with it. As it turns out, the present wheel ratios are very cleverly chosen, and actually result in an accuracy better than that achieved by Giovanni de' Dondi for the *mean* motions of the planets.[f]

The means by which our planetarium was meant to be driven are not known, but it will be argued that it shared one distinctive part, at least, of the anaphoric clock of antiquity; and if this were so, it would be at least probable that both were driven in the same way.

A further manuscript source, which I came across recently, and which is almost certainly copied from an unedited treatise on the design of astronomical clocks by Richard of Wallingford, is further from the present text in its provenance, but closer to it in time than Dondi's *Tractatus Astrarii*. Its discovery was fortunate in one small respect in connexion with the present North Italian text: it tends to confirm what could be no more than a hesitant suggestion on the strength our present text, namely that the astrolabe dial, in the form characteristic of the anaphoric clock, persisted up to the Fourteenth Century.

Finally, a word about the translation. This is fairly literal, and consequently awkward. (It could hardly be other than literal whenever the text is so cryptic that the author's meaning can be understood only because the astronomical parameters he uses are independently recognisable.) The text is written throughout in the subjunctive, and that mood is retained to remind the reader that text has the appearance of a set of *instructions* for making an *Opus rotarum mirabilium*, rather than a

---

origin of clockwork, perpetual motion devices, and the compass, in *Contributions from the Museum of History and Technology* (Bulletin 218), Smithsonian Institution, 1959. The quotations were provided by Professor Loren MacKinney and Miss Harriet Pratt Lattin. Miss Lattin has also singled out for comment a passage in the writings of the Fourth Century writer Macrobius, who apparently added the planets to an armillary, in order to 'perceive by the eyes' their retrograde motions. There is evidence that Macrobius borrowed from Archimedes, but none that the device was automatic, or even geared.

[c] This was found in a sunken treasure ship, off the island of Antikythera, in 1901. For an analysis of it, see D. J. Price, *An ancient Greek computer*, « Scientific American », June 1959, pp. 60-68.

[d] Some of these are mentioned in section 10 of the commentary below.

[e] References to this will be found in the article cited in note b above.

[f] Comparison with Dondi's clock (of 1364) is inevitable, since they are close in both period and provenance. The full text of Dondi's *Tractatus Astrarii*, with introduction and glossary, was published in 1960 by Antonio Barzon, and others, for the Bibliotheca Apostolica Vaticana. See also the last footnote to the commentary, below.

description of such. The ideas involved may not, of course, be the author's own. In two cases the original Latin words ('clavus' and 'signiferus') are retained, since the meaning of each is obscure, and much rests on their translation. Perhaps 'spera' should have been treated with equal caution, for as is well known, this may occasionally be translated 'circle', rather than 'sphere' or 'globe'.

THE MANUSCRIPTS

α - Milan, Ambrosiana, MS H.75 sup., ff 68rb-69ra. A well written manuscript, probably dating from the early fourteenth century.

β - Milan, Ambrosiana, MS 35 sup., ff 42r-42v. Late fifteenth or early sixteenth century.

γ - Oxford, Bodleian library, *Canonici misc.*, Latin MS 61, f 11r-v. Late thirteenth or, more probably, early fourteenth century. By a small margin, this seems to be the earliest and purest text.

Roman numerals are used in α and γ, arabic in β. None of the texts is illustrated. All are Italian and, judging from an associated astrolabe text, γ was copied from a text stemming from the locality of latitude 45°. All are well written, but are either carelessly copied or taken from corrupt or badly written copies. It is possible, of course, that all ultimately stem from a poor translation of a work in some other tongue. There is a surprising coherence between the roman numerals of the two early copies. Inaccuracies were probably introduced into later copies — and several are evident in β — at the stage of turning roman into arabic numerals.

An examination of the variants listed in footnotes to the text shows that, despite the long interval of time between them, α and β are more closely related than α is related to γ. Nevertheless, β does not appear to be a direct copy of α. (Notice that near the beginning 'cclx' is correctly given as '360', and in the penultimate paragraph 'xxiii' is correctly given as '33'. There are several similar mistakes avoided.) A rough classification of the MSS, for what it is worth, is illustrated below. Trivially variant spellings are not noted in the text, in which medieval orthography has been followed as far as possible.

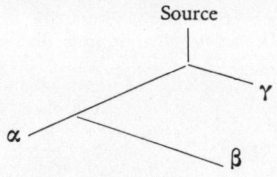

Opus quarundam rotarum mirabilium quibus sciuntur vera loca omnium planetarum et etiam hore dierum ac noctuum.[1,1]

FIAT columpna et locetur in congrua basi. In capite columpne sit clavus concavus substentans[2,1] speram ligneam, in[3,1] qua spera pergameni[4,1] conglutinetur; que habeat extrinsecus, versus columpnam, ccclx[5,1] dentes. Universos illos[6,1] moveat sperula[7,1] xii dentium cotidie circumducenda; cuius axis, clavum concavum transiens, gerat rethe horarum[8,1] sub volvello[9,1] signorum, quod[10,1] eminus[11,1] afixum[12,1] sit, spera[13,1] annuatim cum ea[14,1] circumducenda.[15,1] De centro volvelli[16,1] dependeat filum quo iuxta circumferenciam volvelli[17,1] notande[18,1] sunt hore[19,1] dierum[20,1] ac noctuum.[21,1]

[1,1] α De composicione instrumenti per quod habetur versus motus planetarum β *No title.*
[2,1] α sustentans. *Both are presumed to be forms of* 'sustenens', *rather than* 'subtenens'.
[3,1] α *omits.*
[4,1] α perchameni. β pergramen.
[5,1] α cclx.
[6,1] β istos.
[7,1] γ spertula.
[8,1] γ recte horas. β recte horam.
[9,1] α volvendo
[10,1] β, γ *omit.*
[11,1] β est minus.
[12,1] γ efixum.
[13,1] α, γ spere.
[14,1] α, β, γ *agree. See note 6, traslation.*
[15,1] β circumducendum. α, γ *uncertain endings*
[16,1] α novelli.
[17,1] α novelli.
[18,1] γ notand'.
[19,1] γ hores.
[20,1] γ dies.
[21,1] γ notuum; α, β noctium.

A device of certain remarkable wheels, by which the true places of all the planets are known, and also the hours of day and night.

Take a column on a suitable base. At the top of the column there is to be a hollow *clavus*,[1tr] supporting a wooden sphere. On this a sphere of parchment should be glued, having, outside and towards the column,[2tr] 360 teeth. A wheel[3tr] of 12 teeth, moving round once daily, should drive these.[4tr] The axis of this wheel, crossing the hollow *clavus*, carries the *rete* of hours.[tr5] Underneath a volvelle of signs [of the Zodiac], which is in an elevated position, there must be a sphere, going round annually with it.[6tr] A wire is to project[7tr] from the centre of the volvelle [as a whole?],

---

[1tr] The word is used of the pivot through the centre of an astrolabe. It might mean 'nail' or 'pin'.

[2tr] A sphere is unlikely to have carried teeth. '*Spera*' might be translated 'disk', but it is equally improbable that teeth would be of parchment. It must surely be supposed that the sphere of wood, covered with parchment, is fixed to a metal disk with 360 teeth. Perhaps these were in the form of pegs on the lower side of the disk, or perhaps 'versus columpnam' refers to the teeth as a whole. There may well be a phrase missing here (after '*conglutinetur*').

[3tr] The entire tractate requires '*sperula*' to be some sort of gear wheel, presumably flat. Perhaps the word took more than its share of ambiguity from '*sphera*', which in astronomy is often to be interpreted 'circle'. Du Cange has only one suggestive quotation: '*Amicetus constituit ut clerici coronas portent in modum spaerulae*'.

[4tr] A great deal hinges on the word '*circumducenda*'. As the text stands, it is the wheel of 12 teeth which moves round once daily. This is the driving wheel, not the driven. The driven wheel, of 360 teeth, seems to be fixed to an astrolabe plate, or the like (see on).

[5tr] Taking the reading of γ, we might have 'tells us the hours correctly'. In other words, if there is to be no astrolabe dial, this phrase must be meant to inform us of an ordinary clock-hand. The argument against this is that we are later provided with a pointer for the hours. The reading of γ and β is rejected here for another reason: not only did the Roman anaphoric clocks have a 'rete of hours', that is, a pierced plate with lines corresponding to those of the unpierced plate of an astrolabe, but Richard of Wallingford's clock also had the arrangement, as I shall show in a forthcoming publication. Although no surviving medieval clock has this arrangement, the oldest of these was not made before the end of the Fourteenth Century. Thus if astrolabe dials were at all common in the early century and before (whether water-driven or otherwise), the evidence seems to point to their following the anaphoric clock convention. As knowledge of these clocks became more widespread, it was to be expected that the normal astrolabe arrangement would be preferred.

[6tr] On the nature of the volvelle of signs, see note 8, below, and also section 6 of the commentary. The translation of '*cum ea*' is left vague deliberately, especially as this could be a mistake for '*cum eo*'. I take it to refer in some way to the wheel of 360 teeth, or perhaps to that of 12 teeth, with its concomitant motion.

[7tr] '*Dependeat filum*' might not have referred to a thread hanging, as will become evident when we are given the lengths of certain *fila* later. As it turns out, they represent epicycle radii. Even allowing for a vertical zodiac plane, rather than a horizontal one as suggested here, they might require rigidity, although one way of avoiding this conclusion is outlined towards the end of section 6 of the commentary. However, this first occurrence

In signifero autem, cuius summitatem tenet caput, [22¹] circumducenda est linea Saturni sub tripla diametro; [23¹] et ab eius fine [24¹] dependeat Saturnus filo capiente gradus vi et xv minuta.

In sequenti signifero circumducenda est [25¹] linea Iovis, et ab eius fine dependeat Iupiter filo capiente xi gradus et xv minuta.

In sequenti signifero circumducenda est linea Veneris, sub quintupla diametro et ab eius fine [26¹] dependeat Venus superbi partiente [27¹] quintas linee filo [28¹] cui diameter [29¹] signiferi sit triplex [30¹] sexqualiter.

[22¹] γ cap'; α, β capricornus.
[23¹] α (*here and subsequently*) dyametro.
[24¹] γ finem.
[25¹] β, γ *omit*.
[26¹] γ finem; α filo.
[27¹] γ perficiente.
[28¹] γ f.io (*followed by* c¹ diat); α, β *add* scilicet.
[29¹] β diametri.
[30¹] α (*here and subsequently*) triplicandum; β triplum.

by means of which, against the volvelle's circumference, the hours of day and night may be known.

The line of Saturn is to be carried round on a *signiferus*,[8tr] on top of which the ball is to rest,[8tr] this line extending to one third of the diameter.[10tr] Saturn should be held on a wire of 6°15′, from the end of the line.[11tr]

The line of Jupiter is to be carried round,[12tr] on the next *signiferus*, and Jupiter should be held from the end of this line by a wire of 11°15′.

The line of Venus should be carried round on the next *signiferus*, this line extending to one fifth of the diameter. Venus is to be held from the end of the line by means of a wire. The diameter of the *signiferus* should be 4½ times as great as the length of [the wire].[13tr]

---

of the word 'filum' is translated as 'wire', since the word is suitably ambiguous. In fact this appears to be the first recorded reference to what must be a *clock hand*. Its use suggests that equal, in addition to canonical hours, were being indicated. The fact that they were sidereal hours was probably of no consequence in a machine which was in any case crude.

[8tr] The meaning of this word is not at all clear. In classical Latin, used as a noun, it generally meant 'standard bearer', i.e. a person or thing carrying a sign. It is occasionally used as a synonym for 'Zodiac' (Vitruvius, Pliny, Seneca, etc.). This use is found in the Middle Ages. Du Cange quotes Rabanus Maurus (9th Century): '*Zodiacus vel signifer est circulus obliquus duodecim signis constans...*'. If the word was merely a synonym for 'Zodiac' here, however, the line of Saturn would not, being carried by it, be able to move through the signs. In addition, we have already been provided with a 'volvelle of signs', that is to say, a Zodiac-volvelle (which need not necessarily move, by the way - this term is used of a labelled disk in general). We are to have further *signiferi*. These can hardly be rods, on the other hand (as the translation 'ensign' might suggest), since we later meet the phrase '*circumferenciam signiferi*'. In fact I shall suppose that a *signiferus* is a pierced disk, carrying the planet, or an epicycle wire, at its rim, and just possibly centred at the equant, where appropriate.

Dr Matthias Schramm has drawn my attention to another use of the word '*signiferus*', this time in the commentary on Plato's *Timaeus*, written by Chalcidius. In a recent edition of this (*Plato Latinus*, ed Raymundus Klibansky, vol. IV, *Timaeus, a Calcidio translatus, etc...*, London, 1962) there occurs this passage: '*Etiam illud addendum ceteros circulos id ipsum esse circulos iuxta definitionem... signiferum vero ex multis constare in speciem circuli tympanum... eius circulus qui inter signa medius appellatur, existens maximus... Reliqui extimi duo signiferae latitudinis circuli huius ipsius... breviores sunt*'. (Op. cit. p. 115). It appears that the *signiferus* of a planet was the band contained between the two small circles bounding the planet's motion in latitude. There was thus a different *signiferus* for each planet. It is possible that this (Fourth Century) meaning was a determinant of the meaning found (or rather hidden!) in the present text.

[9tr] Text obscure. (See note 22 of the text).

[10tr] It is not clear *which* diameter is intended. I suspect the writer had in mind the diameter of the portion cut out from the star map of the dial (see commentary, section 7).

[11tr] That is, $6^{15}/_{60}$ parts, where twice the length of the line of Saturn (i.e. the deferent circle diameter) represents 120 parts. It may well be the case that the 'lines' of the planets were threads drawn out from the centre deferent, and were neither rods (as one would perhaps expect), nor even taken from the centre of rotation (which would correpond to the equant, assuming it to be mechanically possible to take eccentricities into account).

[12tr] Notice that the diameter of Jupiter's deferent circle is not specified.

[13tr] The text suggests that the ratio is $4^1/_2$, not $6^1/_2$, as I would suggest. The line

In sequenti signifero circumducenda est linea Lune. Hinc et inde contingens circumferenciam signiferi sit circulus xlv dentium afixus[31¹] signifero, qui[32¹] circumducat sperulam x dentium et, in eius axe, lineam Lune transceuntem. Afixa sperula xiii dentium circumducat sperulam lix dentium. In linea Lune versus eam currentem, a centro huius sperule ad vi gradus et xv minuta fixa, sit linea.[33¹]

In sequenti signifero circumducenda est[34¹] linea Mercurii subdecupla diametro, et ab eius fine dependeat Mercurius filo cui diameter[35¹] signiferi sit triplex sexqualiter.

In reliquo signifero circumducenda est linea Martis cui diameter[36¹] signiferi sit triplex sexqualiter, et ab eius fine dependeat[37¹] Mars filo[38¹] cui diameter[39¹] signiferi sit quintuplex[40¹] sexqualiter, scilicet[41¹] capiente xli gradus et ix[42¹] minuta.

Axes deferentes lineas planetarum transeant[43¹] signiferos in locis augium:[44¹]

Axis Saturni in x gradibus[45¹] Sagittarii, a centro signiferi ad vi gradus et xxx minuta;

Axis Iovis in xxiiii[46¹] gradibus Virginis ad v gradus et xv minuta;[47¹]

Axis Veneris in xxviii[48¹] gradibus Geminorum, a centro ad duos gradus. Ab hoc[49¹] axe dependeat Sol filo equali filo Veneris;

31¹ γ afixum.
32¹ γ *omits*.
33¹ α, β Luna.
34¹ γ *omits*.
35¹ γ diametrum.
36¹ γ diametrum.
37¹ γ dependit ac.
38¹ γ *omits*.
39¹ γ diametrum.
40¹ α, γ quintuplus.
41¹ β filo.
42¹ γ xi.
43¹ γ transceant (*and subsequently*).
44¹ α anguium.
45¹ α gradu (*here and subsequently*).
46¹ β 44.
47¹ γ momenta. *This word could in fact mean 'fortieth parts of an hour'* — *see* Tannéry, *Bibl. Math.* vi (1905) 111 — *but obviously is not used in this sense here.*
48¹ β 24.
49¹ β huius.

The line of the Moon is to be carried round on the next *signiferus*. A wheel of 45 teeth, fixed to it, should touch the circumference of the *signiferus* here and there.[14tr] This is to drive a wheel of 10 teeth, and also, on its axis, the line of the Moon which crosses [?]. An affixed wheel with 13 teeth should drive a wheel with 59 teeth. On the line of the Moon, measuring towards it,[15tr] a line is to be fixed at a distance of 6°15′ from the centre of this wheel [of 45 teeth].

The line of Mercury is to be carried round on the next *signiferus*. The line being one tenth of the diameter in length, Mercury should be held from its end by a wire. The diameter of the *signiferus* is to be $4\frac{1}{2}$ times as long as this wire.[16tr]

On the last *signiferus* the line of Mars is to be carried round. The diameter of the *signiferus* is to be $4\frac{1}{2}$ times[17tr] the length of the line. Mars is to be held by a wire from the end of the line. The diameter of the *signiferus* should be $6\frac{1}{2}$ times the length of the wire, that is to say, the wire is to be of 41°09′ [41°11′?].

The axes carrying the lines of the planets should pass through the *signiferi* at the aux-positions:[18tr]

The axis of Saturn, in the direction 10° Sagittarius is to be at a distance of 60°30′ [from the Earth];[19tr]

The axis of Jupiter, at 24° [44°?] Virgo, is to be at a distance of 5°15′ [from the Earth];

The axis of Venus, at 28° [24°?] Gemini, is to be at a distance 2° [from the Earth]. The Sun is on the same axis, being on a wire equal in length to that supporting Venus;

---

of Venus, i.e. the semidiameter of the deferent, being counted as 60 units, it would follow that the wire (i.e. the epicycle radius) was 600/7 units, or, in the notation of the text, 85°43′, to the nearest minute. This figure, which is much too large, is discussed later, together with the reason for the suggested change.

These calculations rest on the translation of '*triplex sexqualiter*' (which cannot be transcribed as '*triplex sexquialter*' as might be supposed). '*Sexqualiter*' does not occur in any dictionary that I have seen. Perhaps it is a corruption of '*sexquialter*' (for '*sesquialter*') which is invariably translated as 'one and a half'. The prefix '*sesqui*' apparently began by meaning 'one-half more', but in the Middle Ages it seems generally to have meant 'one and a half more'. This meaning is borne out when we come to the case of Mars, where the eccentricity is expressed in both ways On either reading of '*sexqualiter*', the figures ($4^1/_2$ or $3^1/_2$) would give an unacceptable result.

[14tr] The meaning of '*Hinc et inde*' is obscure, but this phrase is presumably hinting at the epicyclic motion which is deduced from this passage in section 6 of the commentary.

[15tr] The Moon?

[16tr] These figures are discussed in section 1 of the commentary.

[17tr] A mistake for '$7^1/_2$ times'? See note 16.

[18tr] A planet's aux position is reached when the wire carrying it is in line with the Earth-equant line ('line of aux'). On the confusion of equant and centre deferent here, see commentary section 5.

[19tr] This entire section of the text is much abbreviated. It can be reconstructed only because the figures are known as constants of Ptolemaic astronomy.

Axis Lune transceat centrum signiferi;

Axis[50¹] Mercurii in xxviii gradibus Libre, a centro ad tres gradus;

Axis Martis in xi gradibus Leonis, a centro ad xi gradus et xxiiii[51¹] minuta.

In clavo concavo intra speram sint[52¹] fixe due sperule, una xvii dentium et altera lxxvii dentium. Habeat[53¹] axi suo afixas duas sperulas unam lxv dentium quibus circumducat sperulam xxvii[54¹] dentium et in axe eius lineam Martis, et alteram xii dentium quibus circumducat sperulam lxxviii[55¹] dentium, et in axe eius lineam Saturni. Altera sperula lxxvii dentium habeat axi suo[56¹] afixam sperulam xxvi dentium, quibus circumducat hinc[57¹] sperulam lxviii dentium, et in[58¹] eius axe lineam[59¹] Iovis, et illinc sperulam xxiii dentium. Et in[60¹] eius foris[61¹] afixa axe sperula[62¹] xiiii dentium circumducat circulum Draconis retrogradi lxv dentium. Centrum[63¹] signiferi Veneris atque Solis habeat parvam sperulam prominentem circa quam circulus Draconis moveatur. Una sperula xii dentium habeat axi suo afixam sperulam xix[64¹] dentium quibus circumducat sperulam lxxv dentium, et in eius axe lineam[65¹] Veneris.

Altera sperula xii dentium habeat axi suo afixam sperulam l dentium, quibus circumducat sperulam xxiiii dentium, et in axe eius lineam Lune.

Tertia sperula xii dentium habeat axi[66¹] suo afixam sperulam xxxiii[67¹] dentium quibus circumducat illam[68¹] sperulam li dentium, et in axe eius lineam[69¹] Mercurii.

Sic habetur verus motus planetarum preter duos circulos breves[70¹] quorum unus uno gradu et xv minutis equat circulum[71¹] Lune, alter uno gradu et xl[72¹] minutis equat circulum[73¹] Mercurii. Deo gratias, Amen.

50¹ α auxis.
51¹ γ xxiiii^or.
52¹ γ sicut.
53¹ β quarum primus hanc *for* habeat.
54¹ α, β lxxviii.
55¹ β 28.
56¹ γ omits.
57¹ β hanc.
58¹ α omits.
59¹ γ linea.
60¹ α omits.
61¹ β fore.
62¹ γ sperulam.
63¹ γ cēt in.
64¹ β 10.
65¹ γ linea.
66¹ α auxi.
67¹ α xxiii.
68¹ β aliam.
69¹ γ linea.
70¹ γ omits.
71¹ α, β cursum.
72¹ *The preceding three words are unintelligible in* γ. β *has* 50 *for* xl.
73¹ β cursum.

The axis of the Moon passes through the centre of the *signiferus*.[20tr]

The axis of Mercury, at 28° Libra, is to be at a distance 3° [from the Earth];

The axis of Mars, at 11° Leo, is to be at a distance 11°24' [from the Earth].

On the hollow *clavus* within the sphere, two wheels should be fixed, one of 17 teeth, the other of 77 teeth. [The former] is to have two wheels fixed to its axis,[21tr] one of 65 teeth, by which it is to drive a wheel of 27 [78?] teeth, and the line of Mars on its axis. The other is of 12 teeth, by which it is to drive the wheel of 78 teeth, and the line of Saturn on its axis. The other wheel having 77 teeth should have a wheel of 26 teeth fixed to its axis. By these teeth it is to drive a wheel of 68, as well as the line of Jupiter on its axis, and beyond, a wheel of 23 teeth. And on its[22tr] axis, a wheel of 14 teeth should drive, in a retrograde sense, the wheel of the Dragon,[23tr] of 65 teeth. The centre of the *signiferus* of Venus and the Sun should have a small projecting wheel[24tr] around which the circle of the Dragon moves.

One wheel of 12 teeth must have, fixed to its axis, a wheel of 19 teeth, by which it drives a wheel of 75 teeth, as well as the line of Venus on its axis.

Another wheel of 12 teeth is to have a wheel of 50 teeth fixed to its axis, by which it drives a wheel of 24 teeth, and also the line of the Moon on its axis.

The third wheel of 12 teeth is to have a wheel of 33 teeth fixed to its axis, by which teeth it drives the[25tr] wheel of 51 teeth, and also the line of Mercury on its axis.

In this way the true motion of the planets is arrived at, apart from two circles, one of which — of 1°15' — equates the circle of the Moon, and the other — of 2°40' [°50'?] — equates the circle of Mercury.[26tr] Thanks be to God. Amen.

[20tr] That is, presumably, no eccentricity is allowed for. The Ptolemaic theory of the Moon is, of course, relatively complicated.

[21tr] A very odd turn of phrase, in view of the interpretation we have found.

[22tr] That is, the axis of the wheel of 23.

[23tr] The line, that is to say, of luni-solar nodes.

[24tr] '*Sperula*' - for 'scale' perhaps.

[25tr] The reason for the emphasis ('*illam sperulam*') is not clear.

[26tr] These last wheels are presumably needed to give moving centre deferents for the Moon and Mercury, as required by Ptolemaic theory. Is 1°15' a mistake for 10°15'? The casual way in which they are mentioned suggests that the instrument, if ever such existed, did not incorporate these wheels. I have ignored them.

COMMENTARY

1. *Equation of the argument - epicycle radii*.

The fact that much of the text is concerned with laying down the ratios between the 'lines' of the planets, and the lengths of supporting wires, leads one to suppose that an epicyclic representation was intended. It will later be shown that there is no obvious provision for driving any planet other than the Moon in its epicycle: only mean *motus* are clearly provided for. Nevertheless, the ratios quoted should not be ignored, and in section 6 below a possible epicyclic arrangement is discussed.

To all intents and purposes the epicycle radii, in terms of the deferent diameter, are the same in the Toledo Tables of az-Zarqellu (c. 1080) and the later (c. 1272) Alfonsine Tables. They differ appreciably from Ptolemy's figures.

	Mercury	Venus	Mars	Jupiter	Saturn	Moon
Toledo	22 02	45 59	41 09	11 03	06 13	
Alfonsine	22 02	45 59	41 10	11 03	06 13	
Planetarium	(see below)	46 09* [1]	41 32* 41 09	11 15	06 15	06 15

Figures given in or deduced from our treatise (the latter figures being marked by an asterisk) are given in the third line — they correspond to wire-lengths. The discrepancy between claimed and inferred values for Mars is probably the result of the author's failure to find small integers whose ratio was 41°09'.

For Venus, the figures given lead to (10/9).60°, which is much too large. Retaining the deferent radius of one fifth of the *signiferus*, we find that altering 4½ to 6½ gives an epicycle radius of 46°09'.

The quantities given for Mercury lead to a nonsensical figure for the epicycle radius, judging by any known version of Ptolemy's theory. It seems likely that the 'one tenth of the diameter' should be kept (see the next section of the commentary), and therefore to restore a reasonable figure we need something like 27½, for the ratio of the diameter of the *signiferus* to the length of the wire holding Mercury. This gives 21°49', for the epicycle radius.

---

[1] Text amended, see below.

On the strength of the evidence of this section, no firm conclusion can be drawn about the tradition in which the instrument lies.

2. *The ordering of the planets by distance.*

If we assume that all *signiferi* have the same diameter, or that there is only one *signiferus* (by and large the text seems to suggest that there are several), then the 'lines' of the planets, which we take to be equal to deferent circle radii, are of the following lengths (the unit being the *signiferus* diameter):

Sun	Mercury	Venus	Moon	Mars	Jupiter	Saturn
—	$1/10$	$1/5$	—	$2/9$	—	$1/3$

(Jupiter and the Sun are given no figures in the text.) Apart from mention of the Moon's epicyclic radius, which is for some reason placed first, the order in which the planets are dealt with on both first and second occasions is Saturn, Jupiter, Venus, Moon, Mercury, Mars.

This order is probably of no significance, even though it is repeated. (There appears to be no mechanical reason for the order — rather the contrary.) The ordering by distance, however, coincides with the order which has been commonplace at least from the time of the Greeks — it is, for instance, the order given in Greek horoscopes. It corresponds, quite simply, to the arrangement of the planets according to their periods of (sidereal) rotation. Many medieval astronomers discussed the distances of the planets, but there are reasons for thinking that it would be unprofitable to pursue the possibility that the above set of 'distances' was meant to reproduce the actual disposition of the planets.

3. *Eccentricities of the planetary orbits.*

For the time being, we shall suppose the instrument to make mechanical provision for non-central equants, as the text, by quoting figures for the planetary eccentricities, suggests. The distances of the equants from the Earth are here collected together. If they were compared with those taken from the Toledo and Alfonsine tables, the agreement would be poor. Instead, they are compared with figures for the greatest equation of centre — often erroneously equated with eccentricities, in the Middle Ages.

	Sun	Moon	Mercury	Venus	Mars	Jupiter	Saturn
Toledo	1°59'	—	3°02'	1°59'	11°24'	5°15'	6°31'
Alfonsine	2°10'	—	3°02'	2°10'	11°25'	5°57'	6°31'
Planetarium	2°00'	—	3°00'	2°00'	11°24'	5°15'	6°30'

For the Sun, the equant is to be identified with the centre deferent. On the Ptolemaic theory the Moon has a moving equant (and centre deferent). Only minimum and maximum values would be of any significance here. The sentence 'The axis of the Moon passes through the centre of the signiferus' is taken to mean that no lunar eccentricity is allowed for. (But see the last footnote to the translation).

As will be seen from the table, the agreement between first and third rows is tolerably good. The crucial figures are those for Jupiter, for it is unlikely that the manuscripts have lost and gained elements of a set of roman numerals so as to change ' lvii ' into ' xv '. It therefore seems highly probable that the planetarium is in the Toledo tradition of az-Zarqellu, for no other well known astronomer quotes an eccentricity for Jupiter of 5°15'. (Most writers follow Ptolemy's 5°30'. Al-Khwârismî has 5°6', and al Kashî 5°28').

We tentatively conclude that the instrument was designed in the pre-Alfonsine period, which in Italy means, perhaps, before 1290 or so. It must be admitted that the reasons for giving this date are not very conclusive. I am not aware of any precise information as to the arrival of the Alfonsine tables in Italy. As J. L. E. Dreyer pointed out, the *Libros del Saber* themselves suggest that the tables were not complete before 1272[2]. Andalo di Negro (pre-1260 to c. 1340), in a set of canons (c. 1323) on the Almanac of Profatius, stated that Profatius used the Alfonsine tables[3]. This famous mathematician, astronomer, and translator from Arabic into Hebrew, flourished in Montpellier, where he died c. 1304. His works were put into Latin almost immediately. Montpellier was within fifty miles of Avignon, and had itself many cultural and commercial connexions with the Italian cities. It seems unlikely that the Alfonsine tables can have been unknown to the leading Italian astronomers after the first decade of the century. (They are thought to have reached Paris by c. 1292 and Oxford before 1320.) But the only conclusive evidence of which I know is that Gueruccius produced an incomplete Italian trans-

---

[2] In this respect he corrects the common belief that they were prepared for Alfonso's coronation in 1252. See *The original form of the Alfonsine tables*, « Monthly Notices R.A.S », lxxx (1920) 243-62.
[3] I think Duhem first drew attention to this statement. See Dreyer, op. cit., and Thorndike, « Isis », x (1928) 52-6.

lation of Alfonsine astronomy in 1341[4], and that Giacomo de' Dondi compiled a set of tables based on the Alfonsine, but for the meridian of Padua[5]. On the other hand, I believe that Campanus of Novara (d. pre-1300?) will eventually be proved to have known the Alfonsine books. He apparently know the Alfonsine theory of procession, but research would be needed to prove that both did not rely on a common source, if this were to be admitted as evidence.

4. *Aux positions*.

Measuring celestial longitudes from the first point of Aries, in the usual way, the auges of the planets are quoted as follows:

	Mercury	Sun and Venus	Mars	Jupiter	Saturn
Planetarium	118°	88°	131°	152°	250°
Approx. Toledo figures for the late thirteenth C.	208°	87°	131°	173°	249°

Since the auges of the planets move at a rate of a little more than 50" per annum (i.e. they were fixed with respect to the fixed stars), and since it is uncertain what theory of precession was used, these rough figures cannot in practice be used to assign a precise date to the instrument. There is another complication: these figures are almost certainly corrupt through miscopying. (One such mistake is very obvious: apart from the substitution of an 'x' for a 'c', the roman numerals expressing the aux position of Mercury in our text agree with the accepted figure.) We must also bear in mind the tendency to drop roman numerals, especially from the end of the group, and the medieval practice of approximating to degrees by dropping *all* the minutes (rather than by going to the *nearest* degree). However, accepting all but the Jupiter figure, correcting the Mercury figure as explained, and working from the tables

---

[4] *Libro di astrologia... Guerruccii filio Cionis Federighi civis Florentini* (Vatican MS 8174). A text of this has recently been published at Saragossa.

[5] These are in at least two MSS, one in the Escorial, and the other Bodleian Library MS Canon. Misc. 436, ff 13 r - 23 v (canons, excluding tables). The catalogue description is: '*Jacobi de Dondis, Patavini, Planetarium, praeviis expositionibus abanonymo quodam confectis*'. (Incipit: '*Cum plures et varie tabule ad celestes motus...*'). The name '*Planetarium*' is misleading to modern eyes, since no planetary instrument whatsoever is considered. The table of contents of the MS is clearer: '*Jacobus de Dondis Tabulae et Canones de Motibus Caelestibus*'.

of az-Zarqellu (I used those in MS Laud Misc 644), I have deduced a date of 1268 ± 40 years. Trepidation was assumed, rather than simple precession, in those tables.

Finally, I should mention one interpretation which the text might just conceivably bear. If the figures quoted were not of aux positions, but of mean longitudes of the planets, we should have what in principle is a very precise means of dating the text. There is, however, no date between the beginning of the Ninth Century and the end of the Fourteenth Century, at which the relative positions of the planets agree at all closely with these figures, apart from one or two in the neighbourhood of the year 1280. Thus on June 27, 1280, the true longitudes of the planets were (with the planetarium figures in brackets underneath):

Mercury	Venus	Mars	Jupiter	Saturn
117	60	134	144	245
(118)	(88)	(131)	(152)	(250)

Unless we allow for miscopying on a large scale — and then there is no case for applying the method in the first place — these figures are not as close as we should expect from the astronomer who designed this instrument. Also the Latin, although highly abbreviated, does not encourage this interpretation of the figures, which are here supposed to be aux positions.

5. *The ratios of the wheels.*

The translation of this highly abbreviated text depends very much upon the ways in which wheels of the given number of teeth can be fitted together to give the requisite motions. Bearing in mind that the ratio of the periods of two wheels, properly meshed, is equal to the ratio of the numbers of their teeth, I shall introduce a new notation for representing gear trains, which saves a good deal of circumlocution, and has many other advantages. Each train of gears is represented by a set of ratios, the product of which must always be unity. This is best illustrated by an example. The expression:

$$\frac{(1^y)}{4} \cdot \frac{88}{(22^y)}$$

denotes a wheel of 4 teeth, with a period of 1 year, driving a wheel of 88 teeth, with a period of 22 years. The expression:

$$\frac{(1^y)}{4} \cdot \frac{88}{77} \cdot \frac{65}{(18\,^4/_7{}^y)}$$

denotes the same arrangement, but with an additional wheel of 77, concentric with and fixed to the wheel of 88, and driving a wheel of 65 teeth with a period of roughly $18^y \cdot 6$. The ease of finding the period of any wheel in a train, almost at a glance, should be obvious: it is the product of all terms, of the continued product up to, and including, the number of teeth it bears, if the number of the wheel's teeth appears on the top line. Otherwise it is the period of the wheel to which it is fixed. (Thus the period of both the wheel of 88 teeth and that of 77 is the same, and equal to $\frac{(1^y)}{4}$ 88) What is commonly known as an 'idle' wheel is not here distinguished from a pair of wheels with equal number of teeth. (Thus $\frac{(1^y)}{4} \cdot \frac{71}{71} \cdot \frac{88}{(22^y)}$ is ambiguous as between the two arrangements.) I shall not introduce a distinguishing notation here, neither shall I give notations for more complex trains (e.g. trains carried on others).

We many now discuss the ratios of the 'remarkable wheels'. Probably the most difficult obstacles to understanding the text, apart from such trifles as miscopied numerals, are (1) the occurrence of two or possibly three wheels with either 77 or 78 teeth; (2) relative pronouns with obscure reference — often so obscure that the MSS disagree on the question.

The second obstacle can never be completely overcome, but once the first is out of the way, it becomes evident that there is an insufficient number of wheels for driven epicyclic movements. With only one set of periodicities to carry in mind (viz. mean *motus*), determining the arrangement of the wheels is then a straightforward exercise in patience. There are viable alternatives to those presented here. To show why they were rejected would be intolerably dull, and I shall simply state what I believe to be the correct ratios.

Ignoring, for the moment, the first set of wheels mentioned in connexion with the Moon, the planets are all driven off one of two wheels having an annual motion, these being of 17 teeth (for Mars, Jupiter and Saturn), and 77 teeth (for Mercury, Venus and the Moon). The text suggests that the wheel of 17 meshes with a wheel of 77, which is not, of course, to be confused with the first. The suggested ratios, requiring not a single amendment to the text, if we choose carefully among the variant readings, are:

Mercury: $\frac{(1^y)}{77} \cdot \frac{12}{33} \cdot \frac{51}{(0^y. \, 2406)}$ $0^y. \, 2406$ (not comparable)

Venus: $\frac{(1^y)}{77} \cdot \frac{12}{19} \cdot \frac{}{(0^y. \, 6150)}$ $0^y. \, 6152$ (not comparable)

Moon: $\dfrac{(365^{1}/4^{d})}{77} \cdot \dfrac{12}{50} \cdot \dfrac{24}{(27^{d}.32)}$ $27^{d}.322$ $27^{d}.33$

Mars: $\dfrac{(1^{y})}{17} \cdot \dfrac{77}{65} \cdot \dfrac{27}{(1^{y}.880)}$ $1^{y}.881$ $1^{y}.885$

Jupiter: $\dfrac{(1^{y})}{17} \cdot \dfrac{77}{26} \cdot \dfrac{68}{(11^{y}.87)}$ $11^{y}.86$ $11^{y}.83$

Saturn: $\dfrac{(1^{y})}{17} \cdot \dfrac{77}{12} \cdot \dfrac{(29^{y}.48)}{78}$ $29^{y}.46$ $29^{y}.58$

In the first of the two columns on the right are tabulated the mean sidereal periods of the planets, to four figures, which are acknowledged to have been known substantially correctly from before the time of Ptolemy. (There is therefore no hope of using them to place the tradition of this instrument[6].) There is little doubt that the Dragon is driven from the final wheel in the Jupiter train:

$$\dfrac{(1^{y})}{17} \cdot \dfrac{77}{26} \cdot \dfrac{68}{68} \cdot \dfrac{23}{14} \cdot \dfrac{65}{(18^{y}.60)}.$$

There are now only two problems left in connexion with wheel ratios. One concerns the wheels of 12 (diurnal period) and 360, mentioned at the very outset. The other concerns an alternative train for the Moon, which the Latin seems to suggest may be represented thus:

$$\dfrac{\ldots}{45} \cdot \dfrac{10}{13} \cdot \dfrac{59}{\ldots} \cdot$$

This ratio is exactly $1\,^{1}/_{117}$, and is not, for example, capable of converting a synodic into a sidereal lunar period (as one might perhaps have expected it to do). The clue to its significance is probably in the quoted 6°15′. This is so like the equation of the argument (epicycle radius) of some versions of Ptolemy's theory (see section 1 of this commentary) that we may suppose the wheel of 45 teeth to carry the Moon's epicycle, the effective equant being at the centre of a fixed wheel of 59, with the 10+13 pair on their (moving) line of centres. This moving line will be carried round the equant by a completely different train, namely that discussed above (see Fig. 4).

---

[6] In the extreme right hand column are listed the mean motions achieved by Giovanni de' Dondi's clock of 1364. Despite the greater complexity of this work, it is worth remarking on the somewhat lower standard of accuracy achieved, so far as mean motions were concerned.

First we notice that the *sense* of rotation of the Moon in its epicycle is given correctly. (The Moon moves in a sense opposed to that of the remaining planets, a point which a great many modern writers completely overlook.) Secondly, we deduce from the tables giving the diurnal motions of the Moon (the different authorities all agree, within sensible limits) the required ratio for comparison with our $1\,^1/_{117}$:

Mean motus	Mean argument	Ratio
13°10′35″	13°3′54″	1·00850 (= $1^1/117.5$)

This good agreement suggests that our interpretation is correct.

As for the wheels of 12 and 360 teeth, I can only suggest that there is a simple train between them, such that the wheel of 360 is driven with an annual motion, and therefore is capable of carrying the wheels of 17 and 77 with such a motion. The wheel of 12 would carry the rete of the dial with a period of one sidereal day. One possible arrangement, with reasonably round numbers, would be:

$$\frac{(366\,\delta)}{360} \cdot \frac{10}{122} \cdot \frac{12}{(1\,\delta)} \cdot$$

(the symbol δ denotes a sidereal day, and of course a year is approximately $366\,^1/_4$ sidereal days.)

This is conjecture only, and there is no point in our commenting on the accuracy, or otherwise, of the representation. We may, however, legitimately comment on the accuracy of the earlier trains, even though we are not sure of the exact sidereal periods which it was hoped to reproduce. So far as it is possible to compare them, the accuracy attained is as good as, if not better than, Dondi's[7]. Only when it comes to the niceties of Ptolemaic theory, with its eccentrics, equants, and so on, does the later device look superior. This is all the more surprising, in view of the astonishing economy of wheels in the present case. (Notice the way in which, twice over, three planets are driven from a single wheel. Notice also the way in which the Dragon train is taken off the Jupiter train.) If we set against this North Italian device all that we known in the way of precise details of its precursors, there is nothing with which it can be strictly compared. The Antikythera instrument perhaps approached it in accuracy, but was intended to give information of a much more restricted sort. Al-Bîrûnî's geared astrolabe is perhaps closer in tradition,

---

[7] See note 6. We have no exact record of Richard of Wallingford's planetary wheel ratios, so far as I am aware at the moment. His representations of lunar, solar, and draconitic motions were, however, much superior to those of the present text.

but it is much less accurate, as well as being restricted to solar and lunar motions. The same must be said for the geared astrolabe in the Science Museum, London, (c. 1300 - see Appendix). But nothing whatsoever is known of the gearing in the two or three true planetaria which there is good reason to believe were made in the Middle Ages. For this reason, one should resist the temptation to heap superlatives on this '*opus quarundam rotarum mirabilium*'.

6. *The practical design of the device.*

The many references to actual eccentricities and epicycle ratios suggest that at some stage it was hoped to make provision for them. This implicit promise was fulfilled only in the case of the Moon. (There is no suggestion that the *full* Ptolemaic lunar theory was reproduced, or even that an attempt was made to do so.) There are several possible explanations for the fact that the text falls short of our expectations, if indeed it does. The author might have been merely describing an object he had seen. The text's astronomical content argues against this, but not conclusively: the description might have been by an astronomer — possibly one in a tradition totally different from that of the maker. The author might, on the other hand, have simply failed to complete his task. Or he might have had his text curtailed by an early copyist or translator. Compatible with the first suggestion is the supposition that the instrument was only partly geared, the final stages in the Ptolemaic constructions being done by means of threads drawn, perhaps, through the circumferences of numbered, but ungeared, epicyclic disks, only the centres of which were driven. Another possibility will be discussed at the end of this section. This requires the dials to be in the more familiar *vertical* plane, the epicycles being described by the planets hanging freely under gravity. There seems to be no known historical precedent for such an arrangement.

If the text is taken at its face value, the equants of the planets (viz. what correspond to the 'axes' of the text) are to be non-central. It is not inconceivable that there should have been a separate dial for each planet, although the fact that three final wheels would need to be driven off each of the two groups of concentric gears, means that these could not be arranged as in the Dondi clock (i.e. at right angles to the horizontal driving wheels). The sizes of the teeth of meshing wheels must be more or less compatible, that is to say, the diameter of a wheel is to be roughly proportional to the number of its teeth. On the hypothesis of separate and coplanar scales, the geometrical problem of arranging the wheels is under-determined. The sizes of the scales over which the planets move may be chosen so as to leave sufficient room for the central astrolabe dial, again assumed to be coplanar. Against this hypothesis is the vaguest of hints, at the beginning of the text, that the Sun at least is to move within the ecliptic ring of a (reversed) astrolabe. This argument is not itself convincing until we realize that the axis of the Sun is shared by Venus. Assuming some sort of symmetry between the planets, we might be inclined to use this as an argument for the planets moving round a common scale.

The near-concentric arrangement of the planets raises its own problems. Chief of these is the need to position the centres of the wheels in keeping with the quoted eccentricities. The necessary geometrical construction is in fact over-determined, assuming the hypothesis of wheel sizes. (It is the problem of making a circle of given size touch each of three circles given in size and position.) It hight be thought possible to reach a compromise solution, changing the sizes somewhat, but an examination of the problem soon shows that this is out of the question. Moreover, the most cursory examination of the minuteness of the eccentricities of most of the planets (see section 3 of this commentary) will suffice to show that either the outer (zodiacal) scale, against which true longitudes were finally measured, was of colossal proportions, or the technology of the miniature Swiss watch was foreshadowed in the Middle Ages.

A possibility remains which has not been mentioned so far: the text might be an expression of a theoretical ideal, which was never realized in practice, and which was perhaps never properly worked out. The ideal can, however, be almost completely realized by the simple expedient of making all equants coincide. The final arrangement is shown in Fig. 2. The figure shows everything apart from the drive, the Moon's epicyclic train (omitted for the sake of clarity), and the epicyclic wires of the planets. The fixed wheel of the Moon's epicyclic train could easily be attached to the fixed zodiac/star plate, and the rest of the train is simply carried on the projecting arm illustrated. Perhaps the strongest argument against the arrangement of Fig. 2 is the large set of concentric sleeves (tubes) in the centre of the device. I can see no alternative, however, unless we suppose there to have been a set of dials grouped round the central astrolabe. An outline of this alternative arrangement, keeping Sun and Dragon within the ecliptic ring of the astrolabe, is shown in Fig. 3. The eccentricities of the planets are allowed for by the simple device of off-setting the zodiac-scales frome the centres of the final wheels in each planetary train. Unfortunately, the Sun's eccentricity is not allowed for: no train being given to take the Sun to an outside scale, its equant was made to coincide with the centre of the rete of hours. Venus is taken outside, notwithstanding the remark about her axis coinciding with the Sun's: it is surely preferable to allow for one eccentric than for none, but this is a weak point in the argument for this arrangement. And why were the planetary 'distances' stipulated so carefully if the planets are to be on different scales?

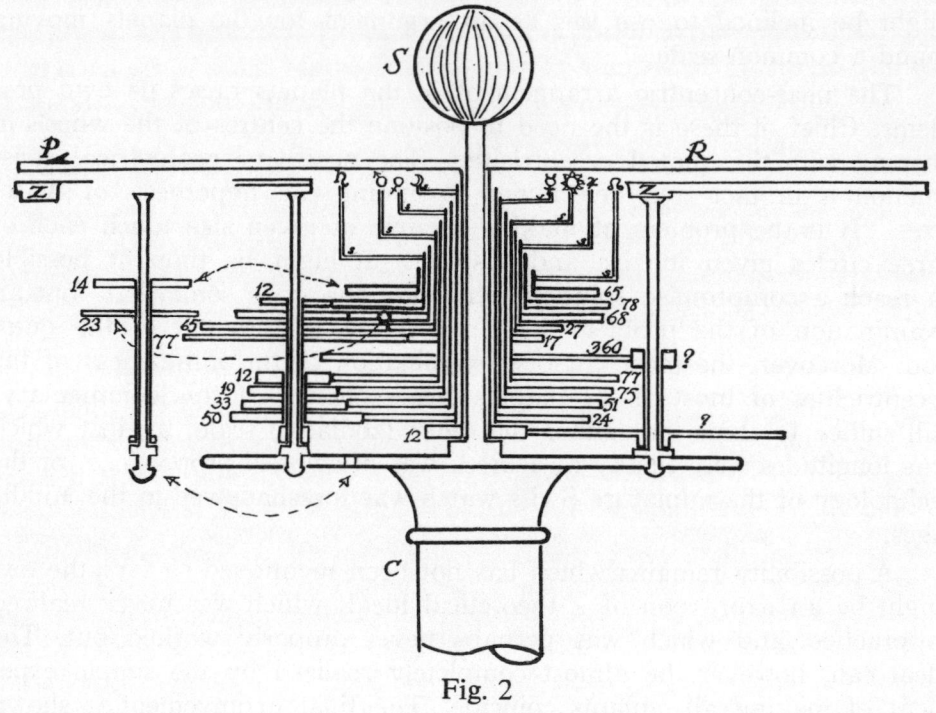

Fig. 2

Lastly, we come to the representation of epicyclic motions for the planets other than the Moon, without the need for gearing. Now it is a characteristic of Ptolemaic planetary theory that in the case of the *outer* planets the direction of the radius of the epicycle which carries the planet is parallel to the line joining the mean Sun to the centre of the Earth. This is not the case with the inner planets. If it could be arranged that each of the outer planets was suspended freely from the end of a wire, the length of which corresponded to the epicycle radius, and the centre of which was carried round the Zodiac with the mean motion of the planet, and if at the same time the Zodiac could be made to move in such a way that the mean Sun was invariably at its lowest point, then the planet would have its correct epicyclic motion *without the need for epicyclic gearing*. By a slight modification of either Fig. 2 or Fig. 3, this could be arranged. Taking Fig. 2, for example, the gears would all be placed in a vertical plane. The ordering of the planets on the central sleeves would need to be rearranged so that the sleeve carrying the Sun could be taken out through the centre. This is what would be fixed to the column. The *relative* motions of the planets and scales (Zodiac, hour-lines, etc.) would be exactly as before, but now the Sun-pointer (*signiferus*) would be the fixed pointing vertically downward. The arrangement and speed of the drive would not be the same; but in any case we known nothing of this. The spacing of the *signiferi* would have to be such as to allow room for the epicyclic wires, but this should not be difficult to arrange. (See also section 9 below.)

Only one consideration prevented me from seriously countenancing this gravity-operated epicyclic scheme: it would not work with the inner planets, and yet the text makes no obvious distinction between inner and outer planets in quoting the lengths of epicycle *fila*. Did the author think that *all* epicycle radii were parallel to the Sun's mean motus? This is unlikely, bearing in mind his otherwise sound employment of Ptolemaic principles. Or was there an even more subtle mechanical principle built into the planetarium, to account for the movement of the inner planets? The text seems to offer no help. In fact the only reason for attempting to complete the scheme outlined in the last paragraph is that it alone has the virtue — if indeed it is a virtue — of making the *volvellum signorum* of the text a volvelle in the truest sense, that is to say, a revolving disk.

In conclusion, there are at least two viable schemes, in accordance with which the planetarium could have been made, without doing undue violence to the text. Neither of these is wholly satisfactory, but of the two it seems that the scheme of Fig. 2 is much to be preferred. Both would need a system of auxiliary alidade(s), pointer(s), or thread(s), not mentioned in the text, to derive true planetary positions with anything approaching Ptolemaic accuracy. The practical difficulties of reproducing the two schemes are not easy to assess, but the greatest difficulty in the first would no doubt have been in regard to the concentric sleeves. The second scheme mitigates, but does not remove, this difficulty.

7. *The astrolabe face.*

The slender evidence for supposing the device to have had an astrolabe dial is, first, that one (and only one) of the MSS, but the best, speaks of a 'rete of hours', and, second, that the face contains a Zodiac volvelle. If there was indeed a revolving rete of hours superimposed on a fixed star map, rather than the conventional rete of stars over a plate of hour-lines, then here is one of the only two pieces of evidence of which I am aware that this arrangement, peculiar to the Roman anaphoric clock, survived into the Middle Ages[8]. The other piece of evidence is that Richard of Wallingford's clock (designed c. 1330) also sported a rete of hours. Of the many late medieval astronomical clocks now surviving,

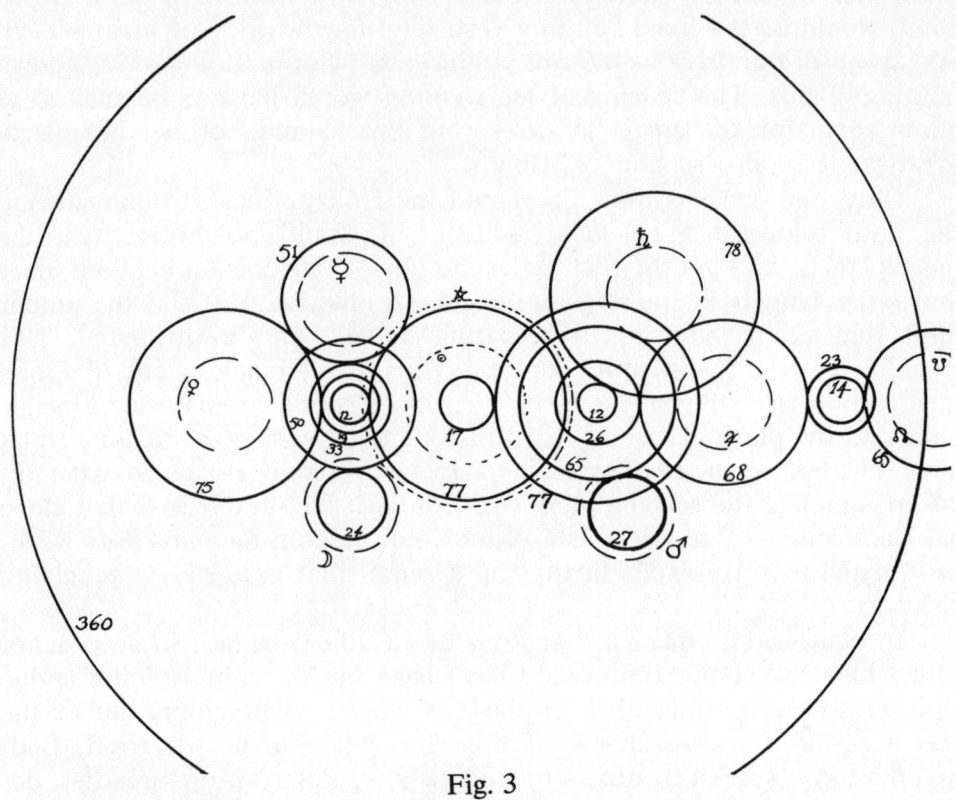

Fig. 3

it appears that none has this characteristic. The Dondi clock, like most surviving astronomical clocks, had one of its faces in the form of the conventional astrolabe — if we ignore the omission of the stars from the rete, which was no doubt intended to make the hour-lines more easily visible. (The well-known Münster clock retains the stars on the rete, but this is unusual.) The reason for the unpopularity of this unfamiliar convention is not hard to seek: the larger the number of hour-lines, the

---

[8] For details, and sources, see Drachmann's articles, cited below. Actually, although the two surviving fragments are from the Roman world (Salzburg and Vosges), and although it is through Vitruvius that its construction is chiefly known, Vitruvius designs his rete for the latitude of Alexandria, and thus 'Roman' may be an unfortunate choice of adjective.

better. But the larger the number of hour-lines in a rete, the more difficult it is to see the star map, the tropic rings, and the ecliptic ring, underneath. If the hour-lines of a rete were formed of thin wire (Vitruvius suggested bronze wire), they would neither hold their shape well, nor be likely to survive for very long. (It is hard to believe that there were no astrolabes made in the same tradition, which failed to survive for similar reasons.)

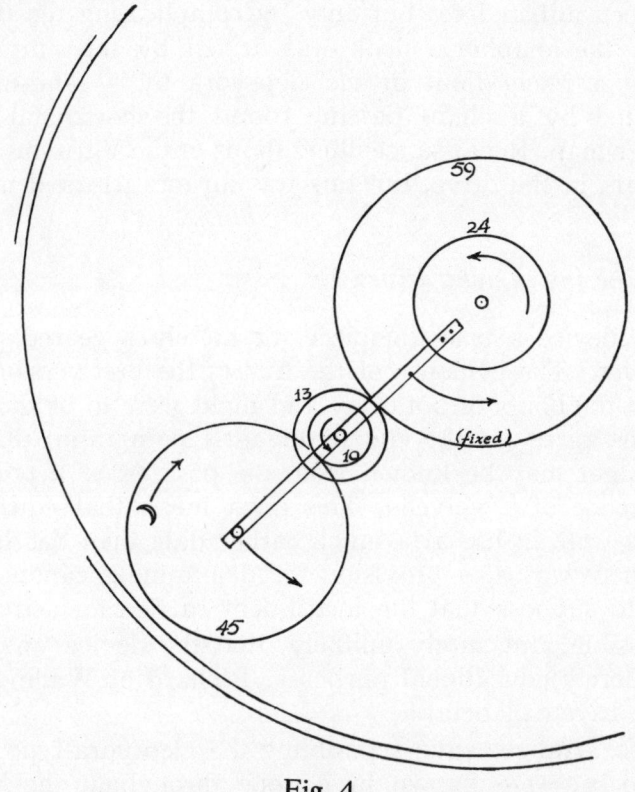

Fig. 4

The relation between the anaphoric clock and the planispheric astrolabe was explored by A. G. Drachmann in *The plane astrolabe and the anaphoric clock*[9]. Drachmann gave a more complete account of the clocks themselves, under the title *Ktesibios, Philon, and Heron*, some years before he had fully appreciated the astrolabe connexion[10]. The stereographic projection of these early clocks was from the North, rather

---

[9] « Centaurus » iii (1954) 183-9. The main points are summarized by D. J. Price in vol. iii of *A History of Technology* (ed. Singer, Holmyard, and Hall), 1960, pp. 604 sq.
[10] « Acta Historica Scientiarum Naturalium et Medicinalium », vol. iv, Copenhagen, 1948.

than the South Pole, as is evident from a reconstruction based on photographs of the Salzburg fragment, illustrated by Drachmann. The Zodiac ring was made with 365 (or 182, or 183) holes, so that a model Sun could be inserted, and moved at daily (or longer) intervals, when the clock was re-set and the clepsydra refilled. It was the star map which revolved, behind a fixed rete, rather than conversely, as is suggested was the case with our *opus rotarum*, with its revolving rete. The older arrangement could have been adhered to, but only by complicating the design somewhat. Lastly, the anaphoric clock was driven by the simple device of counterpoising a rising float in the clepsydra by a sand-bag, the two being connected by a chain passing round the horizontal axle which carried the star-map. Renaissance illustrations of the Vitruvius text showed a train of gears in the drive, but this was an unwarranted interpolation.

8. *How was the instrument driven?*

Was this device a true timepiece, or merely a geared planetarium, manually driven? The evidence of the title of the best version of the text (MS γ) is that the hours of both day and night were to be discerned from it. All versions agree (at the end of the first paragraph) that the hours of day and night may be known from the position of a pointer against the circumference of a volvelle. This must mean that equal (i.e. equinoctial) hours were in use at a much earlier date than has hitherto been suspected. There was also provision for determining canonical hours, if it was right to suppose that the instrument carried an astrolabe face.

It is possible, but surely unlikely, that the device was driven manually, for merely educational purposes. Richard of Wallingford alludes to manually driven planetaria.

Hydraulic arrangements, resembling the clepsydra-type drive of the anaphoric clocks, were known in Europe throughout the Middle Ages, when they were used in conjunction with monastic alarums, if nowhere else[11]. The Islamic world, too, is known to have worked some very remarkable automata, and also time-keeping devices, by hydraulic means. (The almost complete omission of anything resembling gear-work from their automata was a peculiarity of the Islamic world, long after the invention of the mechanical escapement in Europe.)[12] Thus it would be

---

[11] Vatican MS Lat. 5367 (13th. cent.) describes the drive for such an alarum. For further references, see Lynn White's *Medieval Technology and Social Change*, Oxford, 1962, pp. 120 sq. A more venerable example than is mentioned there is that of the 11th (?) cent., discussed by F. R. Maddison et al. in *Antiquarian Horology* iii (1962) 348-53.

[12] E. Wiedemann and F. Hauser, *Über Vorrichtungen zum Heben von Wasser in der islamischen Welt*, Beiträge zur Geschichte der Technik und Industrie », viii (1918).

surprising to find that our planetarium-timepiece did not have a drive of some sort. We should, of course, dearly like to know whether any purely *mechanical* means of driving the device was known to its author. At the moment, the oldest known description of a mechanical movement — albeit tantalizingly vague — is by Richard of Wallingford, and must date from before 1336, the year of his death. There is no earlier reference to a truly mechanical escapement which is not so ambiguous as to be worthless. My feeling is that it was evolved from the striking part of the monastic alarum during the last years of the Thirteenth Century, and in Italy; but the reasons for thinking this are hardly worth repeating. On the whole it is safest to assume that, if the device was driven, water power was used in some form.

9. *The means of representing planetary positions.*

It has been demonstrated that only the Moon was explicitly provided with epicyclic gearing, although for each planet an epicyclic wire was specified in length. In addition, we have the text's reference to a *linea* for each planet, and the mention of what are clearly eccentricities. This all suggests that if the instrument was not completely successful in giving *true* planetary positions, at least it was capable of doing so when supplemented by the residuum of the Ptolemaic constructions. Whether the construction lines comprised drawn lines, wires, stretched threads, or even sliding rules (as in the Dondi clock) it is impossible to say; but it seems fairly obvious that the author was not satisfied with depicting mean planetary positions alone.

Accepting the hypothesis which led to Fig. 2, the planets do not reach to the Zodiac scale; nor can they do so on any hypothesis, if epicycle and eccentric are to be taken into account. The *signiferi*, if disks, probably provided a light background against which the construction lines, whatever they were, could be seen. (If these were sliding rules, with the first hypothesis, they could not all reach to the Zodiac scale without each obstructing the rest.)

10. *On the possibility of identifying the instrument with one mentioned elsewhere in history.*

Although there are records of at least two public clocks in Italy (namely in Ancona and Milan) in the first decade of the Fourteenth Century, the only evidence that they were not of a purely hydraulic design is that one of them was said to possess a 'great wheel', and that the other was of iron — and neither remark is very conclusive. Of relevance to the text discussed here, would be the nature of their dial-work, but this we are never likely to discover. No treatise on a tower-clock, however, is likely to have begun 'Fiat columpna...', or to have required a knowledge of Ptolemaic astronomy for its proper use. A much more likely candidate for identifying with the object described in the text is a gift made to the emperor Frederick II, in 1232, whilst he was in Apulia (in the heel of Italy). In that year, ambassadors of al-Ashrâf, Sultan of Damascus, presented Frederick with some sort of planetarium. (The Sultan was no doubt grateful for the fact that Frederick's exploits in the Holy Land were of a diplomatic, rather than a military, kind.) The planetarium is alluded to, or described in, at least four medieval chronicles. There is a brief mention of the gift (*xenia*), by Ricardus de S. Germano, although he has nothing to say of its nature[13]. Conradus de Fabaria, Abbot of Saint Gall, was told by Frederick that, next to his son Conrad, the 'astronomical heaven, of gold stellated with gems', which had 'within itself the course of the planets', was the possession he held most dear[14]. The Cologne Chronicle speaks of the instrument at greater length (for translation, see the first half of the Trithemius quotation, below, which is essentially the same):

Soldanus Babilonie imperatori mittit tentorium mirifica arte constructum, in quo imagines solis et lune artificialiter mote, cursum suum certis et debitis spaciis peragrant, et horas diei et noctis infallibiliter indicant. Cuius tentorii valor viginti millium marcarum precium dicitur transcendisse. Hoc inter thesauros regios apus Venusium est repositum.

('Babilonia' was the name of a Byzantine fortress near Cairo, and hence a synonymn of 'Egypt'. Venosa is a town in Apulia). It seems likely that this was one of the sources used by the later Benedictine historian Trithemius (1462-1516), Abbot of Saint Jakob at Wurzburg. It does not seem to have been his only source, however, unless it was his imagination which led him to distinguish between two gifts — a *horologium* given by the Sultan of Damascus, and a *tentorium* given by his brother the Sultan of the Egyptians. For this reason, he is worth quoting in full, on the nature of the *tentorium*:

---

[13] *Chronica regia Coloniense, continuatio IV* (Mon. Germ. Hist., Scriptores rerum Germ. in usum scholarum, XVIII, 1880, p. 263). The same text, rendered somewhat less accurately, is in J. L. A. Huillard-Bréholles, *Historia Diplomatica Friderici II...*, vol. IV, p. 369. This work contains the quotation from Ricardus de S. Germano, but without references.
[14] *Conradus de Fabaria casus S. Galli*, cap. 14 (Mon. Germ. Hist., Scriptores, II, 1829, p. 178).

In the same year, the Sultan of the Egyptians sent by his ambassadors as a gift to the emperor Frederick a valuable *tentorium*, constructed with remarkable ingenuity, estimated to have been worth more than five thousand ducats. For internally it was made to resemble the celestial spheres, in which moved likenesses of [or 'signs for', *imagines*] the Sun, Moon and the other planets, fashioned with the greatest skill, and set in motion by weights and wheels. [They moved] in such a way that, describing their courses in fixed and definite periods, they showed the hours, both by day and night, with infallible accuracy [*infallibili demonstratione*]. Moreover, the twelve signs of the Zodiac, with their distinguishing marks [*certis distinctionibus suis*], which move with the firmament, contained within them the course of the planets. [15]

There is no doubt that this description could be applied to the reconstruction suggested in connexion with Fig. 2. *Tentorium* would normally be translated as 'tent' (syn. of '*tabernaculum*'), but any object with a covering plate might have merited this name, and there is no reason why tent-like hangings, attached to the periphery of the star map, should not have been used to hide the wheel-work. Dare we add that tents often had a ball on top of the pole? The 'celestial spheres' does not necessarily mean that globes or armillaries were used: the term was always used very loosely by astronomers, and '*sphera*' is as often as not to be translated as 'circle'. The mention of 'weights and wheels' need not imply a weight-driven mechanical clock: the description could be applied equally to the anaphoric clock. It is, of course, always possible that Trithemius embellished his story to the point where is was more magnificent than the clock itself, but it is not obvious that this was so, even though his writings as a whole are not renowned for their accuracy. The story has a louder ring of authenticity now that the present text has come to light.

The subsequent fortunes of the two gifts cannot be traced after 1248, the year in which Frederick was humiliated by the men or Parma. The town was being besieged by Frederick when the townsmen, taking advantage of his absence on a hunting expedition, made a sortie and stormed the Emperor's camp. His forces were scattered; his treasury, his insignia, his harem, many of his ministers, and all his personal possessions were captured. Frederick's spirit was broken by this disaster[16]. He died less than three years later, without recovering any of his lost treasures.

---

[15] From the Latin, quoted in J. Beckmann's *A History of Inventions, etc.* (4th ed.), London, 1846, vol. I, p. 350, n. 1. The translation given there is much too free.

[16] He need not have grieved for his 'astrologers, magi, and soothsayers', lost in the sortie, for they had predicted the day and month of his own victory. He was persuaded by them to call the wooden township built for the siege 'Victoria', hence the title of the song, hostile to Frederick, from which this information comes: *Carmen de Victoria eversa*, section III (Mon. Germ. Hist., scriptores, XVIII, 1863, p. 796).

What reasons are there for thinking this '*opus rotarum mirabilium*' to have been connected with Frederick, 'immutator mirabilis' as well as '*stupor mundi*'. Next to nothing is known of the frequency of occurrence of this sort of device in the late Thirteenth Century, and nothing could be more ludicrous than to claim the present text as a description of one of Frederick's gifts, merely on the grounds that they were both planetaria with their century in common. If this is a description of the *tentorium* it must have been written by a person with a good astronomical knowledge. He might even have displayed more learning than was possessed by the man who originally designed the instrument, a possibility which might explain away the quoted eccentricities, which on one hypothesis (that of Fig. 2) we had to ignore. The latitude of 45°, quoted in the associated astrolabe text, would, of course, be appropriate to the place where the description was either made or copied. Is it a mere coincidence that Parma is within a third of a degree of this latitude? The evidence is slight, and unless more comes to light this must remain an open question.

Tending to prejudice us against this imperial connexion would be evidence of other thirteenth-century planetaria of a similar sort. One cannot help feeling that, beneath the dust of those thousands of medieval documents which survive unread, there are many other descriptions of similar devices. There are one or two historical traditions that such things were known in the early Middle Ages, but all relatively ambiguous. The eighteenth-century writer Hamberger, who composed the section on clocks and watches in Beckmann's book, cited above, referred to two authors of similar devices. Gerbert, Pope Sylvester II, ' in Magadaburg horalogium fecit' (according to the *Chronicon* by Dithmar). Hamberger rightly questioned that this was anything but a sundial[17]. William of Malmesbury, however, who was not quoted by Hamberger, refers to objects made at Rheims by Gerbert to illustrate his teaching, and in particular to a '*horologium arte mechanica compositum*'.[18] Dr. Harriet Pratt Lattin has drawn my attention to a seventeenth-century notice of the abbey of St Rémy of Rheims, which, perhaps in continuation of the long tradition that Gerbert built a clock at Rheims, mentions a '*horologium ...Gilboti (sic) dictum tinnientium campanularum modulatione, singulis horae quadrantibus concentum edit suavissimum, etc.*'[19]. All this is a far cry from a geared planetarium, and involves nothing incompatible with Gerbert's having built an anaphoric clock. Miss Lattin

---

[17] This fits ill with the desire of later writers to have this timepiece created 'by diabolic art'. See, for example, Wilhelm Maclot, *Metropoli Remensi*, vol. II.

[18] Mon. Germ. Hist., Scriptores X, 1852, p. 462.

[19] The reference is to Paris, Bibl. nat. lat. MS 11819, f. 71.

has in fact collected much information on Gerbert's astronomical instruments. She points out that he made a celestial globe of wood, which he covered in horsehide. This is reminiscent of the parchment-covered sphere of the present text. The stars and circles were painted in colour on this globe, which was one of his aids in teaching astronomy. Gerbert also cites 'the construction of a sphere most suitable for recognising the planets'. The description of this is again intriguing from the point of view of our text, for it runs as follows:

> Within this oblique (Zodiac) circle he hung the circles of the wandering stars (planets) with wonderful ingenuity, whose orbits, heights, and even the distances between them he demonstrated to his pupils most effectually. [20]

Does 'sphere' perhaps refer to a plane projection of the celestial sphere? How else could the planets be within the Zodiac, unless the sphere was an armillary? Were the planets 'hanging' or 'projecting'? (Cf. the dependeat' of our text). For the time being, these questions are unanswerable. Hamberger's remaining reference with any probable relevance to our text was to the anonymous author of the life of William, abbot of Hirsau in the Eleventh Century[21]. This author says of him: '*naturale horologium ad exemplum caelestis haemispherii excogitasse*'. Hamberger, rather extravagently, comments that 'it is evident that it alludes neither to a sun-dial nor to a water-clock, but to some piece of mechanism which pointed out the hours and exhibited the motions of the Earth and other planets'[22]. In fact, so far as this fragmentary passage allows any pronouncement on the nature of the timepiece, it appears likely to have been nothing more than the conventional anaphoric clock, with the lower part of the fixed plate of hour-lines (i.e. that corresponding to the sky below the horizon) covered over, leaving only the celestial hemisphere in view.

In concluding this section a word should be said on Giacomo de' Dondi (pre 1293 to 1359). It is unlikely, but not impossible that he was in some way connected with our text. If so, this would perhaps prejudice us in favour of the arrangement of Fig. 3., on the assumption that this influenced the younger Dondi in his use of separate dials for each planet[23]. Giovanni's *Tractatus Astrarii* nevertheless fails to use these

---

[20] For references to astronomical instruments attributed to Gerbert see Miss Lattin's symposium notes in *Medievalia et Humanistica* (1955) 13-17.
[21] Published by Stengelius, Vienna 1611.
[22] Op. cit., p. 346.
[23] Admittedly Dondi claimed that in using separate dials he was following the practice of Campanus, whose equatorium had a separate plate for each planet. On the Dondi family, for reference

words characteristic of our text: 'volvellum', 'sperula', and 'signiferus'. And above all it is difficult to reconcile the evidence for dating the instrument with even the older Dondi's authorship.

NOTE ADDED IN PROOF

When I first wrote this article I overlooked an important piece of evidence, which as things stand is hidden in note 22[1]. It now seems to me highly probable that MSS α and β have the correcte reading, and that Saturn is to move round a Zodiac with Capricorn at the top. This suggests strongly that the face of the device was in a *vertical plane*, and that the stereographic projection used on the astrolabe face was from the *South pole* (as on conventional astrolabes, and the anaphoric clocks, as well as on Richard of Wallingford's clock). Many clocks in northern Europe surviving from the late Middle Ages show projections from the North pole. There are in fact at least sixteen categories, in any of which a clock with an astrolabe face might be placed. Is it in South or North polar projection? Are the zodiacal signs as seen, or is their order reversed? Is the rete one of hours or of stars? Is it the place or the rete which moves? Probably half of these sixteen possible arrangements have been used in practice, at one time or another, and this fact should eventually prove useful in determining lines of technological influence. It should hardly be necessary to add that horological traditions will never be accurately charted until those who describe the archaeological remains of old clocks also record gear trains in a succinct notation, such as that recommended here.

---

to works discussing them, and especially the history of Giovanni's clock, see S. A. Bedini and F. R. Maddison, *Mechanical Universe*, a forthcoming volume to be published shortly by the American Philosophical Society (*Transactions* lvi (1966) part 5). On the disputed origins of the titule 'dall'Orologio' see p. 19 of this work.

APPENDIX

THE GEARED ASTROLABE IN THE SCIENCE MUSEUM, LONDON.

The only archaeological remains which are in any way close to the device described in the North Italian text are those of the geared astrolabe in the Science Museum, London. This astrolabe (which is French, c. 1300) has been fully described by Gunther[24] and has often been cited in connexion with the early history of the clock, where there is a sad poverty of well-authenticated work of the early Fourteenth Century and before. So far as I know, however, no-one has previously tried to determine the original design, and for this reason some remarks on a possible reconstruction are included here.

The fragment which survives is depicted in the schematic drawing of Fig. 5. (The scales are not shown). The outer circumference bears a calendar scale, and a scale of 1-12 plus 1-12. There is a badly-drawn scale near the centre giving the age of the Moon (0-29, with some remainder). This suggests an instrument to show conjunctions. An arm AB carries the following train (beginning with the fixed outer ring of 180 teeth):

$$\frac{(?)}{180} \cdot \frac{39}{39} \cdot \frac{27}{45} \cdot \frac{14}{} \cdot$$

The arm has, near its pointed end, the word '*nader*'. The other end, which now fails to reach the scale, was therefore presumably meant to mark the Sun's position. Putting a pointer for the Moon's age, and a wheel of 24 at the centre (which measurements show to be compatible with the assumption of roughly equal tooth-divisions, verified for the existing wheels) the following Moon-Sun train results:

$$\frac{(365\ 1/4)^d}{180} \cdot \frac{39}{39} \cdot \frac{27}{45} \cdot \frac{14}{14} \cdot \frac{24}{(29\ ^{11}/_{50})^d} \cdot \text{(Periods relative to the Sun' arm).}$$

The figure of $(20\ 11/50)^d$ which is to be compared with the accepted figure of $29^d \cdot 53$ for the Moon's synodic period, is of an accuracy which is

---

[24] *Astrolabes of the World*, vol. II, p. 347 and plates lxxx, lxxxi.

poor by comparison with the overall accuracy of the North Italian device. (If, as is more than possible, the year is to be taken as 365^d, the figure of 29^d.2 results, which is very slightly worse). The shorter end of the Sun pointer appears to have been broken at a small circular hole, in which another wheel was surely pivoted. If so, the dimensions of the instrument suggest a wheel of about 25 teeth (p, say) meshing with the ring of 180. There is no axis evident for wheels intermediate between this and the otherwise unattached wheel of 32, in the centre; but an offset pair (m, n), as on the other radius, might have been broken off.

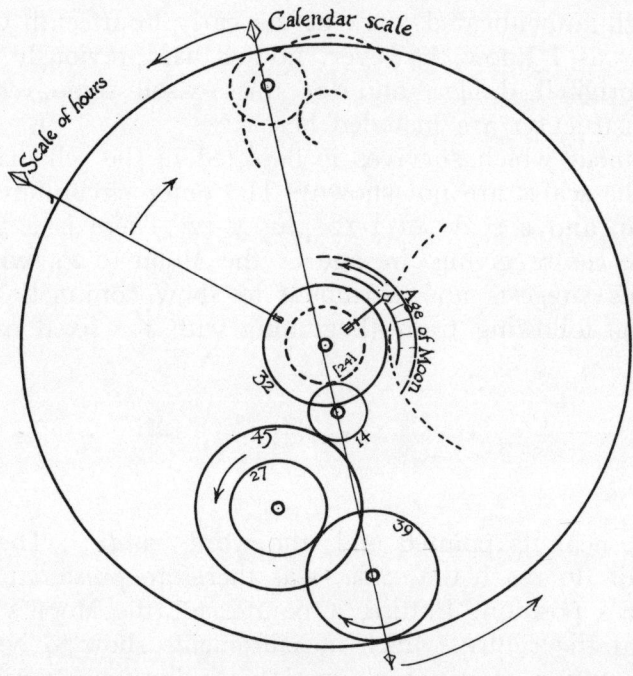

Fig. 5

We should not expect one more pair, or single wheel if we were looking for a clockwise motion for the central wheel of 32; and since we have still to explain the scale of 0-12 plus 0-12, which increases in a clockwise sense, we may assume that this is the direction of motion required. We may have to presuppose three missing pairs or single wheels in all, so long as it is possible to accommodate them in the space permitted.

All periods are here referred to the main pointer which moves round once in a period of T days (roughly 365¼ days, or more probably 365^d on this instrument). In the course of a day, the 24 hour pointer must move through a complete revolution with respect to the main arm, less the angular distance moved by the arm itself, namely 360°. $(1-1/T)$. In other words, it is permissible to write

$$\frac{(T/(T-1))}{32} \cdot \frac{m}{n} \cdot \frac{p}{p} \cdot \frac{180}{(T)},$$

where m/n may, for the moment, stand for one or more sets of wheels. With $T = 365$, this leads directly to the result:

$$\frac{m}{n} = \frac{91.32}{45}.$$

There are many ways of representing this ratio approximately, and there is no hope of discovering the ratio actually used. (One complete train which would suffice is:

$$\frac{(T/(T-1))}{32} \cdot \frac{q}{q} \cdot \frac{64}{8} \cdot \frac{73}{9} \cdot \frac{p}{p} \cdot \frac{180}{(T)},$$

where p and q are idle wheels of any suitable number of teeth. The length of the 'year' is then $T = 366$ days exactly — a very suggestive number!).

It remains a completely open question whether this astrolabe was ever incorporated in a driven timepiece, or whether the idea for it was taken from another timepiece, with similar astrolabe dial. One or the other seems highly probable, a fact which adds to the intrinsic value of the object.

12

MONASTICISM AND THE FIRST MECHANICAL CLOCKS

Had I been speaking in an ancient Athenian law court rather than in modern Japan, my address would have been timed, not by a chess clock but by a clepsydra. Aristophanes[1] and Aristotle[2] both testify to its use in the courts, and the custom was still remembered in the time of Lucian, or a little later, when he or an imitator reported that Demades had made fun of Demosthenes for preferring water to wine. 'Others spoke to water, but Demosthenes wrote to it'.[3] Roman senators timed their discourse by means of the clepsydra, as Pliny, Cicero and others bear out; and Cicero indicates that the very acts of asking and giving leave to speak were described, respectively, as 'seeking the clock' and 'giving the clock'.[4] The clepsydra was used in ancient Greece for timing military watches[5] and for astronomical measurement.[6] It is said that according to Lucian it was used — and if this were not so dubious a reference it would be the oldest historical reference to such a use — for sounding a bell.[7] There is nothing intrinsically surprising about a Greek hydraulic automaton capable of sounding a bell at regular intervals, for Ctesibius had previously, by means described in some detail by Vitruvius, that is, by hydraulic timepieces, caused figures to move, pillars to turn, stones and eggs to fall, trumpets to sound, and other displays *(parerga).*[8] Other winter timepieces described by Vitruvius, driven likewise by water power, required a measure of astronomical understanding if they were to be used to yield the time, having as they did an astrolabe dial.[9]

1. *Vespae,* 93. 857.
2. *Athen. Polit.,* 67. 2.
3. *Demosthenis Encomium,* 15. The work is probably of the second or third century A.D.
4. *De Oratore,* 3. 34. 138.
5. Aeneas Tacticus (4th cent. B.C.) ed. R. Schone, Leipzig 1911, 22-24.
6. See, for example, Proclus: *Hypotyposis astronomicarum positionum,* ed. C. Manitius, Leipzig 1909, 4-74.
7. This statement is made in the *Enc. Brit.,* 11th edn., art. "Bell", by H.M. Ross, who probably took it from the best-selling book by A[lfred] G[atty] : *The Bell; its Origin, History and Uses* London, 1847, p.16. In neither place is a reference given, but the source was very probably chapter 6 of Hieronumus Magius [or Maggi] : *Anglarensis de tintinnabulis,* many editions; but that of Amsterdam 1689 is illustrated with an imaginary reconstruction, at p.31. A rope from a clepsydra float trips a weight which operates a bell once only (through a crank!). Maggi, writing from prison, gives no precise references, but says that his relative Johannes Nicolaus Justus made him a copy from an old book in which the Lucian clock was delineated.
8. *De Architectura,* IX. viii. 4-7. Vitruvius was most probably writing a few years before 27 B.C.
9. *Ibid.,* 8-14.

I mention these early literary references not merely as an introduction to the medieval scene, but as a reminder that a timepiece is much more than a mechanism. To attempt to understand it in isolation from its human setting is to forget that it was made in the first place in response to specific human needs. As those needs altered, its form tended to change, and there might be times when its very survival was in jeopardy. It is doubtful, for example, whether in a simple agricultural society much meaning would be found in an Athenian clepsydra, let alone in a Vitruvian astrolabe dial. However dark or bright a historian might be inclined to find the first few bucolic centuries following the final collapse of the Roman Empire, one thing seems certain: if the Vitruvian tradition was handed down within the West, it was by the chance preservation of text or artefact in a society hostile to the civilization which gave birth to the underlying rationale. In Islam conditions were ostensibly more favourable, and the tradition of hydraulic automata was positively enriched, helped by a religion that encouraged astronomical expertise of a high order,[10] not to say by a climate that seldom encouraged water to freeze. In due course, Europe became more conscious of the need to learn from the past, and something significantly new was added to the ancient tradition of anaphoric clocks, this without much by way of the mediation of Islamic thought. The institution chiefly responsible, both for preservation and for development, was the Church.

This should not be very surprising. The Church was rich and powerful. It controlled almost all academic education. It could afford to employ the best available craftsmen, and numbered among the lay brethren attached to the monasteries must have been some of the most skilful artisans of the time.[11] The monastic orders had played an important part in the development of the many mills and contrivances which could be powered by a waterwheel, which thus made more time available to the monks for their proper vocation of prayer and meditation. (The Cistercian rule enjoined monks to build near rivers so as to make best use of water power.) The administrative machinery for such a vast enterprise as the construction of a large mechanical clock was available nowhere outside the Church and the courts of princes. The transmission of power by means of gearing, rope, and pulley, was admittedly a part of a common stock of craft-knowledge from ancient times. One must be wary of exaggerating the conceptual difficulty inherent in these ideas. The principle of the rope drive, for example, should be immediately evident to anyone who lets slip the windlass as he hoists up a bucket from a well.[12] But the mathematics of gear-trains — and this is especially true of astronomical trains — were the province of the well educated alone. And education of the appropriate sort was the monopoly of the Church.

The Church was a feudal force, and through the close regulation of the monastic day a measure of regularity was imposed on society at large. With or without automatic control, the canonical hours of the monastic life were struck eight times daily on a tower bell which, in summoning the monks to prayer by day and by night, was heard far beyond the confines of the cloister. The rules for the mar-

10. Astronomy served not only astrology, a subject to which Islamic scholars added significantly, but among other things it was required by any who would master the lunar calendar adopted by the faith.
11. The Abbot of St Albans, Richard of Wallingford, whose work I shall shortly describe, was the son of a blacksmith.
12. One might imagine that the principle of the water-driven wheel was arrived at by the same sort of accidental discovery of reversibility, in this case of such a water-raising device as, for example that which appears in Vitruvius: X.4 (the *tympanum*).

ket of Salisbury, for example, in the early years of the fourteenth century, refer specifically to the striking of the cathedral clock,[13] and there are even cases where the Church obliged the townsmen to maintain a church clock. I do not want to give the impression that people outside the churches had no wish to keep their own time: in Cologne, for instance, a guild of water-clock makers was already in existence in 1183, while by 1220 they occupied a whole street, the *Urlogingasse*.[14] Within the Church, however, timekeeping — which had at first been no more than an aid to the regulation of worship — soon became almost a necessary ingredient of ritual. In due course, timekeeping was to encompass the drama of a mechanical cosmos, combined with a wide range of more earthly amusements: striking jacks, jousting knights, wheels of fortune, and in fact all that the Vitruvian word *parerga* might have signified. Not that we should exaggerate the element of drama, or melodrama, to the exclusion of all else. If the concept of precision in timekeeping had been unknown in the fourteenth century, Chaucer could hardly have written, as he did in the *Nun's Priest's Tale,* of the cock Chaunticleer thus:

> 'Wel sikerer was his crowyng in his logge
> Than is a clokke or an abbey orlogge.'[15]

In lands where the Sun does not always shine, and in a community for which the first cockcrow of the day was not early enough for the call to matins, some sort of clock was sorely needed. A classic list of references to early examples has gradually accumulated, and I have no wish to tread more than absolutely necessary on already well-trodden ground, but it is impossible to appreciate the mechanical clock without a knowledge of the role played by its predecessors. There is no good reason to suppose that knowledge of the water clock in the West depended on its importation from Islam. Considerable doubt attaches to the story that there were diplomatic relations between Charlemagne and Harun al-Rashid, as are presupposed by the legend that the Caliph sent the Christian emperor a water-clock, a silk tent, and so on.[16] The oldest detailed account of the construction of a water-driven alarm is in a tenth or eleventh-century manuscript, now unfortunately incomplete, from the Benedictine monastery of Santa Maria de Ripoll, at the foot of the Pyrenees.[17] The text does not appear to be a translation from an Arabic original. There is every indication that the hydraulic driving mechanism, of which the description is lost, did not turn any astronomical dial, but merely a dial to help in setting the alarm. The weight-operated striking mechanism was very simple: an ordinary rope-and-weight drive turned on an axle which acted as a flail on small bells hanging from a rod. This very primitive device had to be re-set after each use. This re-setting was perhaps done by the sacristan, as is explained in the Cistercian Rule which dates from the early part of the twelfth century.[18] In Rule XCIV, the

13. The sources relating to the Salisbury clock are to be found in C.F.C. Beeson: *English Church Clocks, 1280 to 1850,* Antiquarian Horological Society 1971, p.16.
14. Quoted from E. Volkmann by Lynn White Jnr.: *Medieval Technology and Social Change,* Oxford 1962, p.120.
15. *Canterbury Tales,* Fragment VII, lines 2853-4.
16. For the historical controversy see E.A. Belyaev: *Arabs, Islam, and the Arab Caliphate in the Early Middle Ages,* tr. from the Russian, London 1969, p.221. Belyaev is perhaps too extreme in his criticism of the 'traditional opinion inspired by Christian pietism'.
17. A full discussion and translation of the surviving part of the text was given in F. Maddison, B. Scott and A. Kent: 'An early medieval water-clock', *Antiquarian Horology,* 3 (1962), 348-53.
18. I quote from C.B. Drover: 'A medieval monastic water-clock', *Antiquarian Horology,* 1 (1954) 54-8.

sacrist was instructed to set the clock *(horologium temperare)* and cause it to sound *(facere sonare)* on winter weekdays before lauds, unless it was daylight. He was to use it to awaken himself before vigils each day, before lighting up the church. In Rules LXXIV and LXXXIII, the brethren and sacrist were told to ring the larger bell *(signum* or *campana)* on hearing the clock *(horologium)*. These are not the earliest constitutions known that relate to the subject. In the eleventh century, William, abbot of Hirsau, gave similar instructions to the sacristan, using words which echoed the Cluniac rule and also the ancient customs of the monastery of St Victor in Paris, where the registrar *(matricularius)*, the sacristan's companion, was to adjust the clock.[19] Such adjustment was necessitated by the use of unequal hours.[20] Similar commentary on the Benedictine Rule confirms that the same customs were adopted generally.

There can be little doubt that mechanisms were operated by water. The supporting evidence is well known, and has been ably summarized by C.B. Drover.[21] There was the fire of 1198 at Bury St Edmund's abbey, which the clock doubly helped to extinguish, first by rousing the master of the vestry, and secondly by providing water. There are the fragments of slate dating from 1267 or 1268 from Villers Abbey near Brussels, relating to the method of setting a water-clock by the Sun. There are the Alfonsine books of the next decade, which describe both a water-clock and a mercury-clock.[22] There is the clock case of about 1250, drawn in the sketch book of Villard de Honnecourt. And then there is the illustration to which it was the aim of Drover's article to draw attention, an illustration of almost exactly the same period as Villard's, showing a similar clock-case.

The illustration is from a moralized Bible of extraordinary richness.[23] The prophet Isaiah is shown giving the sick king Hezekiah a sign from the Lord that fifteen years will be added to his days. This was done 'that the Sun would be moved back ten degrees in the clock'.[24] I shall discuss this illustration briefly here since it seems to me that it has been misinterpreted by Lynn White Jnr. in a work which is so well known and generally so accurate on points of detail that his version is likely to become canonical.[25]

The medallion illumination illustrates 2 Kings XX.5-11. The same story is told in Isaiah XXXVIII.8.

19. These earlier references and those next following are given in greater detail in John Beckmann: *A History of Inventions, Discoveries and Origins,* tr. W. Johnston, 4th edn., London 1846, pp.346-9. This is a fundamental source of information on the history of the clock. The section was actually written originally by Hamberger.
20. See 33 below. Dante has something to say on equal and unequal hours, and the difficulty of regulating church services, in *Convivio* IV.23, near the end.
21. *Ibid.,* passim.
22. Mr. Francis Maddison, who is preparing an English translation (with commentary) of the text describing the Alfonsine mercury clock, points out that the inspiration for this clock is stated in the text to be a work by 'Iran el filosofo', namely Hero, more specifically where he explains ways of lifting heavy weights.
23. Bodleian Library, Oxford, MS Bodley 270b, f.183v. For a facsimile of the entire MS and its missing parts (B.N. Lat. MS 11560, and B.M. MS Harley 1527) see A. de Laborde: *La Bible moralisée,* Paris 1911-27, 5 vols.
24. '...ut Sol x gradibus retrorsum in orologio reuerteretur'.
25. *Op.cit.,* pp.120-1.

Monasticism and the First Mechanical Clocks 175

The Bodleian manuscript is one of three needed to complete the original Bible, and the Isaiah passage is to be found in the complementary manuscript in the Bibliothèque Nationale, MS Lat. 11560, f.120r, where, sad to say, there is no comparable illustration, owing to the different wording of the Vulgate.[26] Both illustrations show a symbol denoting the Sun, which is emphatically not a 'fan-escapement to slow the action of the chime, at the striking of the hours, by friction with the air'[27] but is a representation in accordance with a perfectly standard convention.[28] The fifteen divisions of the only visible wheel might be significant, but in view of thirteenth-century artistic conventions, it is unlikely that the number does more than pick up the '15' of the years mentioned in the text, as seems to have happened in the Isaiah illustration. It is certainly rash to conclude that since 15 degrees represent an equinoctial hour, therefore the wheel was probably meant to turn once every hour. In any case, the wheel seems to have about 24 teeth, which rather suggests that it might have been meant to turn once in a day.

The clock clearly has some sort of rope drive, but there are several mechanical reasons for thinking that it did not work on the same principle as the Alfonsine mercury clock. (In the latter, a couple created by viscous forces, as mercury flows between radially divided compartments of a wheel, is opposed by a couple created by a rope drive, thus establishing dynamic equilibrium in the wheel, which turns slowly.) Not the least of the objections to this interpretation of the biblical painting is that water is there shown clearly gushing forth from an animal's head spout into a cistern below the clock. There are unsolved problems of interpretation, certainly,[29] but it is difficult to avoid the general conclusion that, however it worked, the clock was mechanically a simple affair, offering little more than encouragement to the men who would make the first purely mechanical clocks.

Within a century, however, two clocks were begun, by Richard of Wallingford and Giovanni de' Dondi respectively, which were so extraordinarily complex that in the sixteenth century the first could be described as even then surpassing all others in Europe,[30] while the second was so intricate that Charles V could find only one technician, Gianello Torriano, who was capable of repairing it, others having failed.[31] How may we explain such a technological advance, for which there were very few parallels in the Middle Ages and few indeed before the industrial revolution of the eighteenth century?

26. There is no mention of a *horologium*, but the mention of *gradus* prompts the illustrator to paint a flight of 15 or 16 stairs!
27. White, *op.cit.*, p.121.
28. The same convention is to be seen on numerous occasions in the manuscripts (Bodley 270b, ff. 10r, 16r, 34r, 57v etc.), especially in connexion with crucifixion scenes. It was still being followed more than a century later. See, for example, Bodleian Library, MS Ashmole 1522, ff. 27r, 39v, 40r.
29. Drover was puzzled at the cranked form of the arm supporting the wheel, but a comparison with the astronomical instruments of ff.11r, 24r and 27r suggests that this was the standard way of supporting a wheel, resulting in a 'handle' in the plane of the wheel.
30. The antiquary, John Leland. No references to the material on Richard of Wallingford are given here. My complete edition of his writings is awaiting publication at the Clarendon Press, Oxford.
31. The authority is Bernardo Sacco, 1565. A somewhat different report comes from the notoriously unreliable Cardano. For the texts, see S.A. Bedini and F.R. Maddison: 'Mechanical universe: the astrarium of Giovanni de' Dondi', *Trans. of the American Philosophical Society*, N.S., 66, part 5 (1966), 37-9.

I must first remind you of that remarkable passage to which Lynn Thorndike first drew attention, in the commentary written by Robertus Anglicus in 1271 on the most widely used of all medieval astronomical textbooks, the *De Sphera* of Sacrobosco.[32] After a discussion of equal and unequal hours,[33] Robert goes on at some length:

> Nor is it possible for any clock *(horologium)* to follow the judgment of astronomy with complete accuracy. Yet clockmakers *(artifices horologiarum)* are trying to make a wheel which will make one complete revolution for every one of the equinoctial circle [i.e. the celestial equator], but they cannot quite perfect their work. If they could, it would be a really accurate clock and worth more than an astrolabe or other astronomical instrument for reckoning the hours if one knew how to do this according to the method aforesaid.
>
> The method of making such a clock would be this, that a man make a disk *(circulum)* of uniform weight in every part, as far as could possibly be done. Then a lead weight should be hung from the axis of that wheel, and this weight should move that wheel so that it would complete one revolution from sunrise to sunrise, minus approximately as much time as it takes about one degree to rise.[34]

This all suggests that no form of mechanical escapement was known to the writer in 1271, and the simple arrangement he describes is not incompatible with the water-clock illustration discussed earlier. Within a few years, however, the number of documentary references to *horologia* grows so very rapidly that we can only suppose the mechanical escapement to have been found at last. C.F.C. Beeson is persuasive in arguing that the earliest of all European records of a clock with such a control is that of 1283, in the Annals of Dunstable Priory in Bedfordshire.[35] This was a house of Austin canons. The clock was set up alongside a great painted crucifixion scene, with attendant images of Mary and John, on the rood-screen and loft, or gallery. Beeson follows with records from Exeter Cathedral (1284), Old St Paul's, London (1286), Merton College, Oxford (1288?), Norwich Cathedral Priory (1290), Ely Abbey, a house of Benedictine monks in Cambridgeshire (1291), and Christchurch Cathedral, Canterbury (1292), all before the turn of the century. Taken singly, the records are easy to view with scepticism, but taking them together, and noting especially that relatively large sums of money are

32. The name Robertus Anglicus points to a man of English family, but the commentary was given as a course of lectures at the university of either Paris or Montpellier.
33. Sometimes called 'equinoctial' and 'canonical' or 'seasonal' respectively. The former are in accordance with our modern convention, each being one twenty-fourth part of a day. The latter are each one twelfth part of day or night. This convention goes back to the ancient world and survived in Japan until the last century at least. A night hour is obviously approximately equal to a day hour only twice a year, near the equinoxes. It should not be thought that the concept of equal hours ('horae de clock', in the late fourteenth century) had to wait for the invention of the mechanical clock, as is sometimes suggested. Astronomers made use of the idea in Antiquity, and it was known in the West in the early Middle Ages from the writings of Martianus Capella, Leontinus, Gerbert and many others. William of Hirsau's *naturale horologium* was probably so called because it showed the equal hours of the 'natural day' (24 hours), rather than of the 'artificial day' (sunrise to sunset).
34. Taken, in a form not significantly altered, from L. Thorndike: *The Sphere of Sacrobosco and Its Commentators*, Chicago 1949, pp.180 (text) and 230 (translation).
35. *English Church Clocks*, Antiquarian Horological Society, 1971, pp.13-14.

involved in payment for the materials used, they persuade us that the mechanical clock had indeed arrived on the scene. When the *orologiarius* Bartholomew drew 281 rations for three quarters and eight days in 1286 at St Paul's, he was surely not building either a sundial or a water clock.

Although it is possible to be reasonably precise as to the time of the invention, the place of origin of the mechanical clock is entirely unknown. Italy has been canvassed, mainly, one suspects, on the grounds that Italy was always a century in advance of the rest of Europe. The earliest acceptable Italian record of which I am aware, however, relates to the year 1309, when an iron clock *(horologium)* was set up in Sant' Eustorgio in Milan.[36] A bell on the bridge at Caen was in 1314 associated with a clock *(l'orloge)* there and according to its inscription served the common people, but this need not have been a purely mechanical clock, and in the absence of further documentation its watery surroundings do not encourage the idea that it was so. The known early English records are at the present time much the richest in Europe, and I am obliged to give most of my attention to them; but I certainly do not suppose that the mechanical escapement was for this reason an English invention. I find it hard to believe, nevertheless, that any early centre of clock-building could have been more advanced than that which took in Norwich and St Albans, and most of what I have to say will relate to the remarkable work done in these two places.

The Sacrist's Rolls of Norwich Cathedral from 1321 to 1325 contain the first extensive financial records concerning the construction and installation of a large mechanical clock.[37] The man in charge of the work was one Roger Stoke, who later worked at St Albans, and who was in both places assisted by Laurence Stoke. The clock had a very large astronomical dial — it was of iron plate and weighed 87 lb — with models of the Sun and Moon, automata, including 59 sculpted images (done by one Adam, a wood-carver), and a choir or procession of monks. There was much colouring and guilding. Smiths, carpenters, masons, plasterers and bell-founders were engaged over a period of three years. The competence of most of the craftsmen concerned seems to have been equal to the occasion, but the making of the main astronomical dial went less smoothly than the rest. In 1323 the fabrication of the large plate was entrusted to Robert of the Tower (Robert de Turri) in London, but in his hands the whole work was ruined. The man was himself ruined *(depauperatus)*, and only 10 of the 18 shillings advanced to him could be recovered. Other artisans proved to be equally ineffectual, ruining the material in their attempts. Men were sent from Norwich to London for news of progress, but at length it was necessary for Roger Stoke himself to ride to London to supervise the engraving of the plate. The total cost of the clock was in excess of £52. There are many ways of working out a modern monetary equivalent, none very satisfactory; but in terms of the salary of the best craftsmen of the time, this amounts to around $250,000, in modern American terms.

36. Mentioned in the Chronicle of Galvano Fiamma. See L.T. Belgrano: *Degli antichi orlogi pubbici d'Italia*, Archivo Storico Italiano, 3rd series. 7, Florence 1868. The clock was restored in 1333 and 1555, and renovated in 1572. See J. Drummond Robertson: *The Evolution of Clockwork*, 1931, p.31.
37. Mr. C.B. Drover kindly provided me with transcripts and photographs of the relevant sections. The Roll for 1323-4 is missing. There are extracts printed in *Archaeological Journal*, 12, 1855, and in the Centenary Volume of the Norfolk and Norwich Archaeological Society, 29, 1946.

Complex as the Norwich clock obviously was, it was as nothing by comparison with that designed by Richard of Wallingford, however similar the two might possibly have appeared outwardly. Before considering the mechanical intricacies of the St Albans clock, I should like to consider that aspect of the clock which gave so much trouble to the Norwich builders, namely the dials. These had the same general appearance as an astrolabe, showing the daily rotation of the sky — the stars, and perhaps the Sun and Moon — against a second 'map' of the observer's coordinate lines and unequal hour lines, the horizon and the meridian being chief of these. We know that the Vitruvian anaphoric clock had such a dial,[38] and it is of some interest to ask whether the ancient tradition was ever completely lost.

I shall offer two tentative arguments for supposing that the tradition had some sort of continuity. The first rests on the fact that there is documentary evidence for the survival from Roman, as opposed to Islamic, sources of the tradition of mapping the sky in stereographic projection in the manner required for the main dial — the most difficult theoretical aspect of the construction. Parts of two discs from anaphoric clocks have been found, one at Salzburg, and one at Grand in the Vosges, both of the 2nd century A.D.[39] Judging by the Salzburg disc, the Sun was not moved mechanically with respect to the stars, but was simply plugged into one of 360 or 365 holes distributed around the ecliptic. Other astrolabe dials are those of the water-clocks of Islam, but here one must be cautious and remember the fallacy of *post hoc ergo propter hoc*. In the Islamic world the dial generally resembled the conventional astrolabe, with its pierced star map placed between observer and the plate of local coordinates. The same seems to have been true of the thirteenth-century Alfonsine dials, which were presumably inspired by Islamic sources.[40] In the European tradition it seems to have been far more often the other way round. In this way, it was possible to paint the constellations in as much detail as was thought desirable. The flimsy overlay of wires representing a horizon, meridian, unequal hour lines, and so forth, offered no great obstacle to vision.[41] Bearing the typical European arrangement in mind, there is a certain class of manuscript illustrations dating from the ninth century onwards which closely parallel the anaphoric clock tradition in the West, as I shall explain in a short digression.

By the ninth century, encyclopaedists like Scot Erigena and Isidore of Seville had brought the contents of several late Roman authors to the notice of the educated few, while Lupitus, Hermann and possibly

38. On the nature of the dial, its relation to the portable astrolabe, and the possibility that it was the invention of Hipparchus and came before the astrolabe (which of course is much the better known of the two), see A.G. Drachmann: 'The plane astrolabe and the anaphoric clock', *Centaurus*, 3 (1954), 183-9.
39. For references to the original literature, and for further details, see article by D.J. de S. Price in *A History of Technology*, ed. C. Singer, et al., vol.3, Oxford 1957, pp.604-5.
40. The fundamental studies of Islamic clocks are by E. Wiedemann and F. Hauser. See especially their *Uber die Uhren im Bereich d. islamischen Kultur*, Nova Acta. Abhandl. d. konigliche Leopoldinisch-Carolinsche deutsche Akademie der Naturforscher zu Halle, 100, no. 5, 1914.
41. The multitude of lines on the plate of an ordinary astrolabe (almucantars, etc.) were superfluous on an instrument which was not meant to be matched to observations, while the fine star pointers of an astrolabe rete would have been difficult to identify on a large dial high above the observer's head in a dark church.

Gerbert, in the tenth century, drew on Arabic sources for their writings on the astrolabe.[42] It is in association with the ancient writings of Aratus (as transmitted by Germanicus), Hyginus and Macrobius on Cicero, however, that we find a certain class of extremely interesting manuscript illustration of this period.[43] One fine example included with scholia to Aratus is in a manuscript of the early ninth century now in Munich.[44] At first sight this is merely a planisphere, showing the constellations pictorially, with one very clearly marked circle. This proves to represent not the ecliptic, but the galaxy, the Milky Way. On closer examination, however, no fewer than eight circles are revealed within the outer pair bounding the planisphere, all in approximate — and not accidental — stereographic projection. Whoever painted this diagram seems to have known the principles of astrolabe projection well. He was certainly not merely painting an ordinary astrolabe rete, since the Milky Way is never — at least to my knowledge — found on an astrolabe, while there is also on the painting an uncharacteristic circle with north polar distance of about 37°. This could be a zenith-track for geographical latitude 53° N,[45] although it is more probably an arctic circle set to a conventional 36° (the height of the Pole at Rhodes). The stars are, moreover, drawn with the signs in a clockwise order, and not as on an astrolabe. I shall return to this point later.[46]

A second, but somewhat different illustration of the same period, also associated with scholia to Aratus, is to be found in the Berlin Codex Phillippicus 1830 at ff. 11 and 12. It is redrawn, perhaps not absolutely accurately, by Georg Thiele,[47] and on the evidence of his drawing it is impossible to know whether the original is as carelessly drafted as it appears to be from Thiele's version. Now, however, we note that this diagram has the sense of rotation of the zodiacal signs as found on an ordinary astrolabe rete, unlike that of the Munich diagram. Yet another example of this type of illustration, which can now be examined in colour in a modern printed source, occurs in a twelfth-century Spanish

42. Gerbert, who was to become Pope Sylvester II, was a highly practical man and it is worth noting that there is mention of a clepsydra in one of his letters (ed. Julien Havet, epist. 153). Mrs. Harriet Lattin has pointed out to me in a note by Oldoin in Alphonso Ciacconius — Augustinus Aldoinus: *Vitae et resgestae Pontificum Romanorum*, Rome 1677, col. 756 — that Gerbert, as Archbishop of Ravenna, constructed a water-clock there: Horologii aquatilis, seu clepsidrae figura est Ravenna in Herculis regione, quam Gerbertus construxit Archiepiscopus tunc Ravennas. This would have been between April 998 and April 999.
43. Aratus of Soli, of the 3rd century B.C., wrote an astronomical poem based on Eudoxus, which was admired by some Roman writers. Cicero, Caesar, Germanicus and Aviennus translated it, and much still survives. There are at least four English translations. The circles which interest us (galaxy, equator, ecliptic and the tropics) are discussed at p.202 of the Loeb edition Gaius Julius Hyginus was a prolific Latin author, living in Spain or Alexandria, whose elementary astronomical treatise drew heavily on the poem of Aratus. The commentary by Macrobius on Cicero's *Somnium Scipionis* is, of course, much better known.
44. CLM 210, f.113v. This is reproduced in several places, perhaps the best illustration being that in D. Bullough: *The Age of Charlemagne*, London 1965, plate 50 (in colour).
45. There is a very different diagram, but one based on an astrolabe plate for approximately this latitude, in Bodleian MS F.1.9, f.88r. The latter was drawn at Worcester, lat. 52° 10', in circa 1130.
46. Further notes on the circles will be found with the diagram, in which they are redrawn.
47. *Antike Himmelsbilder*, Berlin, 1898, p.164.

Fig. 6. Construction lines taken from the constellation map of C.L.M. 210, f. 113v.

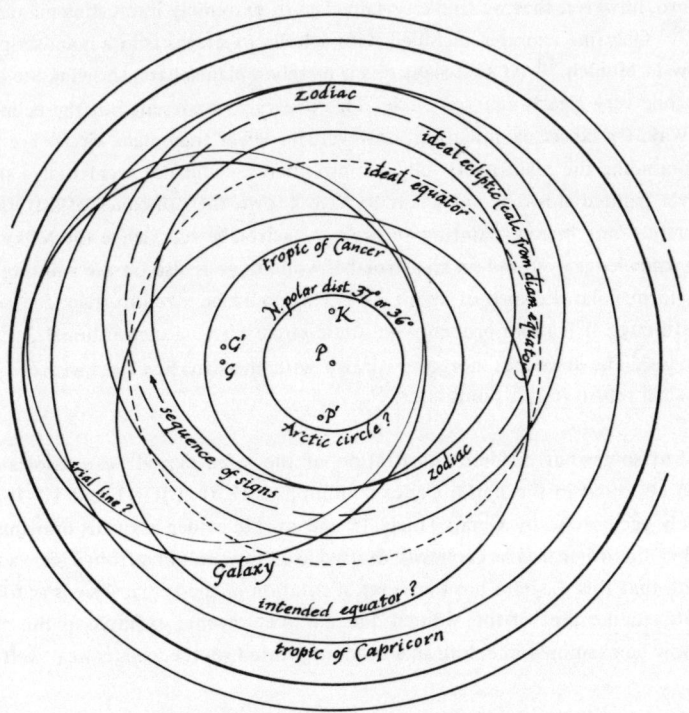

Notes on the constellation map of C.L.M. 210, f. 113v.

1. The projection is from the *south* pole, as on an astrolabe, but the stars are depicted *as seen*, and not with the order to the signs reversed, as on known astrolabes.

2. The tropics are not drawn with our convention, for one touches the northern limit of the zodiacal band, the other the southern limit.

3. The hesitant way in which the circle centred at G^1 is drawn, together with the equality of its radius and that of the circle centred at G, and their proximity, suggests that the former is a trial line, or a mistake. Neither can be the projection of a great circle, whether from P, P^1 or K. See below.

4. The circle centred at P^1 appears to be the intended equator, since no other line will serve, and since the diameter is correct. In stereographic projection, however, the equator would be centred at P; see the dotted line. (The tropics must be concentric with the equator.)

5. The small circle centred at P has a north polar distance of about $37°$, and could mark the arctic circle in keeping with one ancient convention ($36°$ = latitude of Rhodes). Alternatively, it could be the locus of the zenith point for an observer in a latitude of approximately $53°$. (The MS was for some time in the library of St Emmerammus, in Ratisbon.)

6. The catalogue entry on C.L.M. 210 has

> 'f. 114 Excerptum de astrologia . Adiecta est picta caeli tabula. Inc.: Duo sunt extremi vertices mundi quos appellant polos ... '

This text was edited by E. Maass in 1898.

manuscript.[48] Thiele's *Antike Himmelsbilder* is a useful source of precise information on the texts with which these illustrations are associated, and the book is in fact prefaced with a list of 27 manuscripts (but not all of them illustrated).

I have no wish to suggest that these diagrams were copied from an anaphoric clock plate, but they show that at least from the time of the ninth century the necessary skills for making such a plate were once again available in Europe, and that Pacificus, Archdeacon of Verona, could well have had on the clock mentioned in his epitaph a 'song of the heavens' in the form of an astrolabe dial of one sort or another.[49] It is quite possible, of course, that the medieval illustrations were copied from ancient exemplars which were themselves taken from anaphoric dials, or were even done by craftsmen skilled at both arts.

Coming down for a moment to the thirteenth century and another record of an astrolabe dial, this time from northern Italy, we find what is probably the oldest extant detailed description of the sequence and counts of the wheels of a planetary model.[50] This was probably water-driven, if it was ever built, and if so it must have demanded many advanced technological skills.[51] The device had an astrolabe dial in a vertical plane,[52] and it exhibited a number of Ptolemaic planetary motions. It is not at all unlikely, however, that the description related to a non-European device, perhaps that given to the Emperor Frederick II, in 1232, by the ambassadors of al-Ashraf, Sultan of Damascus. This highly valued gift was presumably lost in the seige of Parma in 1248, with the Emperor's treasury, his insignia, some of his ministers and, no less significantly, his harem.

I now come to my second argument — and one which is, I think, much stronger than the first — for supposing that there was a tenuous European continuity in the transmission of the ancient tradition of putting astrolabe dials on clocks. If we look at the astrolabe dials of the great European cathedral and abbey clocks of later centuries, we find that they are for the most part in stereographic projection from the north pole, rather than — as with almost every known portable astrolabe — from the south. This is true of the clocks at Valenciennes, Münster, Prague, Bourges, Doberan, Lübeck, Lund, Stralsund and Ulm, for example. Notre Dame at St Omer, and Berne, were relatively rare specimens in south projection, and like the Alfonsine mercury clock and the De' Dondi clock, they had a rete of stars which was turned by the mechanism. In the typical anaphoric clock, however, and in the Salz-

48. The illustration is at f.92v. of an Osma cathedral MS of Ciceronian pieces. It is reproduced in colour in G. de Champeaux and Dom. Sebastian Sterckx, O.S.B., *Le Monde des symboles*, 2nd edn., no place of publication, 1972, p.66.
49. Beckmann, *op.cit.*, p.344, gives as part of the epitaph of this man: 'Horologioque carmen spherae coeli optimum,/ Plura alia graviaque prudens invenit.'
50. I gave the text and translation of the work, with detailed discussion and two potential reconstructions, in 'Opus quarundam rotarum mirabilium', *Physis*, 8 (1966), 337-72.
51. One of my suggested reconstructions required it to have no fewer than ten concentric arbors (tubes). This might seem improbable, and yet we do know that the St. Albans clock used multiple tubes (*caligae*), which were therefore not beyond 14th century technological resources.
52. See the note added in proof, *ibid.*, p.368. Note that on p.362 I was wrong to repeat a claim that the stereographical projection of the Salzburg fragments is from the north pole.

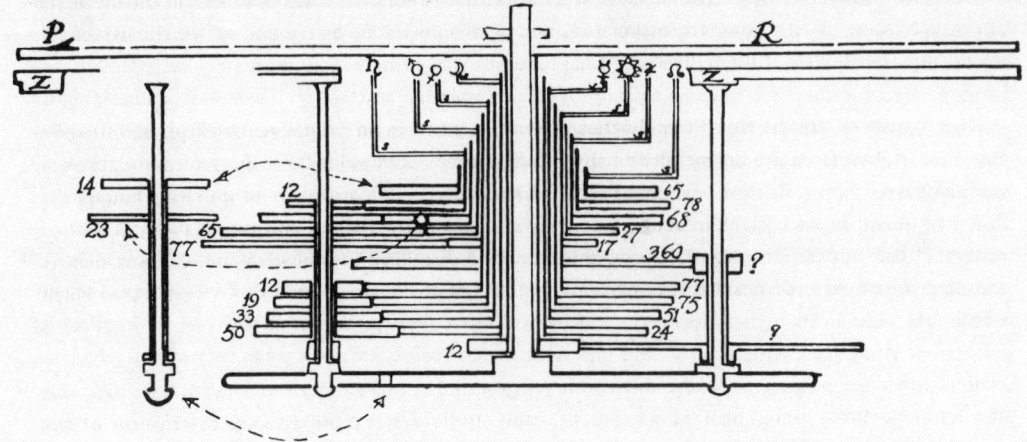

Fig. 7. One possible reconstruction of the 'device of certain remarkable wheels' described in the manuscripts from northern Italy (13th cent.).

R is the rete, Z the zodiac, P the pointer, and C the supporting column. The models of the planets are indicated by their symbols, and the numbers indicate the numbers of teeth on the adjacent wheel. For clarity, one train of gears has been drawn out of position. It is to be inserted in a way indicated by the dotted arrows.

burg fragment, and in Richard of Wallingford's clock at St Albans, while the projection was from the south pole, the rete was fixed and was a rete of hours, rather than of stars. If the Phillippicus and Osma manuscripts had been drawn from a moving plate, this would have been in the same tradition.[53] I would therefore like to suggest, very tentatively, that Richard of Wallingford was following an ancient tradition perpetuated, whether by artefact or document, within the monasteries of northern Europe. And lest it seem that the probabilities are high that a fourteenth-century author should hit on the ancient arrangement by chance, it should be noted that there are sixteen possible arrangements of a rete and plate, and that no fewer than five of these are actually found in use among the clocks I have mentioned.[54]

The first English clocks seem to have been made almost wholly of iron, and to have been of large dimensions: a frame three or four feet across was not unusual, and the frame of the St Albans clock was probably more than twice as great as this. De' Dondi's clock, of the mid-fourteenth century, was of brass and much smaller, while in the inventory of Charles V it is recorded that Philippe le Bel, who died in 1314, possessed a clock of an even more costly metal, silver, 'une reloge d'argent, avec deux contre poix d'argent empliz de plomb'.

53. A plate in the style of the Munich MS would merely have been required to turn in an anti-clockwise sense.
54; The projection may be north or south; the stars may be as seen, or as they would be seen by an observer outside the star sphere, as it were; the rete may be of hours or stars; and it may be fixed or moving.

The Church cannot perhaps claim much of the responsibility for the rapid improvement in the quality of metalwork or for the rapidly growing numbers of metal-workers at the end of the thirteenth century and the beginning of the fourteenth. Chain armour was giving way to intricately hinged plate, and cannon were for the first time being employed in European warfare.[55] There was social mobility enough, however, for ecclesiastic and artisan to come together on such a venture as that of making the most elaborate device in metal designed at that time. Some clocks, to be sure, were relatively simple devices, but a high proportion of those early clocks of which we have records were much more. Some of them showed jacquemarts to amuse or astonish the onlooker, and others perhaps a simple astrolabe dial, but far more significant were those which displayed the daily and annual movements of the Sun, the Moon (and possibly the planets), the phases of the Moon and other astronomical phenomena. At least in its highest form, the mechanical clock was largely the product of an intellectual movement going back to antiquity, and one with which Professor Price's paper very effectively deals.

The tradition of the geared planetarium simply could not have begun or have been maintained without considerable astronomical and mathematical knowledge. The single diurnal movement of the water-clock was something which could be altered at will by adjusting the water-flow; but to interrelate the complex movements of the planets and the cosmos requires a far from trivial ability to compute trains of gears. Since the same ability is needed to compute even the much simpler going and striking trains of a mechanical clock, it seems not unlikely that the first mechanical clocks were the product of the academic world — and here it is impossible to draw a dividing line between Church and University — rather than of the unschooled artisan. We need only recall the difficulties encountered in the much simpler matter of engraving an astrolabe dial for the Norwich clock.

Richard of Wallingford was in an excellent position to unite the craft tradition with the academic. As I mentioned earlier, he was the son of a blacksmith of Wallingford in Berkshire. Orphaned at an early age, he was sent by the Prior of the Benedictine house at Wallingford to study at Oxford, where he stayed with one short break from about 1308, when he was roughly 16, until he took the monastic habit at St Albans at the age of 23. After his double ordination, first as deacon (1316) and then as priest (1317), he was sent back to Oxford by his abbot. There he studied philosophy and theology for nine more years, in the course of which he wrote some remarkable pieces on trigonometry and astronomy, as well as astrology — which was of course almost *de rigeur* at that time. In 1327 he visited St Albans at a time which coincided with the abbot's death. He was himself elected abbot, duly visited Avignon for the papal confirmation, and at length returned to rule his monastery, which was in several ways the most important in England. He had a difficult time. He contracted leprosy on his overseas journey; his abbey was saddled with numerous debts; the townsmen were in revolt against the feudal privileges of the abbey; and there was an inquisition to investigate his fitness to rule. Despite his firmness towards them, however, he appears to have inspired enormous confidence in most of his monks, and when he died after only nine years in office, he left behind him a far greater secular inheritance than did his predecessor. At St Albans he was remembered above all else for his clock.

Leaving aside, for the present, the nature of the going and striking trains, I will first say something of

---

55. There are actually many late medieval examples of men who were both armourers and clockmakers.

the general character of the astronomical section, which was built into an adjacent frame. It seems, judging from the manuscripts, that Richard began to write a thoroughly formal treatise on the whole subject of making an astronomical clock, that he completed a first section on the arithmetical techniques to be used in calculating trains of gears for general astronomical use, but that he died before editing his several drafts of chapters describing the actual construction of a clock — the St Albans clock. A single manuscript of these drafts survives, not in the hand of the author, but copied out for the sacristy at St Albans by a man who had a very poor idea of the best order, and who mixed the drafts with other quite irrelevant texts. The manuscript was later misbound, and a whole gathering is missing, presumably with details of the actual planetary trains hinted at by certain surviving data.

In successive parallel drafts we can see how Richard improves upon his first thoughts until the accuracy of the gear trains he advocates is exemplary. By using a transported correction train to drive the Moon, for example, the theoretical error in its mean motion is only 7 parts in 10^6. But accuracy is not only achieved by the careful determination of gear ratios. The problem of driving the Sun round the zodiac at the correct rate is simply that its speed should be variable. Richard solves this problem by computing an oval gear wheel, with contrate teeth, linked to a long pinion the axis of which is parallel with the plane of the oval wheel. The result is highly accurate, but more to the point, there is absolutely no known historical precedent for gears of this design, which until the discovery of the St Albans manuscript would have been generally thought beyond the potential of medieval technology. (I should say here that the oval wheels of the later De' Dondi clock are of an entirely different sort, and serve a different purpose.) The intricacy of the gear work of the St Albans clock may be judged from another type of wheel, which transmitted power from conchoidal contrate teeth to a worm gear. This offers an excellent solution to the problem of smooth transmission, and — unlike the oval wheel — is still much used.

Among the clock's other refinements, I may mention a half-blackened Moon-globe which, rotating on its axis as the globe moved round the dial, showed the lunar phase. But this was not all. The lunar globe was so placed that it was automatically drawn under a small eclipsing disc whenever the time was ripe for an eclipse of the Moon. I hardly need say that this was an extraordinary feat of design.

Very little about the arrangement of the planets of the clock is now known, but when John Leland saw the clock at the time of the dissolution of the English monasteries, he reported that it included planetary trains and a tidal dial. (English manuscripts of the period are frequently found which tabulate the time of the tides at London Bridge, and a tidal dial would have been a relatively simple affair, following the lunar motion.) And this is a suitable point at which to record the placing of the clock. Like so many abbey clocks, it was put into the south transept of the church, and was therefore rather better protected from the weather than the average tower clock.

The St Albans clock was not completely finished during Richard's lifetime, but there is no reason to suppose that those who finished the work altered his design.[56] Judging by a remark made in chroni-

56. Abbot Thomas de la Mare (ruled 1349-96) saw to it that 'the upper dial and the wheel of fortune were perfected' by Laurence Stoke and a monk who was skilled in woodcarving, William Walsham by name. Richard of Wallingford 'first arranged' them, but they were left off the clock on account of his early death. Almost all of the clock was therefore completed before 1336.

cles of the abbey after his death, there was no one capable of doing comparable work — as he had indeed foretold, in excusing his lavish spending on the work on the occasion of his being upbraided by the king who pointed to the poor state of repair of the fabric of the abbey. There was, however, an established clockmaking craft, and it is of the greatest interest that the very Stoke family who had worked at Norwich were also engaged over a long period of time in work on the St Albans clock. Roger and Laurence Stoke were professional *horologiarii*, but they were lay brethren,[57] who must have done other monastic work in the area of England north-east of London. How tight the monastic hold on their profession was, at this period, is difficult to determine for lack of evidence, but the only stylistic parallels I have found with the St Albans clock are with the west country series at Exeter, Wells, Wimbourne and Ottery St Mary. All of these, however, are much later. The resemblances are slight, and mostly concern the simplest aspect of the lunar mechanism.

Finally, I come to the mechanical escapement, that characteristic of the truly mechanical clock which distinguishes it from what went before. It has always been supposed that the first mechanical escapement was the familiar verge and foliot, that is to say, the sort found on the De' Dondi clock, and found on large church clocks throughout Europe, until at length it was replaced by the pendulum. It cannot be said with certainty that this was not the first; but the first escapement of which we have certain knowledge, namely that of the St Albans clock, was substantially different. The clock had two similar escapements, one to control the going train and one to sound the bell at the hour — and on a twenty-four hour system, with a number of strokes equal to the hour. A double-edged pallet was thrown first one way and then the other, each edge being acted upon by one of several radially disposed pins on one or the other of a pair of similar wheels fixed together in parallel planes a suitable distance apart. (The pins on one wheel came mid-way between those on the other.) A vertical verge with cross-bar, carrying weights to adjust the moment of inertia of the verge assembly, and hence the period of swing, oscillated accordingly, and on the striking side of the clock it was these oscillatory swings which caused the bell to be struck. Needless to say, by working with one pallet it was not necessary to traverse the arbor of the crown wheel (or pin wheel) with another arbor. On the other hand, the St Albans mechanism required an extra wheel, and it might well have quickly fallen out of favour on the grounds of expense. I am convinced, however, that it was mechanically more efficient than the common verge and crown wheel. To distinguish this type of escapement, I shall use the word 'strob', which is that given to it in the manuscript. In dictionary language, strob is 'etym.dub.' The pallet is simply called a 'semicircle'.

There is no drawing of the strob wheels and verge in the St Albans manuscript, and the reconstruction is one which was achieved by painstakingly piecing together a whole series of measurements of things given in words, many of which were unknown to the dictionaries. Having, after many attempts, arrived at a successful solution, its correctness seemed to be confirmed when I found drawings of the very same escapement in Carlo Pedretti's *Studi Vinciani*,[58] reproduced, need I say, from the work of Leonardo da Vinci. These drawings are from Codex Atlanticus and date from about 1495. There are similar drawings in Codex Madrid. It is not for me to correct here those who have ascribed to Leo-

---

57. Laurence was important enough to accompany abbot Thomas to the papal court when he sought confirmation of his election.
58. Geneva 1957, pp.103-4.

nardo the invention of this ingenious device, but rather to suggest that if an escapement is known from two places at opposite extremes of Europe and from periods 160 years apart, there is a strong likelihood that the escapement was once widely diffused. There is in fact one other manuscript, of the early fifteenth century, in which the same escapement appears to be the subject of discussion.[59]

The two types of verge and foliot escapement — they may be distinguished as single- and double-wheel types — are obviously functionally related. The double-wheel is mechanically superior and more expensive, and on both scores it might therefore be thought wise to place it historically earlier than the single-wheel. It is the earlier, as far as extant records go, and moreover when Richard of Wallingford gave some numerical data for it, between 1328 and 1336, he wrote as though its construction was generally known. In the customary scholastic manner, he went into minute detail on the astronomical and arithmetical side of his draft treatise, but when he came to the going and striking trains, he took their general construction for granted and contented himself with measurements and gear counts. Some of the most technologically interesting parts of his work are thus the most difficult to disentangle, and this is especially true of the part which concerns the striking mechanism. I hasten to add, however, that I have not the slightest doubt that the clock had hour-striking of the sort mentioned earlier, with some sort of locking barrel, even though this directly conflicts with the well-known statement by a certain monk of Malmesbury, who dates the invention to 1373.

On this subject of the oscillatory mechanism for striking the bell, I wish to end. This, in the St Albans clock, works exactly as the main escapement, and it is not inconceivable that such an oscillatory striking device triggered at suitably chosen intervals by a hydraulic clock, pointed the way to the first mechanical escapement proper. (Perhaps my title should have echoed Professor Price's, and read simply 'The escapement before the escapement'.) The bell to be struck in this manner could have been large and the noise impressive, unlike the tintinnabulation to be expected of a water clock of the type illustrated in the thirteenth-century moralized bible. Can it be that the more ponderous tones of a clock sounded in this way were those which, somewhere between 1235 and 1260, it was ordained should be heard at the installation of every new abbot of St Albans? There are water-clocks (*orologia ollarum*, 'clocks of earthernware vessels') briefly mentioned in the St Albans chronicles under the rule of John de Maryns (1302-8), but the earlier reference is more revealing by far. The prior shall enter the church from the chapter house, the rule of John of Hertford tells us, to be presented to God and the Holy Martyr Alban at the high altar, with the striking of a summons, the shawms[60] sounding with the horologe, the tapers lighted round the altar, and the throne uncovered. Very few of those treatises from the centuries following, in which are given precise and complete descriptions of actual clocks, can convey quite as much about medieval humanity as can this passing reference to one way in which a great abbey clock of the thirteenth century was absorbed into the high ritual of the church. Nothing could have been more fitting.

59. Cracow, MS 551, ff.44v -49r. There is a further class of Italian double-wheeled escapements used in alarm clocks, as well as cognate devices from later periods. These are cited in my forthcoming edition of the works of Richard of Wallingford.
60. The chronicler adds that they are sometimes called 'mules'!

HIERARCHY, CREATION, AND *IL VELTRO*: THREE FOOTNOTES TO DANTE'S *INFERNO*

> Three tentative suggestions are made, all relating to Dante's *Inferno*. First, that the hierarchical structure of hell reflects in one of two ways that of the heavens. Second, that Dante's allusion to the Sun's position (and through it to the date in the year) is unlikely to have been meant to be precise, but that he might nevertheless have intended this as the Sun's own (astrological) exaltation, following an earlier creation convention. Third, that "il Veltro" might have been a veiled allusion to the republic of Venice, regarded as having been born at Torcello, between Feltro and Montefeltro.

1. *Hierarchy in the Inferno*

The first thesis that I shall tentatively advance is such a small one and the literature on Dante so vast, that I fear I am likely to be repeating what has been said elsewhere — and said, no doubt, with more authority. The thesis can be put very simply, and if indeed it has been proposed before, Dante scholars have certainly not given it the prominence I think it deserves. It is that the ninefold structure of the Inferno in some sense mirrors that of the material heavens, and that by adding Lucifer to the first, we complete the symmetry, or antisymmetry (in a moral, if not mathematical sense), with the tenth, immaterial heaven, namely the Empyrean, the heaven of light and love, and the home of God and his angels.

Before I go any further, I want to emphasize that I am not chiefly concerned with simple numerical equivalences, with nines and with tens, for example — although inevitably these come into the picture. The correspondence I want to indicate is a simple cosmological one. Its very simplicity is the greatest of my problems, for it means that Dante might have taken it into his poem unawares, from his sources. I shall try to show that he did not do so.

As everyone who has read Dante knows, there are at least three important sorts of harmony in his *Divine Comedy*. There are cosmic harmonies, reflecting those of the Aristotelian universe; there are metrical and other broadly "poetic" harmonies inspired by number symbolism, and what we might call a Pythagorean view of the world; and there is the view that human fortune is somehow ordered, linked harmoniously, that is to say, with nature and the universe of nature, not only in its contingent aspects but in its very forms. To take the last point first: Beatrice, answering Dante's question as to how it comes about that he can fly through the spheres of the heavens, tells him that

> ... Le cose tutte quante
> hanno ordine tra loro, e questo è forma
> che l'Universo a Dio fa simigliante... [1]

Higher beings are ordered by form, which is the impress of God's image. She goes on to explain that this form draws the fire up to the Moon, binds the Earth and gives it unity, controls the heavens generally, and moves creatures, irrational and rational alike. Indeed she adds ironically (if that is the right word to describe the pitying sigh, *un pio sospiro*, prefacing the lecture) that had Dante, being free from constraint, stayed on the ground, that really would be a marvel. There is magnificent ambiguity here, for she speaks as a natural philosopher, but is aware that her audience will understand by *privo d'impedimento* an untrammelled *will*. Her harmony is always a consequence of conformity to *two* sorts of law, natural and moral. Aristotle had been Christianized, in cosmology in particular through the writings of, as Dante among others thought, Dionysios the Areopagite himself. The pseudo-Dionysius was the fifth or sixth century neo-Platonist who wrote works with the titles *Divine Names*, *Mystic Theology*, *Ecclesiastical Hierarchy*, and *Celestial Hierarchy*. The last of these was translated into Latin in the ninth century by John Scotus Erigena, after which it quickly became the standard textbook of angelology. (How Dionysius influenced Erigena's highly intellectual scheme of a hierarchy of beings that parallels a hierarchy of logical orders of universality is outside the

---

[1] *Par.* I. 103-5.

scope of my simple thesis; but a third parallel, the hierarchy of degrees of value or perfection, is obviously not without relevance to Dante's grand conception. Note that the infinitude of Erigena's hierarchies has a counterpart in *Paradiso* XXIX. 130-5, where Beatrice talks of an indeterminate number of degrees.)

Now in *Paradiso* XXVIII, after (in canto XXVII) looking down from the Primum Mobile at the degenerate family of man on Earth, Dante has what might be called a "Dionysian" vision. He looks into Beatrice's eyes and sees a vision mirroring what is behind him. Turning, he sees a point radiating light of unbearable brightness — unbearable despite the fact that the point is so minute that even the smallest star would be a moon by comparison. Turning around the point are nine concentric haloes. Since the canto opens with a mirror metaphor, it seems as though Beatrice's eyes must be merely reflecting the celestial spheres below; but although there is indeed in her circles an indication of decreasing speed with increasing size, what Dante sees is clearly representative of God and His angels. The innermost ring — a ring of fire — turns more rapidly than "quel moto che più tosto il mondo cigne"[2]. The scholastic Aristotelian commonplace that God is without parts and indivisible, makes the analogy with the *point* clear enough. As Beatrice explains, "Da quel punto depende il cielo e tutta la natura", adding that the circle nearest to it (which Dante reasons is brightest because sharing most in the essence of the central point) is kept in motion by a burning love[3]. The analogy with the world of sense is not, however, complete, for as Dante observes, in a world of sense the revolutions are more divine the further they are from the centre. Why do model and copy not agree? His guide explains at length that things are not always what they seem, and that virtue is not to be measured by circumference alone, but by its intrinsic qualities. In fact the angelic orders — for that is what the rings of light are at last revealed to be — *are* congruent with the heavens, and more *can* follow from less and less from more. In short, we are not to expect Dante the poet to present us with "perfect" reflections of world in world. I shall later consider an explanation of the sort Beatrice gives here, that is to say, one that reconciles an order that is the inverse of what for some people might be the "obvious" order. It is perhaps worth adding that the seventh

---

[2] *Par.* XXVIII. 27. Presumably the reference is to the daily motion of the fixed stars. In lines 44-5 of the same canto we are told that the swift movement of the circle of fire is caused by burning love, presumably of God at its centre. This is a reflection of a typically medieval (albeit mistaken) exegesis of Aristotle, for whom the unmoved mover acts as a final cause of the motions below. Final causes, according to Aristotle, were not of course *desires*.

[3] *Par.* XXVIII. 40-44.

sphere in the vision merits Dante's special attention. It is the only sphere after the first that is properly described:

> Sopra seguiva il settimo sì sparto
> gia di larghezza, che'l messo di Iuno
> intero a contenerlo sarebbe arto [4].

It is perhaps not irrelevant that the corresponding celestial sphere, counting down from the Empyrean, belongs to Venus, and earthly love.

Beatrice identifies the circles with the three hierarchies or nine orders of angels as named and distinguished by Dionysius, rather than Pope Gregory ("the Great"). In fact, she adds, Gregory, when he reached heaven, was obliged to smile at his error, for he had mistakenly put Principalities in place of Powers. In his *Convivio* [5] Dante had followed a third system, namely that suggested by Bruno Latini, according to which a few angels of each order preside over the movements of each sphere. Since there was to be (following Aristotle) a separate intelligence for each movement, an attempt had to be made to identify the individual movements (epicycle, deferent, and so on) in such a way as not to leave one with a shortage of intelligences. Does every planet's daily motion come from *its own* intelligence, for example, or is *one* such, allocated to the Primum Mobile, enough for all daily movements? Dante, in the *Convivio*, had not been able to decide [6]. In the *Paradiso* it is simply a question of one order of angelic intelligences for each sphere.

The second, "Pythagorean", form of cosmic symmetry in Dante's work is not totally unrelated to the first. The idea was that mathematics could furnish all cosmic secrets: as Plato said, mathematics "draws the soul upwards" from the unstable world to the pure forms — stable blueprints of the world, as we might loosely describe them. The Pythagoreans had ascribed a divine quality to the *tetraktus*, the number 10 expressed as a triangular number $(1 + 2 + 3 + 4)$. Great importance was attached to the discovery that the parts of musical harmonies could be expressed in terms of ratios of those first four numbers. The Pythagoreans also attached importance to the number 9, not only because it could be related to astronomical models, but because it had the perfection of the number 3 twice over. The literature is endless, and I have no intention of even trying to begin to do justice to it. I will mention one or two key texts, and simply remark in passing how fond

---

[4] *Par.* XXVIII. 31-3. The messenger of Juno is the rainbow, Iris—too small to contain this seventh circle. (Cf. *Par.* XII. 12.).
[5] *Conv.* II. 7-11.
[6] *Ibid.*, II, 5 and 6.

Dante was of the number 3, and how common it has always been in all forms of ritual. It was commonly said that 9 brought the mystery of the Trinity closer to our understanding. "Dionysius" and his like had managed to introduce the angels into the scheme of nine, as we have seen, and yet they always had to contend with the apparently greater perfection of the number 10. (Thus Augustine, Isidore, Hrabanus Maurus, and Hugh of St. Victor — to name four influential writers — all believed the perfection of 9 to fall short of that of 10.)[7] The lesser number had strong biblical credentials, to be sure. There were the thrice nine sieges of Jerusalem, God's call for nine judgements, the nine graces, and the nine gifts of the spirit. Outside the scriptures there was the institution of the nine Muses. There were nines built into the calendar, the most conspicuous being that in the Roman calendar — the nones comes 9 days, taken inclusively, before the ides. Martianus Capella, not clearly appreciating the character of Ptolemy's system of geographical zones, says there are nine of them too[8]. The game was endless and obsessive, and none was more obsessed than Dante. To Beatrice the number nine was said to be a friend, a miracle at the source of which, Dante tells us, was the Trinity. There are at least eight occasions in the *Vita Nuova* when she is associated with the number nine[9]. The number can be found in the arrangement of the 31 poems which, with their explanations, go to make up the *Vita Nuova* (10 + 1 + 9 + 1 + 10). The extraction of number symbolism from poetry has led many a scholar to force the evidence, perhaps unwittingly; but in the case of Dante we have his own testimony. We might look at some of this more closely.

After an astonishing piece of slanted calculation, in which the poet invokes successively the Arabic and Syrian calendars, yet again throwing out nines in connexion with Beatrice, he explains to us what might be described as her "cosmic nature":

> Since, according to Ptolemy and according to Christian truth, there are nine moving heavens, and since in keeping with common astrological opinion these heavens affect the Earth below according to their conjunctions, this number was associated with her in order to show that at her generation all nine of the moving heavens were in perfect conjunction one with another...[10]

---

[7] Augustine, *Quaestionum Evangeliorum*, P.L. 35, 2.40, col. 1356; Isidore, *Liber numerorum*, P.L. 83, 10.52, col. 190; Hrabanus, *De universo*, P.L. 111, 18.3, col. 4910; Hugh, P.L. 176, 2.6.3, col. 449.
[8] *De nuptiis*, ed. A. Dick, p. 375. For Ptolemy, see *Almagest*, II. 8.
[9] III. 1,8; VI. 2; XII. 9; XXVIII. 3; XXIX. 1.
[10] *V.N.*, XXX.

Dante then goes on to give a deeper, religious, reason, but I am concerned here with the nine spheres. There is no mystery as to Dante's use of the nine celestial spheres, which added to the Empyrean, God's own abode, makes for perfection in the decad. Once Beatrice has explained it for Dante, there is no mystery in the vision he saw reflected in her eyes in canto XXVIII, representing the angelic orders in inverted sequence, and encircling a point representative of God Himself. (Although this might have been inspired by no outside source, there is such an arrangement beautifully illustrated in a ninth-century breviary of Hildegard of Bingen. There nine choirs of angels encircle a small white disc.) [11] But then nine occurs too in the arrangement of the circles of hell, which Dante's guides do *not* systematically explain. On what principle was that arrangement achieved?

First, a reminder as to the overall structure of the place. Dante enters hell with Virgil as his guide, and in due course finds that the place is, broadly speaking, a conical funnel (with vertex at the centre of the Earth and hence of the universe), ridged in nine circles. There is of course considerable minor structural irregularity. At the lowest point of the funnel is Lucifer, and each level is ruled over by a different demon. The nine circles are grouped into two main lots. In upper hell there is first Limbo, and then the four circles in which the sins of incontinence are punished. Here Dante draws on Aquinas, Aristotle, Cicero, and Virgil, at the very least, but not for the *patterning* of the place. There is little subsidiary structure in the first five circles. As the descent proceeds, Dante meets first the worthless ones — the unbaptized — and the virtuous pagans, whose only torment is deprivation of God's presence. The other circles of the first five are filled with those whose failure was to act positively for the good. They had given way to appetite — the lascivious, the gluttons, the avaricious and prodigal, and the wrathful. The sharpest division in hell comes on entry to circle Six, just within the walls of the City of Dis. Beyond, beginning with the intellectually so, come the obdurate. In circle Seven come the violent, in circle Eight the simply fraudulent, and in Nine the positively treacherous. Three types of violence are considered in circle Seven, and in the last two circles there are respectively ten types of fraudulence and four of treachery.

Although there is an explicit reference to Aristotle, who in the *Ethics* (ch. 7) is said to have made incontinence less serious than malice and bestiality [12], there is no obvious candidate for precisely Dante's

---

[11] Reproduced in R. HUGHES, *Heaven and Hell in Western Art*, Weidenfeld & Nicolson, 1968, p. 23.
[12] *Inf.* XI. 79-84.

scheme in any earlier author. There is nothing, for example, like the seven deadly sins that lend their structure to the terraces of purgation in Purgatory. In the doctrine of the *fall* of angels we might imagine that there would somewhere be discussion of a hierarchy of fallen angels precisely mirroring that of the good angels, but if so, I have not been able to find it. Gregory the Great, who, as we have seen, had modified "Dionysius" slightly (if that is the right chronological order — it was what Dante thought) on the nine orders of angels, deals in another place [13] with evil spirits, but not in the same hierarchical terms. The doctrine of future retribution had emerged gradually in Hebrew thought. Apart from the usual Gehenna as a place of punishment, there is the fiery abyss in *Enoch* XVIII.11 and XXI.8, for evil angels, but no hint of any place with an intricate structure. I cannot judge whether relations between Jewish eschatology and the Avesta support the thesis that the latter influenced Judaism, or the reverse, but it seems that the Later Avesta divides hell into parts — in fact into four parts of increasing severity [14]. Pali scripture apparently knows a system of eight hells below ground in descending order corresponding to the ascending heavens [15]. Certain Hindu writings enumerate seven hells, and others multiples of seven [16]. None of this, however, brings us one step nearer to Dante's source of the idea underlying the topology of his Inferno.

The progression of good through the ninefold heavens, and of evil through the ninefold hells, is obvious at once. The ninefold nature of hell was *not* emphasized by the poet, but perhaps his audience would have thought him perverse had he done so, bearing in mind the celestial "Trinity twice over" which was offered as a reason for the great goodness of the heavens. The clearest aspect of the relationship between the two places is the *contrast* — of the heavens with loathsome hell, of love with hate, of sympathy with hostility, and of pity with cruelty. We would hardly expect a strong correlation. What I want to emphasize here, in our search for symmetry or some weaker sort of correlation between the two realms, is that there are two essentially different ways of attempting to associate the two sets of circles: 1) starting with Lucifer at the centre and the Earth's surface at the outside, we can match each of the circles of hell with one of the spheres in the series beginning with the Earth's surface in the middle of the universe and ending with the Empyrean at the outer extreme; or 2) we may match

---

[13] *Moral.* III, *passim*.
[14] J. HASTINGS (ed.), *Encyclopaedia of Religion and Ethics*, XI, p. 847 (article by L. C. CASARTWLLI).
[15] *Ibid.*, p. 830 (article by E. J. THOMAS).
[16] *Ibid.*, p. 845 (article by A. B. KEITH).

Earth	—	Lucifer	—	God
Moon	—	9 Treachery	—	Primum mobile
Mercury	—	8 Simple Fraud	—	Fixed stars
Venus	—	7 Violence	—	Saturn
Sun	—	6 Heretics	—	Jupiter
		[castle wall]		
Mars	—	5 Wrath	—	Mars
Jupiter	—	4 Avarice	—	Sun
Saturn	—	3 Gluttony	—	Venus
Fixed stars	—	2 Lasciviousness	—	Mercury
Primum mobile	—	1 Limbo	—	Moon
God	—	Earth's surface	—	Earth
		(1) (2)		

The two correspondences between the circles of heaven and hell, numbered (1) and (2) in the text.

God, and the Empyrean, with Lucifer and the Earth's centre, and so correlate the two hierarchies until we finish with a correspondence between the circle of the Moon and Limbo, and finally of Earth's surface with Earth's surface. The first of these alternatives might be called the "geometrical", and the second the "moral".

The geometrical alternative has much to recommend it to modern ways of thinking. There is a sense in which the universe is a progression of increasing good from Lucifer at the evil and material centre to God at its perfect and immaterial Empyrean extreme; but the view of evil as a deprivation of good, the mathematician's way, as we might regard it, is not characteristically medieval. Evil was a positive thing, and it does not do justice to the typical medieval outlook on evil — perhaps not even to our own — to think in terms of a linear graph, with the surface of the Earth at the "zero-point" of good, and hence of judgement. I must confess that the alternative (1) to which this view gives rise has a certain appeal. The Earth partakes of (corresponds to) the character of God and Lucifer. The least good inhabitants of the heavens, those in the lowest three spheres (of the Moon, Mercury, and Venus), are paired off with the least good, or rather most evil, inhabitants of hell (circles Nine, Eight, and Seven respectively). The match is not what one could wish it to be, however, to the extent that whereas the circle of the Sun forms a natural division in the heavens, and the wall around the castle of Dis encloses the worst sinners in hell, this wall takes in one ring too many, for it includes that of the heretics (circle Six). Note, by the way, that the Sun cannot, on this argument, be plausibly paired off with the heretics, for it does not fall within the zone of the Earth's shadow, used to explain away weakness in the three inferior celestial spheres. It creates the circumstances of the shadow.

Taking the second alternative, the "moral" correspondence, does not provide any startling new insights into the meaning of the poem, so far as I can see, but there are one or two points which might be construed as support for it. For the sake of clarity I will number them:

(I) The numbering of the circles of hell, through the sentencing procedure of Minos, who coils his tail round his back, for example, eight times to send a sinner to the penultimate circle [17], agrees with that generally used for the corresponding heavens. Of course this is a very slight point, since the circles are numbered in the order in which they are encountered. Notice, however, that Minos is himself situated in the second circle, and is thus using a sort of "absolute" numbering system. More significant is the fact that the sinner in the *eighth* circle — Guido da Montefeltro — tells how on his death he was claimed by St Francis, but carried off to circle Eight by one of the black *cherubim*. It is the cherubim who have charge of the *eighth heaven*, on all three schemes of angelic hierarchy.

(II) God and Lucifer are likened implicitly when they are separately described as emperors — which is hardly relevant to our thesis — and as *points* — which might be thought relevant. On God's pointlike nature we have seen the account of Dante's vision within Beatrice's eyes. Lucifer is more obviously "'l punto dell'universo" [18].

(III) There is a definite tendency for the time spent in the different circles of hell, as measured by the numbers of lines devoted to them, to increase regularly with increasing proximity to Lucifer [19]. I cannot see any exact rule of increase, although it is, loosely speaking, "exponential", and bears a rough qualitative resemblance to the increase in size of the spheres of astronomy, on the system of Ptolemy later taken over by Alfraganus. For those who are not aware of the system, however, I should add that the distances to the outer planets — in that system and in reality — are so enormous by comparison with distances to the nearer ones, that there can be no question of a straightforward proportionality between line numbers and distances. Ironically, the times Dante allocates to the visits to the several heavens do not show the same reasonably smooth pattern of increase as times in the circles of hell.

---

[17] *Inf.* XXVII. 124-5.
[18] *Inf.* XI. 64-5.
[19] The approximate points of entry to the different circles of hell come at the following line numbers in *Inferno*: (One) 438; (Two) 569; (Three) 714; (Four) 838; (Five) 898; (Six) 1188; (Seven) 1540; (Eight) 2289; (Nine) 4280; (centre) 5058.

(IV) There is what seems to me to be a common element in the approach to Lucifer in the *Inferno* and the approach to God in the *Paradiso*, namely in the image of crystal in both places. The frozen lake in hell, in which traitors were embedded, looked more like glass than ice, "un lago che per gelo / avea di vetro e non d'acqua sembiante"[20]. Although other sources (in particular the Apocalypse of St Paul) refer to punishment by cold, it seems to be agreed that Dante's structural use of the image is his own. Turning to *Paradiso*, one finds that Dante tends to refer to the "first mover", rather than (as in *Convito*) to the "cristallinum", as might have been expected; but there is at the *corresponding* point in *Paradiso* — that is, just before entry to the Empyrean — the account of the creation with this crystal and glass metaphor:

> E come in vetro, in ambra od in cristallo
>   raggio risplende sì, che dal venire...[21]

The addition of amber, and the language generally, have added a certain warmth to the image.

(V) In the heavens, the space between the Sun and Mars represents a place of stellar symmetry in the sense that there are four spheres below and four (of stars — that is, of Mars, Jupiter, Saturn, and the fixed stars) above. There is no similar place of symmetry between circles Four and Five of hell, which correspond, on hypothesis (2), with the spheres of Sun and Mars; and yet just prior to the descent to circle Five in the *Inferno* there is a short lecture on God's provision at the creation of intelligences for the guidance of heavenly spheres:

> Colui lo cui saver tutto trascende,
>   fece li cieli e diè lor chi conduce
>   sì, ch'ogni parte ad ogni parte splende,
> distribuendo igualmente le luce[22].

Is there here a hint of symmetry in poetic structure corresponding to that in the exchange of light? If so, it seems that it might have been consciously repeated at what would be the corresponding point of *Paradiso*. The effulgence of the Holy Ghost is there described, in a protracted song that positively burns with light, as "equally bright all round":

> Ed ecco, intorno, di chiarezza pari,
>   nascere un lustro sopra quel che v'era,
>   per guisa d'orizzonte che rischiari[23].

---

[20] *Inf.* XXII. 23-4.
[21] *Par.* XXIX. 26-7.
[22] *Inf.* VII. 73-6.
[23] *Par.* XIV. 67-9.

One last potential analogy. As Virgil explains to Dante in the seventh circle of hell, they have always circulated to the left, as they have descended [24]. (Here he overlooks the sixth circle, where there has already been some retrograde movement.) [25] It is tempting to suppose that the sense of their circulation was meant to agree with the daily rotation of the heavens, but it must be admitted that another explanation can be found without difficulty. The typical movement of the planetary spheres is in a sense opposed to the daily rotation of the stars, and thus the circulation through hell might have been seen as in *contrast* with *that* rotation, that is, as a contrast of evil with good. Whatever the answer to the question of interpretation of the sense of rotation, the more important and prior question is whether Dante was viewing the rotations in heaven and hell as analogous at all (or negatively so). The poet retrogresses again, later in the seventh circle [26], and attempts have been made to find theological explanations for the fact. They are to my mind strained, but I have no wish to counter them with what would be even more strained explanations in terms of *planetary* retrogression.

Why, if the "moral" analogy was in Dante's mind, did he not make it more evident? As I have already suggested, there is an element of blasphemy implicit in the analogy, which might thus have made Dante naturally reluctant not only to explain it, but also to develop it in all its potential ramifications. His reluctance to introduce the Sun into the *Inferno* is very obvious — an exception I shall discuss shortly comes before entry to the first circle. Likewise his avoidance of the use of God's name, except when uttered blasphemously. In canto v, for example, God is simply referred to as "Another" [27]. Under the circumstances, perhaps we are entitled to say about the analogy I have been suggesting that Dante, while tempted to use it, was afraid to press home the point.

2. *The season of Creation*

The very first line of the *Inferno* refers to the time of the journey in terms of human allegory. It takes place at the mid-point of the journey of the poet's life. Shortly thereafter there are astronomical references to the time — to daybreak, through an oblique reference to the rays of

---

[24] *Inf.* XIV. 124-9.
[25] *Inf.* IX. 132.
[26] *Inf.* XVII. 31.
[27] Line 81: "venite a noi parlar, s'altri nol niega!".

the Sun on the hillside, in the sixth tercet, and to the time of year in the thirteenth, through mention of the Sun. Its position is given, but only by reference to the stars that accompanied it "when the divine love first set those beautiful objects in motion":

> Temp'era dal principio del mattino,
> e'l sol montava'n su con quelle stelle
> ch'eran con lui quando l'amor divino
> mosse di prima quelle cose belle... [28]

Intertwined with this cosmic theme there is the moral and perhaps political allegory with which I shall eventually be concerned. Before looking to the latter, I should like to comment on the first — the *only* clear reference to the time of year of the journey — simply because no commentary I have seen explores the whole range of its possible meaning for Dante.

The first point to bear in mind is that he might not have been thinking very deeply about the astronomical implications of the remark. He *is* capable of loose astronomical comment — as for instance when he says in the seventh canto that every star is setting that was rising when he commenced his journey [29]. Now to make an accurate association of the Sun's position and the positions of stars at a time in the remote past — in this case at the Creation — requires a certain calculation, namely of the movement of the stars with respect to the vernal equinox, the movement we now refer to as "precession", but which was for pre-Copernican astronomers a "movement of the eighth sphere". Dante knew of this movement, for which he acknowledged the Ptolemaic figure (got from Alfraganus) of one degree per century. He not only used it for his oblique allusion to Beatrice's age, at the beginning of his *Vita Nuova* — which we have already mentioned — but in the *Convivio* he actually calculated the movement of the stars from Creation to his own time. Since he reckoned the interval to be about 6400 years, the movement was a little over 60 degrees [30]. The precise details are not important for our purposes. Had Dante changed his mind about either the value for the age of the world, of that for the rate of change of stellar longitude (say by taking better data than Ptolemy's, as were indeed readily available in works stemming from the Arab world) [31], or about the choice of model for its evaluation [32], he would

---

[28] *Inf.* I. 37-40.
[29] *Inf.* VII. 98-9. This is astronomical nonsense, and has no obvious allegorical purpose.
[30] *Conv.* II. XIV. 15-17. Cf. II. VI. 16.
[31] A degree in about 66 years would have been a good "modern" value.
[32] The trepidational models of Thābit ibn Qurra and others, giving rise to a

have been obliged to revise that figure. Fortunately we shall not need to look very deeply into these possibilities. Let us first consider some fourteenth-century traditions as to the time of the creation of the world, and the Sun's whereabouts at that time.

There are, roughly speaking, four conventions to be investigated. Some scholars argued — usually on very general scriptural grounds, considering, for instance, the availability of food for Adam — that the Creation took place when the Sun was at the autumnal equinox; others that the vernal equinox was the appropriate time [33]. Others followed a more overtly astrological tradition, the most important source of which was perhaps Macrobius, who said that at Creation the Moon was in Cancer, Aries was in mid-sky, the Sun rose in Leo, Mercury in Virgo, Venus in Libra, Mars in Scorpio, Jupiter in Sagittarius, and Saturn in Capricorn [34]. These are the astrologers' "domiciles" of the planets, as accepted virtually universally. No specific positions within the domiciles were quoted by Macrobius, but there is a typical statement on this subject by John Ashenden, a late contemporary of Dante, to the effect that at the Creation each planet was at the 15th degree of its domicile "according to Julius Firmicus and Macrobius" [35]. There was a different astrological tradition, however, according to which some or all of the planets were in their "exaltations" at the time of Creation. Thus the Sun was supposedly in Aries, the Moon in Taurus, Jupiter in Cancer, Saturn in Libra, Mars in Capricorn, and Venus in Pisces. (Mercury cannot be as far from the Sun as is its exaltation, and tends to be put with Venus in Pisces.) A possible variant on this is one where *precise* degrees, rather than whole signs, are assigned to the exaltations. The Sun's exaltation, which is the one that concerns us here, was taken as the 19th degree of Aries.

Since I am inclined to favour the hypothesis that Dante was referring to the Sun's having been near its exaltation on the morning before Good Friday, as it had been at the creation of the world — a hypothesis I shall shortly defend — a few brief remarks as to the sources of this hypothesis are in order. As David Pingree has pointed out — and the following details are drawn from his writings — the "planets in their exaltations" is the horoscope of a mahāpuruṣa ("great man") in

---

non-uniform change of longitude with time, are what I have in mind. For more information see my *Richard of Wallingford*, Oxford University Press, 1976, vol. III, Appendix 38.

[33] For further information about these beliefs, see my "Chronology and the age of the world", chapter 20 of W. YOURGRAU & A. BRECK (eds.), *Cosmology, History, Theology*, Plenum, 1977, p. 307 ; see above, chapter 9, pp. 112-13.

[34] MACROBIUS, *Comm. in somm. Scipionis*, I. XXI. 23.

[35] See p. 114 above.

Indian astrology, starting with Sphujidhvaja's *Yavana jātaka* (written 269/70)[36]. This was taken over by the Sasanians as the horoscope of the first man, Gayomart, whose birth coincided with the beginning of planetary motion. In this horoscope note that Mercury is sensibly removed from its exaltation, as already explained, and placed in its own dejection, Pisces. The Sun is in its Greek exaltation (19° Aries) and not its Indian exaltation (10° Aries). The convention is completely different from the normal thema mundi, with the planets in their domiciles, as found in the authors already mentioned and others reflecting Greek sources (Vettius Valens, Rhetorius, and so on). The "exaltation" horoscope nevertheless found its way to the Byzantines; but what literary source Dante could have seen is not yet clear. It is quite possible that his source was not literary but pictorial. Some but not all of the planets are in their exaltations in the damaged first panel, depicting the creation of the world, of a series of frescoes by Bartolo Battilori in the Collegiate Church of S. Gimignano. The painter was born shortly after Dante's death, but the tradition is likely to have been alive in other places Dante visited.

It is quite wrong of commentators to conflate these different traditions — as when the "spring" tradition is linked with that set down by Macrobius, for instance. But which was Dante's? Let us first suppose that he was calculating reasonably carefully. For the stars to have moved from the autumnal equinox, or from Leo, to the sign of Aries, from the time of the Creation, is something he could not possibly have countenanced, within any reasonable medieval view of the available time interval. On the other hand, stars that were in the sign of Aries in his own day he would have reckoned to have been in or not far from Capricorn at the Creation — not, I think, a likely belief, involving as it does a wintertime creation. Thus every plausible alternative is excluded, without the need for us to calculate the matter precisely. The only alternative that remains is that Dante was speaking loosely — perhaps deliberately so, as might have been the case had he wanted popular appreciation of the point he was making — namely that the Sun was in Aries at the Creation of the world. In short, he was not following the theory of the autumnal Creation, or Macrobius, or the more "precise" version of that "domicile theory", but was acknowledging either the "spring equinox" view, or the "exaltation" view.

There are important implications here for an understanding of Dante's technique. Was he the devious calculator he has so often been

---

[36] See especially "Māshā'allāh: some Sasanian and Syriac sources", in *Essays on Islamic Philosophy and Science*, ed. G. F. HOURANI,Albany, State University of New York Press, 1975, pp. 5-14.

made out to be? I think not. The "spring equinox Creation" hypothesis will not fit with the precise position of the Sun near Good Friday, either in 1300 or 1301. To say that Dante meant Aries only in a loose sense is to sacrifice the very principle of exactitude that is at stake. But where, precisely, was the Sun on the morning before Good Friday in 1300 and (for generosity's sake) 1301? In fact it was reasonably close to the precise degree of the Sun's exaltation, 19° Aries; for in 1300 (when Good Friday was 8 April) the Sun at daybreak was at 25° Aries, while in 1301 (when Good Friday was 31 March) it was at 17° Aries, to the nearest degree. With an almanac he might have slipped out by a degree or so, but the generally favoured year of 1300 will not give us a precise 19°, without undue distortion of the evidence. It might be argued that we are near our target, and that Dante was only using astronomical doctrines in an approximative sense, as indeed has already been acknowledged in the assumption that he ignored the adjustment necessitated by the movement of the eighth sphere (precession). The trouble is that as soon as you make such a concession, you sacrifice the principles which will allow you to decide between the only two hypotheses left in the running. The most that can be said, I believe, is that Dante believed in a Spring Creation, and that *either* he thought it remarkable that the Sun was near its exaltation and place of Creation, on the day his journey through the universe began, *or* he was speaking very loosely. (Twenty degrees of sky is a great deal. Even so, one may say with Dante that on Good Friday the Sun was rising with all the stars in that twenty degrees, although they would not have been visible. As for the equinox, be it noted that the Sun had not been there for two or three weeks — that is, since about 12 March. It is high time that we eradicated from Dante commentary those references to 21 March as the date of the equinox.)

3. Il Veltro

From two aspects of cosmic symbolism that must strike most readers as quite trivial, I pass to a sphere where angels fear to tread. Who, or what, was intended by "il Veltro"? I have no firm answer to give to a question that has already given rise to a vast literature of its own [37], but I have some provocative and not entirely implausible observations to make which — so far as I can see — have not been

---

[37] Perhaps the most accessible source of reference is the entry under this word in the splendid *Enciclopedia Dantesca*, edited by UMBERTO BOSCO and others, Rome, 1971. It should not be necessary to repeat what is so readily available.

made before; and that is generally thought reason enough for adding to the existing stock of commentary.

At the very beginning of his journey, Dante finds himself lost in a dark wood from which he attempts to escape by climbing a steep hillside. He is immediately turned back by a fleet and light footed leopard (*una lonza*) with a dappled pelt [38]. No sooner has the sight of the rising Sun raised his hopes than he is confronted by a ravenous lion that causes the air to tremble, followed by a lean and hungry she-wolf that had caused many to live in misery (*grame*) [39]. As he flees back to the wood he is halted by the shade of Virgil, who tells him that he cannot hope to climb the hill, but must follow another path. The she-wolf, with insatiable greed, mates with many beasts and will continue to do so until the Hound (*il Veltro*) shall come who will deal her a sorrowful death. He will feed on neither land nor money, but on wisdom, love, and virtue, and his birthplace (or birth, or people; *nazion*) shall be between Feltro and Feltro:

> Molti son li animali a cui s'ammoglia,
> e più saranno ancora, infin che'l Veltro
> verrà, che la farà morir con doglia.
> Questi non ciberà terra nè peltro,
> ma sapienza, amore e virtute,
> e sua nazion sarà tra feltro e feltro [40].

The Hound will save that low-lying Italy for which died Camilla and Euryalus, Turnus and Nisus. He shall hunt the wolf back into hell, whence she was first sent forth by envy [41], and his coming will be decided by the stars [42].

So much for the words of the prophecy, which has been seen by some as one of celestial (stellar) influence, by others as a prophecy of the second coming of Christ, and by most early commentators as pointing to the coming of a great leader — perhaps Can Grande della Scala. Others have favoured a pope (as, for instance, Benedict XI), or an emperor (Henry VII of Luxemburg being the front runner in the imperial stakes).

The mystery of *il Veltro* is compounded by Beatrice's prophecy of a "five hundred, ten and five" sent by God to slay the harlot and the giant in the mystical procession witnessed by Dante in the terrestrial paradi-

---

[38] *Inf.* i. 31-3.
[39] *Ibid.*, 45-51.
[40] *Ibid.*, 100-105.
[41] *Inf.* i. 106-11.
[42] *Purg.* xx. 13-15.

se[43]. Most commentators take this to denote a leader (DXV = DVX), but some argue for a pope (Domini Xti Vicarius), and others for Christ himself (whether Dominus Xtus Victor or the subject of a monogrammatic VXD commonly found among twelfth-century and thirteenth-century liturgists). It is easy to see why DXV and il Veltro have so often been identified, although Dante gives us no clear license to do so. Of course, as far as cryptic religious and political prophecy is concerned, he was very much a product of his times.

In Canto xx of *Purgatorio* he refers again to the one who will slay the wolf, and addresses himself to the heavens:

> O ciel, nel cui girar par che si creda
> le condizion di qua giù trasmutarsi
> quando verrà per cui questa disceda [44]?

This astrological allusion could be of many sorts. It could be of a very general nature, linking stars and nativities in customary ways; or it could relate the second coming of Christ to the Ages of the World, which were sometimes reckoned as a thousand years apiece, but often calculated in terms of the motion of the eighth sphere, within traditions connecting with the Creation topic discussed in my second section; or the allusion could be to the relationship between religious change and so-called "great conjunctions" of the planets Saturn and Jupiter [45]. (Even ordinary conjunctions of those planets are rare. There were six in the thirteenth century, one of them perhaps within a couple of months of Dante's birth.) As far as this new and indirect reference to il Veltro is concerned, any of these interpretations can be justified in terms of medieval belief.

The problem of il Veltro begins with what at first seems the most secure hint of all — a birthplace between Feltro and Feltro. Even early commentators considered the possibility that these were not names of places, but words for a poor fabric (felt), or even for the sky. The felt interpretation has led to some astonishing astronomical readings, ending with Dante himself as the Hound, in one case, while it has led others to see in the opening canto a reference to the poor stuff of the friars' habit.

My interpretation requires a geographical reading. There have been

---

[43] *Purg.* xxxiii. 43-5. See the entry "Cinquecento dieci e cinque" in the *Enciclopedia Dantesca*.
[44] *Purg.* xx. 13-15.
[45] In addition to my chapter referred to in note 33, see my "Astrology and the Fortunes of Churches", *Centaurus*, xxiv (1980), pp. 181-211|; see above, ch. 8, pp. 59-89.

several contenders for the title to "Feltro", chosen according to the birthplace of the favoured candidate for the Hound. San Leo Feltrio and Macerata Feltria have between them Faggiuola, birthplace of Uguccione della Faggiuola. Feltre in Venezia Euganea and Montefeltro in the Romagna have between them both the birthplace of Benedict XI (Treviso) and the country in which Can Grande della Scala (lord of Verona) worked in the imperial cause. The last of these attributions seems more forced than that of Benedict and Treviso — but of course Dante might have been doubly obscure in this matter. I shall accept the two places, Feltro in Venezia and Montefeltro, but propose *Torcello* as the place between.

Torcello is a small island in the Venetian lagoon. Like Feltro and Montefeltro it was the seat of a bishop. (Treviso was not.) Cunizza, in the third heaven, mentions Feltro in connection with a crime perpetrated by its bishop [46]. Montefeltro was the name of a whole district (the country ruled by Guido da Montefeltro, encountered in the eighth circle of hell), a name taken from that of its chief town, now San Leo.

The chief glory of Torcello is its basilica, and the glory of its basilica is its twelfth-century mosaic of the last judgement. This enormous mosaic contains five rows of figures below Christ crucified. Christ as a saviour in limbo comes in the top row. Christ in glory, as a judge — with saints and confessors — comes in the row below. In the middle row come angels blowing the last trump, while sea and land give up their dead in fragments regurgitated by animals. In the fourth row there comes a segregation of damned and saved, while in the lowest row, to the left-hand side of the door piercing the mosaic (and thus on Christ's right), are the elect in Paradise, and to the other side are the damned, in a compartmented hell [47].

The mosaic is not strongly reminiscent of the *Divine Comedy*, apart from inevitable themes common to Christian iconography generally, *and the depiction of the beasts in the middle row.* On the left side, to the front of the dead who are rising whole, in their grave clothes, from their tombs, come (from front to rear): a lion, an elephant, a wolf, a leopard, a canine animal with a spotted muzzle, and a griffin. There are also two crows. To the right hand side of the wall there are fishes, and a creature it would be reasonable to describe as a "sea-leopard", spot-

---

[46] *Par.* IX. 52-60.
[47] An illustration of the entire mosaic, together with some details, is to be found in R. HUGHES, *Heaven and Hell in Western Art*, Weidenfeld & Nicolson, 1968, pp. 12-14. More profusely illustrated, and in colour, is A. NIERO, *La Basilica di Torcello e Santa Fosca*, Venice, Ardo, n.d., pp. 32-45.

ted, and in a sphinx-like posture, its front paws being visible. It has a canine head. On its back rides a human form, perhaps not representing Antichrist (who was taken to ride the back of Leviathan)[48], for he carries a trumpet, like the angels on shore.

Now I have nothing of any importance to add to what has already been written about the religious symbolism of Dante's various creatures. That the leopard conventionally stood for incontinence, and the sins suited to the first circles of hell; that the lion stood for violence, and the second region; and the she-wolf for malice, and the final part of hell; that the griffin in the mystical procession in the terrestrial paradise was commonly understood as representing, by its nature — it was half lion and half eagle — the human and divine characters of Christ; that Geryon, guardian of the fraudulent in the eighth circle of hell, and named after the *tergeminus Geryones* of the *Aeneid*, was believed to entice strangers and slay them; all these things one can accept, with the enormous weight of supporting scholarship supplied for them over the centuries. It is impossible to deny those biblical and other texts that provide what it is now felt safe enough to call the "recognised sources" of Dante's beasts. Thus in a single verse of *Jeremiah* (v.6) we have a lion, a wolf, and a leopard — in that order — to tear in pieces the unrighteous, on God's behalf. (No doubt this text is in ancestry of the essentially Byzantine iconography of the Torcello mosaic.) Surely, though, the point is not that it should be possible, with great effort, to assemble suggestive literary sources for Dante's use, but rather that we should try to reduce those sources in number and extent. At Torcello, in a single work of art, we have what could well have stirred his imagination on three key passages, and more. We do not have to suppose that he actually saw the mosaic, although that is not out of the question. A copy, or a description, is all that he might have had — and of course either might have been associated with biblical and other texts.

There are many differences between Dante's Geryon and Torcello's, but in any case there is no really adequate single *textual* source for the creature. The *Aeneid* is good only for the name and the monstrous character[49], although — as if to justify his choice of name — Dante

---

[48] Leviathan is one of several names for sea monsters occurring in Jewish and related Babylonian writings. For a description of the beast associated with the sea, see *Revelation*, XIII. 1-10. Although it is there said to have seven heads and ten horns, which Torcello's does not, it is likened to a *leopard*, with feet as of a bear, and the mouth as of a lion. It carries on its heads the name of blasphemy, and was consequently later associated with Antichrist.

[49] *Aen.* VIII. 202.

gives his monster three natures (of man, beast, and serpent) in place of the three heads and three bodies mentioned by Virgil [50]. Dante's extended account, in cantos XVI and XVII of the *Inferno*, presents an image of a fairly conventional dragon (apart from the man's face), with a serpent's tail, which Torcello's sea creature seems not to have, although it is aptly described in the words:

> lo dosso e'l petto e ambedue le coste
> dipinti avea di nodi e di rotelle [51].

Dante's swims through the air, but is likened to a swimmer in the sea [52]. Torcello's swims, but above all looks like a leopard, at least at first glance; and Dante makes much of the fact that the cord with which Virgil attracts the monster from the ravine is one with which he had earlier hoped to capture the leopard [53].

I am not suggesting that the Torcello mosaic wrote Dante's story for him, but that it might well have put ideas for characterization into his mind. The lion, the wolf, the leopard, the griffin, and the swimming leopard are all there. Is the Torcello elephant responsible for the allusion to elephants and whales (*elefanti*, *balene*) in canto XXXI of the *Inferno*? There are large fishes in the sea section of the mosaic, and Dante is as likely to have seen the mosaic as he is to have seen a whale. Did the mosaic put him in mind of dolphins, in canto XXII? And what of the canine creature whose head and neck can be seen between the leopard and the griffin? Can this be il Veltro? Probably not. It is canine in appearance, but not excessively greyhound-like, and it seems to have tiny wings sprouting from its lower neck. What matters, of course, is not what the artist meant it to be, but what Dante — if this has anything to do with his source — believed it to be. I can only leave this as an open question; but I think there may be a better candidate for il Veltro, on the assumption that the Torcello mosaic is Dante's source of inspiration. Might he not have intended *the Venetian state itself*?

It is necessary once more to emphasize that the most one can expect of a representation such as that at Torcello is that it put the seeds of ideas in the poet's mind, and not that it forced his hand with a relentless logic. Even scriptural texts were powerless to do that. Take the

---

[50] *Inf.* XVII. 10-15.
[51] *Inf.* XVII. 14-15. The biblical references to serpents' tails usually offered by way of explanation for Dante's description are acceptable without being very compelling.
[52] *Inf.* XVI. 134-6.
[53] *Inf.* XVI. 106-8.

lion, the wolf, and the leopard of *Jeremiah*, for instance, which were the agents of God's judgement. It would be hard to propose them as candidates for il Veltro's vengeance, although I suppose that if Christ were implied, his having offered redemption through repentance as an alternative to irreversible judgement might be the required exegesis. (It seems very forced to me, but perhaps that is because I am not completely medieval in outlook.) I will do what I can to give the Torcello case its own logic. The place lies between Feltro and Feltro. In ecclesiastical terms, it was the most important place between those two bishoprics. It was, so to speak, the place where the lion, the wolf, and the leopard were to be found; but who was to be born there? There is no obvious candidate — and presumably Dante would have referred to someone or some institution in which he saw current promise, at the time of writing. Implausible as the solution may seem from what we know of Dante's political hopes, however, the republic of Venice can easily be regarded as having been born in that place.

The ancient story of the migration to the islands in the lagoon, under pressure from incessant barbarian harrassment, is told in the Altino Chronicle [54]. Paul, Bishop of Altinum, having been alerted to an impending invasion by the movements of birds, on the third day heard a voice bidding him rise and look to the stars — which he did from a high tower, an action commemorated in the name of Torcello itself. Duly enlightened, the people of Altino sailed forth to the islands of Torcello, Mazzorbo, and Burano. The present basilica goes back, not to this sixth-century event, but to 639 and 864-7. The nave was restored, and an octagonal baptistery built, in 1008, while the large mosaics were put up in the 12th and 13th centuries. Two curious — but perhaps quite irrelevant — observations might be made about the church and its name. Its very foundation, and its name, have to do with *heavenly* events, events thus appropriate to the *Divine Comedy*. And yet the arrangement of the seats for the priests around the semicircular apse is such that they rise in steps, in a form somewhat suggestive of the ridges of hell! (Dare I add that the bishop's throne is in the centre?)

This reading of "il Veltro" as "Venice" would make sense of the reference to "quella umile Italia", which is thought to echo Virgil's *Aeneid* III. 522-3, taking "umile" in the sense of low-lying, rather than humble. (The Latin *humilis* can mean both things, and as a variant on the first, and clearly meaning "near to sea level", is found elsewhere in

---

[54] *Chronicon Venetum (vulgo Altinate)*, in *Monumenta Germaniae Historica Scriptorum*, tom. XIV, Hannover, 1883, pp. 5 ff. I am here concerned only with medieval tradition. Recent archaeological work has revealed first and second century remains of Roman type at Torcello.

Virgil.) The low-lying land would then be all that part of Italy between the Apennines and the Dolomites. A number of allusions to Venice have been claimed from the *Divine Comedy*, some of them more easily given an alternative interpretation. It is mentioned by the eagle in the heaven of Jupiter in connection with the counterfeiting of the Venetian *grosso*. Cunizza, in the heaven of Venus, mentions the island Rialto as the eastern limit of the March of Treviso. There is a three-tercet description of the Venice Arsenal in canto XXI of the *Inferno*, where an analogy is drawn between activities in the eighth circle of hell and the shipyards. The allusion to boiling glass in canto XXVII of *Purgatorio* has suggested to some a knowledge of Murano. But none of this can be counted as offering any support for the Torcello reading. There is perhaps even counter-evidence in the prophecy that the saviour will hunt the wolf through every town until he has driven her back to hell:

> Questi la caccerà per ogni villa
> fin che l'avrà rimessa nello 'nferno [55].

A sea-going power does not naturally hunt from town to town. But did Dante perhaps consider the Venetians the people best fitted to recover the *Holy Land*, which the pope, "Principe dei novi Farisei", was certainly not doing, busy as he was with war near the Lateran [56]? It is not for me to say that the association of this passage with Venice is plausible, for I have a vested interest in making it. I merely note that V. Zanetti holds that it alludes to Venice, and he presumably has no axe to grind [57].

I can take the prophecy of the first canto of *Inferno* no further, but what of the DXV prophecy? If we take this to mean a leader, DVX, how can we identify it with a republic? The answer is that there is no reason whatever for taking DXV and il Veltro as absolutely identical, although it seems that they must be closely associated — as the republic of Venice was associated with its doge. The word *doge*, of course, comes immediately from the Latin *dux*.

How probable is this interpretation? It would be foolish to pretend that Dante's monarchical aspirations can be easily fitted to a declaration of faith in a chief magistrate of fourteenth-century Venice. One would have to argue that he was attempting genuine, rather than pseudo, prophecy, and that he thought he foresaw, not Lepanto, but a situation where Venice could recapture Acre and other cities of the

---

[55] *Inf.* I. 109-10.
[56] *Inf.* XXVII. 85-90.
[57] See the article "Venezia" in *Enciclopedia Dantesca*.

Holy Land. It has long been accepted that the harlot and the giant whom the DXV shall slay [58] represent respectively the papacy and the French royal house, and it is hard to see how Venice can have been thought capable of such an act, except perhaps in the long term. It is easy to imagine a situation in which Dante, visiting the Venice Arsenal, was overwhelmed by the thought of the potential might of the Venetian state; but it is not easy to find any independent evidence for such feelings on his part. It is not even certain when, if ever, he was there — before, that is, the mission of 1321 that led to his illness and death. But if my argument is not overwhelming, it does at least hang together. Torcello, the heavenly tower, lies between Feltro and Feltro. The wolf is to be found there, which is to be hunted by il Veltro, Venice. Torcello is the birthplace of Venice. The Venetians will drive the wolf back to hell and recover the Holy Land. The DXV, the doge, will slay the harlot and the giant, and the stars will bring this about. If my argument seems to have a circular quality, dropped from Bucentaur into the waves of Dante scholarship it will at least find itself with many comparable rings for company when it finally reaches the sea bed.

Addendum

Note that Aquinas, *Summa Theologica* 1a 68.2.2, has a passage reminiscent of the crystal and glass metaphor (p. 196). There it is said that the waters above the heavens (cf. *Genesis* 1.7) are not necessarily fluid but are rather crystallized round the heavens in a state similar to *ice*.

---

[58] *Purg.* XXXIII. 43-5.

14

THE ASTROLABE

The astrolabe was the most widely used astronomical instrument of the Middle Ages. It originated in antiquity and was still not uncommon in the 17th century. One purpose of the instrument was observational: it was employed for finding the angle of the sun, the moon, the planets or the stars above the horizon or from the zenith. It could also be used for determining the heights of mountains and towers or the depths of wells, and for surveying in general. Far more important, however, was the astrolabe's value as an auxiliary computing device. It enabled the astronomer to work out the position of the sun and principal stars with respect to the meridian as well as the horizon, to find his geographical latitude and the direction of true north (even by day, when the stars were not visible), and it allowed him to indulge in such prestigious and lucrative duties as the casting of horoscopes. Above all, in the days before reliable clocks were commonly available, the astrolabe provided its owner with a means of telling time by day or by night, as long as the sun or some recognizable star marked on the instrument was visible.

A more precise name for the instrument I am describing is the planispheric astrolabe. There are three other types of astrolabe: the linear astrolabe, the spherical astrolabe and the mariner's astrolabe. The linear astrolabe was an instrument that was difficult both to use and to understand, and it was rarely made. The spherical astrolabe was also rare; it was in the form of a globe, although it had much in common with the flat planispheric astrolabe. The mariner's astrolabe was a relatively late instrument; as far as is known it was first used only a little before the time of Columbus. It was a crude device, serving chiefly to find the altitude of the sun, moon or stars above the horizon, and it was used for much the same purpose as the sextant of later centuries. Basically it consisted of an alidade, or straight rule, pivoted centrally on a single pin on a circular plate. On each end of the alidade was a vane pierced with a hole. The mariner hung the instrument from his thumb and adjusted the alidade so that he could sight the celestial object through the holes in the vanes. He then read the altitude of the object on the scale of degrees around the edge. (In working with the sun he would have allowed one vane to cast a shadow on the other in order not to injure his eyes by direct observation.) The mariner's astrolabe was made of heavy brass so that it would hang steadily from its ring and shackle, and it was also pierced so that the wind would affect it as little as possible.

The planispheric astrolabe I shall henceforth call simply the astrolabe, since it was by far the commonest type. In order to fully understand even its simple uses, it is necessary to examine its outward form and trace how it acquired that form.

Both sides of the astrolabe bore valuable information. Generally speaking, the alidade was pivoted on the back. The back was a repository for information that could in principle have been stored elsewhere. It usually carried a number of scales and tables whose precise nature tended to change from century to century.

A scale that is found on almost all astrolabes is the calendar scale, which represented the days and months and correlated the position of the sun with the date within the year. If we could only see the stars by day, we should find it easier to appreciate the apparent movement of the sun against the stellar background. This movement is of course a consequence of the earth's motion around the sun; as the earth proceeds along its orbit the sun appears to shift with respect to the stars. It is therefore often convenient to speak as though the earth were at rest in the centre of a vast sphere on which all the celestial objects are situated. The stars and even the planets are at such immense distances in comparison with the size of the earth that the celestial sphere is a reasonable convention, as long as one is concerned only with the direction of the celestial objects from the observer.

The path of the sun on the celestial sphere is the ecliptic, and the sun completes one circuit of the sky along this path in a year. The planets appear to travel in a band of sky several degrees on each side of the ecliptic; this band is the zodiac. It is possible to give the approximate position of the sun on the ecliptic (its place on the zodiac) for any date of the year. Leap years present a small problem, but it is not a very difficult one, since the accuracy expected is only a relatively large fraction of a degree.

The calendar scale of the astrolabe has engraved on it the days and the months. There is also a zodiac scale, usually concentric with the scale of dates, which correlates the dates with the sun's position on the ecliptic. The sun's position can be given as a celestial longitude from zero degrees to 360 degrees, reckoned from some suitable point of origin. In the Middle Ages a variant of this system was used: the zodiac was divided into 12 signs. Each sign was 30 degrees in length and had been named after a prominent constellation. In actuality the constellations had long previously moved into neighbouring signs as a consequence of the slow precession of the equinoxes, which in turn is due to a conical movement of the earth's axis. Partly because of precession, and partly because the time it takes the earth to go around the sun once is not exactly 365¼ days, there are small shifts in the sun's position for any particular date as the years pass. These shifts can be taken care of without much difficulty by the rules of the calendar. Note that a medieval calendar scale is likely to be 10 or 11 days out of register with the Gregorian calendar we use today.

The front of the astrolabe is more important than the back. It has two principal parts. One, the rete, is a fretted plate, which like the rest of the astrolabe is usually made of brass. It overlies an unperforated circular plate. The rete (from the Latin word meaning "net") is a representation of the heavens. The tips of small pointers mark the positions of the brightest stars, an off-centre circle represents the ecliptic, and there are also parts of three circles representing the celestial equator and the tropics of Cancer and Capricorn. Through the centre of the rete is a pin around which it can rotate. The pin, which also holds the alidade on the back, is kept in place by a wedge passing through a slot in the shaft of the pin. The thicker end of the wedge was traditionally in the form of a horse's head, and thus the wedge was often called "the horse". If any durable transparent material had been readily available at the time, the rete would probably have been made of it; anyone today who wanted to build a simple astrolabe could use a sheet of plastic to make the star map.

The other principal part of the astrolabe is the plate under the rete. It is graduated with a series of circles and straight lines representing coordinate lines that are fixed with respect to a given observer. The centre of the astrolabe, around which the rete turns, represents the north celestial pole, around which the stars appear to turn. Concentric with it are the Tropic of Cancer, the celestial equator and the Tropic of Capricorn. These circles can be represented both on the rete and on the plate below it. On the plate there is a line representing the observer's horizon and a point for his zenith. There is a set of almucantars, or circles of constant altitude, above the horizon and encircling the zenith. There are also lines of constant azimuth, which appears as arcs of circles radiating from the zenith and running down to the horizon.

Clearly the distance separating the pole and the observer's zenith on the astrolabe plate depends on the geographic latitude of the observer. If he lived at the North Pole, the two points should coincide, whereas if he lived at the Equator, the two should be separated by whatever represents 90 degrees on the astrolabe plate. The necessity of having a different plate for every latitude at which the instrument was to be used was always a source of chagrin to the astrolabist. He would have a plate for his own latitude, and he might have as many others as he was likely to need on his travels.

Such plates were often called climates. An astrolabe might have as many as four, five or even more climates, each plate being engraved on both sides and all being stacked in the mater, or main body, of the astrolabe. They fitted under the rete and were secured by the pin and horse [*illustrated in Fig. 9*]. There were astrolabes that could be used at any latitude with a single plate, but they were not easy to use nor were

Plate 1. Persian astrolabe, signed Muhammad Muqim, and dated A.H. 1067 (= A.D. 1647/8). This was made for Shah Abbas II. Note how cleverly the maker has given an illusion of symmetry in the rete.

(Museum of the History of Science, Oxford University)

Plate 2. The back of the astrolabe depicted in Plate 1. The scales above the shadow-square are characteristic of Indo-Persian instruments. They could be used for various trigonometrical purposes, including the inter-conversion of equal and unequal hours.
(Museum of the History of Science, Oxford University)

Plate 3. English astrolabe of the late 14th or early 15th century, with a Y-type rete closely resembling those in the manuscripts of Chaucer's treatise on the astrolabe.

(Museum of the History of Science, Oxford University)

Plate 4. The back of the astrolabe shown in Plate 3. The calendar scales are complete, and even include the names of feast days.

(*Museum of the History of Science, Oxford University*)

they ever common.

How are the stars and the coordinate lines on the celestial sphere mapped on to the rete and the climates? Suppose the observer was at the centre of a large hemispherical dome on which the almucantars and the coordinate lines of constant azimuth were drawn at intervals of five or 10 degrees. Through this series of lines he would be able to see the stars of the night sky, which would move with respect to the lines because of the daily rotation of the earth. If the observer were to take a long-exposure photograph, the pinpoints of starlight would trace out arcs of concentric circles rotating around the north celestial pole. (In the true medieval manner we shall overlook the needs and prejudices of those living in the Southern Hemisphere.)

Just as it is possible to make a flat map of a terrestrial globe, so it is possible to map the two spheres introduced here: the fixed network of coordinate lines and the moving sphere of the sky. There are certain necessary practical requirements if the maps are going to be made of brass and are to serve at all times. If the two maps are to be arranged so that one pivots around a fixed point of the other, as with an astrolabe, then this point should be one of the poles, preferably the north pole if the instrument is to be used in the Northern Hemisphere. Furthermore, the projections of both maps should be alike for all positions of the rete and the plate with respect to each other; a map projection would be no good at all if it meant that the rete had to be distorted as it rotated.

The stereographic projection was admirably suited to the needs of the astrolabist. It has the property that circles on a sphere remain circles when they are projected on to a flat plane, and that the angles between intersecting circles on the sphere remain unchanged when they are projected. Although there are reasons for suspecting that other conventions were used in earlier times, the convention that was almost universally followed with small astrolabes was to project stereographically from the south pole of the celestial sphere on to the plane of the equator. A line was extended from the south pole to the desired object on the celestial sphere; the point where this line intersected the plane of the projection was the location of the celestial object on the map. A series of such points was mapped to yield the coordinate lines.

With this stereographic projection, the closer a southern star is to the south celestial pole, the farther it will be from the north celestial pole on the plane of the projection, that is, on the rete. The projection of the entire celestial sphere is infinite in extent. In practice the rete is almost invariably made only a little larger than is necessary to accommodate the Tropic of Capricorn. Stars on the rete are represented by the tips of brass pointers. In principle these could be bent after a time to allow for the precessional movement of the earth's axis (although such allowance is not worth bothering with over periods of half a century or less). The bending is more likely to happen by accident than by design, however, and the pointers were usually made as rigid as possible. On the rete the circles of the tropics and the equator are not much needed, since they also appear on the plate below, and so they simply serve largely as supports for the star pointers.

The equator and the tropics are at right angles to the axis of the projection. As a result they turn out in projection to be circles that are concentric with the rete and centred on the north pole (represented by the pin). Moreover, if any degree graduations were to be put on the equator of the celestial sphere, they would lie uniformly on the projected equator. Neither of these properties belongs to the most important circle on the rete, namely the ring that represents the ecliptic. The centre of the ecliptic ring differs from the centre of the equator and the tropics because the plane of the earth's equator is inclined at an angle of 23½ degrees with respect to the plane of the earth's orbit. Longitudes are measured along the ecliptic from the vernal equinox, one of the two points where the ecliptic crosses the equator. This is the beginning of the sign of Aries; when the sun is at the "first point of Aries", day and night are of equal duration.

At the vernal equinox the sun is passing from south of the equator to the north and is heading into Aries and thence into the sign of Taurus on its progression around the ecliptic. When it reaches its most northerly point of the ecliptic at the summer solstice, 23½ degrees north of the equator, it leaves Gemini and passes into Cancer, hence the name of the tropic in the Northern Hemisphere at the latitude of +23½ degrees. As the sun continues its course along the ecliptic it eventually enters Libra as it again crosses the equator, although this time it is passing from north to south. This it does at the autumnal equinox, when again day and night are of equal duration. The sun reaches the winter solstice as it enters Capricorn 23½ degrees south of the equator, hence the name of the tropic of the Southern Hemisphere at a latitude of −23½ degrees. The sun's annual path as a whole is indicated by the outer edge of the ecliptic ring on the rete.

How is the ecliptic ring on the rete constructed? All that is needed is to plot the points of the summer and winter solstices [*illustrated in Fig. 14*]. Since in the stereographic projection circles remain circles on the map, these two points define the diameter of the ecliptic circle. The geometric centre of the ecliptic will lie midway between the two points. The ecliptic circle, when constructed, will cross the equator at the points corresponding to the equinoxes. (It so happens that the geometric centre of the ecliptic always falls at such a point that the angle made at the equinoxes from the centre of the ecliptic to the centre of the rete is twice 23½ degrees or, more precisely, twice whatever value is accepted for the angle the ecliptic makes with respect to the equatorial plane.)

The almucantars are drawn on the astrolabe plate below the rete in much the same way. The horizon of the observer is inclined to the celestial equator by 90 degrees minus the geographic latitude of the observer [*illustrated in Fig. 11*]. To find the two points determining each of the almucantars, it must be remembered that the almucantars are no longer great circles in planes passing through the centre of the earth; they are small circles parallel to the horizon. When they are drawn, the result is a series of circles around, but not concentric with, the observer's zenith. All their centres lie on the meridian.

The lines of equal azimuth are much more difficult to construct. They are a series of great circles stretching from the horizon to the zenith, and cutting the horizon circle and the almuncantars at right angles. Since the stereographic projection leaves angles unchanged, the lines of equal azimuth on the astrolabe plate will be arcs of circles that retain this property. In general, astrolabes show only those parts of the lines of equal azimuth that appear above the observer's horizon [*illustrated in Fig. 12*].

Before we turn to some uses of the astrolabe, what of its history? The theory of the stereographic projection can be traced back to one of the greatest of Greek astronomers, Hipparchus. He was born about 180 B.C. in Nicaea, not far from modern Istanbul, and he made observations from Rhodes and Nicaea. Unfortunately most of what we know about him comes from secondary sources. One of the most important of these sources is the Alexandrian astronomer Ptolemy, who was writing some four centuries later. Ptolemy was perhaps the greatest astronomer of the ancient world. His most important book, now known as the *Almagest*, makes no mention of the planispheric astrolabe. There are, however, references in his *Planisphaerium* to the "spider" of the "horoscopic instrument", suggesting that an instrument with something like the later form of the astrolabe was known in his day. The *Planisphaerium* is a treatise not on the astrolabe but on stereographic projection. It is known only from a Latin translation by Hermann of Carinthia (A.D. 1143).

Other scholars besides Ptolemy refer to the astrolabe, but many of the references are cryptic. The oldest surviving account of the instrument's construction and use was written in the sixth century by John Philoponos of Alexandria. A century later Severus Sebokht wrote on the subject in the Syriac language. After this time the instrument became moderately well known, judging from the many different treatises devoted to it in both the Islamic world and the Christian. Perhaps the first European treatise was one written by Hermann von Reichenau, or Hermann der Lahme (the Lame), a monk of Reichenau who died in 1054.

Much better known in medieval Europe was a work originally written in Arabic by Māshāʿallāh, who is thought to have been an Egyptian Jew. It was translated into Latin by 1276, and was the basis of the only good early treatise on the astrolabe in English, namely the one written a century later by none other than Geoffrey Chaucer.

His work, titled *A Treatise on the Astrolabe*, is dated about 1393. It

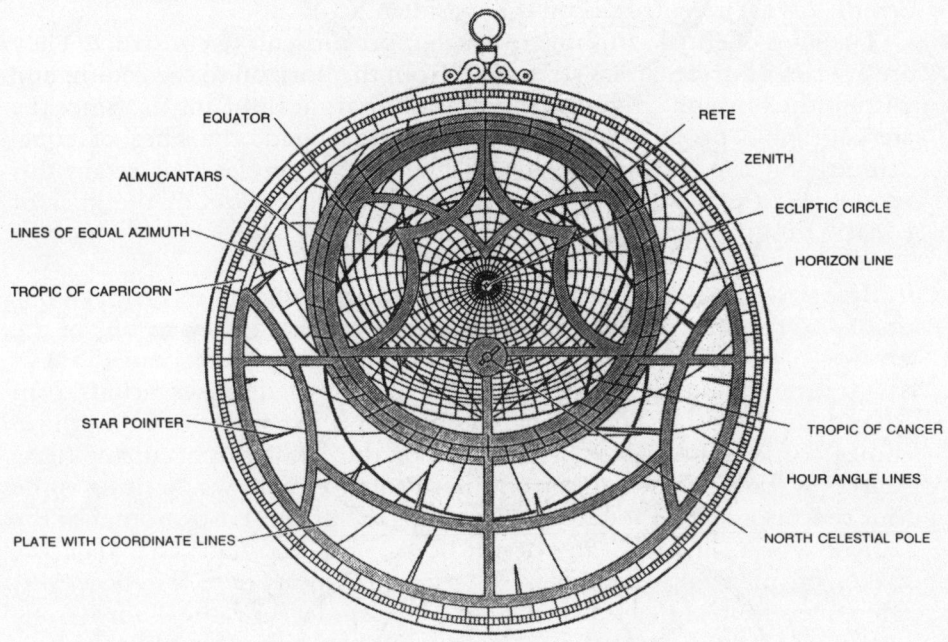

Fig. 8. **The Front of an Astrolabe,** showing those parts that were central to its function as an instrument for calculation. The fretted network, known as the rete, is a reproduction of the heavens. The tiny pointers indicate the positions of the stars. The eccentric circle at the top is the ecliptic: the yearly path of the sun through the sky. The rete pivots around a pin that holds in the plate behind it. The pin's position corresponded to the north celestial pole. The lines on the plate represent coordinate lines and are fixed with respect to an observer on the earth. The turning of the rete showed the daily motions of stars in relation to observers.

(Figs. 8-16 based on drawings by George V. Kelvin)

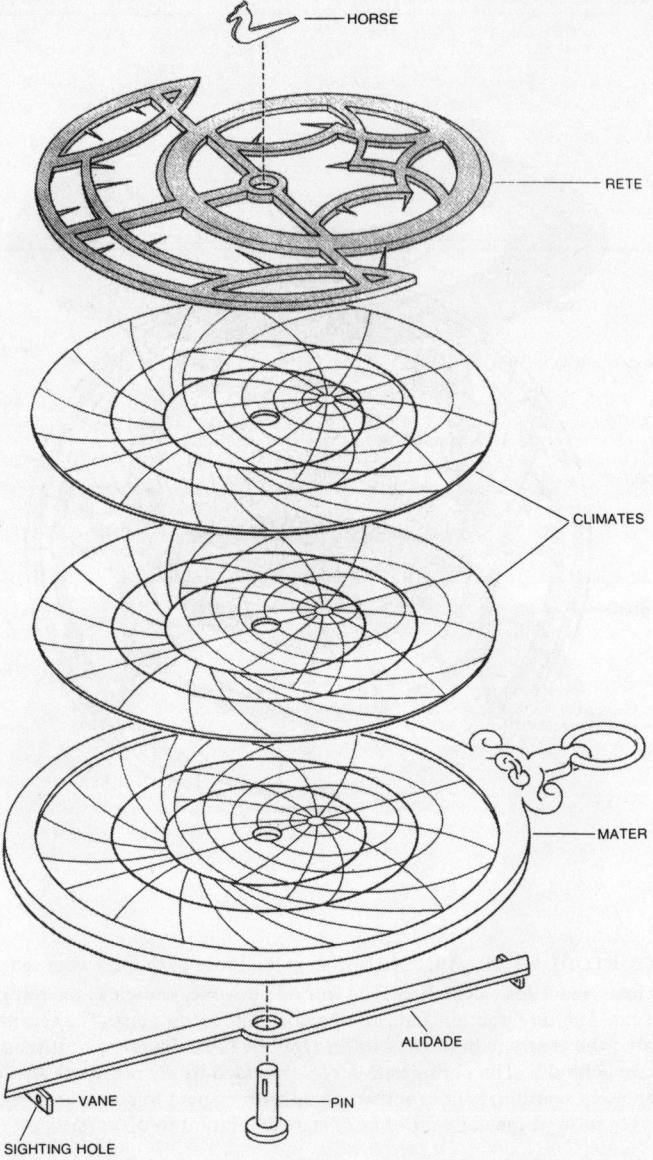

Fig. 9. Exploded View of an Astrolabe, showing the relationship of its various parts. The mater ("mother") is the main body of the astrolabe. The climates are plates engraved with coordinate lines for different latitudes, usually those to which the observer might travel. The alidade is a straight rule that was used for sighting celestial objects and finding their altitude. It was held to the back of the astrolabe and was free to rotate like the rete. The rete fits over all the climates, which are contained within the mater. The pin slides through the centres of all the plates and is secured by the horse, a wedge whose thicker end was traditionally in the form of a horse's head. Some astrolabes had no loose plates; in such instruments the mater was engraved as the one and only climate.

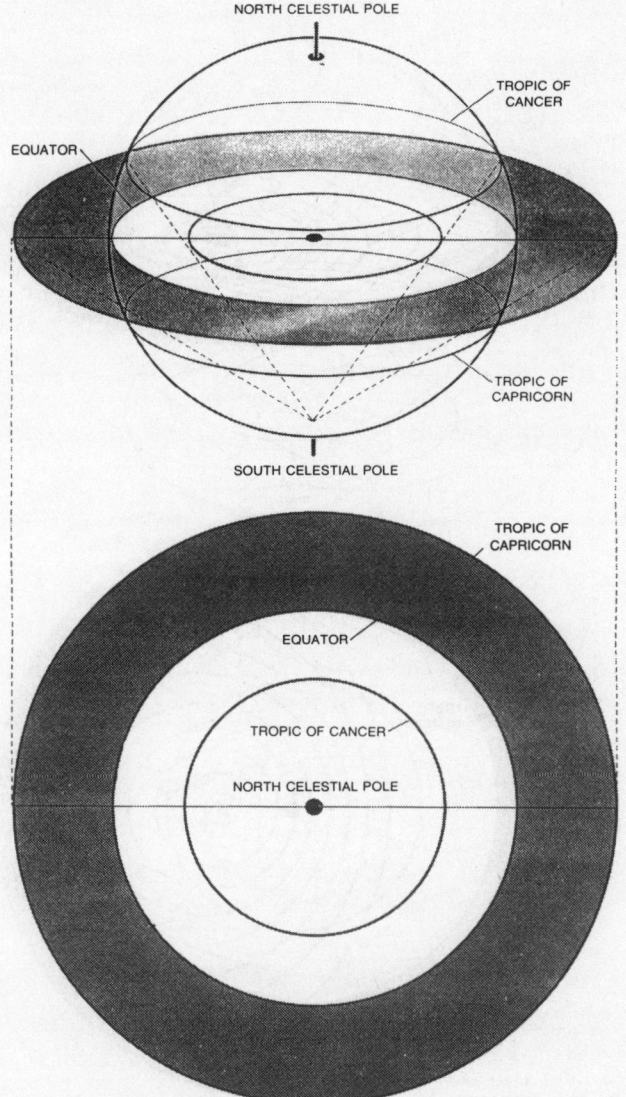

Fig. 10. **Stereographic Projection of Equator and Tropics**, showing how these circles on the celestial sphere (*top*) are projected on to the astrolabe plate (that is on to the mater or one of the climates) or on to the rete. On most astrolabes the plane of the equator (or a plane parallel to it) is taken to be the plane of the projection. A line is extended from the south celestial pole to the desired point on the celestial sphere (in this case one of the tropics or the equator). The point where this line intersects the plane of the projection is the location of that celestial point on the map. A series of such points is charted to yield the coordinate lines. The equator and tropics are at right angles to the axis of the projection. As a result they turn out to be circles that are concentric and centred on the point representing the north celestial pole (*bottom*). The astrolabe pin goes through the north celestial pole.

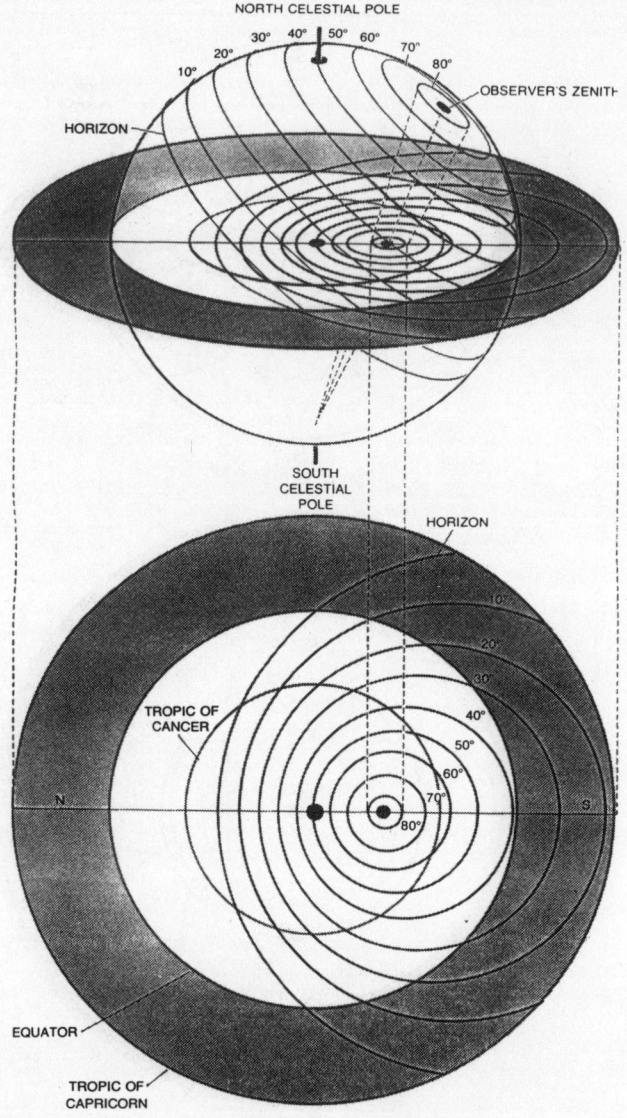

Fig. 11. Stereographic Projection of Almucantars, or circles of equal altitude concentric with the observer's zenith and parallel to the horizon, making circles on the plane of of the projection. They do not, however, have a common centre. In the illustration the observer's zenith is 50 degrees north of the equator. His horizon and almucantars are first shown as they appear on the clestial sphere (*top*). The stereographic projection has the property that circles on a sphere remain circles when they are projected on to a flat plane. In projection all the centres of the almucantars lie on the line (*NS*) that runs through both north pole and observer's zenith (*bottom*). This line is the projection of observer's meridian.

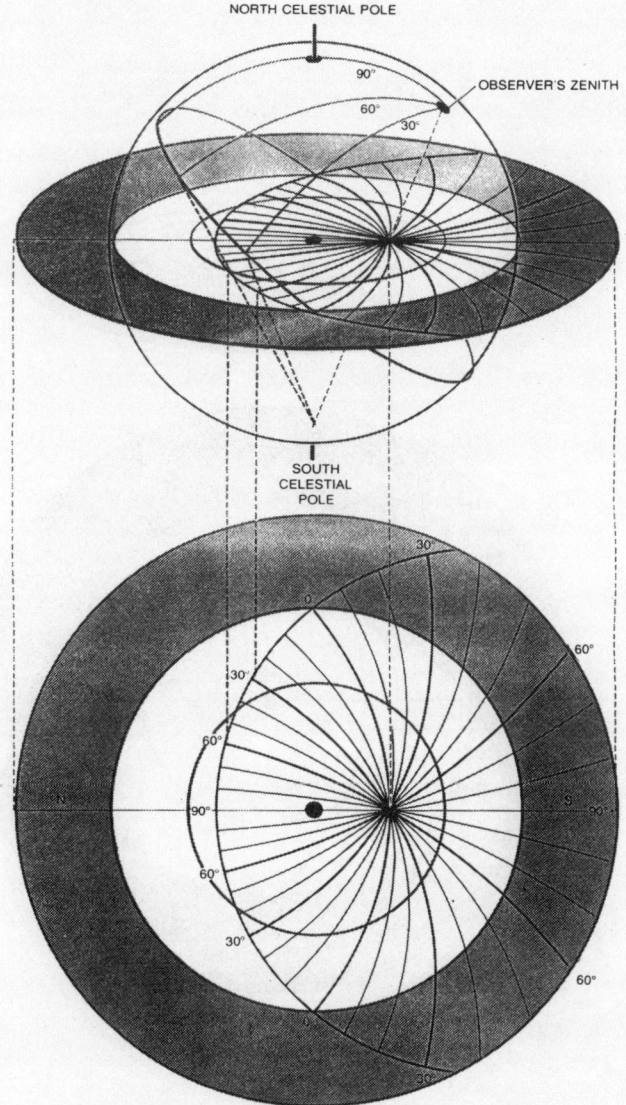

Fig. 12. **Stereographic Projection of Lines of Equal Azimuth,** is a series of great circles that stretch from the horizon to the zenith. Hence they cut the horizon circle, and the almucantars (*not shown*), at right angles (*top*). Angles between intersecting circles on a sphere remain unchanged when they are stereographically projected on to a flat plane. Therefore on the astrolabe plate the lines of equal azimuth will be arcs of circles that cut the lines of the horizon, and the almucantars (*again not shown*), at right angles. Most astrolabes show only the lines of equal azimuth that would appear above observer's horizon (*bottom*).

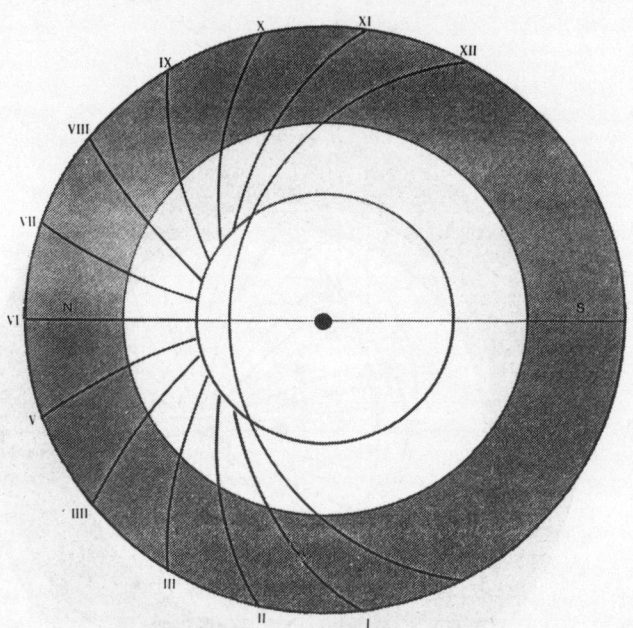

Fig. 13. **The Unequal Hour Lines.** The entire cycle of one day is divided into 24 hours. When the time was reckoned in unequal hours, as it is here, the period of daylight and the period of night, regardless of their duration, were both divided into 12 equal parts. Thus the hours of the day were not in general equal in length to the hours of the night. The hour lines were usually drawn only below the horizon line. Those portions of the concentric circles of the equator and the tropics are divided into 12 equal parts, beginning with the points of intersection with the west horizon. Corresponding points are connected then with smooth curves.

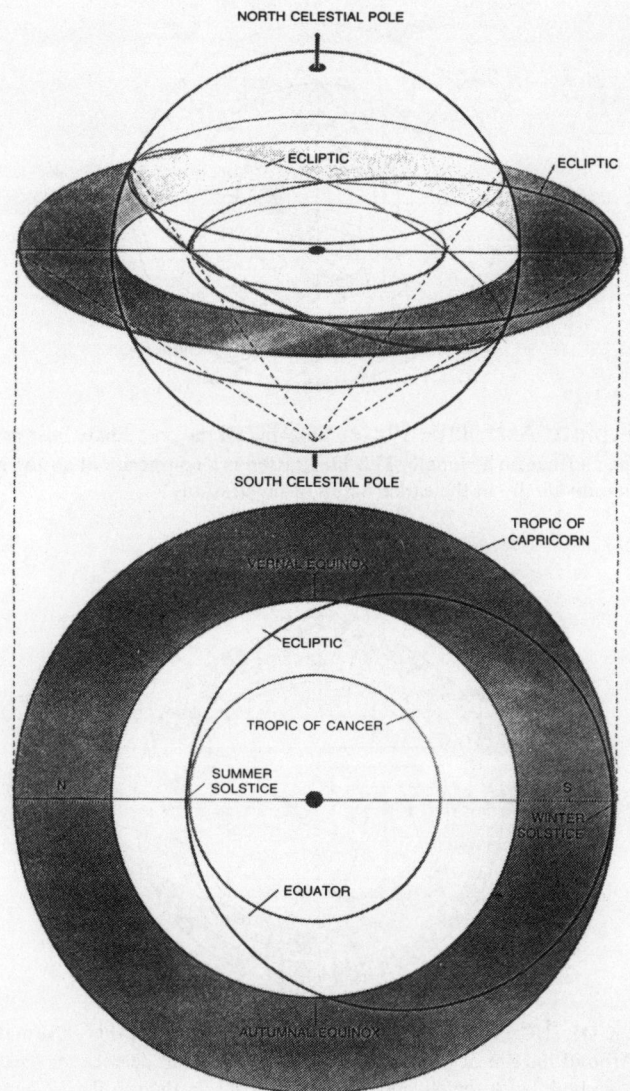

Fig. 14. Stereographic Projection of the Ecliptic as used for the Rete.
The ecliptic is the apparent annual path of the sun on the celestial sphere as seen from the earth. The Equator of the earth is tipped at an angle of 23½ degrees from the plane of the ecliptic, so that this angle is preserved on the astrolabe rete. All that is needed to draw the ecliptic is to plot the point of the summer solstice on the Tropic of Cancer and the point of the winter solstice on the Tropic of Capricorn. These two points define the diameter of the ecliptic circle, whose centre lies midway between. Ecliptic crosses equator at points corresponding to vernal equinox (first day of spring) and autumnal equinox (first day of autumn). Ecliptic is divided into 12 signs of zodiac starting at the point representing vernal equinox. The lines dividing the ecliptic on the astrolabe rete radiate from the north celestial pole.

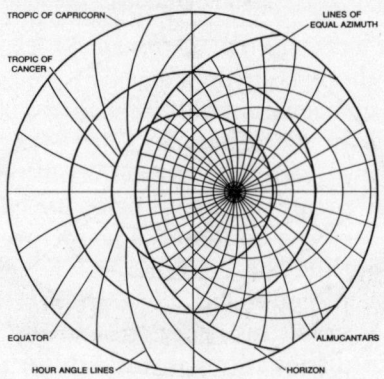

Fig. 15. Complete Astrolabe Plate, showing all the coordinate lines as they appear with respect to one another on a climate. This illustration is a composite of all the stereographic projections shown individually in the earlier series of illustrations.

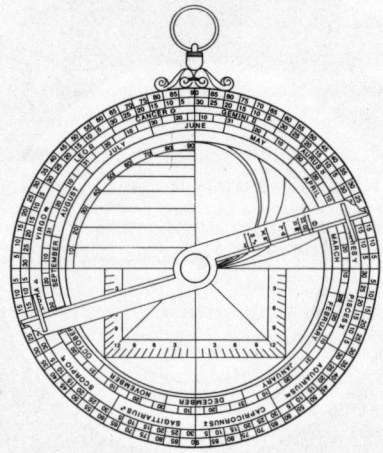

Fig. 16. Back of the Astrolabe, carrying the alidade and other information necessary to the observer. Around the rim of the example shown is a scale of degrees for measuring the altitude of a celestial body with the alidade. Immediately inside the rim the 12 signs of the zodiac are listed and divided into 30 degrees each. The scale of the months and the days inside the zodiac scale correlates the position of the sun on the ecliptic with the correct date. It is not concentric with the other circles to allow for the sun's nonuniform motion along the ecliptic. The design of the interior portion of the astrolabe back varies widely with the individual instrument. Here the quadrant at upper left contains horizontal lines from the degree markings; their distances from the horizontal diameter of the astrolabe correspond to the sine of the altitude of an object above the horizon. The quadrant at upper right contains lines for computing the time in unequal hours directly, independently of the front of the astrolabe. These lines are used in conjunction with the graduations on the alidade. The two quadrants at the bottom contain the "shadow squares". These could have been used in conjunction with a gnomon to get the cotangent or the tangent of the altitude of an object above the horizon. If they were accurately and completely divided, which they rarely were, they provided a means of measuring altitudes more precise than sighting with alidade.

The Astrolabe

survives in more than two dozen early manuscripts. In some of them it has the subtitle *Bread and Milk for Children*. The subtitle was probably provided by a scribe who was surprised at Chaucer's opening remarks in a work that would have been generally regarded as hard-tack for adults. In a modern rendering the work begins:

"Little Lewis my son, I well perceive signs of your ability to learn the sciences of number and proportion, and I also have in mind your earnest request especially to learn the contents of the treatise on the astrolabe." Chaucer goes on to outline the contents of the treatise, which in fact seems never to have been completed. He explains the need for a work in English, and he mentions his debt to earlier astronomers. It is unfortunate that his English is about as difficult for the ordinary modern reader as Latin was for Lewis.

By the 16th century the advent of printing and the steady improvement in techniques of engraving for publication had given rise to a number of magnificent new treatises on the astrolabe. These were in turn partly responsible for some striking advances in the art of the instrument maker. Astrolabes became larger, more decorative and more finely and accurately engraved. Nevertheless, allowing for differences in the language of the inscription, there was little or nothing about the typical astrolabe of the early 17th century that would not have been immediately familiar to an astrolabist of a thousand years earlier. The oldest surviving dated instrument is believed to date from A.D. 927/8. This particular astrolabe also carries a signature that is difficult to decipher, but which could be an Arabic form of a Greek name (Bastulos or Nastulos).

Before the end of the 13th century the planispheric astrolabe was known and used from India in the east to Islamic Spain in the west, and from the Tropics to northern Britain and Scandinavia. Variations in the general style of decoration are usually characteristic of the county and period of origin. The star pointers of the earliest retes, for instance, are usually of a simple dagger shape engraved only with the name of the star. At the other extreme later Indo-Persian astrolabists would often work the rete into an intricate and highly symmetrical foliate pattern, a difficult thing to do with what is essentially a star map having an asymmetrical natural arrangement. Astrolabe makers throughout the eastern world often damascened their instruments with silver and gold. It is interesting to trace from their surviving signed work successive generations of the same family. The family might all, for example, have worked at a centre such as Lahore, and thus perhaps have had connections with the Mogul court. Persian instruments tended to be extremely ornate, filled with fine ornamental engraving.

In the West the style of the rete is usually reminiscent of contemporaneous styles of church architecture. The style of written inscription is similar to the style of Western manuscripts in general, and is highly characteristic of the period in which it was done. There is good evidence that many astronomers the world over made their own instruments, although there was scarcely any important centre of learning that did not at one time or another have its specialist workshops turning out instruments professionally. European instruments were rarely signed with the maker's name during the Middle Ages, a time when anonymity was considered no vice. By the 16th century European astrolabes were often signed.

The physical size of most astrolabes is between three and 18 inches, although much larger ones are found in a number of rather different forms as the dials of astronomical clocks. The use of the astrolabe as a clock dial goes back to classical antiquity, when the rete was made to rotate once daily by waterpower. After the invention of the purely mechanical escapement at the end of the 13th century, astronomical clocks were to be found in most large European cathedrals. In a typical arrangement the star map and the map of coordinate lines of the conventional astrolabe change places, the coordinate lines being made into the rete and the stars being painted on a plate behind it. Usually the stars were made to rotate and the rete was fixed, but sometimes these roles were reversed. A model of the sun is occasionally found on the ecliptic of the star map; it is moved along the ecliptic manually, or by a mechanism, so that it completes one circuit of the ecliptic in a year. In order to judge the time from such a dial one must be familiar with at least the basic principles of the use of the astrolabe.

The chief purpose of the astrolabe was for telling the time. First the altitude of the sun or of a star was found by employing it as an observing instrument. Then, assuming that the observer knew where the sun or the star was on the rete, the rete was revolved until that point coincided with the almucantar for the appropriate altitude. (It is assumed that the observer knew which climate to choose for his latitude and on which side of the meridian line the object fell.) The refraction of the atmosphere, which changes the apparent position of objects in the sky, and which is greater the nearer they are to the horizon, was ignored. The sun's approximate position on the ecliptic for any day of the year is found from the calendar scale on the back of the astrolabe.

Once the rete is in the correct position, the observer can find his local time according to any one of several conventions. If the circumference of the astrolabe is marked in degrees, 15 degrees correspond to an hour. Noon will be when the sun is towards the top of the instrument, midnight when it is towards the bottom, 6:00 A.M. when it is to the left and 6:00 P.M. when it is to the right. Imagine now a great circle joining some object in the sky to the north celestial pole. The angle that this great circle makes with the meridian is the hour angle of the object. As a consequence of the stereographic projection, a rule lying on a line passing through the centre of the astrolabe and the point of the instrument representing the object makes an angle with the vertical diameter (the meridian line) equal to the hour angle of the object. The hour angle is so named because it can provide one with a measure of time through its change as the earth rotates (or as the stars rotate with respect to the earth). It is usually quoted in hours, minutes and seconds rather than in degrees of arc.

A number of different kinds of time can be told from an astrolabe. The first is sidereal time, or time by the stars, which is defined as the hour angle of the first point of Aries. If the day is counted from zero hours to 24 hours beginning at midnight, 12 hours will have to be added to the count for sidereal time, because at the vernal equinox the first point of Aries (which is at that time the position of the sun) will cross the meridian at local noon.

A second kind of time is true solar time: the hour angle of the sun regardless of its position with respect to the stars. There is another and more familiar type of solar time called mean solar time, which postulates a "mean sun" moving around the equator (rather than the ecliptic) at a uniform rate throughout the year, and making one complete circuit in exactly a year like the true sun. The earth moves around the sun in an ellipse with the sun at one focus, and it travels faster in its orbit the closer it is to the sun. Therefore from the earth the true sun seems to speed up and slow down in its course around the ecliptic. Thus the true sun and the mean sun move not only along different paths but also at different rates. In order to convert from observed true solar time to the more useful mean solar time, one must apply a correction known as the equation of time. It is based on knowledge of the earth's motion in its orbit and it can be found in reference books. The correction for the equation of time was scarcely ever applied before the 17th century. In order to convert mean solar time to the local time at some standard location such as Greenwich, the observer needs to know his geographic longitude, and again that adjustment was seldom made in astrolabe work.

A third kind of time is time measured in unequal hours. The ordinary man in the Middle Ages divided the period of daylight into 12 equal parts and the period of night into 12 equal parts regardless of the actual length of day and night. The length of the day-hours would obviously equal the length of the night-hours only when the sun was at one of the equinoxes. Many astrolabe plates include unequal-hour lines. In order to avoid their being confused with the almucantars, the unequal-hour lines were drawn only below the horizon line [*illustrated in Fig. 13*].

Time in the Middle Ages was often reckoned from sunrise or sunset even when measured in ordinary equal hours. Many astrolabe plates show lines similar in appearance to the lines of unequal hours but which are in fact for measuring the time in equal hours from sunrise or sunset.

Although the astrolabe was primarily an instrument for determining the time, it was an extremely useful adjunct of the astrologer's art. To cast a horoscope for a particular moment of time, an astrologer needs to know the degree of the ecliptic that is on the eastern horizon ("the ascendent"), the degree of the ecliptic that is on the western horizon ("the descendent"), the degree of the ecliptic where it crosses the meridian ("the degree of mid-heaven") and the degree of the ecliptic where it crosses the northward continuation of the meridian, once called the midnight line ("lower mid-heaven"). These degrees are easily read off the ecliptic ring once the rete is correctly positioned for the moment of time that is of interest: perhaps a moment of conception, of birth, of death or of some other important event such as a coronation. Once the four key points of the horoscope have been found, the 12 astrological houses (which are not to be confused with the signs) can be ascertained and the planets can be assigned to them. There are, however, many systems by which the division can be made. The most important one of these can be found in Chaucer's treatise on the astrolabe.

Like a modern electronic computer, the astrolabe in the Middle Ages was a source of astonishment and amusement, of annoyance and incomprehension. Imprecise as the astrolabe may have been in practice, it was undoubtedly useful, above all in judging the time. The instrument might have been used, more often than not, in the dark, but "dark" is hardly the word to describe the age in which it was so widely known and so well understood.

ASTROLABES AND THE HOUR-LINE RITUAL

IT SEEMS TO BE COMMONLY BELIEVED that a standard part of the engraving of the back of an astolabe is a set of hour-lines forming, as it were, a double horary quadrant. Although I have made no systematic study of the extant astrolabes of the world, I have examined 132 astrolabes in the Museum of the History of Science in Oxford for unequal-hour lines in the form of circular arcs, with rather surprising results.

Out of a total of 57 European astrolabes from before the year 1800, 25 have these unequal-hour lines, whereas only 16 of a total of 75 eastern astrolabes have them. Of the 25 European astrolabes, 15 have the lines symmetrically arranged as between the two upper quadrants, whereas only two of the eastern 16 have the lines in two quadrants. More significant is the empty ritual in accordance with which the lines are included on almost all of these 41 astrolabes. At best, the lines can give the (unequal) hour with an accuracy only about half as great as that given by the conventional astrolabe itself. At worst, the lines are carelessly drawn, unnumbered, very small indeed, and – worst of all – not associated with an auxiliary scale of solar positions.

This auxiliary scale may be included in at least three different ways:

1. Through graduation of the alidade.
2. Through concentric arcs, crossing the unequal-hour lines, marking as many solar positions during the course of a year as possible.
3. Through a scale of solar positions (mid-day altitudes) on the rim of the astrolabe.

The third possibility is never found on the Oxford astrolabes, although one might have imagined that the idea would have occurred to at least one astrolabist in history, for it is the alternative found on the 'old' quadrant–with–cursor. (On that instrument the date scale is movable, as it should be if the observer's geographical latitude is to be taken into account.)

Graduation of the alidade is found on only three of the European instruments, and on only two of the eastern – in both cases ignoring graduations with a separate purpose. Out of 41 instruments, the alidades of 5 are lost, and of three or four are possibly modern. Even so, it appears that, at best, about one in six of the 41 instruments is likely to have left the workshop with a graduated alidade.

Three European instruments, and two eastern, have concentrics, one of each set having been counted previously for its graduated alidade. The concentrics are at best for the divisions between zodiacal signs. In one European case, the concentrics are drawn as though for all solar altitudes between 0° and 90°. The concentrics on both eastern astrolabes are quite useless, being for the wrong latitude (see below). In only one case out of 41 has the maker given any indication that his unequal-hour lines are – as the graduations stand – of value at specific latitudes. (I say latitudes rather than latitude, since in one of his quadrants the graduations are for latitude 52°, and in the other for latitude 49°.)

Thus out of our 41 instruments, only six are at first sight of any value whatsoever as horary instruments, and of these, only one is earlier in date than the sixteenth century. This solitary medieval exception is a Fusoris-type instrument, IC 192, and on closer inspection it appears that the graduations on its alidade are worthless. The others, with their IC numbers where appropriate, are: IC 165 (Flemish?, 1558); acc. no. 73-11/2 (Italian, 1558); IC 274 (German, c 1580); IC 211 (French, 1595); IC 276 (German, 1609 + pasteboard); IC 19 Persian, AD 1641); acc. no. 57-84/164 (Indo-Persian, AD 1666/7).

In summary : out of 132 astrolabes examined, 41 instruments have the unequal-hour lines, and yet only four could have been used in at best a rough and ready way to find unaided the unequal hour. At a season well removed from equinox or solstice, only one of these (57-84/7, with its scale of mid-day altitudes) could have given the time with an accuracy approaching that of the main astrolabe, without the curious technique of using the astrolabe as an auxiliary instrument. Not a single medieval instrument has survived in a form which would suggest that the unequal-hour lines were used meaningfully. But finally, we note the possibility of our using the graduations associated with the unequal hour lines (either the graduations on the alidade, or the concentrics) as a means of deducing the geographical latitude for which the astrolabe (*if properly constructed*) was intended.

Thus on IC 19, the best eastern example, the six o'olock line intersects the concentric for the summer solstice at a point P for approximate altitude (shown on the rim, when the alidade passes through P) 57°. Subtracting 23½°, the approximate geographical colatitude emerges as 33½°, making the latitude 56½°, a nonsensical result. (The four plates now with the instrument range from latitude 21°40' to latitude 37°. The instrument was made for a man in Mashhad, where the geographical latitude is 36° 21'). As a European example: on the instrument IC 211, made for Paris (48° marked) or Lille (51° marked), the horary quadrants (one of equal hours) prove to be of value at geographical latitude 50°, a very reasonable figure.

SUMMA RATIONE CONFECTUM:

AN ASTROLABE DRAWN BY COMPUTER

(With O. Østerby and K. Møller Pedersen)

Refert Abraam Iudaeus in libro super opere tabularum de duobus astrolabiis summa ratione confectis, tum magnitudinis tantae, ut utriusque diameter novem palmis extenderetur, et cum duo fratres Bersechit instrumentorum compositores simul, ingrediente Sole in Arietem, Solis altitudinem observarent, non idem utrumque retulit instrumentum, sed duobus minutiis invicem variarunt.

Pico della Mirandola, *Disputationes adversus astrologiam divinatricem,* IX. viii

Two astrolabes, each about a metre in diameter, and yet they differed by two minutes of arc. Our brief guide to a way whereby an astrolabe of comparable dimensions may be charted automatically is meant for readers less fastidious than Pico. The conventional planispheric astrolabe, even when accurately drawn, cannot rival ordinary calculational procedures, correctly carried out; but it has, as many will testify, certain intuitive advantages over the more accurate alternatives.

The greatest practical problem in constructing an astrolabe from first principles is that of drawing the coordinate lines, for apart from the difficulties of finding their centres, an instrument of even modest dimensions will include circles with radii of several metres. A method which many people have been driven to use is to photograph separately the rete and plate of an old instrument, and to print the one image on an opaque background and the other on a transparent material, such as litho film, suitably reinforced for the pivot at the centre. Our real concern is not with such practical matters, which are left to the ingenuity of the individual, but with a relatively painless way of plotting the coordinate lines. For this it is necessary to have access to a computer and plotting equipment, but such are probably more easily found than a photograph of an astrolabe plate for your particular latitude. We plot two charts, one essentially of local coordinates (altitude and azimuth). This, for simplicity, will be called the *plate*, regardless of its final use. The other chart, which for reasons of tradition we call the *rete*, is actually nothing beyond a set of ecliptic coordinate lines on which astrolabe stars may be plotted as required. Since star coordinates change in time, as a result of precession and proper motion, and since some are looking for a historically useful instrument while others have the present sky in mind, we have made no attempt to add stars to the rete. Ours in therefore no more than the bare bones of an astrolabe, but at least it is the part hardest to draw. Lists of star coordinates are scattered and numerous, but one which will be found very useful for the middle ages is Paul Kunitzsch, *Typen von Sternverzeichnissen* (Wiesbaden, 1966); but note in this connection that the medieval words *latitudo* and *longitudo* should be interpreted with care. Most national astronomical ephemerides carry revised star lists with coordinates. Incidentally, it will probably be found most convenient to put the stars on the opaque background, since it is more likely that information will be added to the star map than to the chart of local coordinates. Note that equatorial coordinates may be easily superimposed on the rete, for they simply require a set of suitable radius vectors, drawn to a uniform circumferential scale, and a series of concentric circles at radial distances corresponding to the constant declinations they represent. These are easily drawn with the usual stereographic construction. (For an elementary introduction to the construction of the lines on an astrolabe, see above, chapter 14.)

It is not, of course, essential to make either chart rotate above the other. Most astrolabe operations can be done with a rule, a pair of compasses, and a graduated outer scale; but then the visual advantages of the astrolabe are lost, and one might as well use a desk calculator.

The equations of the curves

Although it is not necessary to know the equations which are used in the following programmes, they are included for reference purposes. The equatorial radius is denoted by R, and the plate radius by P. A and a denote azimuth and altitude respectively, and φ is typically geographical latitude; but these variables have other meanings which should be obvious enough, when the rete is plotted, since the required equations are fundamentally the same.

Altitude lines: circles $\quad x^2 + (y - d)^2 = r^2 \quad$ (condition: $x^2 + y^2 \leqslant \varrho^2$)

$$\text{where} \quad d = \frac{R}{2} \left\{ \cot \frac{\varphi + a}{2} - \tan \frac{\varphi - a}{2} \right\}$$

$$\text{and} \quad r = \frac{R}{2} \left\{ \cot \frac{\varphi + a}{2} + \tan \frac{\varphi - a}{2} \right\}$$

These are plotted for $a = 0, 1, 2, \ldots 89$ degrees and, in our example, $\varphi = 51° 50'$ (the traditional figure for Oxford). We also plotted for $a = -1, -2, \ldots -18$ degrees, to allow for twilight calculations.

Azimuth lines: circles $\quad x^2 + y^2 - 2 Rx \sec \varphi \tan A + 2 Ry \tan \varphi - R^2 = 0$

(conditions: $x^2 + y^2 \leqslant \varrho^2$ and $x^2 + y^2 - 2 Ry \cot \varphi \leqslant R^2$).

These circles are plotted for all integral values of A, and for the same value of φ as before.

Lines of ecliptic latitude: circles as for the altitude lines, but with φ taken as the complement of the obliquity of the ecliptic, and circles taken equally above and below the ecliptic.

Lines of ecliptic longitude: circles as for azimuth lines, but with φ now the complement of the obliquity, and the second condition removed.

Notes to the programmes

The following peculiarities of the programmes should be observed, so that necessary changes can be made to meet special needs. The most important of these are (1).

The Plate

1. The latitude for which our plate was drawn was 51°.83. See line 4.
2. The radius *(P)* of the completed disc was in our example 30 cm. See line 4.
3. The stereographic projection of the coordinate lines extends, as is customary, only as far south as the tropic of Capricorn. This can be changed by altering the 23.5 to whatever maximum southerly declination is required. Note: If the rete is going to be drawn for an obliquity differing from 23°.5, then the 23.5 in line 5 should be altered accordingly.
4. The lines of constant altitude (almucantarat) are first plotted at five-degree intervals, and then the programme is completed by adding intervening lines at intervals of one degree. This offers an opportunity for the two sets of lines to be drawn in different colours, or with different thicknesses of pen. The instructions as to spacing, for those who wish to alter them are da := 5 (line 4) and da := 1 (last line). The eighteen lines of the second set, mentioned above, are plotted without interruption below the horizon. The instruction (last line) reads starta := −18.
5. Azimuth lines are for five-degree intervals only, the instruction being dA := 5 (line 4). They are drawn above the horizon only.

The Rete

1. The obliquity accepted here is 23°.5, its complement being entered in the programme (line 4).
2. The radius *(P)* is again 30 cm.
3. A limit corresponding to that in (3) above is again set to the disc. To extend the disc to latitude (celestial) $x°$ south, replace instruction delta := 23.5 by delta := x.
4. Similar instructions to those under (4) above relate to the spacing of the two sets of parallels of celestial latitude, except that there is now no distinction between north and south sides of the ecliptic. Instructions as to spacing are as before (again in the fourth and last lines).
5. Longitude lines are at five-degree intervals as before, but these no longer cover only the upper half of the sky.

Correct orientation of the finished charts is simplified by our inclusion of an equatorial circle and a line through the centre (colure of the equinoxes) on both charts.

Note that for both plate and rete the pen must first be positioned on the plotter at the centre envisaged for the final disc. Computation time for each chart, with the 30 cm radius used, was approximately one hour on the Gier computer at Aarhus, and plotting time was two hours. The times taken are roughly proportional to the radius. Much time can be saved, of course, by omitting the lines at a spacing of one degree (see (4) above).

Three plotting procedures have been used in the programme, and a short description of their function is given here. The unit is 1 cm.

plotstop   interrupts the plotting for change of pen (see (4) above);
plotline   (xo, yo, x1, y1) draws a straight line from (xo, yo) to (x1, y1);
plotcurve   (x, y, t, t1, t2, n) draws a curve given by (x(t), y(t)), with the parameter
            t going from t1 to t2. The curve will actually consist of n straight line
            segments.

Programme for the plate

begin integer n; *boolean* thin;
   real R, P, delta, phi, starta, dda, da, a, ar, A, dA, pi, ro, v, z1, z2, tan, co, si,
      se, tanA, d, r, theta, theta1, theta2, k, omega1, omega 2, omega;
   P := 30; phi := 51.83; da := 5; dA := 5; starta := 0; thin := *false*; dda := da;
   pi := arctan(1)×4; ro := pi/180; phi := phi×ro; delta := 23.5;
   plotline(−P, 0, P, 0); z1 := 2×pi;

   if delta > 0 *then*

   begin delta := delta×ro; omega := (pi/2 + delta)/2;
      R := cos(omega)/sin(omega)×P; n := P×pi×20;
      plotcurve(cos(v)×P, sin(v)×P, v, 0, z1, n)

   end else R := P;
   n := pi×R×20; plotcurve(cos(v)×R, sin(v)×R, v, 0, z1, n);
   z1 := sin(phi/2)/cos(phi/2)×R; plotline(0, P, 0, −z1);
   co := cos(phi); se := 1/co; si := sin(phi); tan := si/co;

   for A := −90 + dA *step* dA *until* 89 *do*
   begin tanA := sin(A×ro)/cos(A×ro); r := sqrt((tanA×se) ↑ 2 + tan ↑ 2 + 1)×R;
      k := (tanA ↑ 2×se + tan×si)×R/r/si − (P ↑ 2 − R ↑ 2)/2/R×co/si/r;
      z1 := tanA/si; theta1 := arctan(z1);
      z2 := sqrt(z1 ↑ 2 − k ↑ 2 + 1); omega1 := arctan(k/z2);
      k := (tanA ↑ 2 + 1)×se×R/r×si;
      z1 := tanA×si; theta2 := arctan(z1);
      z2 := sqrt(z1 ↑ 2 − k ↑ 2 + 1); omega2 := arctan(k/z2);
      theta := theta1 + omega1;
      if theta < theta2 + omega2 *then* theta := theta2 + omega2;
      omega := theta1 + pi − omega1; omega2 := theta2 + pi − omega2;
      if omega > omega2 *then* omega := omega2;
      z1 := tanA×se×R; z2 := tan×R; n := (omega − theta)×r×10;
      plotcurve(cos(v)×r + z1, sin(v)×r − z2, v, theta, omega, n)

   end;

again : *for* a := starta *step* da *until* 89 *do*
   begin if thin ∧ a ⩾ 0 ∧ entier(a/dda)×dda = a *then* a := a + da;
      ar := a×ro; z1 := (phi − ar)/2; z2 := (phi + ar)/2;
      tan := sin(z1)/cos(z1); co := cos(z2)/sin(z2);
      d := (co − tan)×R/2; r := (co + tan)×R/2;

```
        if d + r < P then theta := pi
        else begin k := (d ↑ 2 + r ↑ 2 – P ↑ 2)/d/r×.5;
                if abs(k) < .1 then theta := pi/2 – arctan(k/sqrt(1 – k ↑ 2))
                        else theta := arctan(sqrt(1 – k ↑ 2)/k);
                if theta < 0 then theta := pi + theta
        end;
        theta1 := 3×pi/2 – theta; theta2 := 3×pi/2 + theta; n := theta×r×20;
        plotcurve(cos(v)×r, sin(v)×r + d, v, theta1, theta2, n)
    end;
    if ¬ thin then
        begin plotstop; da := 1; starta := –18; thin := true; go to again end
end
```

Programme for the rete

```
begin integer n; boolean thin;
    real R, P, delta, phi, starta, dda, da, a, ar, Á, dA, pi, ro, v, z1, z2, tan, co, si,
        se, tanA, d, r, theta, theta1, theta2, k, omega1, omega 2, omega;
    P := 30; phi := 66.5; da := 5; dA := 5; starta := –45; thin := false; dda := da;
    pi := arctan(1)×4; ro := pi/180; phi := phi×ro; delta := 23.5;
    plotline(–P, 0, P, 0); z1 := 2×pi;

    if delta > 0 then

    begin delta := delta×ro; omega := (pi/2 + delta)/2;
        R := cos(omega)/sin(omega)×P; n := P×pi×20;
        plotcurve(cos(v)×P, sin(v)×P, v, 0, z1, n)

    end else R := P;
    n := pi×R×20; plotcurve(cos(v)×R, sin(v)×R, v, 0, z1, n);
    plotline(0, P, 0, –P);
    co := cos(phi); se := 1/co; si := sin(phi); tan := si/co;

    for A := –90 + dA step dA until 89 do

    begin tanA := sin(A×ro)/cos(A×ro); r := sqrt((tanA×se) ↑ 2 + tan ↑ 2 + 1)×R;
        k := (tanA ↑ 2×se + tan×si)×R/r/si – (P ↑ 2 – R ↑ 2)/2/R×co/si/r;
        z1 := tanA/si; theta1 := arctan(z1);
        z2 := sqrt(z1 ↑ 2 – k ↑ 2 + 1); omega1 := arctan(k/z2);
        theta := theta1 + omega1;
        omega := theta1 + pi – omega1;
        z1 := tanA×se×R; z2 := tan×R; n := (omega – theta)×r×10;
        plotcurve(cos(v)×r + z1, sin(v)×r – z2, v, theta, omega, n)

    end;
again : for a := starta step da until 89 do
```

```
begin if thin ∧ entier(a/dda)×dda = a then a := a + da;
    ar := a×ro; z1 := (phi − ar)/2; z2 := (phi + ar)/2;
    tan := sin(z1)/cos(z1); co := cos(z2)/sin(z2);
    d := (co − tan)×R/2; r := (co + tan)×R/2;
    if d + r < P then theta := pi
    else begin k := (d ↑ 2 + r ↑ 2 − P ↑ 2)/d/r×.5;
        if abs(k) < .1 then theta := pi/2 − arctan(k/sqrt(1 − k ↑ 2))
                       else theta := arctan(sqrt(1 − k ↑ 2)/k);
        if theta < 0 then theta := pi + theta
    end;
    theta1 := 3×pi/2 − theta; theta2 := 3×pi/2 + theta; n := theta×r×20;
    plotcurve(cos(v)×r, sin(v)×r + d, v, theta1, theta2, n)
end;
if ¬ thin then
begin plotstop; da := 1; starta := −46; thin := true; go to again end
end
```

Summa ratione confectum

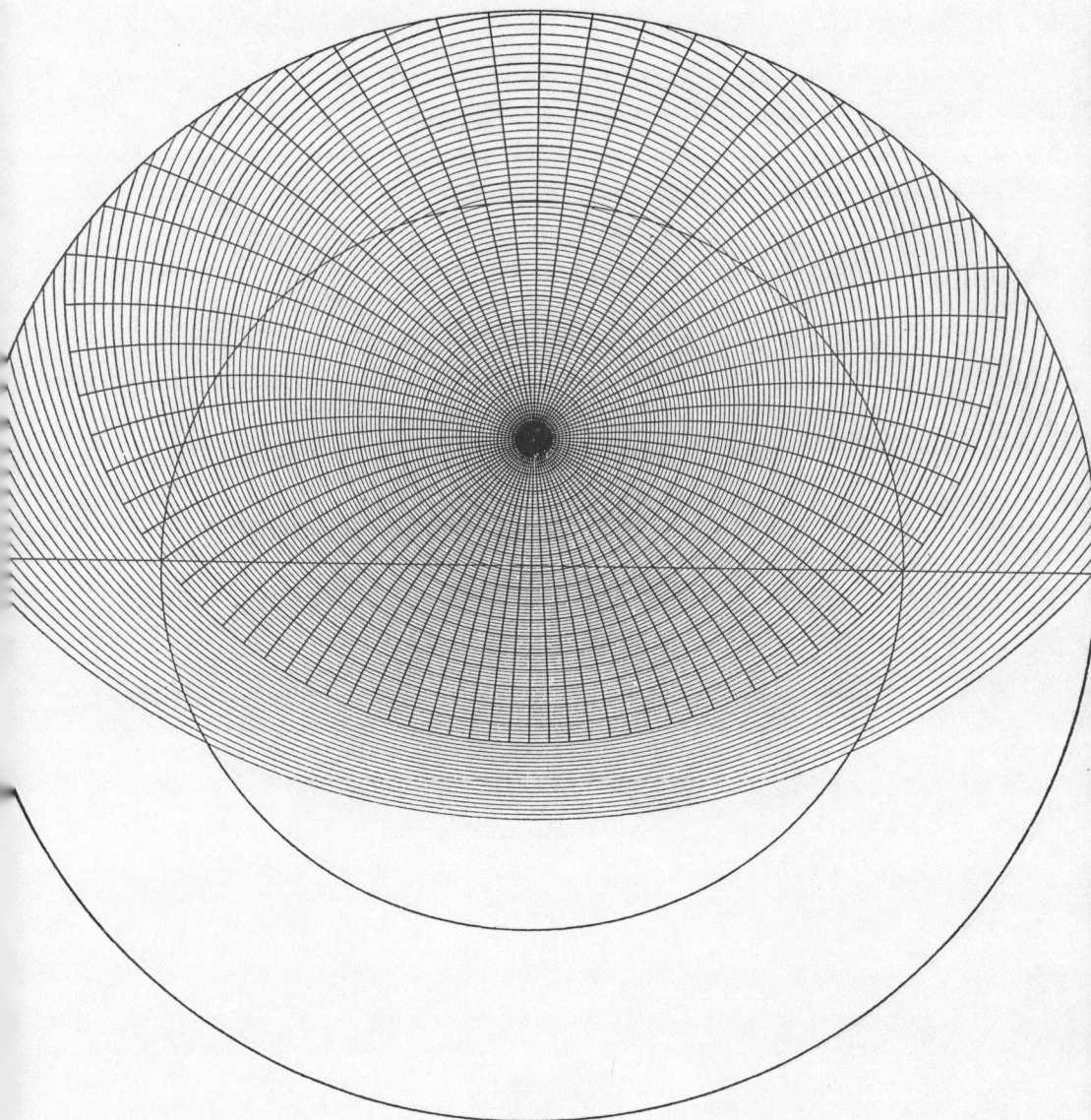

Fig. 17. Plate for latitude 51°.83 (the medieval figure for Oxford).

Figs. 17-19: Specimen plates, plotted using the above programme at the University of Aarhus in 1974.

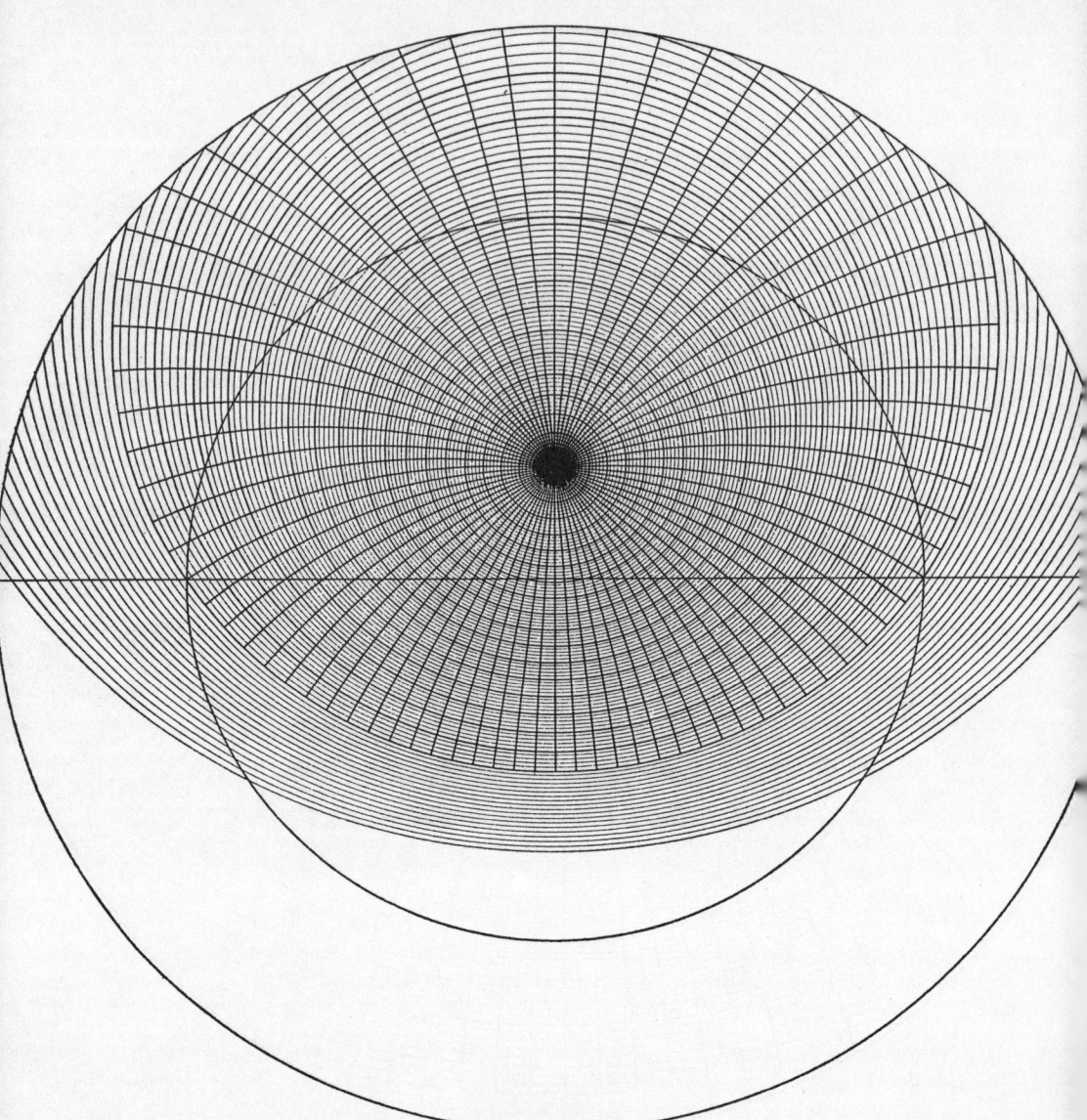

Fig. 18. Plate for the latitude of Aarhus (56°).

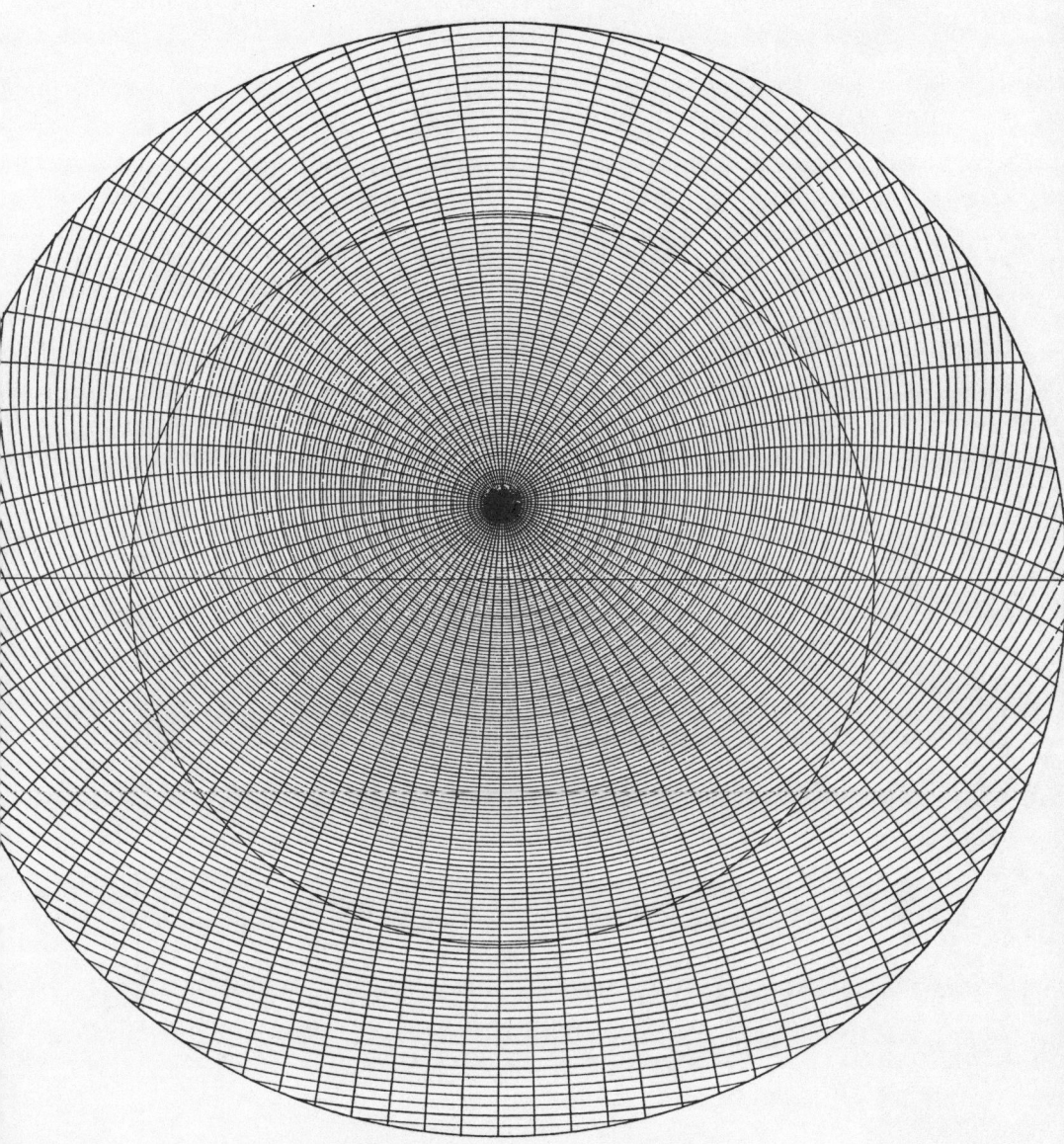
Fig. 19. Plate of ecliptic coordinates (rete).

17

ETERNITY AND INFINITY IN LATE MEDIEVAL THOUGHT

Among the many topics from Aristotle's *Physica* that stimulated debate in the late Middle Ages, those concerning the infinite, in its various senses, occupy an important place, since they had repercussions not only in natural philosophy, but in theology too (¹). Their history, however, has become mixed with a strongly mythological element. It is well known that Georg Cantor, architect of the most influential modern mathematical theory of the infinite (that is, the theory of transfinite sets), paid many compliments to scholastic philosophers. As a result, we often meet with empty comment of the sort «Grosseteste, in proposing that some infinite quantities could be greater than others, took the first step on the road to modern set theory». On the other side, we have those who make the casual claim that the actual infinite in a strong sense presented scholastic philosophers with no important problems, since Christian belief forbade any rival to God's infinity, and since Aristotle's arguments against the actually infinite were generally accepted. It is true that Aristotle's ideas and vocabulary occupied the centre of the stage — as they were to do to the nineteenth century, and do in some quarters even in our own day — but there were many variants on his fundamental theses, and in any case there was always a strong neo-Platonic element in the theological backdrop. It is hardly possible to provide a brief summary of two thousand years of active debate, but a few general remarks are in order.

Aristotle's central definition of 'infinite' was «that which cannot be gone through», and his paradigm of a sequence that cannot be gone through was the sequence of natural numbers (²). He distinguished further «infinite with respect to addition» (as number and time are) and «infinite with respect to division» (as in the case of an endlessly divisible geometrical line, or the time continuum). In *Metaphysica* K. 10, he made it clear that he was not concerned with two other meanings: the infinite as a separate, independent, thing; and as the Indivisible, the One. He asked questions rather about the limitation (the possibility of 'going through' in appropriate senses) of, for instance, number, magnitude, time, and body. He made much use

---

(¹) This paper is written from a largely Oxonian point of view — and this is no accident. It supplements, and is in turn supplemented by, my two chapters in the second volume of the *Oxford University History*, ed J. Catto (forthcoming).

(²) The Aristotelian doctrine is mostly to be found in books III and VIII of the *Physica*.

of the elimination of concepts on the grounds of hidden inconsistency. Thus if 'body' means «something limited by a surface», then the concept of 'unlimited body' is inconsistent. And if «a quantity is infinite» means «it is always possible to take a part outside what has already been taken», and if 'whole' and 'complete' mean that there is nothing more to be taken, then a completed infinite, an actual infinite, is a contradiction in terms. Without some sort of revision of these fundamental definitions, no one was likely to make any more headway with the problem of the actual infinite than Aristotle had done.

After the powerful onslaught on Aristotle by the sixth century Christian (Monophysite) theologian John Philoponos of Alexandria, a number of Arab theologians were inspired to develop his style of argument for the existence of God in particular. The so-called kalām argument, in the hands of al-Kindī and al-Ghazālī, derived the existence of God using the lemma that the series of past events in the world must be finite. From there, it was a question of arguing that the world could not have originated without a cause; and God is the cause. This type of argument was handed on to the Latin West, by routes that have not yet been fully charted, but the discussion became heated with the attempts made by the ecclesiastical authorities in the thirteenth century to control the rising tide of Averroism. When Aquinas gave an account of the standard arguments for a world of creatures that has always existed, he was able to draw on Avicenna, Averroes, and Maimonides. From the point of view of influence, rather than originality, Bonaventure had already become a spokesman for the Christian's belief in a world with a beginning in time. His commentary on the *Sentences* of Peter Lombard, composed around 1250-1251, could not, however, escape the criticism of those Christian philosophers who wanted their faith to be defended by cogent argument rather than argument that was merely comforting. Aquinas is the classic example of one whose honesty obliged him to admit that he could not give intellectual reasons for rejecting the thesis of the eternity of the world. This is the context of the condemnation of Averroistic and Avicennian (and other) theses by Etienne Tempier, bishop of Paris, in 1277. At least ten of the theses concern the eternity of the world. The spirit of that condemnation was opposed to the prevailing 'scientific' tendencies of the time, as manifested especially in Paris and Oxford. It was a plea for a free, undetermined, world, ruled by a free God; for a world that is free even though we are in possession of a knowledge of necessary laws. And this idea touches on the more reputable arguments both for and against an eternal world.

Rational proof, conceivability, and faith were not always easily reconciled, added to which the scholars concerned rarely appreciated the many possible combinations of doctrinal stances on the eternity question (the same is true of most modern commentators). One might hold that a rational proof of (non-) eternity is already available; or that it is conceptually admissible, even if the proof is not available; or inadmissible; or something to be held true by faith alone; or something in principle unprovable. Some of these positions may obviously be combined in an interesting way; others in a way that is uninteresting ([3]). Both in the *Summa theologiae* (c.

([3]) See my *The Eternity Question; Some Medieval Arguments*, (forthcoming), for a more detailed discussion.

1266-8) and again in a *quodlibet* (Easter 1270) that came shortly before the earlier batch of condemnations by Tempier (December 1270), Aquinas claimed that it is impossible to give a demonstrative proof of the fact that the world was not always in existence: as Giles of Rome apparently found, to defend this position was to lay oneself open to misinterpretation and could be positively dangerous ([4]). Aquinas, in other words, sat on the fence; and some of his commentators have tended to be alarmed by signs that he might have been looking too sympathetically at the 'eternity' side of it. His view on an actual infinity of created objects certainly shifted between the writing of *Summa theologiae* and *De aeternitate*, and in the latter he is to be found arguing that there is no inconsistency in the two notions 'to be created' and 'to exist always'. In effect, he thought something *could* have been produced from eternity by the world's Creator. Most scholars took eternity to be a possibility as soon as they had scanned existing arguments and found none against it convincing. The later Aquinas saw the need to demonstrate the consistency of the language used to describe a world as eternal or non-eternal. Naturally he saw the heretical side of the defence of eternity for the world, rivalling God's eternity. Oddly enough, in saying (like his followers Giles and Godfrey of Fontaines) that the old arguments against this heretical view were unsatisfactory, in his *Summa contra Gentiles* he even made Aristotle out to have been no more than a critic of *actual* arguments against eternity, rather than a positive believer in the idea. Grosseteste in *Hexaëmeron* had earlier scorned William of Conches for reading Aristotle in this way ([5]).

In his *Summa theologiae* Aquinas collected together most of the previous arguments for and against the world's eternity — in the form of theses, counter-theses, syntheses, and emendations in general, all without much attention to order. Confronted with such a collection — and there were many others of much the same scope — the average medieval scholar must have felt cheated, much as Laplace felt when he considered how unlucky he was to have been born after Newton. There were a few, nevertheless, who did manage to add appreciably to the stock of relevant argument, and the first text I shall consider is one by John Duns Scotus, from his *Sentences* commentary in its *ordinatio* version, known as the *Opus Oxoniense*.

Duns Scotus' final purpose was to prove the primacy of God, and to derive God's absolute properties. As regards primacy, he must show that there is something 'first with respect to efficient causality'. He also tried to show that there must be a first in the order of final causes; and a first by way of pre-eminence. But he began by asking whether there is an actual infinite in the realm of beings generally ([6]).

Aristotle had already discussed the possibility of a compound infinite body with

---

([4]) After the 1277 condemnations, he refused Tempier's request that he retracted, and left Paris in 1278. He returned in 1285, as holder of the first Augustinian chair in theology there.

([5]) Aquinas quoted from the *Topics* (Leonine ed., pp. 33ff). For the Grosseteste criticism, see R. C. Dales (ed.), *Roberti Grosseteste... in VIII Libros Physicorum* (etc.), Boulder, Col. 1963, p. XX.

([6]) The *Opera omnia* of Scotus being produced by the Vatican Scotus Commission (ed. C. Balić et Al.) began with the *Ordinatio* (prol.-dist. 48, vols 1-6, 1950-58). The relevant arguments are largely available in John Duns Scotus, *Philosophical Writings, a Selection*, translated with introduction by A. Wolter, Edinburgh 1962, pp. 39-79.

elements finite in number [7]. He was concerned with an opposition or contest, so to speak, between two elements, where the potency of one element (say fire) can overcome that of another (say air) in a certain ratio. In this case, he thought, the fire would succumb to the air if the air were infinite in amount. The fact that we have fire in the world means that the air is finite. In this, its original scientific dress, the argument had an obvious rationale: it was based on the contrariety presupposed in the elements — air is cold, fire is hot, and so on (having nothing to do with the logical theory of contraries). Later Christian theologians were exercised by another problem this type of argument seemed to pose: God's infinite goodness should counteract evil; yet evil exists, tempting to lead one to the conclusion that God does not [8]. A version of this sort of puzzle is found in Duns Scotus, who considered whether it militated against actual infinite being of any sort. The argument he quoted went as follows: *If the actually infinite were one of two contraries, nothing of the other could exist in nature*. For all the outward physical or metaphysical form of the argument, the problem of evil was always lurking in the background. Henry of Ghent, and Aquinas before him, had argued that the major premiss in such an argument holds only of formal contraries, and that evil is not formally contrary to God. But it makes no difference, said Duns, whether the contrariety be formal or virtual (to do with powers). If something be infinite «it will not tolerate anything contrary to its effect, since by reason of its infinite power it will destroy anything incompatible with its effect». His example: if the Sun were to be infinitely hot, either virtually or formally, in either case it would leave nothing cold in the universe. The argument is badly flawed: even on its own terms, it cannot prove that there is *nothing* infinite until *every* potential contrary in the world has been reviewed. He had no theory worthy of the name, telling him what was incompatible with what, outside the (here trivial) bounds of logic. It is noteworthy, though, that the principle of annihilation of contraries that Duns involved was not an all-or-nothing affair, but involved *degrees* of compensating contrariety, very much in the spirit of the new physics. Of course, he was not a product of the mathematically orientated Oxonian groups, and he made a poor showing in his handling of another, quasi-mathematical, argument against an actual infinite in the realm of beings, evidently relying too heavily on an anthropomorphic analogy [9]. His volitional physics is still a far cry from the Mertonian kinematics that was just around the corner; but he did rescue the Anselmian argument for the existence of God, and revitalise the Augustinian tradition. Without going into the details more closely, it has to be said that — like Augustine — he saw the dependence of the world on God as a *continuing* dependency. Creation and conservation are concepts, he thought, distinguished only by their different sorts of time-reference. This doctrinal position, bearing in mind the great influence of Scotus, must have smoothed the way for those who, already half convinced by Aquinas that there is nothing irrational in

---

[7] *Phys.* III. 5 (204b. 10-18).

[8] Thus Aquinas in *Summa theologiae* 1.2.3. Cf. [ps.-] Siger of Brabant, *Questions sur la Physique d'Aristote*, ed. P. Delhaye, Louvain 1941, pp. 193-4.

[9] He tried to show how instantaneous motion need not be self-contradictory, as Aristotle had said it was. He phrases the argument in terms of powers, finite and infinite, and gets very confused when considering the cooperation of both sorts of power. He was perhaps led astray by Averroes, in *Met.*, Book XII. See note 3.

the concept of an eternal world, wished to live dangerously and investigate it more closely.

Duns Scotus had some original things to say in his proof of God's existence concerning the relations between two different sorts of series of 'causes', accidentally and essentially ordered series. Without pursuing the argument, note that his aim was to show a *producing* cause outside the series of contingent beings. Ockham wished to make the external cause a *conservator*, in addition. Briefly, the difference between an efficient cause and a conserving-efficient-cause is that the latter has to be simultaneous with what it conserves while the former does not have to be simultaneous with its effect. A string of conserving causes of something in existence *now* would be — if infinitely long — a present and actual infinity, a much more challenging 'actual infinity' than a series of causes disappearing into the remote past. Ockham accepted Aristotle's arguments against this actual infinity, concluded that the conservation series must be finite, and made God the first term, the unconserved conservator as we might say [10]. He found his God in the present, and not by looking into the distant past; and he did so using Aristotle's physics.

It is commonly said that the effect of the 1277 condemnations was to reduce the number of topics in theology where rational demonstrations were to be expected. Whatever the evidence — and it does not seem to me to be very great —, Oxford scholars, at least, do not seem to have been afraid to debate the 'endless hierarchy' question. Of course half a century had passed by the time Bradwardine (and for that matter Ockham) touched on the subject, but in its geometrical style alone his argument is revealing. When he wrote on God — and he tells us that the structure of the universe interested him only inasmuch as it throws light on the being of man and God — he began with two postulates: «(I) God is the highest good, in comparison with whom nothing is better or more perfect; (II) There cannot be an endless sequence or hierarchy in things. There must be a first cause in the chain of causes» [11]. He followed, in this Euclidean-style presentation; with a forty-part corollary! (God can only act where he is, ...etc.). Admittedly, most of the corollary is filled with orthodox truths, directed against atheists, image-worshippers, and so forth, and although the outward dress of his argument is axiomatic, it is in reality often impressionistic and weak, relying on the use of concepts that led his audience to think analogically. He saw the danger: anthropomorphism could only work to suggest limits to God's infinity (he even thought that taking God as First Cause was a dangerous step, inasmuch as all heresies were traceable to the idea!). He made a number of points with a bearing on Ockham's argument, and one wonders whether perhaps Leibniz had read the passage where Bradwardine asked whether God leaves the world to run on, after creation, without interfering in secondary causes. Leibniz, be it noted, was one of the intermediaries in that long (but finite)

[10] *Quaestiones super libros Physicorum*, qq. 135-7; discussed by D. Webering, *Theory of Demonstration according to William Ockham*, St. Bonaventure, N.Y. 1953, pp. 105-106, whose interpretation differs from mine. See note 3. Reminiscent of Ockham here is an argument by Fitzralph that can be summarized: what God can do successively, He can do simultaneously. He cannot create an infinite in act simultaneously, therefore neither can He do so successively. See for this argument A. Maier, *Ausgehendes Mittelalter*, vol. I, Rome 1964, p. 78.

[11] *De causa Dei*, 1.1.1.E and 1.1.2.A. For more circumstantial detail, see H. A. Obermann, *Archbishop Thomas Bradwardine, a Fourteenth Century Augustinian*, Utrecht 1958, pp. 51-54.

series of scholastic writers on the eternity question running up to Kant, whose First Antinomy retraced much conventional ground.

Appraisals of the 1277 condemnations have varied widely. For some, they represent only a last ditch stand by conservative secular theologians against the Faculty of Arts and the regulars trained in the new sciences; for others, they represent the birth of modern science. A more balanced view of their impact on science is that they encouraged scholars to look beyond the limits of Aristotelian and Averroistic philosphy for new interpretations of nature, limited only by the demand that God be not required to do the impossible, «if 'impossible' be understood according to nature» ([12]). Often it is hard to decide whether the wording reflects a desire to be free from Aristotle, or merely ignorance of his work, and the reasoning behind it. It is condemned to imply that there are limits to God's power by saying «that God *could not* move the heavens in a straight line, the reason being that he would then leave a vacuum» (art. 49). Granted Aristotle's understanding of the meaning of 'place', such a condemnation was nonsensical; but one could make some sense of the prohibition by simply seeing the world in a new light, namely as a spherical region in unending space.

In short, the very incompetence of the condemnations — from an Aristotelian point of view — might well have encouraged infinitistic thought. Although infinite space was much discussed by Oxford men — Richard of Middleton, Walter Burley, Robert Holcot, and others — Thomas Bradwardine is a rare case of a conscious defender of the reality of infinite space. Alexandre Koyré suggested that his *Euclideanism*, essentially anti-Aristotelian on the nature of space, was responsible ([13]). By comparison, Richard of Middleton was a prudent and timid spirit of little originality; but one should not forget that the two were more than a generation apart. Richard was a finitist in the sense that he followed Aristotle's general line, rejecting, however, eternity, and the necessity of the world's uniqueness. He thought that God could have created a *finite* number of other worlds; or that he could have constantly enlarged the world ([14]). In age more or less mid-way between them we find Walter Burley, with his much more impressive reasoning — for instance on the possible creation of place simultaneously with matter ([15]). This no doubt helped Bradwardine on his way: God acts where he is; he must be in the site prepared for what he is to create there, namely in a void place. Since *a limited void is not imaginable*, the argument goes, God's effective and eternal presence is in all of infinite space extending outside the universe. When Bradwardine implied that God is extended, he was skating of thin ice, as he well knew, and he went to some lengths to stress that God is not extended in an ordinary sense, but in a metaphysical sense, «inextensively» and «indimensionally» ([16]).

([12]) For a modern edition of the condemned articles, with an attempt to locate sources (there are many earlier editions, some rationalised by subject) see R. Hissette, *Enquête sur les 219 articles condamnés à Paris le 7 mars 1277*, Louvain 1977.
([13]) *Le vide et l'espace infini au XIVe siècle*, in «Arch. d'hist. doct. et litt. du moyen age», 1949, pp. 45-91. See p. 52.
([14]) *Ibidem*, pp. 69-74.
([15]) *Ibidem*, pp. 75-80.
([16]) *Ibidem*, pp. 88-90.

While Oxford's affection for Euclid clearly did no harm to Bradwardine's frame of mind, Koyré should perhaps have paid more attention to a not irrelevant place in the *Sentences* commentaries of the time. In I.44 it was asked whether God could make this world better. Thus Ockham, pointing out that God could make a world better in kind, better in accidents, or better *and distinct in number*, gave much attention to this last possible plurality of worlds ([17]). He considered, in other words, different sets of natural places — that is, separate Aristotelian-type worlds. Strange to say, Copernicus played with a not wholly different idea in his *De revolutionibus*, the centuries later ([18]). As soon as the Earth's centre ceased to be identified with that of the universe, it was natural for anyone brought up on the Aristotelian theory of natural place to consider a multiplicity of such places, *i.e.* planetary centres. Ockham had to face the same problem, with his hypothesis of a plurality of universes. Others included the discussion, in a slightly different form, in their comments on II.35, where the inquiry is into the nature of sin. Could God make another world, with men exactly as we, but without sin? Aristotle and Averroes were cited against, but the Church Fathers did not rule out the possibility, and therefore the topic was open to free discussion ([19]). Note that this had nothing to do with 1277 and all that. It was to become a common topic for discussion: Nicole Oresme later brought his wits to bear upon it, and it did not escape the attention of Paul of Venice, John Mair, or Leonardo, to name but three of many who seem to have been influenced directly or indirectly by Oresme.

The question as to the potential existence of infinite worlds seems to have had little to do with the abstract treatment accorded to the various concepts of infinity — less, for instance, than did such theological questions as whether God himself is potentially infinite, whether one may contemplate infinitely, whether angels are eternal, whether truth is eternal, or God's wisdom formally infinite. From two manuscripts originating in late thirteenth century Oxford, A.G. Little and F. Pelster list *quaestiones* that include more than a dozen such themes ([20]), but intermixed with natural philosophical subjects, such as infinite and eternal motion and quality and dimension. It is hard to say whether this eclecticism, still more evident from *Sentences* commentaries, was cause or effect of the general feeling for the unicity of truth. If theology was marked off from the rest of knowledge, it was perhaps as much by its use of revelation and the principle of God's omnipotence as by its subject matter. Even within theology there were some (for example Ockham and Aquinas) who wanted to get as far as possible without theological principles, for they felt that they would otherwise bring the faith into disrepute. This sort of urge does not, however, explain a certain measure of success to which late scholas-

---

([17]) *Opera plurima (Super 4 Libros Sententiarum, in lib. I*, vol. 3), Louvain 1962, *distinctio* XLIV. I have given the three arguments in a different order from Ockham. Note that other writers of the time make use of his example of fire rising from Oxford and Paris (sometimes changing the towns), and it is quite possible that there is no originality worthy of the name here.

([18]) See my *The Medieval Background to Copernicus*, in *Copernicus Yesterday and Today*, in «Vistas in Astronomy», 17, 1975; see below, chapter 24, pp. 401-14.

([19]) Thus Baconthorpe in l. 44: «*An theologiae repugnet pluralitas mundorum? Ubi Catholici dicunt quod non, Chrysostom super illud Ioannis, 'Omina per ipsum facta sunt'*».

([20]) *Oxford Theology and Theologians c. 1282-1302*, Oxford 1934 [O.H.S.]. My examples are on pp. 104 (qu. 1), 110 (60), 127 (176), 311 (26).

tic writers attained in their attempts to go beyond Aristotle in an abstract analysis of relevant concepts.

Aristotle, as we have seen, ruled out the actual infinite using arguments as to inconsistency (he also invoked his doctrine of the elements and natural place, but less successfully). He did not make use of one particular inconsistency argument much used in the late Middle Ages, that might be called the paradox of unequal (actual) infinites. It is probable that this paradox has ancient roots, and that it was obtained from Philoponus through Simplicius. The best early medieval exploitation of the paradox is perhaps that to be found in al-Kindī, the ninth century Baghdad philosopher responsible for introducing much Greek philosophy to the Islamic world. Unlike most later Islamic philosophers, he rejected the notion of an eternal (emanative) creation, and for his proofs of creation *ex nihilo* drew on arguments already sketched by Philoponus. Al-Kindī used a postulational method which makes his assumptions stand out very clearly. One of these was at the heart of his arguments, and may be written

$$\text{If } B_1 = B_2 \text{ and } B_3 = B_1 + B_0 \text{ then } B_3 > B_2,$$

where 'B_0' etc. denotes a body in the sense of any extra-mental quantitative entity ([21]). Putting aside what we might call 'negative body', and for that matter zero body, the axiom makes no allowance for the behaviour of the infinite. This remark in a sense begs the question: al-Kindī would have said that his axioms were self-evident, and they showed the infinite to behave inconsistently, and therefore to be worthy of rejection. Thus (and this is not as he expressed himself, but it makes his point briefly) if B_0 represents an actually infinite pile of stones correlated with the odd natural numbers, B_1 a pile correlated with the even numbers, then B_3 is one correlating with the natural numbers. B_2 might be a pile numbered by the natural numbers, since each may be correlated with an even number (its double). The two hypotheses are satisfied, then, but the conclusion nevertheless has the set of natural numbers greater than itself. There are many other arguments relying on the same sort of assumptions — boiling down to the assumption that infinites have to be assumed equinumerate in the sense of one-to-one correlatable. One very common argument concerned the revolutions of the Sun and Moon (often the planets were introduced to add variety). In an eternal world, the revolutions of the Sun (the argument goes) would be infinite in number up to the present time; but the revolutions of the Moon would be twelve (or more) times as many, implying that one infinite is greater than another. The absurdity shows (it was said) the absurdity of the eternity hypothesis. The argument was used by Phyloponus (Jupiter, Saturn, Sun, Moon, and stars), al-Ghazālī (Jupiter, Saturn, stars), William of Auvergne (Saturn and Sun), Bonaventure, Henry of Harkeley, William of Ockham (Sun and Moon), Buridan (Sun and Mars), and many others. It has occasionally been ascribed to Galileo, but it was available in many of the books widely read in Galileo's time.

In retrospect we now value more the attempts made to resolve the paradox than those of men who wished to use it to refute an eternal world. A number

---

([21]) See W. L. Craig, *The Kalām Cosmological Argument*, London 1979, pp. 23-27, for one analysis of the argument as a whole.

of fourteenth century Oxford writers in the first category acknowledged with Aquinas that the intuitive uses of the concepts of equality and inequality were not applicable to infinites. The approach by Henry of Harkeley, who believed in the composition of continua out of indivisibles, was quite different: he admitted that one *can* compare different infinites (of the same kind [*eiusdem rationis*] — for example a line and a line, a multitude and a multitude) and that it *is* possible to use the concepts of whole and part in this connection (it being of the essence of quantity that whole is greater than part). He believed, though, that one must avoid the use of the simple Euclidean axiom asserting every whole to be greater than its part [22]. This axiom was said to fall under a more general axiom applying to infinites, and it was in terms of the latter that the inequality of infinites must be understood. His words «*Illud quod continet aliud et aliquid ultra illud vel praeter illud est totum respectus illius*» have been rendered as follows by J. E. Murdoch, who is preparing an edition of the *Questio de infinito et continuo* from which they are drawn: «that which (*e.g.* an infinite set) contains another thing (*e.g.* an infinite proper subset) and something else beyond it, or in addition to it, is a whole with respect to that other thing» [23].

Harkeley's opinions were soon to be challenged by the Franciscan William of Alnwick, who noted a difference between «beyond» — which he thought entailed having more absolutely — and «in addition to», which among infinite things he thought entailed diversity but not a greater plurality [24]. It is tempting to see in this distinction an awareness of the difference between the set-subset relation (which infinites may obey) and the relation of unequal cardinality. Whether or not he saw this very clearly it is hard to say, but it is certainly clearly appreciated in the *Sentences* commentary of Gregory of Rimini, given as lectures in Paris in 1342 [25]. As far as basic mathematical structures were concerned, one should not ask too much: this insight was not likely to lead very far. Its context was the series of *distinctiones* (42-4) in the first book of the *Sentences* dealing with the absolute power of God — can God, for instance, using his infinite power, create an actual infinite? The aim was to circumscribe (in this case) God's powers by ever more careful conceptual distinctions. Once this was done, the writer passed dutifully on to the next *distinctio*. Gregory, for instance, broke the problem down into problems of infinites according to multitude, magnitude, and quality or intensive perfection; and in doing so he was following a standard procedure. John Baconthorpe, for example, separated the problem of the 'simple infinite' (*infinitum simpliciter*), which does not admit of greater and less, from the rest. It was supposedly infinity of the first kind we ascribe to God — illustrated by the universal relationship of paternity [26]. As soon as this sort of decision had been made, the theologian's urge to explore the implications of the other meanings was bound to be diminished.

---

[22] Henry was apparently encouraged in his views by Grosseteste (*super 4 Phys., cap. de tempore*). See A. Maier, *Problem des aktuell Unendliche*, in *Ausgehendes Mittelalter*, vol. I, pp. 51-55.

[23] *Cambridge History of Later Medieval Philosophy*, ed. N. Kretzmann *et al.*, Cambridge 1982, p. 571.

[24] *Ibidem*, pp. 571-572.

[25] *Ibidem*, p. 572.

[26] *Op. cit., In Primum Sent.*. d. 23; see esp. pp. 255-257. Note the reference to «Cogmiton» — the Franciscan Richard of Conington, perhaps.

Fortunately there were many whose philosophical curiosity went further, although for the most part it was a question of accepting Aristotle's definitions with minor modifications. «*Infinitum in fieri*» was a typical phrase for the potentially infinite: it is «always possible to go further»; of extensive magnitudes, they are «not so great that it is impossible to have greater»; and of multitudes, «they are not so numerous that you cannot have more». What was most appealing about the actual infinite («*in actu*», «*in facto*») was that it could be spoken of as a definite thing — «so great that there is no greater», «so numerous that there is no more», and so on. It was *that which* exceeded (in some appropriate sense) every finite whatsoever, a *unique* maximum of its kind; and it was no doubt this feeling for its uniqueness that made most philosophers believe that inequalities between infinites (as in the planets argument) were nonsensical. One could have one's non-unique infinite, however, if need be, by suitable rephrasing — the infinite «exceeds any finite whatsoever», or «is beyond all determinate proportion with the finite». Differences of definition led to misunderstandings, and a failure to achieve true confrontation in some disputes ([27]), and yet even then there were useful points to be made, as may be illustrated from a mid-fourteenth century Oxford controversy.

Something of the controversy is known from casual remarks made by Adam Woodham in his *Sentences* commentary ([28]). It involved Richard Fitzralph and Richard Kilvington, of Balliol and Oriel colleges respectively. (Fitzralph lectured on the *Sentences* in 1328-9, and Kilvington within two or three years of this. Their commentaries on the *Physica* of Aristotle seem to be no longer extant, but we do have a *quaestio* by Kilvington on whether creatures are of their nature circumscribed by certain limits — in short, a consistency question) ([29]). Somewhat as Henry of Harkeley had done, Kilvington accepted unequal infinites; in short, the infinite for him was no unique maximum. Fitzralph took a totally different standpoint. Ordered things that are actual must have a first term and a last; which is to say that they are finite. This is to speak categorematically. The categorematic-syncategorematic distinction as applied to infinity is one that has been explained in many different ways. Thus Knuuttila and Lehtinen explain that Kilvington was following the usual distinction in saying that the term 'infinite' was used syncategorematically (*i.e.* having no signification taken by itself) if it came at the beginning of the sentence. It 'consignifies' with the subject and predicate ([30]). When preceded by the term it modified, it was supposed to have a categorematic sense, and to give a determinate supposition to the term it modified (one must beware of taking the statements about the relative positions of terms as the definitions of the words; in the last analysis they were rather forced grammatical, or rather logical, conventions). More simply expressed, a categorematic use of the word 'infinite' was when it was thought to

---

([27]) And quite understandably so, as may be well illustrated by two modern writers' insistence that Gregory of Rimini and John Kilvington (for whom see below) were using the concept 'infinite' in the potential sense, when their contemporaries took them to be radical exponents of the doctrine of actual infinity. See S. Knuuttila and A. I. Lehtinen, *Plato in infinitum remisse incipit esse albus*, in *Essays in Honour of Jaakko Hintikka*, ed. E. Saarinen *et Al.*, Dordrecht 1979, pp. 309-329, esp. at pp. 311-312.
([28]) A. Maier, *op. cit.*, pp. 75-81.
([29]) *Idem*, *Die Vorläufer Galileis im 14. Jahrhundert*, Rome 1949, pp. 174-176, 210-215.
([30]) *Op. cit.*, p. 311.

function as a common term, as though there were no problem of meaning. Syncategorematically, on the other hand, it was used in a way that conveyed information about the behaviour of terms of a general class. In the first case, said Bradwardine, it was *«quantum sine fine»*; in the second *«quantum et non tantum quin maius»*. Now Fitzralph seems to me to have appreciated an important property of such statements (syncategorematically expressed) as that a line is composed of an infinity of points, that a continuum is an infinity of proportional parts, even that God can know an infinity of things, namely that there is in them no reference to individuals. Since there seems to be such a reference they are dangerous, and one should, he thought, use 'infinite' only of a class (set) of objects rather than of individual members (of lines rather than points, etc.) ([31]). One might almost call this a Fregean observation, but I think it could be prised out of the writings of many others of the time, and it does, after all, have a certain Aristotelian ring to it. On the question of intellectual allegiances note that Fitzralph is said to have been a friend and pupil of John Baconthorpe, who (with much exaggeration) was later described as «Prince of the Averroists» ([32]). And that Adam Woodham took an argument *verbatim* from Fitzralph on God's ability to make infinite equals; and that he strongly influenced the similar arguments of his fellow Franciscan Robert Halifax, of Cambridge, and possibly too the Dominican Robert Holcot ([33]). There is a certain sameness about so many of these writings on the infinite, and, as with wines, it is often difficult to judge provenance from labels. 'Aristotelian' and 'Averroist' were epithets often applied with polemical intent. Such labels should not be allowed to obscure the genuine originality of at least some of those to whom they have been applied.

([31]) A number of relevant extracts from Fitzralph's work are given in G. Leff, *Richard Fitzralph*, Manchester 1963, pp. 111-116.

([32]) See the article on Baconthorpe by T. A. Archer in *D.N.B.* John is not mentioned in the work by Leff (see note 31), although Archer seems to imply that he had a part to play in influencing jointly with Fitzralph Wyclif's teachings on dominion and grace.

([33]) A. Maier, *Problem des aktuell...*, pp. 79-81.

CELESTIAL INFLUENCE –
THE MAJOR PREMISS OF ASTROLOGY

Although I use the word 'influence' in a modern sense, I hope that its older meaning will not be forgotten. When the English translators rendered the Lord's question to Job as 'Canst thou bind the sweet influence of Pleiades, or loose the bands of Orion?' (xxxviii.31), the Latin *influere* or *influentia,* with connotations of an in-flowing of something or other (let's keep it vague), would have been at the back of most educated readers' minds.[1] Here I want to consider the more general problem of ways in which the celestial bodies were supposed to produce their effects, whether by an influx or infusion, or some other means yet to be specified. I shall have nothing to say about influenza.

Those who wrote about celestial action were rather like those who wrote about the unicorn – that is to say, they were more concerned with its behaviour than with its existence. As to the question of existence, just as there were traces of the animal of legend (in the forms of authentic horns), so there were signs that one of the planets at least, namely the Sun, produces changes on the Earth. This was the starting point of Aristotle's doctrine of efficient causation in the *De generatione et corruptione* (II.10), the fundamental text for those who tried to justify astrology in the scholastic tradition. He had, as he there explained, shown in the *Physics* that *motion* is the primary cause of change. (This was to take issue with those who said that coming-to-be was the cause of change). How then can a generation *and* destruction be explained? As two different processes, there must be two different motions – different, that is, in speed or direction. The answer is to be found in the Sun's motion along the inclined circle, the ecliptic, which implies in a sense a certain advancing and retreating. It seems that Aristotle thought that this motion implies a variation in distance from the Earth. This is not the only problem for those who would make sense of the passage in question. He was in difficulties over the periodic time of the process of generation and decay, which on this theory ought surely to be one year for everything affected by the Sun. He said, lamely, that 'there is an Order controlling all things, and every time (every life) is measured by a

---

[1] It is of no consequence that the original Hebrew talks only of binding the cluster of the Pleiades.

period. Not all of them, however, are measured by the same period . . .'.[2] He seems here to be suggesting that some of the responsibility lies with the changing thing. He needed the Sun's movement to explain something else that had obviously given him trouble: how, in a sublunary world that had an eternal existence, are we to explain that the element fire, for example, that rises naturally, has not been lost to our Earth's surface? The answer given was that the Sun's movement produces a cyclical interchange between elements. Even time itself is said to get its continuity from the circular movement of the Sun. Notice that there is no mention here of the planets – although some modern writers have somehow managed to substitute the planets for the Sun, in explaining the passage in question.

The second relevant Aristotelian source is that section of the *Meteorologica* in which he discussed phenomena in the upper reaches of the sublunar region. Not only did this include wind, rain, thunder, lightning, and so on, but also comets and the milky way. The predominant efficient cause of these phenomena was presented as the Sun, while the material causes were supposedly the elements earth, air, fire, and water. Near the beginning of the work there is the casual remark that 'we must assign causality in the sense of the originating principle of motion to the influence of the eternally moving bodies'.[3] Curiously enough, the importance of the Moon is played down, and even on the question of tidal phenomena, Aristotle assigned main responsibility to the Sun.[4] Others in the ancient world appreciated the Moon's importance, and Aristotelian commentaries in the middle ages were replete with lunar associations. This was a subject for introducing into commentaries on the *Meteorologica,* for instance, where it was sometimes coupled with a theory of critical days in the course of human disease – thanks to the quasi-tidal behaviour of the humours. I shall return briefly to these traditions later. In passing, let me mention three quite distinct medieval traditions concerning the tides: one was a philosophical tradition, in which tides were accommodated to the general physical principles of Aristotle; one was an empirical, astronomical, tradition, in which the monthly movements of the Moon were correlated with the tides at a particular port; and the third was an astrological tradition with an admixture of Aristotelian philosophy.[5] I will introduce the last of these, and do so in connection with the highly influential views of Ptolemy, as expressed in the *Tetrabiblos,* but first I shall look at some influential strains in Stoic cosmology.

[2] 336b 12–14.
[3] 339a 31–32.
[4] Thomas Heath, *Aristarchus of Samos* (Oxford, 1913) p. 307.
[5] See my *Richard of Wallingford,* vol. 3 (Oxford, 1976), pp. 206–8.

The three main representatives of the early Stoa, namely Zeno of Citium, Cleanthes, and Chrysippus, were all from outside Greece proper – respectively they came from Cyprus, Assos in Asia Minor, and Soli, likewise in Asia Minor. Partly for this reason, there has been a tendency to look outside Athens for the main influences on the philosophy first given there, on the Painted Porch *(Stoa Poikile),* by Zeno around the year 300 B.C. Whether or not Heraclitus, or Semitic sources, were more influential than Aristotle and Plato, the fact is that Stoic doctrines became extremely popular, and that for half a millennium they were current at many different levels of society, not only markedly intellectual ones. In their physics as in their theology, the Stoics harmonised all with God, even to the extent of supposing God and the universe to be qualitatively identical. In proving that the universe is intelligent, Cleanthes stressed the identical character of heat in the Sun and in living things.[6] The heat of the cosmos is not moved by another force, for there is none stronger; it is self-moved, and thus of the nature of soul. The heavenly bodies are moved 'by their sensation and divinity', not by nature nor by force, but by *will.*[7] There are many echoes of this belief in medieval writers. The Sun was supposed by Cleanthes to be the ruling principle of the cosmos – perhaps this is a principle somehow related to Aristotle's remarks in *De caelo* about the true centre and *archê* of the universe being at the periphery, just as an animal's heart is not at its true centre.[8] The biological parallels in Cleanthes are too involved for us to follow here, but note his analogy between the renewal of the human soul by exhalation from the blood, and the sustaining of heavenly bodies by exhalation from the oceans, and other liquids of the world's body, which thus circulate, for they are radiated back to the Earth in due course.[9]

Whereas to Cleanthes the world soul was a heat permeating the cosmos, for Chrysippus it was a permeating *pneuma* – that which also makes a man an organic, living, whole. The biological parallels are even more marked than in his predecessor's theory. As the third century B.C. wore on, the preference shown towards the *pneuma* theory by physiologists led to the gradual decline of its predecessor. Both theories had lives of their own in medicine, of course, independently of the Stoics, in the shape of a well known antagonism between haematists and pneu-

---

[6] D. E. Hahm, *The Origins of Stoic Cosmology* (Athens: Ohio State U. P., 1977), pp. 136ff.
[7] Cicero, *Nat. D.* 2. 44; the argument goes that, following Aristotle, all motion of living bodies is due to nature, force, or will, and that the Sun and Moon are not moving round by nature (the implication is that they would in this case move up or down), nor is there any force greater than theirs to move them contrary to their nature.
[8] 239b 6–15; Hahm, ibid., pp. 150–1.
[9] Note Cicero's language, as used of the Moon: *multaque ab ea manant et fluunt (Nat. D.* 2. 50; for other material on circulation see 2.118 especially).

matists, beginning long before the period I am discussing. (Aristophanes, in the *Clouds,* for instance, indicted Socrates' pneumatism before his Athenian audiences). These traditions with their different 'motive principles', as we may call them, cannot be very directly connected with the reasoned defences of astrology offered at a later date, but their language has much in common with the language of astrological influence. The same goes for the language of a related work *De mundo,* that for long went under Aristotle's name, but that if we are to follow Wilhelm Capelle derives from two works by the Stoic Posidonius (c.135–c.51 B.C.). It does seem to share something with the arguments of Posidonius preserved for us by Ptolemy, two centuries later, and known through Cicero's critical account in the *De divinatione* (II.43) – a work written close to Posidonius' in time.[10] The *De mundo* is no more a work of astrology than were the works of Aristotle I cited, but like them, the pseudo-Aristotelian work offered a slender pseudo-physical basis for celestial influence; and since it was several times translated between the thirteenth century and the fifteenth, its arguments are worth summarizing.[11]

The God of the *De mundo* preserves the ordering and arrangement of things within it.[12] The upper regions of the world are 'called Heaven, the abode of the gods'. The planets seem to be supposed to move to greater and lesser distances from the Earth,[13] but the general Aristotelian patterning of spheres and elements is clear. After the ethereal and divine (celestial) element 'there follows immediately an element (fire) which is subject throughout to external influence and disturbance, and is, in a word, corruptible and perishable'.[14] He then proceeds through the lower terrestrial elements, discussing ways in which they are illuminated and moved. 'And since the air too admits of influence and undergoes every kind of change, clouds form in it, rain-storms beat down . . .', and so on.[15] What follows is an account of meteorology without much by way of further reference to the heavens, although 'Many tides and tidal waves are said always to accompany the periods of

---

[10] Capelle, *Neue Jahrbücher,* vol. 15 (1905), pp. 529–68, according to E. S. Forster's introduction to the Oxford translation (1914), assigns it to the second half of the first century A. D. I shall quote from this translation.
[11] It was translated by Bartholomew of Messina, c. 1260, and again by Nicholas of Sicily, Grosseteste's collaborator, at about the same time. In the fifteenth century it was translated by John Argyropulos and by Rinuccio Aretino. There is a translation given to one Engelbert, and another, anon., listed in Thorndike and Kibre, which see for further references.
[12] 391b 11–12.
[13] 392a 14–19.
[14] 392a 31–35.
[15] 392b 9–13.

the Moon at fixed intervals'.[16] Everything, in fact, in the universe is seen as 'ordered by a single power extending through all, which has created the whole universe out of separate and different elements – air, earth, fire, and water – embracing them all on one spherical surface and forcing the most contrary natures to live in agreement with one another...'.[17]

If the author of the *De mundo* had gone no further, we might have supposed the harmony to have been of a familiar Pythagorean sort – but it is rather a *forced* harmony, as promised by the last quotation. In the sixth chapter, we are told of an untiring God at the outermost limit of the heavens, the nearest body to him being that which most enjoys his power, then the next nearest, and so down to the Earth, where the effects of God's ordering are so feeble that we meet with incoherence and confusion. *Proximity to God determines the degree of orderliness.* This, in brief, is the 'law of force' of the *De mundo*. An interesting aspect of the account offered is the human analogy drawn between God and such great rulers as Cambyses, Xerxes, and Darius, all of whom knew how to *delegate* organisational tasks to those below them. So it is with God, a great chorus-leader, bestowing his purpose on the heavenly bodies singing and dancing together. The principle of decreasing orderliness is forgotten as the chapter wears on, waxing ever more lyrical. God is a principle of organisation, to all intents and purposes. He has an irresistible binding action, governing all things' destinies.[18] This is ostensibly not astrology; but no-one with a deep-seated wish to justify astrological belief could have preferred any other Aristotelian text to this, blending as it did the imagery and sentiment of Plato with the authority of the supposed Aristotle.

There is in fact evidence that the 'Chaldean' astrologers picked up these largely meteorological ideas at much the same period as that which saw the rapid rise of a more mathematically orientated astrology – that is, in the century or more before Christ. Cicero again, in his *De divinatione* (2.42), tells us that the Chaldeans spoke of a force in the zodiac such that different parts of it transform the atmosphere. (The word used is '*caelum*', but the meaning here seems to be fixed by a later reference to '*tempestates caeli*'). The force is set in motion by the planets; and just as the Sun's revolutions affect our atmosphere, so the 'Chaldeans' thought that children at their birth are differently animated and formed according to the state of the air, while indeed the soul, the body, and all our actions are likewise fashioned. (Cicero himself thought that celestial bodies beyond the Sun and Moon were *too distant* to affect human lives. Note that he did not take a stand against a faulty princi-

[16] 396a 26–28.
[17] 396b 25–33.
[18] See esp. 401b 8–end.

ple).[19] Two centuries later we find the greatest apologist for astrology in ancient times, Ptolemy himself, setting out arguments that Boll believed to have been drawn from Posidonius.[20] Whether or not this was their immediate source.[21] Ptolemy falls comfortably into the Aristotelian-Stoic meteorological tradition. A certain power (δύναμις) emanates from the aether, causing changes in the sublunar elements, and in plants and animals. Effluence from the Sun and Moon – especially the Moon, *by virtue of her proximity* – affects things animate and inanimate, while the planets and stars also have their effects. If a man knows accurately the *movements* of the celestial bodies, and their *natures* (perhaps not their essential but at least their potentially effective qualities), and if he can deduce scientifically the qualities resulting from a *combination* of these factors, why should he not judge both weather and human character?[22] These sentiments are, scientifically speaking, perfectly laudable – at least, as an expression of an ambition. Needless to say, they were constantly attacked, especially by those who feared for human freedom; but that is not my theme. One last point before passing on to other writers: there are strong similarities between the way many astrologers, like Ptolemy, speak of the Aristotelian aether, the fifth element or quintessence, and the way others speak of the World Soul, *anima mundi*. It would be easy enough to bring the alchemists into the picture, too, with their attempts to infuse this superior spirit into lower forms; and likewise the musicians, who tried to identify the celestial spirit with the harmony of the universe, considered as a musical harmony. Considered in this last sense, it was supposed to exert an influence analogous to the unquestioned influence that music has, in moving human spirits – perhaps even in moving God's.[23] Following Boethius, the middle ages regarded musical forms as arranged hierarchically, with *musica instrumentalis* below *musica humana, musica mundana* above that, and *musica caelestis* higher still, or rather an all-embracing harmony.[24] Harmony was sometimes spoken of as a governing instrument or even as a moving power, but to try to take this doubly aetherial influence into account would be to wander too far from specifically astrological doctrine.

[19] Ibid., 2.43.
[20] Franz Boll, *Studien über Claudius Ptolemäus* (Leipzig, 1894), pp. 131 ff.
[21] Boll was able to draw parallels with not only Cicero but Philo Judaeus, Cleomedes, and Manilius – Posidonius providing the putative link. The argument continues that, like Ptolemy, Posidonius correlated geography and astrology, but K. Trüdinger attacked the hypothesis; see his *Studien zur Geschichte der griechisch-römischen Ethnographie* (Basel, 1918), pp. 81 ff.
[22] This long justificatory passage is in *Tetrabiblos* I.2. For actual specimens of Ptolemy's reasoning along these lines – falling somewhat short of his ideals, however – see *Tetrabiblos* III.11, 'Of bodily form and temperament'. See also p. 56ff. below.
[23] Cf. II *Kings* 3.15.
[24] *De institutione musica*, I.2.

Implicit in the doctrines to which Ptolemy gives careful expression is the view that the planets bestow their gifts according as (so Plotinus puts it) they themselves are agreeably or disagreeably affected at the various parts of their course.[25] This Plotinus thinks as absurd as that evil stars should (in certain situations) be of help, and benevolent stars a hindrance. These are revealing comments, for they show that ideas of a celestial force that depends on (a) position pure and simple of a planet in the zodiac; (b) relative separation of a planet and the subject it is supposed to affect; and (c) the rising and setting, ascending and declining, of planets in relation again to the subject – that such ideas, intuitively acceptable to Ptolemy and his like, were quite otherwise to Plotinus, and presumably many more. We, with our consciousness of Newtonian and other accounts of how forces may operate, should find no difficulty with (b) and (c). (Absolute position as a determining factor might be a little harder to swallow). Plotinus never really tried. He could, though, vaguely appreciate laws of algebraic addition of forces, or laws of interaction (he criticised as illogical a particular doctrine of aspects). One might go so far as to say that his arguments made sense by the lights of his age; but of course at the back of his mind there is always the presumption that the stars, being in a divine place, and thus divine, will do us no hurt.[26] It is not that the stars are lacking in influence – they are constantly spoken of as pouring out warmth – but that they are individually only *signs* in a harmonious whole, a sovereign unity, under the control of the One. We are to take them as 'letters inscribed (and yet moving) in the heavens',[27] and from them divination is possible, as it is possible – so he believed – from birds and other animals.

This point of view, as explained by one who is usually regarded as the founder of neo-Platonism, does not seem to have influenced the main stream of astrologers, although I can see no intrinsic reason why it should have done so – apart from the fact that it was allied to a criticism of an interpretation of astrological procedures that tended to be prefixed to the procedures themselves. Like those who offered the procedures criticised, Plotinus seems to have been strongly influenced by the Stoics. Neither the wills of deities, nor mechanical causation, however, have any place in the new scheme. Divination is the art of reading the signs given to us by parts of a harmonious whole, parts between which there exists the sympathy to be expected within any organism. Cicero had long before hinted at the teaching of certain Stoics that sounds

---

[25] *Enneads* II. 3.1. This third tractate, 'Are the stars causes?', is the source of all the following remarks about Plotinus. It has 18 short sections, and I shall not distinguish them further.

[26] A nail he puts firmly in the coffin of doctrine (a) is that the planets are not in the zodiac itself, but much lower. See II.3.3.

[27] Ibid., II.3.7.

rather as though it resembled the Plotinian doctrine.²⁸ It presented the same sorts of problems of determinism and freedom of human action as the teaching it was meant to rival, but they are not my present concern.

The number of classical references to the 'meteorological' tradition, starting usually from solar radiation and later taking in the Moon and tides, is potentially huge. One could quote, for example, Aulus Gellius citing the hostile remarks of Favorinus, or Strabo, Priscian, Seneca, and St. John Damascene who all referred (without exactly agreeing) to Posidonius. There are numerous references in the Church Fathers, not to mention the alchemists; but the law of diminishing returns is as crucial as the laws of astrology, and I shall leaf through the centuries in a way that hardly does justice to intervening civilisations (particularly Indian and Islamic), in order to bring my story into the Latin West. There, without any doubt, the most influential Islamic writer was Abū Maʿshar (786–866). Fashions came and went, in astrology, but he seems to have been read and quoted constantly from the time of the translations of John of Seville and Hermann of Carinthia in the twelfth century to the decline (or at least turning native) of the subject in the seventeenth.²⁹ Richard Lemay has perhaps gone too far in his remark that 'during the thirteenth century [in the West], the authority of Abū Maʿshar on astronomy-astrology, and on cosmology disputed the first place with Aristotle himself', but the *Introductorium* undoubtedly provided the link scholars felt was needed between astrology and (Aristotelian) cosmology.³⁰ Lemay could even quote a remark made in the fourteenth century to the effect that Aristotle had himself made astrology out to be a part of physics.

In the broadest of outline, Abū Maʿshar's astrology, like that of Islam generally, can be seen as an amalgam of Hellenistic and Indian astrology. The first, with neo-Platonic and Aristotelian accretions, was transmitted through the Ṣabians of Harrān. Pre-Islamic Indian genethlialogy took over the techniques of its Hellenistic sources, rather than the underlying philosophy. Much came by way of Arabic and Persian translation, and Greek and Roman cultural application were easily brushed aside in favour of Indian (the caste system, metempsychosis, and so forth).³¹ In all systems lacking the Plotinian type of subtlety, the planets are dispensers of power in some sense. In Hermetic philosophy,

---

28 *De div.* I. 51–2.
29 For a comparison of the translations and an assessment of the Aristotelian content of the *Introductorium,* see the first two sections of Richard Lemay, *Abū Maʿshar and Latin Aristotelianism in the Twelfth Century* (Beirut, 1962). Note the sub-title of the work: The Recovery of Aristotle's Natural Philosophy trough Arabic Astrology.
30 Lemay, *ibid.,* p. xxxvi.
31 One should not exaggerate the Hermetic ingredient in Abū Maʿshar. As Richard Lemay points out, only four out of 103 chapters in his main work are Hermetic. For a

for instance, the planets are intermediaries between the One and the sublunar world. Through them the divine will manifests itself, and the demons act on their behalf below. This gave rise to Talismanic methods of approaching the One through the demons and planets. For the Gnostics, the planets were not so much parts of the plan of a cosmic ruler as instruments of a spirit of evil, to be resisted by purification of the soul – again often using Talismanic methods. In India, however, the system was different again: the planets dispensed *their own* power, and very complicated rules were devised for ascertaining their relative strengths in different astronomical situations. (There were chains of causality, not only from planet to man, so to speak, but linking acts of a former existence to the present!)[32] The Islamic compound of Hellenistic originals with Indian innovations was then further complicated. In view of Hermetic, Gnostic, Indian and Harranian precedents it is not surprising that Islam too developed rituals to avert or alter the influences of the planets. Here, it is the nature those influences were supposed to have that concerns us, and I shall make only passing reference to ways in which they were meant to be controlled.

Albumasar (as I shall spell his name henceforth, whenever I am taking his doctrine from the two rather different Latin translations) began like Aristotle in noting the obvious changes on the Earth wrought by the Sun and Moon. Like Aristotle, he needed more than the constantly rotating primum mobile to explain the seasonal fluctuations; and like Aristotle, he settled on the obliquity of the ecliptic to explain them. Unlike his predecessor, he wrote the planets explicitly into the story, in order to explain the enormous variety of change over and above the seasonal changes.[33] He added that experience bears out the postulated involvement of the planets, over a period of years, and that philosophy, with its logical handling of systems of causes, is capable of fortifying the astrologers' discoveries.[34] Albumasar summarized what are essentially Aristotle's views on generation and corruption, adding the extra factor of a *receptivity* to planetary influences on the part of terrestrial things.[35] He then turned to the doctrines of *De caelo* (had he known it, II.7-12) to explain how *motions* in the heavens are the key to *alteration in the elements* which combine to make up natural things. The heavenly bodies themselves have *natural* motions. (I note that Adelard of Bath, in chapter 74 of his *Quaestiones naturales,* talks as though natural motion

convenient short account, see D. Pingree, 'Astrology', in the *Dictionary of the History of Ideas,* vol. 1 (New York, 1973), pp. 118-26.
[32] Ibid., p. 122.
[33] Lemay, ibid., quotes some relevant passages from Hermann's and John's translations, pp. 50-1, 85.
[34] Ibid., p. 52.
[35] *Introductorium* I.2. There was an Aristotelian precedent even for this, see p. 78 below.

can only be up or down, and thus rejects it for the heavens. We have already seen that Cicero has the same view, and it seems to me that Adelard is taking the whole of the relevant passage from him)[36]. The heavens exist in order to produce the necessary changes – one is again reminded of Adelard's remarks to the effect that there would be not harvest if spring were permanent – and the terrestrial world is joined with the celestial and its motions necessarily, by God's command *(iussu Dei)*. The supreme circumscribing sphere, that is, the Primum Mobile, in circling the universe produces heat in it, which somehow makes the world more subtle, and changes in bodies begin to take place.[37] The question remains: how, precisely, does the circling of perfect heavenly bodies cause inferior, imperfect, terrestrial bodies to move and change?

Albumasar acknowledges two sorts of action, by contact and through a medium. In the second case, there might be voluntary action bringing contact about, through a medium, and there might be natural action transmitted through a medium, as heat through a vessel to water within. A third possibility, which he thinks applies in the case of heavenly bodies moving terrestrial ones, is where a *natural action is transmitted through an invisible intermediary,* as when a magnet attracts iron through the (invisible) air. Again there is the idea of the attracted thing being suited to receive the attraction, and perhaps here we have the source of the notion of a natural receptivity on the part of terrestrial things for the influences of celestial movements. Both translators have a tendency to cut out examples, but both include a reference to the extremely interesting (but illusory) experience of sensing the magnetic influence by means of a piece of copper between the magnet and the iron.[38] An uncharitable reader might see in all this the reduction of one mystery to another. One thinks of the attempts made in the nineteenth century to devise explanatory models of magnetic action, mostly mechanical hybrids of one sort and another; but Albumasar fell back rather on a rather neo-Platonic explanation, when he tried to go deeper, an explanation in terms of 'emanation' rather than mechanism – after all, we have seen how he by-passed action by contact. This, at least, is the judgement of Lemay, who contrasts Aristotle's view that form originates from a potency in matter itself, with neo-Platonist ideas of a descent of forms from a superior order of beings in the hierarchy of existence,

---

[36] See p. 247 above.

[37] Lemay, *ibid.*, pp. 59–61, seems to be so anxious to find Aristotle here – quite rightly, of course – that he does not draw attention to the obvious additions.

[38] I doubt whether anyone would be prepared to analyse this in terms of electromagnetic induction, paramagnetism, or whatever. On the score of sources, Lemay, p. 64, again speaks of the Aristotelian origin of the classification; but his numerous references (n. 1) do not seem to suggest more than that Aristotle was the starting point of the scheme, e. g. in *De gen. et corr.,* I.6.

introduced without alteration into matter to inform it and give it existence.[39]

There are many different ways of reviewing the Albumasar material. It was of great influence in the West, and yet there it was supplemented by texts, in due course, that were ultimately seen as underpinning it – namely the texts of Aristotle, Plato, and other Greek writers. It was of influence on twelfth century scholars who did not have the advantage of a good untersdanding of those proto-texts – Hermann of Carinthia is a good example. But then one may ask for the routes by which Abū Maʿshar himself came to know what he retailed in his 'Great Introduction', and here it has been shown by David Pingree that he was not using as immediate sources 'the Arabic translations of the *De caelo,* the *Physica,* and the *De generatione et corruptione,* but the purported writings of the Ḥarrānian prophets, Hermes and Agathodemon. That the "Sabaeans" of Ḥarrān depended on Aristotle's *Physica, De caelo, De generatione et corruptione,* and *Meteorologica* for their theories regarding the material universe is clearly stated by [a student of al-Kindī].'[40]

Here I am concerned with this influence on the West, and it will suffice to say that he passed on a doctrine of the formation of individual beings in terms of (1) the properties of the elements of which they are formed; (2) their forms – which do not admit of degrees (one is not less or more a man on different days); (3) all else that is of consequence. This last category is what is produced by the heavenly bodies (ultimately by the command of God). It includes the differentiation of genus from genus, species from species, and individual from individual, and the production of harmony of soul and body and other innumerable accidents such as 'sex, distinctness of form and habit, beauty or ugliness, variation in size, variety of colours, different customs, and so forth'.[41] Apart from a theory of heavenly influence in individuation, Albumasar helped to fix the idea that the planets are intelligences of some sort; and more specifically, he passed on doctrines concerning the tides, and the functioning of celestial influence. I will consider these last in turn.

Broadly speaking, the neo-Platonists argued that, since the entire *world* was, as they supposed, animated by a divine Intelligence and Will, the planets were for that reason animated. In Albumasar, the effects of planetary motion, being in part, at least, rational in nature, the causes must be rational. Another line was taken by, for instance, Raymond of Marseilles: the planets are self-moving, and therefore rational. Of course their ultimate mover and creator was admitted to be God. The all-

---

[39] On Albumasar's 'emationism' in this sense, see Lemay, ibid., pp. 65–7, 231–43.
[40] *Dictionary of Scientific Biography,* vol. 1, art. Abū Maʿshar, p. 33.
[41] See Lemay, ibid., p. 83, n. 1: 'Id autem est ut generis a genere . . . . atque id genus'. From the Hermann edition, fol. a7r.

pervading World Soul was often equated with the Holy Ghost – although this was regarded with theological suspicion in the twelfth century. Raymond, like Abelard and William of Conches in their different ways, vacillated in their opinions. William, in his *Compendium philosophiae,* went so far as to say that the World Soul was neither God nor creature, but a third thing, although present in God's mind when the universe was created.[42] It was not the Sun itself, nor a virtue emanating from the Sun. In saying this, it does seem probable that he was treading warily in a minefield newly created by notions taken from Albumasar and dropped into older theology. The Sun as a source of heat was one thing; the Sun and the seven planets as ensouled by the Holy Ghost was quite another.[43] In view of attacks on such ideas by William of St Thierry, it is not surprising that William of Conches should have decided to make no mention of the World Soul in his *Dragmaticon philosophiae,* his masterpiece of the late 1140s.

Albumasar's remarks on the tides are perhaps the most relevant to our theme, and to introduce them it will be useful to turn back to Ptolemy's *Tetrabiblos* III.11, where he discusses the form and character of the human body. Abstracting from his various pronouncements on the subject, we can say that he believed that when the planets are morning stars they make a body large; when at first station powerful and muscular; when they move with a direct motion, of poor proportions; at second station, somewhat weak; and at setting, without repute, but able to bear hardship.[44] The idea at the root of this, and accepted tacitly by many later writers, is that the virtue of a planet depends on its motion with respect to the fixed stars, and this in turn depends on its position on the epicycle, being greatest in a direct sense at the far point of the epicycle, stationary near the two tangents from the observer to the epicycle, and least (even retrograde for all but the Sun and Moon) at the near point. According to Bouché-Leclercq, the first to adapt a simpler qualitative scheme to the system of epicycles and eccentrics was Adrastus of Aphrodisias.[45]

Ptolemy ascribed powers to the planets, related, as explained, to their (apparent) motions, but he also wished to take into account their positions relative to the horizon, 'for they are most powerful when they are in mid-heaven or approaching it'.[46] So far so good. No doubt we may see an analogy with the case of the Sun. (It had long been considered a paradox that the summer heat coincided with the Sun's being near

---

[42] Lemay, ibid., p. 190, following H. Flatten.
[43] Raymond of Marseilles, quoted by Lemay, n.1, p. 191.
[44] Cf. similar remarks at I.8, I.24, II.6.
[45] *L'Astrologie grecque* (Paris, 1899), p. 116.
[46] Ibid., I.24.

apogee. Ptolemy realized that the key factor was proximity to the zenith – even though he made the seasonal weather depend also on the qualities of the zodiacal signs)[47]. But alas, the parallel with thermal radiation or its like breaks down in the very same sentence, which continues: 'and second [most powerful] when they are exactly on the horizon or in the succedent place . . .' (i.e. about to rise). In other words, we have here a blend of the Aristotelian ideas already encountered, from the *Meteorologica,* making terrestrial heat depend on the Sun's motion, with what might possibly be a thermal analogy (mid-day Sun), but also with a sort of horoscopic astrology that it would be very difficult to relate to any recognisably physical theory of radiation.

Turning again to Albumasar: we saw how, in the *Introductorium* (the third book), it was argued, in much the same way as in *Tetrabiblos,* that all that is born and dies is subject to the stars, although the planets, with their varying velocities, are better fitted to producing movements in the world. The more rapid their movements, in fact, the more powerful their effects, and for this reason the Moon is especially important to us. (The 'fixed stars' – which he knew to have a slow displacement from the eighth sphere – produce the *stable* aspects of life through their daily motion, and extremely slow changes corresponding to that slow displacement, which we think of as a precession of the equinoxes)[48]. There are no fewer than six chapters on the effects of the Moon, and almost all of this material deals with tidal phenomena.

We can learn a great deal from this material of the physical models that were in the writer's mind – or at least, were transmitted to his Western audience through the translations we are considering. Of these translations, Hermann's is the more detailed here – although I shall later show influences from both versions. There is a nice symmetry outlined in III.4, with fire and air made subject to the Sun, and water and earth to the Moon. Sun and Moon are said to receive their movement from the superior world and *by their rays* to transmit influences to the sublunar world. (Albumasar cited Hippocrates, on the Moon as a mediator between heavens and Earth. It is not clear to me how the present passages were supposed to link up with earlier ones, in which heavenly bodies were supposed said to affect the Earth by actions transmitted through an invisible medium, like magnetic actions in that respect)[49]. As already explained, their efficacy consists in their *movement*. The stars and other planets, in fact, are said not to radiate at all, although they shine – an interesting indication of an attitude to light that a modern reader would not understand. As for the tides, the variety, diversity, and magnitude of

---

[47] Ibid., I.10 and II.1 respectively.
[48] A post-Copernican idea. One should use the word carefully.
[49] See p. 254 above.

tidal movement is such that there are some who do not believe that the Moon can be their cause, says Albumasar. Later (III.7) he opposed the opinion of those who think the tides to be *natural* movements. Water naturally falls: the tides rise and fall. He insisted, rather, that the Moon, that repeatedly rises and sets over the sea, *draws the waters by a cognate force*. The sea follows this tractive force *of itself (sponte)*.

For the insights it gives into the near-intuitive elements in the growth of a physics of tidal force (and let us not dismiss all this as rationalistic, for there was a strong empirical element), one chapter in particular is invaluable (III.5). Here eight causes are listed for tidal inequalities. Perhaps one should also add an idea that is aired in a later chapter (III.8) to the effect that the magnitudes of the tides are functions of how *receptive* the sea is to the Moon's influence.[50] (Note how most laws of force include a term concerning the object on which the force acts). It hardly needs to be said that the tractive force was said to provide two tides, one following the Moon, one opposing it; but Albumasar could not say how the second was to be explained.

The tidal inequalities, in brief, are caused by the Moon, and to a lesser extent the Sun, and are functions of:

1. The (angular) separation of Sun and Moon, which affects the lunar radiation (which we recall was a vital factor). The Sun also has its effect on the tides, we are told (an afterthought?), as indeed do certain humid stars, and in conjunction with the Moon these may reinforce it.

2. How different the lunar motion is from its mean motion.

3. The position of the Moon on its eccentric.

4. The position of the Moon on its own orbit, i.e. in declination. (Its orbit is at roughly 5° to the ecliptic.)

5. Whether the Moon is to the north or south of the equator.

6. Whether the days are what the Egyptians call 'sea days' and Westerners call 'days of increase and decrease'.

7. The length of daylight (roughly expressed, in spring and summer, when the Sun is in the northern signs, it will give more assistance to the Moon in moving northern seas).

8. The action of winds, which may cause exceptionally high water.

There is a great measure of good scientific sense in all this (most of it having ancient roots, needless to say), and it is not without significance that Albumasar gives relatively little attention to the Moon's influence on other phenomena, such as the production of growth in plants, changes in medical conditions, etc., when they are simply not quantifia-

---

[50] Cf. remarks about receptivity, on p. 253 above.

ble. He might not have had the inverse square law of gravitation, but he was emphatically in direct and not retrograde motion towards that end.

The great influence of this work of Albumasar, in these 'neo-Aristotelian' respects, was perhaps nowhere more evident than in the work of one of its translators, namely the *De essentiis* of Hermann of Carinthia (1143).[51] Hermann referred to Albumasar more than to any other author, by name, apart from the arch-astrologer Ptolemy. It seems unlikely that Hermann knew any of Aristotle's physical works directly, and it is hardly surprising, therefore, that he managed to work into his writings so much neo-Platonism that seems to us to run counter to the spirit of Aristotle. It seems likely that he used the so-called *Theologia Aristotelis*, which – despite its name – is a selection from Plotinus' *Enneads*, translated into Arabic.[52] This might well explain Hermann's having accounted for the bond between the heavens and the sublunar regions, not in terms of motion, radiation, and a fitting receptivity, as Albumasar had done, but in terms of *musical proportions*. The textual influences are hard to assess, because there were other neo-Platonic works that might have served this purpose. One absolutely central doctrine is that of the trio Same-Different-*medium*, which is based on Plato's account of the World-Soul, as in *Timaeus*.

Hermann's universe looks to us something of a hybrid, therefore. In the zodiac 'was constituted the whole power of efficiency, destined to flow through the middle as far as the lowest part' (63rC).[53] When God had finished distributing the elements and parts of his creation, the lower world was encircled by the higher, 'the perpetual power of its natural motion' (63vB). The language is not that of the analytical natural philosopher. In trying to translate Hermann's into such a language, we should not be taken to be claiming that his readers could have abstracted the meaning. Take, for example, this sentence: 'Since, then, the whole power of efficiency was to descend in order from the Same through the Different – just as the breath from one mouth bursts forth through different pipes – the Same, being simple, remained undivided; the Different and complex demanded a many-formed division.' (63vH)

With the text in hand, this piece of almost Plotinian vagueness can be seen as nothing more than a statement that a single sphere (on which the stars will eventually be put) with a constant motion (i.e. the Same) was needed by God to explain what is stable in the world, while the complex of spheres (the Different) is that having to do with the Sun's oblique path, that is, along the ecliptic. Through it, the power from the first is transmitted, and by this means we obtain variety. The Different is

---

[51] I refer henceforth to the critical edition by Charles Burnett (Leiden, 1982).
[52] Burnett, *ed. cit.*, p. 41.
[53] This and the following references are to Burnett's edition.

thus what Aristotle, and all who followed him, invoked to explain generation and corruption in accordance with the seasons. Later, Hermann introduces the *medium* in the form of the planetary spheres, the 'strong bond' that is 'filled with every musical proportion' (65rC). He spoke of this as a mediating bond, reconciling extremes by means of love.

This will seem very unpromising material if we are looking only for signs of celestial influence of a generally radiative kind, but it does at least show us that any typology of such 'influence' must be drawn more generously. The connection with Hermeticism is clear. He was citing the *Tabula Smaragdina* indirectly, via ps-Apollonius, *De secretis naturae,* when he wrote of the Sun and Moon that they were established as the 'parents' of the universe – that is, were involved in generation going on in it (65vC). He referred to Trismegistus as having assigned one principle to birth and the other to growth *(loc.cit.).* He went on to say how in secondary generation and corruption, two principles were needed for each; and he related these to the planets, and their order. (Saturn is a principle of corruption contrary to the Sun, and Mars contrary to the Moon. The entire passage gives a very obscure rationalisation of the favourable and unfavourable, co-operative and opposing, characters of the planets.) The drift of the planetary theory was to show how, by their double movements (on epicycle and deferent – Hermann rarely hints at further complexities of motion), the planets 'might provide sufficient variation for all the movements of subject nature' (68rE). He reiterated the doctrine that the harmony in the celestial music ('every vigorous movement gives forth a sound') is 'the one bond of society holding everything in the universe in an indissoluble knot', before ending the first book with his 'physical' explanation of influence, as we might describe it: *sympathetic vibration in the sublunar world.*[54]

If the emphasis in the first book of the *De essentiis* is interestingly idiosyncratic, the second only serves to prove that one may not blend one's sources with impunity. It would be unfair to put too much emphasis on his references to the calling down of celestial spirits – introduced in connection with his distinction of mortal from immortal animate being. It is no easier to square with his theory of sympathetic harmony, however, than is the account he offered of the tidal movement of the sea. Drawing on Albumasar, he must of course now think in terms of *directed* influence of Sun, Moon, and stars. (I know that we found some inconsistencies in Albumasar, but the overall pattern of his explanation corresponds reasonably well with our own ideas of directed forces.) This he cannot allow himself to do. He wrote of the Sun's effect

---

[54] His sources are uncertain – see Burnett's commentary, p. 293 – but he might have been influenced by some remarks in the Albumasar *Introductorium*. The uncertainty stems from sheer superfluity of 'Pythagorean' material.

(effectus) not always being the same, when it is at the same place in its orbit, and – following Albumasar – ascribed this to the participation of the stars and planets, the former for 'singular' guidance, the latter for 'universal' (74rBC). But it was the image of a *bond (nexus)* between the dragging Essence (the heavens) and the dragged Substance that he used at the appropriate place in his text (74rBff.). At all events, he leaves us in no doubt that the influences are there, whatever their explanation, and he derides 'the naïve medical man who thinks that nothing has an effect, unless it is of the lower nature alone' (74rG).

It is doubtful whether there is any better or more complete example than Hermann of Carinthia of the reception of the 'Albumasar-Aristotle' in the twelfth century. We have already made brief reference to Raymond of Marseilles and William of Conches. Bernard Silvester, author of *Cosmographia,* the most important cosmic myth between Lucretius and Dante, is in no doubt: the lower world was governed by the all-embracing laws of celestial motion.[55] God's effective power is, by celestial mediation, translated into causal terms that allow of prediction – such as in the anticipation of the virgin birth (an example taken from Albumasar, it seems).[56] Bernard's world was redeemed 'from complete and senseless mechanism ... [by] not only the humanistic alternative of free will but the idea of progress', information about the reform of the cosmos being foretold in the stars (Stock).[57] The problem of determinism and human responsibility is not, however, central to my theme. It would be as churlish to expect Bernard to have achieved great philosophical depth on that subject as to look a closely argued rationale of celestial involvement in earthly affairs; but what he does say on this last subject serves to remind us that it is a mistake, when following the history of a series of ideas, to overlook more colourful, and surely more deeply felt, literary and spiritual allusions – especially in an age that had not learned the obsessive delights of natural philosophy. For Bernard, all forms of terrestrial life draw this life from the god of life, and where his friend Thierry of Chartres 'expressed this conceptually – a "shaping fire" *(ignis artifex)* is the moving cause of things, and earth is their material cause – Bernard goes back to the archaic mythical image of the marriage of Heaven and Earth, Ouranos and Gaia. The eternal Fire is the lover and husband who puts himself into the womb of his bride Earth.'[58] The cosmos portrayed is one marvellously harmonious, but peopled with daemons of many kinds, much of the daemonology evidently drawn

---

[55] The most convenient text and commentary is Peter Dronke, *Bernardus Silvestris Cosmographia* (Leiden, 1978). See also Brian Stock, *Myth and Science in the Twelfth Century. A Study of Bernard Sylvester* (Princeton, 1972).
[56] Stock, *ibid.,* pp. 132-3.
[57] Ibid., p. 161.
[58] Dronke, *op. cit.,* pp. 37-8.

from Chalcidius and Martianus.[59] This is hardly Aristotle talking, but is rather the scholar whom the newly arrived Aristotle had to convert to a new mode of thought.

Some took more readily to the new mode. Daniel of Morley is a good example of one who tended to reject the Latin Platonic tradition in cosmology, and in drawing heavily on William of Conches to add the astrologized Aristotelianism of Albumasar. He took over many texts from the *Introductorium,* especially noteworthy being those on the theory of the tides, drawn from III.5, I.4, and III.9, in that order.[60] The persistence of this tidal theory through the thirteenth and fourteenth centuries may be illustrated through a treatise which I take to be by Richard of Wallingford (c.1292–1336), although it has some thirteenth century characteristics, *Exafrenon pronosticacionum temporis.*[61] It appears that this draws on the John of Seville translation. It includes a none too detailed account of the dependence of the weather on the luminaries and planets, this being decided by position (e.g. in the epicycle) and hence motion – which last is the crucial factor, as in Albumasar.[62] The other factor emphasized is the native character of the planet (the Sun being hot and dry, Venus being warm and moist, and so on, as laid down in *Tetrabiblos* I.4). There is mention made of a factor that may *diminish* the efficacy of the planets thus far assigned, namely their proximity to others. This is the doctrine of rays – not rays extending from the planets to the Earth, but rays extending along the zodiac for a short distance – for example, 8 degrees either way for Mars, 9 degrees for Jupiter, 15 for the Sun, and so forth. A planet falling under the Sun's rays in this sense was burned, *combustus,* ὕπαυγος. This classical doctrine persisted throughout most of astrology's long history, in slightly variant forms, but does not seem to connect with our theme in a significant way. I note that in many optical writers of the middle ages a distinction was drawn between *lux*, the brightness one sees an object to have, *lumen,* the light it passes to other objects to illuminate them and make them visible, and *radii,* the rays that appear *around* bodies, as in a glory, aureole, or nimbus. This fits the astrological usage. (It has to be admitted that *splendor* and *radiositas* are sometimes distinguished – as by Grosseteste, for example – but as explained, this can be put aside as irrelevant to our theme).

The Albumasar treatise was by no means the only Arabic work on celestial, and particularly tidal, influence to help create a Western tradi-

---

[59] Ibid., p. 43, n. 11.
[60] Lemay, *op. cit.,* p. 332, n. 6.
[61] This is included in my *Richard of Wallingford,* 3 vols., (Oxford, 1976). See the commentary on cap. 4, in vol. 2 at pp. 113 ff.
[62] For the text, see esp. vol. 1, pp. 210 ff.

tion, but others did so in a more piece meal way, and I shall mention them only as the occasion arises. By the early decades of the thirteenth century, scholars were looking at their textual sources more critically than hitherto, and none set a better example than Robert Grosseteste. One of the chief historical problems in this connection relates to the authenticity of a work on the tides that is widely accepted as his: *De fluxu et refluxu maris*. Attention was first drawn to this in 1926 by F. Pelster, who settled for Grosseteste's authorship; others later suggested Adam Marsh; R. C. Dales, who has given it closest attention, first argued for Grosseteste[63] but later had doubts, especially since an early work by Grosseteste, as well as a late one, accept al-Biṭrūjī's views on the tides, while *De fluxu*, having otherwise the appearance of an intermediate work, criticises that theory.[64] For these reasons I shall consider those views Grosseteste expressed in works that are generally accepted as unproblematical, and only then consider the *De fluxu*.

In his early writings Grosseteste accepted Islamic astrological material at its face value, more or less, and there is no reason to suppose that he ever rejected the basic premise of celestial influence on terrestrial bodies. Perhaps before 1209 (he died in 1253, at the age of 84 or thereabouts) he wrote a work under the title *De impressionibus aeris seu de prognosticatione,* concerned mainly with astrological methods in meteorology.[65] Tides were a traditional topic in this sort of literature, but the impression given by Grosseteste is that he has thought the subject through, to his own satisfaction, and is not simply copying out other men's sentences. He placed great emphasis on the *motion* of the luminary or planet, but where 'motion' for most of his predecessors had been a rather vague concept, for Grosseteste it was the *daily* motion of the planet across our sky that is the decisive factor in settling the influences of the planet on the Earth. When the Moon is at its aux, it is going at its slowest as seen from the Earth with reference to the stars; and therefore its daily motion across the sky is at its greatest, since these motions are opposed. By the rules we have seen in Albumasar, therefore, the Moon can be thought to be at its most potent, as far as this factor is concerned.[66] (Grosseteste gave Ptolemy as his authority for the influence of motion). The same goes for the Sun. The scheme works very nicely, since aux coincides nearly enough with summer solstice. In the case of the planets, they are strong at their apogee (aux) as regards the deferent

---

[63] 'The authorship of the *Questio de fluxu et refluxu maris*...' *Speculum,* 37 (1962), pp. 582–8.
[64] 'Adam Marsh, Robert Grosseteste, and the treatise on the tide', *Speculum,* 52 (1977), pp. 900–01.
[65] *Die philosophischen Werke des Robert Grosseteste, Bischofs von Lincoln* (Beitr. z. Gesch. d. Phil. d. Mittelalt.; Münster i. W., 1912), pp. 41–50.
[66] Ed. Baur, pp. 47–8.

circle, and at the *near* points of their epicycles, since there they have the slowest direct motions.[67] Other things being equal, the tides, he said, will be greatest when the Moon is at its aux; and he was a good enough astronomer to know that this occurs when the Sun is either in conjunction with, or opposition to, the Moon.[68] *'Ideo fluit mare et refluit bis in mense lunari'.* As for the daily tides, he saw that the light of the Moon, their cause, would bring them round to us over periods of roughly once a day; but why, as experience taught, should there be a tide twice a day? His handling of this problem is interesting: on rising above the horizon the Moon draws our sea waters up, and this continues until the Moon has passed our meridian, after which the waters subside, 'and the waters flow to the opposite place and a similar peak *(tumor)* is generated there'. This, he added, is why there are two ebbs and flows every day.

There is in all this a very interesting sleight of hand, and some modern commentators have gone along with the way of thinking that Grosseteste spells out in the first half of his paragraph. He has concentrated on our local waters, and on the Moon as it comes up above *our* horizon. Our waters subside, and somehow they magically slip away to form a swell on the other side of the Earth, with the implication that there are always two peaks in the Earth's seas – for we do need a second peak to explain the second tide. Now there is another way of looking at our tides: one can think of the peak in the sea pointing towards the Moon, and brought to our port as the Moon comes to our meridian. Not all tide treatises show evidence that this mode of thought had been appreciated, but in the second half of his paragraph, Grosseteste makes it clear that *he had appreciated it*. *'Et sicut luna totam terram circuit in die et in nocte, similiter circuit tumor ille.'* The trouble was, he could not say why, with this way of talking, there should be a second peak; but of course he did not draw attention to the fact.

Grosseteste had not yet developed his 'light metaphysics', which arrives with the *De luce* (between 1225 and 1230?). In this earlier period, in his *De artibus liberalibus,* and in his *De cometis,* for instance, he had taken what might be distinguished as the 'common astrologers' view', that the planets and fixed stars act on the lower world in virtue of qualities that they share (hot, cold, wet, and dry). They were assumed to differ in constitution from the quintessential spheres on which they were supposedly carried. This view had implications for alchemy that were not in keeping with orthodox Aristotelianism. With the more

---

[67] Ibid., p. 49. He does not mention the epicycle and talks of *'zenith capitis'* (zenith) rather than of its far point. This might be thought to suggest a misunderstanding of a source (cf. 'zenith' and 'highest point').
[68] For medieval terminology, and a brief account of the Ptolemaic theory of the lunar motion, see my *Richard of Wallingford,* vol. 3, (Oxford, 1976), pp. 171ff.

mature Grosseteste of the *De luce,* his position on this important point changes. He there gave as a reason for the centrality of the Earth in the universe that it may be a natural focus, so to say, for the light from the spheres. From astrology he had learned to treat the elemental qualities as a common denominator, shared by the cause and the affected thing, the heavens and the sublunar region; but now he was convinced that every natural body has in it a celestial, luminous, nature, and a fiery luminosity at that. How he used light in such works as the commentary on the *Posterior Analytics* and *Hexaëmeron* to explain a whole range of physical phenomena is well known. What is less well appreciated is that light has stepped into a role played earlier by the qualities of the elements. One might even see traces of this in his *De cometis* (early 1220s?), written in ignorance of *Meteorologica*. There he wrote of the powers of the stars and planets to sublimate from the earth spirits of the nature of the Sun. This power *(virtus)* is clearly another to be added to our list of celestial influences.[69]

The light metaphysics first came into its own in a two-part treatise, *De lineis – De natura locorum*. There he laid down a theory for explaining the effects of light according to general principles of perspective familiar to the select few who knew Ptolemy's *Optics*. Grosseteste has a primitive sort of 'photometry', based on the 'pyramid' of light propagation: the power (e.g. of the Sun) increases as the angle of incidence (measured from the normal to the plane on which it falls) decreases. It is all rather impressionistic, and I will not got into the details, but note something of no less interest than his crude geometry of forces: Grosseteste had found an explanation for the piling up of water on the side of the Earth away from the Moon! He rejected the idea put forward by philosophers that 'opposite quarters of the Earth are of the same mixture [of qualities] and therefore subject to the same effects'.[70] This, he thought, was nonsense, since the Earth blocks the way for the rays. The solution is that the rays are *reflected,* 'quoniam radii lunares multiplicantur ad caelum stellarum, quod est corpus densum'.[71] Because of its density we cannot see the slight luminosity of the heaven, but reflected rays were supposed, somehow, to get round by reflection to the other side, and to produce the second peak. He went on to talk of concentrating solar rays by refraction through a flask; and of a reference by Albumasar to Hippocrates' statement to the effect that the density of the air would be

---

[69] Baur has two texts. See esp. *op. cit.,* p. 39. S. H. Thomson later published another, and then supplemented it. For a bibliography of Grosseteste's writings, see Thomson's excellent *The Writings of Robert Grosseteste, Bishop of Lincoln, 1235-53* Cambridge, Mass., 1940); and as a supplement, see the bibliography in J. McEvoy, *The Philosophy of Robert Grosseteste* (Oxford, 1983).

[70] Ed. Baur, p. 70.

[71] *Loc. cit.*

so concentrated that we should be unable to live, were it not for its irradiation by the stars at night.[72]

In the *Hexaëmeron* (c.1235) Grosseteste's final position on astrological belief is stated at some length. Superficially, he is hostile: 'We wish to warn that astrologers are both seduced and seducers, and their teaching is impious and profane, and written at the dictation of the devil. Therefore their books ought to be burned.'[73] The reason for the hostility is theological, and has to do with free will rather than celestial influence, in which Grosseteste's belief was as strong as ever. As far as the human body is concerned, he made statements to the effect that, for example, the planets could arouse the blood so as to stir up ire; but the soul has a potentially greater power, so all is well. Besides which, medicines can counter these bodily affections![74] Dales rightly remarks his 'growing awareness of the inescapable accuracy of all human measurements [that] led him to realize the inherent inability of astrology to do what its practitioners claimed, since the accuracy they pretended to was unobtainable'; but from my point of view this is not very significant. More to the point is the fact that he thought astrologers had failed, and must fail, because the influences they sought were so precisely focussed (if I may use this expression) in accordance with the momentary stellar configurations, that even the most accurate astronomer would not find them. They were *real* enough.[75]

What, then of the *De fluxu et refluxu maris?* Was it by Grosseteste? Consider, first, a few of the arguments to be found in it. The elements, says the introduction, are not moved by themselves or by another element. The efficient cause must either be a power of the heavens or a star in the heavens. And then: 'In this connection, Alpetragius [al-Biṭrūjī] says that all the lower spheres, as far as the sphere of water, are moved from east to west by the power of the outermost sky. But the lower any particular sphere is, the less power it will receive from the outermost sky, because this power is the power of a body, and diminishes with distance.'[76]

The writer holds this to be false, and simply *ignores* the premiss concerning the diminution of the power of the body with distance. This

---

[72] Ibid., p. 72.
[73] R. C. Dales, 'Robert Grosseteste's views on astrology', *Mediaeval Studies*, 29 (1967), pp. 357-63. The work in question has recently been edited by Dales and Servus Gieben, *Hexaëmeron* (Oxford, 1983).
[74] Ibid., pp. 362-3 (1967).
[75] Ed.cit., fols 214-5.
[76] The best edition (which also has a translation, from which this first excerpt is taken), is R. C. Dales, "Robert Grosseteste's 'Questio de fluxu et refluxu maris'", *Isis*, 57 (1966), pp. 455-74; text at 459-68.

is not the main prong of his attack, however. Even on the assumption that it is the heavenly spheres that cause the tides, the theory will not work; for the fastest moving part of the fastest moving, the equator of the outermost, is operating on a complete circle of sea round our globe, and ought to pile up the sea uniformly all the way round our equator. It does not, ergo . . .

This was only marginally dishonest of the writer – who smuggled in the assumption that *motion* is the key to the power. He affirmed – as Grosseteste would have done – that the Moon is the cause. The sea follows its motion more closely than any other celestial motion. And next comes the account of the Moon's rising over the sea, culminating, setting, and so forth, all more or less as it was found in *De impressionibus aeris*.[77] There are one or two small differences. There is no talk of the sea slipping away to the other side of the globe when it recedes on our side. The writer substitutes a sleight of hand of different sort: he says that when the Moon is below the horizon the sea will rise (etc.), but he does not say *which* sea. He hopes the reader will take it to be *ours* ('and thus in one lunar revolution there are two high tides' is in essence his conclusion), whereas surely the intelligent reader would have expected him to say that there will be a high tide *where the Moon is*. In view of my remarks about Grosseteste having seen this other way of looking at things, I am inclined to doubt whether this is his. Note that it presupposes that 'heavenly bodies do not act on lower bodies except by their light'.[78] Admittedly, at a later point, the text repeats this, with some ideas that smack of the *De natura locorum,* but that might simply have been inspired by Grosseteste. Part II of the text, however, is something we have met before – but nowhere in the attested works of Grosseteste: it is a section almost wholly based on Albumasar's *Introductorium*.[79]

Like Albumasar, the writer gave eight reasons for tidal variation, but he found it difficult to reproduce the older reasons, founded as they had been on other principles than his. The first reason given was that when the Sun and Moon are in conjunction, the Moon's power is augmented; at the quarters it reaches a minimum, and in opposition a second maximum. So far so good. This is a consequence of the argument from *motion* that we saw in Grosseteste; but when the Moon is in opposition (the only case explained) we are told that 'its power is not because of its aspect to the Sun, but on account of the increase of *light* in

---

[77] See p. 263 or 264 above.
[78] *Ed. cit.,* lines 110–11. The earlier parts of this paragraph correspond to lines 89 ff. Note that Dales' rather free translation might give the impression that high tide occurs whilst the Moon is rising, i.e. before it culminates. That is not the writer's intention.
[79] There is some Albumasar material in the spurious *Summa philosophiae*. See Baur, p. 627.

it'.[80] Now this seems to me to be a mark of a compiler of doctrine, rather than a careful thinker, such as Grosseteste has at least tried to be on this subject. Note, in passing, that when the Sun and Moon are in conjunction, the Moon's light is least!

The second reason is, like its source, an unspecific reference to variation of the Moon's motion around the mean. There is no hint – any more than in the original – of Grosseteste's interpretation of the motion as a daily motion. The third reason has to do with the variation of the Moon's distance by virtue of its motion on the *epicycle*. This conforms with Grosseteste's reasoning, and does not seem to be Albumasar's, which had reference to variation by dint of the eccentric motion on the epicycle. The new rule is that the Moon will be weakest when it is at the farthest point of the epicycle, and strongest at the nearest point.[81] The fourth, fifth, and sixth reasons are more or less Albumasar's, but the seventh involves yet another misunderstanding. It concerns the help the Sun gives to the Moon, which in northern waters (tacitly understood to be at issue) is greatest when the Sun is in northern signs. The help, we may say, is always positive, but varies sinusoidally, following the pattern of daylight. The writer found the verbalization of this beyond him, rather as in the case of the quarters of the Moon's *day*, and wrote as though the assistance was increasing in autumn.

The eighth reason, as before, has to do with the winds. There follows a short third section of a vaguely practical sort, but adding nothing to the topic of celestial influence.

Whether this treatise could have been written by Grosseteste is a matter for Grossetestians to decide. I am inclined to think – for reasons I have given – that it is not consistent enough to pass as his; but one should not readily overlook the fact that two manuscripts assign it to him. One way or another, he did contribute a great deal to the general subject of celestial influence, rather inadvertently, through his writings on light, and his new paradigm, explaining light propagation and vision in terms of species. An alternative to the ray optics that was accepted alongside it, and that is still familiar, it is now usually regarded as having been somehow the 'wrong' alternative. The concept of 'species' has been variously understood: in Augustine's *De trinitate,* for instance, it was part of a theory of human perception. The object seen gives off a

---

[80] *Ed. cit.,* Pt. II, lines 9–10. The word I translate as 'aspect' is given by Dales as *'respectum'* which he translates as 'nearness' – a meaning which more or less coincides only in the case of conjunction. I note that he records two MSS as having uncertain readings: – *'aspectum'(?)*

[81] The word 'epicycle' is not used, and there is reference to the 'aux of its circle', but the technical terminology used in the passage (*ad longitudinem propinquiorem,* and so on) makes the meaning clear.

likeness (species) which is perceived. Another species is given off by the first to the senses; that gives off a third species to the memory; and the third gives off a fourth to the recollection if called upon.[82] With Grosseteste and his followers the meaning of the term was extended. It now meant the likeness of the object, regardless of the presence of a perceiving being, but it also meant the *force* or *power* by which an object acts on its surroundings.[83] For Bacon, even more broadly, it was the first effect of any naturally-acting thing, and *all* natural causation, for him, was attributed to the multiplication of species. This whole theory, or rather set of variant theories, gave rise to lengthy debates about what actually happens in the medium. Do the forms or species exist there? To mention only three philosophers: Averroes had said that there are forms in the medium, with an existence somehow between corporeal and spiritual existence. Albertus Magnus settled for corporeal existence. And Ockham rejected the idea of any such intermediary.

It is hardly surprising that mention of celestial influence tended often to be couched in the language of the propagation of species. Consider Bacon, for instance: 'Although terrestrial things cannot resemble the heavens in their complete natures, they agree at least to the point [of being linked] through the reception of species'.[84] We have seen already how there was a belief that the heavens and the sublunar world must share something, or some property, for it to be possible for the former to influence the latter. For the early Grosseteste, it was the conventional sharing of elemental qualities; later for him it was the binding of light in matter. Bacon went back to Aristotle's *Metaphysics*, Book V, saying that it is evident that lower things can be influenced by higher things, 'since they share the same matter, because things of the same genus have the same matter'.[85] A passing reference to the part played by the obliquity of the Sun's path in generation – straight from our classical Aristotelian text, of course, now so well worn that he hardly needed to fill in the details – and then the following argument: the Sun and other celestial bodies do not exist in terrestrial bodies in substance, only in power, *virtus*. In this sense, celestial nature may be generated in celestial matter.[86] This is not a work of astrology, but rather takes the basic fact of astrological influence for granted, and tries to provide it

---

[82] P. Michaud-Quantin, 'Les Champs sémantiques de *species* . . .', cited, and supplemented by D. C. Lindberg, *Roger Bacon's Philosophy of Nature. A Critical Edition*, etc., of *De multiplicatione specierum* and *De speculis comburentibus* (Oxford, 1983), liv-lv.
[83] Lindberg, *ibid.*, p. lv.
[84] Ibid., p. 71 *(De mult. spec.)*.
[85] *Loc. cit.;* cf. *Met.* 1024b.
[86] Bacon, loc. cit. For the record, he added that the Moon and stars receive the virtue and species of the Sun, 'bringing light into being there . . . out of the potentiality of the matter'. And even terrestrial things may produce a species in celestial bodies.

with a theoretical basis. No doubt Bacon would have said, had he been asked, that the chapter in question (I.6) was what Aristotle's *De generatione et corruptione* would have included had Aristotle been fortunate enough to have lived in the thirteenth century. Without reproducing the argument in great detail, we can say that for generation to take place, (1) matter had to be shared by agent and recipient, and (2) the agent has to have greater power than the recipient, to prevail over it. (There were other conditions)[87]. Not only substances but also accidents (such as the 'tangible qualities', warmth, cold, dryness, and humidity), can also 'complete their species' in certain suitable recipients. His first example is of *light* affecting bodies,[88] for instance the light from the Sun. This whole theory of species is double-layered, and of considerably greater ingenuity than anything found in straight astrological texts; but it can rarely be said to have been connected with experience in any clear way.

There are some noteworthy exceptions. Bacon touched on the problem of the diminishing effect of the agent with distance. Is distance the cause, as Alhazen had said?[89] The power of a magnet is stronger, noted Bacon, in distant iron than in adjacent air, yet it passes through the air to the iron! The medium comes into the calculation, he decided; but discounting this factor, 'the species of light or anything else must be weakened with distance, since as far as sense can discern (or at least according to nature) density always increases as one proceeds from the heavens to the lower regions . . .'.[90] In other words, the law (if that is not too strong a word for it) of variation of force with distance is not meant to be universal, but to apply to forces along radial directions in the universe. Bacon's is not the best of arguments, but then, the whole theory of species is unsuited to laws of photometry or force in general, at least until blended with a theory of geometrical rays, after which the species are hardly of any significance. Bacon certainly knew the power of geometrical methods, and the praises he so often heaped on them were largely due to their value in 'speculative and practical' astrology and astronomy, by which we learn 'the celestial causes of the generation and corruption of all lower things'.[91] There were, though, places where he obviously thought a species-explanation highly appropriate. To single out an example with relevance to our present theme: in the *Opus tertium,* again, he discussed the miracles wrought by words, from the time of the world's beginning. The good or ill power of the soul generates a corresponding species or power that is multiplied in the voice, and

---

[87] Ibid., p. 83.
[88] Ibid., p. 89.
[89] Ibid., p. 209.
[90] Ibid., p. 211.
[91] *Opus tertium,* ed. J. S. Brewer in the Rolls Series, 1869, cap. xxx, p. 107.

in the air carrying it. There is also the species or power of the body being acted upon. But fourthly, there are the heavens, and the species or power of the celestial configuration at the time the words are uttered. After sketching briefly his doctrine of how charms, fascination, and certain sorts of magic function, he was here content to refer the reader back to *De multiplicatione specierum*.[92] At the end of the fragment of the *Opus tertium* first published by A. G. Little, Bacon made some cognate remarks about celestial involvement in alchemy, which need not detain us, for they add nothing to the underlying theory of influence.[93]

Bacon, and those who were likewise prepared to give a rationale of celestial influence, did not forego an opportunity of retailing traditional astrological teaching. He touched on astrological medicine in many places in his writings, on planetary dignities, aspects, properties – and the effects of compounding them –, major conjunctions and their effects on religious history, and so forth, drawing on standard authors in a standard way. He reported the tidal doctrines of Albumasar, and was obviously proud of his own explanations of the tides, with his analysis of their causes 'by the multiplication of the rays of the Moon according to lines and angles, and by the reflections of rays'; but there seems to be nothing extant from him that merits such enthusiasm.[94]

When considering Albumasar, we saw that he likened celestial influence to magnetic powers, acting as it seems to do without contact. This is not the same thing as supposing that celestial influences are by nature magnetic, although there might have been some connection in Abū Maʿshar's mind, if the direction-finding properties of the magnet were known in his day – which seems rather unlikely, at the moment. (The Chinese had this knowledge, admittedly, before 1088).[95] It was debated whether the north- and south-pointing properties of the magnet were due to deposits of the stone at the Earth's poles, or whether the key to the mystery was the pole star. The famous letter of Petrus Peregrinus on the magnet (dated 1269), which hints at earlier discussions no longer extant, settles on a celestial solution, since the stone is found at many places well away from the poles.[96] Bacon, writing at almost exactly the same time, does the same,[97] but neither will agree with the common philosopher, who ascribes a power to the *star*. It is the heavens themselves that matter. As Petrus explains, the poles of the magnet derive

---

[92] *Op. ter.*, ed. cit., cap. xxvi.
[93] *Part of the Opus Tertium of Roger Bacon* (Aberdeen University Press, 1912), pp. 80–9.
[94] The claim is at Brewer, ed. cit., p. 120.
[95] Joseph Needham, *Science and Civilisation in China*, vol. iv, pt. 1 (Cambridge, 1959), p. 249.
[96] The letter can be conveniently found in translation in Grant, *Source Book* (op. cit.), pp. 368–76, together with a useful bibliography of relevant editions, etc.
[97] For example in *Op. ter.*, ed. Brewer, p. 384.

their virtue from the poles of the world (and the pole star is not precisely there), while the rest of the magnet receives virtue from appropriate parts of the heavens. Both he and Bacon allude to a way of illustrating this dependency by experiment: a spherical lodestone perfectly balanced and smoothly pivoted will turn as the heavens turn – serving, incidentally, as a timepiece! (A design for a perpetual motion machine ends Petrus' letter). William Gilbert and Galileo rejected the idea: as Galileo said, even when it is stationary with respect to the Earth, the magnet has a diurnal rotation, and there is no reason for a second movement.[98] This would not have been acceptable to a believer in a geocentric universe who was also convinced of a correlation between the heavens and the parts of the lodestone.

This question of a correlation between the heavens and points of the lower regions is one to which Bacon, like Grosseteste before him and John Pecham after him, paid some attention, producing in the course of writing on optical matters an argument with some astrological value. I will quote from John Pecham, whose writings draw heavily on Bacon's, perhaps following upon personal contacts.[99] In his *Perspectiva communis* Pecham made ample use of the theory of species, assuming like Grosseteste and Bacon before him that they proceed from the eye as well as from the object. Exactly as they had done, he noted the significance of the Earth's centrality in the universe:

> Light from a concave luminous body is received most powerfully at the centre. The reason for this is that, for every point of a concave body, perpendicular rays, which are stronger than others, converge in the centre. Therefore the virtues of celestial bodies are incident most powerfully in and near the centre of the world...[100]

One has only to think of the context, a work on optics, with its ray diagrams, and its Earth at a focus of the world, so to speak, to see how appealing must have been the idea of an 'optics' of astrological influence. It is what Grosseteste had been practising in a rough and ready way when he offered an explanation of the peak in the seas to the far side of the Moon, and what Bacon had been hinting at, in speaking of the tidal theory he was developing. They were all three agreed that every place on the Earth receives a different 'radiant pyramid' from any celestial object. To quote Pecham:

---

[98] Note 15 to the Petrus Peregrinus letter, ed. cit.
[99] Whether in the 1240s in Paris, in the mid-1250s in Oxford, or in the 1260s in Paris, is uncertain. Both were of course Franciscans, in the wake of Grosseteste's teaching to that order, in Oxford.
[100] D. C. Lindberg, *John Pecham and the Science of Optics. Perspectiva communis* (Madison, 1970) p. 99.

Light departs powerfully from any point on a luminous body, and the more nearly perpendicular, the stronger it is; therefore as pyramids proceed more obliquely from the surface of a luminous body, they are without doubt correspondingly weaker, and as they are more perpendicular they are correspondingly stronger...[101]

This, he went on to say, is why plants are so diverse. (As we might say, each place has a different perspective on the heavens – but that turn of phrase is to think of the 'pyramids' in the other direction). Maimonides, he added, had no need to suppose that individual stars correspond to individual plants. Bacon had added another illustration, which must have seemed like manna from heaven to the astrologer who had been embarrassed by the age-old argument of the twins.[102] Their aggregates of pyramids of celestial influence are subtly different! *(In eodem tempore veniunt diverse figurationes virtutum celestium ab eisdem stellis...)*

One last remark, in connection with John Pecham's *Perspectiva communis*. He offered there another solution to the problem of tides on the side of the Earth away from the Moon, a solution quite different from Grosseteste's. He settled for the *penetration* of planetary influence, and turned Bacon on his head. 'Because every body is susceptible to celestial influence *(influentia)*, it is certain that no body lacks transparency altogether... Hence lynxes are said to see through walls'.[103] Although not mentioned here, this was an answer to the question as to how metals can be formed under astral influence – as metallurgists so often supposed them to be – in the darkness within the Earth.

Grosseteste clearly changed the ways in which scholars spoke of celestial influence, in as much as he persuaded them to accept a model of light rays incorporated in a dense transparent medium, and producing heat there (especially by reflection back on themselves) as a result of the breaking up of matter. (This idea is to be found in four or five of his short, later, works.) Even in Grosseteste's own affections, however, this did not entirely replace the old Aristotelian principle that somehow it was 'motion' in the heavens that caused the effects on Earth. The trouble with this type of explanation was that it was difficult to see how the motions of planets on perfectly frictionless spheres could transmit motion inwards. On the face of things, the homocentric system of al-Biṭrūjī, the twelfth century Sevillian astronomer, seemed to have the quality of an inward transmission of movement – at least the text claimed it, albeit in a suitably vague way. In this respect it was the answer

---

[101] Ibid., p. 65; and n. 10 for cross references to Bacon and Grosseteste.
[102] Why, with identical horoscopes, do they not have similar fates? See *Op. ter.*, ed. Little, p. 5; and *Opus maius*, ed. J. H. Bridges, vol. 1 (London, 1914), p. 138.
[103] Ibid., p. 133. Sight occurs only through a transparent medium, because species are multiplied only through transparent bodies. (So begins the proposition).

to the Aristotelians' prayers. (Whether or not it was inspired by the system of Eudoxus, or by that of al-Zarqālī is beside the point). Its popularity in the thirteenth century in the West probably had nothing to do with the demand for a model for the transmission of astrological force, but that it lent itself to explaining this could have done it no harm. Grosseteste often cited it, and obviously welcomed it; Albertus Magnus publicized it, but tended to stick with Ptolemy. Bacon took it very seriously, and like the author of the *De fluxu,* reported the theory of the tides presented by al-Biṭrūjī (Alpetragius). The best astronomers, if we may use the phrase to distinguish the mathematically minded astronomers from the natural philosophers and cosmologists, usually stood by Ptolemy. Among the second group there were many whose instincts were to tell them to do likewise – Giles of Rome, for instance, and Pietro d'Abano, and John of Jandun. By the fourteenth century, the fashion for homocentric spheres was fading fast. The hesitations of the thirteenth are beautifully mirrored in the thought of Thomas Aquinas. He started his career as an adherent of Ptolemy, later treated the theory of Eudoxus as on a par with Ptolemy's, went on to speak grudgingly of Ptolemy, then for long treated both systems as unproved hypotheses, before finally showing rather more sympathy for Ptolemy.[104] That the Ptolemaic system occupied the centre of the stage between his time and the sixteenth century was due in large measure to the demand for accurate planetary information that it alone could provide – now through the medium of such tables as the Toledan, the Alfonsine, and their derivatives. This demand, of course, was frequently from the direction of astrology.

It has been as presupposition of all the writers so far discussed that there is a pervasive causality operating in the world – not, of course, necessarily of a single and universal type. All assumed a Prime Mover, God, usually along the lines of Aristotle's Unmoved Mover, but problems appeared as soon as it was asked how God moved the *primum mobile,* the first moving sphere. It was generally accepted that Aristotle was right when he wrote in the *Metaphysics* that the final cause 'produces motion as being loved, but all other things move by being moved',[105] and evidently God moved the *primum mobile* at least in this way, as a final cause, loved and disired. There was, furthermore, no harm in assuming that God acted as an efficient cause of motion in the *primum mobile.* We must obviously not allow more than *one* such efficient cause: if too much emphasis were to be placed on the system of

---

[104] For further details of these changes of heart, spanning the period from his *Sentences* to his *De caelo* commentaries, see T. Litt, *Les corps célestes dans l'univers de Thomas d'Aquin* (Louvain, 1963), ch. xviii.
[105] 1072b 3.

Eudoxus and Aristotle, where it was tempting to treat each planetary motion as having its own Unmoved Mover, we should be risking a polytheistic world picture. As is well known, this risk was exchanged for another, when scholars decided in favour of the idea that celestial bodies are moved by angels souls, or intelligences, peculiar to themselves. There were many doctrinal positions adopted here. Some treated the intelligences as stationary, others as though they moved with the bodies whose movement they caused. Some explicitly rejected the idea that the soul could be sensitive or vegetative, and insisted that it is only intellective. Some philosophers placed great emphasis on the wills of the movers, as against their intellects, for in the absence of resistance in the heavens it was thought that the planetary movers, to avoid producing infinite velocities, would have to use voluntary restraint. Some rejected intelligences or angels in favour of some sort of intrinsic principle, perhaps an impetus given to every orb by God at the time of the creation of the world. In view of the multiplicity of belief on these fundamental motions in the universe, it is rather surprising that there was not greater variety in the theories of causation by the celestial bodies on the inferior world. What was usually lacking was a systematic vision of the relationship between the various sorts of causality operating in the world as a whole. In Aquinas we have a philosopher who, more than most, was aware of this need.

Aquinas was in no doubt that the celestial bodies are moved by created spirits, or separate substances, with a purely intellective life. He remained undecided over the question of whether they are united with the celestial\bodies as are souls with living bodies, or are only motors to the inanimate bodies they move.[106] Of their causal influence over the inferior world he was in no doubt, and Litt has noted more than a hundred and thirty places in Aquinas' writings where this influence is affirmed.[107] The first comes from the *De regimine principum,* a work discussing the advantages and disadvantages of an absolute monarchy, where there is an analogy drawn between the state with its ruler and the universe with its prime mover – both being examples of the ruling of a multitude by a single power.[108] In *De Trinitate* he spoke of celestial bodies as 'complete natures', with a science by which they may be considered in themselves, even though they may also be treated as principles of other, inferior, bodies. In this, he said, they are like the elements, which are also the principles of mixtures; and in another place he spoke of the firmament as 'causa mixtorum activa', which is remi-

---

[106] Litt, ibid., ch.V.
[107] Ibid., chs VI, VII and VIII.
[108] Ibid., p. 110. From *De reg. princ.,* I.i.

niscent of a doctrine we have seen already, according to which the planets control the adjustment of the balance between the elements in mixtures. Litt was probably reading too much into the relevant passage, when he said that the comparison with the case of the elements 'ne peut viser que l'universalité des deux causalités'. The two situations were simply taken as examples, to be contrasted with other situations where it does not make sense to talk of principles as complete natures when they are divorced from those things of which they are components (as units of a number, points of a line, form and matter of a physical body).[109] In fact Aquinas on several occasions decided that certain activities could not be explained in terms of elementary qualities, and *therefore* they could only originate with the celestial bodies,[110] although in *Summa theologiae* I^a.110.1 ad 2 there is talk of operations on inferior bodies that are neither natural nor explicable in terms of the powers of heavenly bodies, but under the *direct* control of angels. As he had written a few lines earlier, 'just as the lower angels who have less universal forms are controlled by the higher ones, so also material things are controlled by angels.'

There is in many of the relevant passages from Aquinas the notion of a ranking of explanation, one might almost say according to the mystery attaching to whatever is to be explained. If something is reducible to an explanation in terms of elementary qualities, it is least mysterious, if the heavenly bodies are needed, then it is more so, and direct intervention by angels is more mysterious still. There is a ranking of angels as there is of forms: the nobler a form, the more it dominates physical matter.[111] In the case of the angels, if not of the stars, he had the comforts of a long theological tradition. In a short sketch of the views of earlier philosophers in the same article in *Summa theologiae,* he grouped the doctors of the Church with the Platonists. Briefly, Plato was said to have argued that certain non-material beings, namely *rationes* and species of sensible bodies, have direct control *(praesidentia immediata)* over perceptible things; but that Aristotle had made out these non-material beings to be something higher and more universal. They control, said Aristotle, not particular things 'but only those with a universal activity, namely the heavenly bodies'.[112] Augustine, on the other hand, wrote that 'Each visible thing in this world has an angelic power set over it'; Damascene spoke of the Devil as one of the angelic

---

[109] Ibid., p. 111.
[110] Ibid., p. 114, ref. to text no. 2.
[111] *Summa theol.* I^a. 76.1.c.
[112] In the *Metaphysics* at 1073a 31–33, Aristotle says that the movement of each of the planets must be caused by a substance both unmovable in itself and eternal. Cf. *De gen.* II.10, 336a 15.

powers over and above the terrestrial order; and Origen, noted Aquinas, remarked on the need of angels to watch over beasts, trees, plants, and the increase of all other things. Aquinas went on to speak of the ranking of angels. Here we need not follow him, but only emphasize how very central to his thinking was the whole subject of angelology. It provided his account of causation with what looks to us like a thoroughly arbitrary component – arbitrary, that is, to the extent that we are not privy to the workings of angelic minds.

This does not seem to have been a cause for regret to him. In a sense, the angels are a structurally superfluous refinement to his philosophy, the main lines of which have a stark symmetry: God is the incorruptible and unmovable cause; the heavenly bodies are causes incorruptible but movable; and the inferior bodies are corruptible and movable. He often spoke of the second class as though they governed *all* changes in the third, and Litt reports this as his view,[113] having overlooked what must again seem to most modern readers a very arbitrary division of subject matter. Aquinas had spoken in the *De veritate* (22.13.c.) as though bodies have some of their tendencies decided by the nature of the elements of which they are constituted (and these include, for instance, falling downwards, heating, and cooling, he added), while they get others from the influence of the heavenly bodies, *ex impressione caelestium corporum* – and here has that old favourite the magnet attracting iron, but it is an analogy no longer, rather an *example* of an effect, a tendency *(operacio)* arising from the impress of celestial bodies.[114] What he never explains is how we are to know where to draw the line dividing these two sorts of causality. By his example, though, we are inclined to see the heavens as a sort of back-stop, to catch whatever elementary explanations (i.e. in terms of elemental qualities) cannot. A body can fall simultaneously under both sorts of 'force': thus water may be affected by the Moon, but it will not thereby lose its natural downward motion. One feels that for Aquinas, as for so many lesser philosophers, the theory of heavenly influence was a theory of last resort, and this feeling is not diminished when we read a statement in the *Contra Gentiles* (III.92) that extends the example: not only is the attraction of a magnet for iron subject to the heavens, but also certain stones and herbs have other 'occult virtues' from the same cause. It is not enough to distinguish – as is often done – only between Aquinas' view of natural powers as God's vicarious actions, and supernatural actions as God's direct intervention in the world. There is a large area of middle ground, as I hope is now clear, and this was by no means without its own subdivisions. Not that these were all of Aquinas' making. Litt has an

---

[113] Op.cit., pp. 146–7.
[114] See Litt., pp. 117ff., for this and related texts.

interesting text from Albertus Magnus, included because it seems to be reflected in what, as it happens, is a somewhat notorious doctrine by Aquinas that 'the form of a mixed body has activities not belonging to the elements of which it is composed'.[115] The idea that the qualities of the elements give rise to new, intermediate qualities participating in the extremes of the qualities combined, became popular in scholastic alchemy. Albert had said much the same; but he too had gone on to say, in the text quoted by Litt, that the attraction of a magnet *(adamas)* for iron has nothing to do with the nature of any dominant element, but arises from the influence of the celestial bodies.[116] As we have seen, this was becoming something of a commonplace. One could have pointed out to Albert that it is logically independent of the doctrine he was denying.

Another intimation of uncertainty crept into Aquinas' account of celestial influence when he described the activities of the heavenly bodies as 'variously received in the lower bodies according to the differing disposition of their material contents *(secundum diversam materiae dispositionem)*'. This is one of Aristotle's four conditions to be met before generation can take place – there must be pre-existing matter capable of losing its existing structure and of taking on another.[117] There is nothing in principle wrong with making change a function of the qualities and dispositions of both agent and patient – on the contrary – and yet when one considers the boldness of the enterprise of astrological prediction and explanation it is easy to understand why Aquinas cut the relevant passage down to a minimum. The material conditions of human conception might not be wholly disposed to maleness, so they might receive a male or female form; and Augustine used this uncertainty as a reason for rejecting divination by the stars. That is all; and it is all too easily said! Anyone who, like Aquinas, believes in celestial generation, within an Aristotelian philosophical frame, is intellectually obliged to try to create a 'science of material conditions'. No-one took this task very seriously. For the Aristotelian astrologer, Augustine's argument should have been no more than a challenge to improve an art that took into account only the 'celestial influence' side of the equation.

---

[115] *Summa theol.,* I^a.76.1.c. T. Suttor, in the Blackfriars edition translates *'forma mixti corporis'* over-generously as 'form of a chemical compound'. For further historical details concerning the fate of doctrine (Duns Scotus, nominalists such as William of Ockham and Gregory of Rimini, terminists such as Buridan, Oresme, Marsilius of Inghen, Albert of Saxony), see A. Maier, *Studien. III: An der Grenze*... (Roma, 1952), pp. 89–140.

[116] Op. cit., p. 128.

[117] The other conditions are that there be a generating agent, already possessed of the form to be imposed on the receiving matter; the two must at some moment be in contact; and there must be an extrinsic active principle – such as heat, in the case of living beings, and heat and cold for inanimate ones. See *De gen. et corr.* I.6–10.

The views of the 'Aristotelian astrologer', as I have called Aquinas, are found an unself-conscious application to astrology, in the *Summa theologiae*. He holds, he tells us, 'that demons are spiritual substances that are not joined to bodies, and thus it is clear that they are not subject to the action of the heavenly bodies either essentially or accidentally, either directly or indirectly'.[118] Why, then, do they harass men at certain phases of the Moon? They find their task easier because the *Moon* has disposed the brain, an organ that readily receives the Moon's influence, to be easily disturbed during the phases in question. (The demons also come then in order to trick men into believing in the divinity of the stars!) Why is the brain so susceptible to the Moon's influence? Because, 'as Aristotle says, the brain is the moistest of all the parts of the body'.[119] This is a rare fragment of the sort of 'scientific' thinking that I said an Aristotelian astrologer should have felt obliged to pursue.

One of Aquinas' favourite examples of phenomena depending on celestial influence is that of animals (and on one occasion plants) engendered by putrefaction. Litt notes twenty-nine texts, and from them it is evident that the basic principles of Aristotelian instrumentality are always at the back of his mind. The celestial body is in possession of universal (not particular) active powers, *virtutes universales agentes*, and causes life by acting on the passive virtue of the four elements, while they have also a certain active virtue. Once more the heavenly bodies may produce different results according to the different dispositions of matter; and always they already contain the perfection of the animals they engender *per modum eminentiorem, quodam excellentiori modo,* or with other similar qualifying phrase in the Aristotelian tradition.[120] It seems that Aquinas thought generation by putrefaction to be the *most obvious* case of celestial causality. Litt was able to supplement his Twenty-nine texts with fifteen on generation in general, but they add nothing of principle to the metaphysics of the subject. This metaphysics may be summarized very briefly by saying that the celestial bodies are equivocal causes (as opposed to inferior bodies, which are univocal); they are universal causes, of the species rather than the individual (inferior bodies are particular causes of the individual); they are instruments of their moving spirits, but use, in turn, inferior bodies as their instruments, guiding these inferior bodies to their ends. Chance events, as well as acts of human intelligence and will, are not subject to them.[121]

Where, in all this, do the traditional theories of influence belong? What of motion and light, for example, and the 'physics' of astrology?

---

[118] II^a. 115.5.c.
[119] *De partibus animalium* II.7, 652a 27.
[120] Litt., op. cit., pp. 130–6. Cf. n. 117.
[121] See further chs VII and VIII, ibid.

What of the traditional qualities of the planets, and their transmission to Earth? Aquinas was not untypical of his century in the attitudes he struck towards astrological tradition and innovation. In his *Sentences* commentary he followed something like the Grossetestian line: light *(lux)* is the active quality of celestial body just as heat is the active quality of fire, and the rays *(radii)* of different planets have different effects.[122] They have different effects according to *their own* natures – Averroes was wrong to suppose that all stars are of the same species in nature, for if this were so, they would have the same effects, which is evidently not the case. Aquinas never thought to repeat the standard lists of planetary qualities that he could have found in any one of a dozen astrological texts, but from occasional remarks he makes about the planets one gets the impression that he had no reason to reject conventional wisdom. He accepted the basic Aristotelian ideas, which we took at the outset from *De generatione,* and they of course carried with them the idea of *motion* as a cause of generation and corruption. He never wrote his commentary on this work, but the same ideas are found in a more cryptic form in the *Metaphysics,* which he did comment, and he referred to the *De generatione* text in his *De caelo* commentary.[123] (He benefited from Simplicius on *De caelo* for an understanding of the varying heating effects of the Sun with season, not an obvious piece of physics.)[124] He picked up Aristotle's idea of matching planet with subject according to their natural cycles – so that Saturn's causality, for example, is exercised on permanent or slowly changing things, this being the slowest of the planets. (Aristotle did not extend the principle beyond the Sun and Moon). And of course there are many texts (at least ten) attributing the tides to the causality of the Moon: thus *'fluxus et refluxus maris non consequitur formam substantialem aquae, sed virtutem lune'*.[125] Many though they are, it has to be said that they are always so brief as to suggest that Aquinas had very little interest in the physical problem, as against the metaphysical. As for the problem of astrological prediction, it is hard to say whether he had a stronger interest, but he certainly spoke on many occasions of *true* predictions – even though it was often only to attribute success to help from demons. He quoted Ptolemy's *Tetrabiblos,* but very rarely.

Viewed from the point of view of the development of physical concepts, Aquinas' pronouncements on the functioning of celestial influence might at first seem disappointing. He was an eclectic, who evidently took on board at least some physical ideas that to some of his

---

[122] Ibid., pp. 220ff.
[123] II.4, text 341.
[124] Litt, op. cit., p. 227.
[125] *Summa theol.* I^a. 110.3 ad 1.

contemporaries must have seemed irreconcilable, or at least uneasy bedfellows. (The theory of rays and the theory of motions, for instance, are such ideas). What he did achieve, or at least tried hard to achieve, was a listing of boundary conditions for determining what sort of concepts should be used in what situations. Terrestrial bodies are to be treated as instruments of celestial bodies. Do not ask *how* the instruments come to be manipulated! (If you ask, you will be given snippets of non-systematic information.) The causality of celestial bodies is irreducible to a causality of the sort we know from the physics of the elements. That, at least, is what it is *not*! You can now rule out, it seems, the sort of physical speculations as to its nature that we have seen being tentatively put forward by earlier writers. The sense of mystery surrounding the idea of celestial influence is heightened. St Thomas Aquinas was, from our point of view, rational and systematic, but not in the most promising tradition of natural science.

There is little point in pursuing every minor variation in thirteenth century belief on the subject of celestial influence. All believed in its existence in some shape or form, and most scholars can be treated as composites of ingredients we have already encountered. Not only is this so, there is a danger in taking anyone as a stereotype. The most famous of 'Averroists', namely Siger of Brabant, inserted so many ideas from Avicenna in his philosophy that he cannot be described as Averroist or Aristotelian in any simple sense. His God, like Averroes' (but not Aristotle's) was a first creative cause, acting from the outside inwards, and making use of the intelligences associated with the spheres, but doing so successively, since 'from one simple being [the First Cause] only one effect can immediately follow. The First Cause determines the motions and positions of the celestial bodies, and they in turn necessarily cause what happens in the sublunar world, although the arrangement of matter here below can under certain circumstances interfere with this causation. In *De necessitate et contingentia causarum* – a work which, if not his, can be paralleled almost throughout by passages from his accredited writings – this necessary causation is more or less taken for granted.[126] What exercised Siger's mind was another question, namely whether future events are necessary before they happen. He said not, but only because he distinguished between the necessary and the accidental. It is not that events are free, but that the chain of causation can be

---

[126] P. Mandonnet, *Siger de Brabant et l'Averroisme Latin au XIIIème siècle,* Part 1 (2nd edition Louvain, 1911); Part 2, containing the texts (2nd edition; Louvain 1908). See Part 2, pp. 91 ff. for work by Siger on the topic of future contingents, and pp. 111 ff. for the central points of relevance to us from *De necessitate.* Other works have been brought to light since Mandonnet's study. Note the *quaestiones* on the *Metaphysics* (editions include C. A. Graiff's, Louvain, 1948).

broken, and free will is manifested in our decision-making about 'future impedibles'. Although this is not my theme, it is an excellent example of the cast-iron grip that the notion of celestial causation had on thirteenth century thought.

Another example is Dante, a man cast in a less severely scholastic and much more interesting mould, from our point of view. (We may safely side-step the old controversy over why he put Siger of Brabant in Paradise, and made Thomas Aquinas praise him, when the two were to a great extent adversaries). In the *Convivio,* Dante actually makes use of the idea of celestial power (*virtù celestiale*), the influence of the stars over material things, when the things are suitably placed, to explain by analogy the passing of varying degrees of perfection to us from the sciences.[127] Putting the analogy this way round is, to say the least, paradoxical. In the *Convivio,* Dante acknowledged that his ideas on generation were Aristotle's, but of course he knew them by many different routes. (He quoted from *De generatione, Metaphysics,* and *Physics,* on the subject.) In the *Divine Comedy,* he speaks of the 'informing power' of the animals and plants as originating in the movement and *light* of the stars, and there are many other touches of neo-Platonism. The Empyreum is a heaven of pure light, that will at length reveal the Creator to man, and that passes on both power and movement to the *primum mobile.* This in turn moves the whole universe, the power passing through the spheres, the living light of God diminishing as it passes from one level to the next.[128] It is the angelic intelligences in the spheres that are the causes of their motion and power: they use the heavens as their instruments in order to realize forms that they have themselves intuited in the mind of God, but the degree of likeness they manage to impress will depend on a number of factors, largely Aristotelian in spirit, but coming perhaps indirectly, from Albertus Magnus and Aquinas. The special emphasis Dante placed on these may be most easily discussed in relation to the Latin work *Quaestio de aqua et terra.*

The authenticity of the *Quaestio,* first printed by Padre Moncetti in 1508 and evidently unknown before then, has been powerfully defended by Edward Moore, and is now generally accepted.[129] It is in the form of a philosophical question determined on 20 January 1320 in the presence of the clergy of Verona, as the colophon tells us, the question being very simple: Can water be anywhere naturally above dry land? (Note that Dante, who died in 1322, wrote parts of the *Divine Comedy* as late as 1319

---

[127] II. 14, in the Florence, 1893 ed.
[128] *Paradiso,* i. 1–3; ii.55 ff.; xiii.55 ff.; xxviii.64 ff.; xxx.39–42 and 100–02. Cf. *Conv.,* II.vi.9; III.xiv.2–3; IV.ii.6–7; etc. It should go without saying that Dante's universe is strongly hierarchical.
[129] *Studies in Dante, Second Series,* (Oxford, 1899), pp. 303–74.

or 1320.) More precisely, the question was 'Whether water, in its own sphere, that is in its natural circumference, was in any part higher than the earth which emerges from the waters, and which we commonly call the habitable quarter'.[130] Dante's surprising answer is No! His approach was first to dismiss five standard arguments from his predecessors. To those, for instance, who had said that the seas would be eccentric to the Earth, because moved by the Moon which itself moves in an eccentric orbit, he answered that imitation does not have to be in all respects; and that falling earth would fall towards a different point from falling water – which would make Aristotle laugh to hear it![131] He argued that neither could there be a piling up of waters (even though this might leave the centre of mass, as we should say, at the centre of the Earth), for any humps in the water would soon disappear, as the water flowed naturally down. The elevation of the sea-shore above the sea was a problem, though, for being heavier than the sea it should 'seek the centre with greatest force'. It is, he thought, no answer to say that earth has different densities in different places, for as an elementary body it must possess gravity uniformly in all its parts.[132] Putting this less than adequate reasoning aside, we find his positive explanation extremely interesting. The Earth is elevated in part so that the elements can come together and be mixed there. How does it come about, and why is it so important that the elements be mixed? 'Since all forms which are ideally within the potentiality of matter are actualised in the mover of heaven, as the Commentator says in the *De substantia orbis,* if all these forms were not continuously actualised, the mover of heaven would fail of the complete diffusion of his excellence, which may not be uttered.'[133]

This somewhat notorious doctrine of the actualisation of all possible forms (not, incidentally, to be found in the work cited) is thus made the key to the problem. It is given a religious rather than a purely philosophical cast: this is what is meant when we talk of the diffusion of God's power. The heavens are the instruments; and the elevation of the land is a small detail in God's total plan. We can now afford to forget this small detail, but not the mechanism by which it is brought about. It is elevated in part by the virtue of heaven

> as the obeying by the commanding; just as we see in the case of the appetitive and resenting nature in man, which, although their proper impulse urges them to obey the affections of sense, yet insofar as they are susceptible of obedience to reason, are some-

---

[130] I follow the P. H. Wicksteed translation in *A Translation of the Latin Works of Dante Alighieri,* (London, 1904).
[131] Sections 7, 12 and 23.
[132] Sections 16 to 18.
[133] Section 18.

times restrained from their proper impulse, as appears from the first of the *Ethics*.[134]

This use of a human analogy to explain what may be broadly called 'celestial influence' (a phrase used elsewhere rather differently) is absolutely typical of Dante. For him, every natural movement is activated by love, a 'force that unites', whether it is the love minerals may have for the place of their generation, the love of plants for their places, or the more complex love of animals, or the love that makes the magnet turn towards the pole star. It is this that marks out Dante's theory of celestial influence, if 'theory' is not too strong a term, from those that are more recognisably physical, in the modern sense. He is quite happy with a loose, two-way, causation, always drafted on the basis of essentially Aristotelian elements, but always strongly humanized. He was fond of the metaphor of a 'drawing out'. As he explained in the *Convivio*, precious stones are formed by the Sun's first drawing out what was base in the material from which they are formed, after which a star gives the stone its special form.[135] The souls of plants, the leaves they bear in spring, the souls of animals, all are drawn out by the heavens as a goal.[136] His love for the teleological did not, however, prevent him from discussing efficient causation, in the *Quaestio*. The elevation of earth is not caused by earth, he said there, since earth moves naturally down, nor is it by water, air, or fire, since they are distributed (on his view) uniformly. The tides are conveniently overlooked. That leaves the heavens – but which? The Moon is ruled out, since she would not give the distribution of land we observe. (Note that he rejected the Moon's eccentricity as an explanation, but accepted the axiom that 'agents operate with greater power the nearer they are').[137] All the other planets are excluded for the same reason; and so is the ninth sphere, which is simply too uniform! This leaves the eighth sphere, the sphere of the stars.

The sphere of the stars, Dante tells us, has unity in substance but multiplicity in virtue, or power (*virtus*).[138] The magnitudes, luminosities, and arrangements of the stars in constellations, as well as the side of the equator to which they fall, are what decide their powers. Again and again we see celestial influence being taken for granted, so that no-one contradicts the basic premiss of its occurrence; and here Dante adds that these things are perfectly clear to all who have been nurtured in philosophy. As he wrote in the *Convivium, all* philosophers agree that the

---

[134] Loc. cit.
[135] III,iii,3 and 6–8; IV, xviii 4; etc.
[136] *Rime* C. 40–1; *Purg.* xxx, 109–111.
[137] Section 20.
[138] Section 21. Cf. *Par.* ii.115–6.

heavens are a cause.¹³⁹ We need not dwell on the special pleading by which he escapes from the difficulty of explaining why the land raised by the stars is not in a uniform circular band, since the stars rotate. (The answer was that there is not enough matter to go round!) What is more interesting is the fact that he was obviously unclear in his own mind about the nature of the celestial power. It is manifest, he wrote,

> that the lifting virtue is in those stars which are in the region of heaven contained between [the equator and the ecliptic], whether it elevates it by way of attraction, as the magnet attracts iron, or by way of impulsion, by generating vapours that force it up, as in the case of special mountain ranges.¹⁴⁰

Drawing up of vapours, generating vapours that impel, powers of attracting, powers effected through a formal similarity – these were never worked into anything like a coherent 'celestial physics' (as one might call it) by Dante. But then, the same could be said of the typical university philosopher. Of the broad 'geometrical' principle of influence, however, Dante was in no doubt: 'the aspects of things below are like to the aspects of things above, as Ptolemy asserts'.¹⁴¹ This suitably imprecise axiom is the fundamental premiss of the astrologer, although one did not, of course, have to accept the rest of the conceptual paraphernalia of astrology to accept it.

There were university scholars with a more carefully worked out conceptual scheme, philosophers who had clearly given more thought to the inconsistencies in the ever changing mixture of doctrine. A good example is John Buridan, perhaps the most influential teacher of natural philosophy at the University of Paris in the fourteenth century, a man who dominated the so-called terminist movement in formal logic in his time, a thoroughgoing nominalist (although his dependance on Ockham is a matter of controversy), and now perhaps most widely remembered for his theory of projectile motion and impetus. He was twice Rector of the university (1328 and 1340), but remained in the faculty of arts, that is, did not graduate in theology. He does not appear to have been expert in the technical aspects of astronomy, but his *Quaestiones super libris quattuor de caelo et mundo* (c. 1340) show him to have been one of the best of commentators on the Aristotelian tradition, and it is there that we find his views on celestial influence.¹⁴² They were not, needless to say, primarily directed by astrological considerations.

---

¹³⁹ II.xiv.
¹⁴⁰ *Quaestio,* section 21.
¹⁴¹ Loc. cit.
¹⁴² Edited under this title by E. A. Moody, (Cambridge, Mass., 1942; repr. 1970).
¹⁴³ See the lengthy note on this by W. K. C. Guthrie, in the Loeb edition.

As a suitable point of entry to the problem we may take *quaestio* 15 of Book II, concerning *De caelo* II.7, and the question of whether celestial bodies generate heat by their light (*lumen*). Aristotle's account of how the stars generate light, and, at least in the case of the Sun, heat also, had long been considered less than satisfactory. Heat and light is emitted by them, he said, because the air is chafed by their movement. Alexander of Aphrodisias and Simplicius had, in their commentaries, objected that there were two problems here: if the stars are of the same material as the spheres that carry them, why do they produce so different an effect? And why did Aristotle say that the air, rather than fire (the uppermost element) is ignited by the stars?[143] This is not the place to conjecture what Aristotle might have answered. Buridan's question was different, and concerned the heating effect of light, taken, along with motion, as a candidate for bringing heat about. He could accept it as a cause of heat, but not as an astrological cause. The evidence of which he made use was more textual than empirical. Aristotle, in *De partibus animalium* IV, had noted that the Moon had a warming effect, and yet the astrologers said that in its nature the Moon was cool and moist. The same could then be said of the others stars, receiving, as they all do, light from the Sun – light that is admitted by everyone to have a heating effect. (Here 'Aristotle' was again the authority, or rather the author of *De proprietatibus elementorum*). Under certain circumstances, though, according to the astrologers, the stars could have cooling powers. Averroes contradicted Avicenna on this point, but why should we believe the one rather than the other? When clouds obstruct the Moon, or when it is lost to view in a conjunction, its powers are not diminished, claimed Buridan – on the contrary, they are greater in the case of conjunction and opposition than at the quarters, as is shown by the tides and the size and number of shell-fish.

It is clear, therefore, that for Buridan celestial influence was not to be confused with light. He would not go along with those who had said that light from the Sun picks up heat as it passes through the sphere of fire,[144] or with others (we recall that Grosseteste was one) who spoke as though rays emitted to the Earth's atmosphere and intersecting themselves in the air, cause it to be dispersed and rarefied and thus warmed. Against this, he noted that the weather may be at its most tranquil when the summer Sun is most powerful. He would not accept that light carries the powers of the stars, warmth in the case of a warm star, cold in the case of a cold one. This was contrary, he said, to the 'proved' principle that all the stars have a warming effect, besides which he has already said

---

[144] Ed. Moody, p. 195. He argued that since, following Aristotle, light is not a body, it cannot incorporate anything in itself.

that the powers of the stars are propagated other than as light.[145] He said very little on the subject in this *quaestio,* but he did make the claim that there are many qualities, celestial and belonging to the influx from the heavens, that are more fundamental than heat, cold, moistness and dryness, which last are only called first in relation to the tangible and elementary qualities.[146] Earlier in the work he had made clear his agreement with Aristotle on the question of whether the qualities of the elements could be ascribed to the heavens. Some philosophers had said that since the stars rise and fall on their eccentrics, they must be heavy and light: but no, said Buridan, they do so with circular and not rectilinear motion. We say that the signs are fiery, aery, and so on, but this is a circumlocution, meaning only that they, or certain stars, cause these effects here below. The heavenly bodies do not have these qualities formally, but only virtually, in their powers, *virtualiter.*[147]

Buridan was too cautious to say much on the score of how precisely the celestial influences work. (Note how at one point he expressed agreement with Alexander of Aphrodisias – who had been cited by Averroes – that just as it was not necessary (*non opportet*) to try to say why (*propter quid*) gravity moves things down, only that we are here faced with a natural property of it, so it is not necessary to say why light warms, etc.)[148] He did, though, note that in their acting on things here below the heavens are not fatigued, just as the intelligences do not tire in the act of moving the heavens.[149] This tells us a great deal about his attitude, and no doubt that of many of his contemporaries. Influences from the heavens were somehow disengaged from physical forces such as we know here on Earth. Of course it is a platitude to say that the Aristotelian philosopher thought of the celestial and terrestrial regions as fundamentally different in most respects, but the influences were, after all, supposed to pass from one to the other; and it seems that for Buridan their activity was taken to be intellectual.

In the end, however, Buridan had to come back to the question of motion in the heavens, and the difficulties that Aristotle had faced. Here, in questions on the *De caelo,* the problem was primarily that of finding a mechanism for linking motion in the heavens with heat here below, and not with the more general connections between celestial motion and generation, much less the astrological problem. The first problem reflected on the others, even so. It was accepted that the Sun's

---

[145] Ibid., p. 196.
[146] Ibid., pp. 198–99.
[147] Ibid., p. 43, namely Lib.I, q.9.
[148] Ibid., p. 97.
[149] Ibid. pp. 131-2. Note: 'Fatigatio est diminutio virtutis activae cognoscitivae, propter longam eius actionem ...' It occurs on account of an exhalation of spirit. We do not say that a man is fatigued, he added, if he is impotent, old, ill, or has a defective limb.

sphere causes more heat than any other. Why, when there are larger spheres than the Sun's and closer spheres than it? How, when the Moon's sphere, and others, gets between the Sun's and the Earth? It was all very problematical, he admitted; but he was sure that the motion of the Sun's sphere did not act of its own accord (*proprie et per se*), but by the way such agents do.[150] If the motion is nothing other than the mobile thing, why should it need to move to get its effects? Would a stationary sphere do just as well? Buridan offered two explanations, one from Averroes, the other from Peter of Auvergne, and asked us to choose for ourselves, for want of anything better. Averroes had said that the *aggregate* of the celestial spheres should be looked upon as an animal, the spheres themselves corresponding to its parts. Its action, like that of an animal, is that of the totality.[151] This analogy is obviously in the macrocosm-microcosm tradition. The alternative solution seems to me to be much more interesting, in the present context.

Peter of Auvergne, we are told, maintained that the Sun's sphere causes in the lower spheres not heat – because they are not heatable – but a quality that we cannot sense, but which, when propagated as far as the lower heatable things generates heat in them.[152] One might regard this as highly sophisticated and thoroughly scientific, or deeply supernatural, according to taste. It was certainly not entirely original, although the example by which it was illustrated was given different twists by different scholars. According to Peter, there is an analogy with a certain fish called *'stupor'* (the torpedo fish) which, when caught in a net causes numbness (*stupor*) and tremor in the hands of the fisherman holding the net. This it could not do unless it altered the medium in some way, yet the medium is not numbed, for it is not numbable. Some other quality must be generated by the fish in the medium, neither sensible nor manifest to us, and this, propagated (*multiplicata*) to the fisherman causes numbness in him.

This analogy (not, it should be said, called that) had a long and vigorous life. The qualities generated in the medium are the occult secondary qualities against which so many self-righteous seventeenth century philosophers directed their polemics. There is a somewhat lengthy rationalisation of the sort of explanation offered here, in Henry of Langenstein's *Tractatus contra astrologos* (1373), where primary qualities give rise to accidental and substantial change, and to secondary

---

[150] Ibid., p. 207. 'Sed quicquid de hoc sit verum, una conclusio potest poni satis concessa: scilicet quod motus localis caeli agit in ista inferiora et caliditatem et frigitatem et humiditatem et siccitatem, non proprie et per se, sed quia per ipsum applicantur nobis sol et stellae agentes in ista inferiora et caliditatem et [etc.]'.
[151] Averroes, *De caelo,* comm. 42.
[152] Buridan, ed. cit., p. 208, Book II. q.17.

qualities, some of which may be occult, hidden from our senses. Secondary qualities do not affect the medium through which they pass. Occult secondary qualities may be responsible for the transmission of disease to Earth.[153] This is the end of the story; the beginning might be taken to be Aristotle's discussion in *De caelo* (I. 3) of the nature of the 'first body', of which the heavens are made. It is unalterable, he said, for it is not susceptible to growth or diminution; and this means there is no change in quality in it. Buridan, in his *Quaestiones,* went along with this then standard view,[154] although he conceded that the Sun might alter the lower spheres with an alteration *communiter dicta*. When he gave Peter of Auvergne as his source for the torpedo analogy, he was evidently not aware that Thomas Aquinas was perhaps Peter's source: he used the analogy at least four times, and on one occasion reinforced it with that of the magnet, which attracts the iron through the air, while the air itself is not attracted.[155] Thomas, however, refers us back to his source, Averroes' commentary xxxvii on the eighth book of the *Physics* (which Peter of Auvergne might have borrowed); but it goes back further, at least to Simplicius, who uses it in the same context (*De caelo* II. 10) as he, so that Litt regarded this as his source.[156] There is no reason why he should not have been aware of both. At all events, this interesting explanation of the transmission of an effect through a medium must be treated as a commonplace by the mid-fourteenth century.

As the fourteenth century progressed, we find the old familiar ingredients in the many references to celestial influence, with many differences of emphasis and degrees of philosophical caution, but little positively new. For their different elaborations on earlier ideas, however, two further Parisian masters are noteworthy, namely Nicole Oresme, who studied with Buridan, and Henry of Langenstein, younger by a few years than Oresme.[157] I shall refer in particular to two works by each scholar, and since the work of the two men overlaps in date, it will be as well to list the four works here:[158]

1. *Tractatus de configurationibus qualitatum et motuum* (perhaps written between 1351 and 1356). By N.O.

[153] H. Pruckner, ed., *Studien zu den astrologischen Schriften des Heinrich von Langenstein* (Leipzig, 1933) pp. 197-9 (cap. 3).
[154] I.10, ed. Moody, pp. 48-9.
[155] Litt, op. cit., p. 41, n. 4 for the references.
[156] Ibid., p. 392. The translation *stupor* is William of Moerbeke's at the appropriate point; elsewhere in the same work he translates the word (νάρχη) as *torpedo*.
[157] Oresme (c. 1320-1382) had ceased active teaching in arts before Henry of Langenstein (1325-97) arrived there (c. 1363). Buridan died in or shortly after 1358.
[158] Text 1 was edited (etc.) by M. Clagett in *Nicole Oresme and the Medieval Geometry of Qualities and Motions* (Madison 1968). Texts 2 and 3 are in the Pruckner volume already cited. See n. 153 p. 85. Text 4 is edited by A. D. Menut and A. J. Denomy in *Maistre Nicole Oresme. Le livre du ciel et du monde* (Madison 1968).

2. *Questio de cometa* (1368). By H. L.
3. *Tractatus contra astrologos coniunctionistas* (1373). By H. L.
4. *Le livre du ciel et du monde* (1377). By N.O., for the French king, Charles V, to accompany a French version of *De caelo*.

Oresme had earlier composed a *Quaestiones de caelo,* but his later commentary on the *De caelo* (that is, text 4) does represent some changes in attitude on his part – for example, he now admits the possibility that the heavens are sustained in motion by an impetus given them by God at their creation (Buridan's idea), something he had earlier rejected.[159]

The configuration doctrine of Oresme is now too well known to need more than cursory introduction. It was a geometrical (graphing) system for representing abstract quantities of quality and motion. It was also found a use for the physical explanation of a number of different phenomena, physical, biological, medical, even psychological. (The 'configuration' can be thought of as the total graph, the total picture, of the thing; or it can be regarded as the internal relations between the parts of the graph.) Bearing in mind Dante's recourse to the concept of love, to explain celestial influence[160], and parallels with the magnet, it is interesting to note Oresme's use of the configurations doctrine to explain beauty, natural friendship, and hostility. It was, he said, all a matter of compatibility, or harmony of configurations.[161] A magnet, for example, was supposed to attract iron because of a certain natural bond they possess as the result of '(1) a fitting accord between the ratio of the [natural] qualities of one with the ratio of the natural qualities of the other, and (2) of the fitting accord between the configurations of the qualities of this sort in each of them'.[162] Note that this sort of causation (which is described as probable no more), is said to be *occult*.[163] It is clear that Oresme thought that he had hit on an invaluable principle, giving him an insight into what previous writers had loosely described as 'harmony'. (This, I imagine, is what he would have claimed). It is hardly surprising, therefore, that he extended the principle to the classic case of harmony – as he called it, 'the figuration of celestial velocity'.

The planets – even the fixed stars, if we accept the doctrine of trepidation – move with a difform velocity, as Oresme noted, and one must suppose that the velocities are 'figured properly' and 'assimilated to beautiful figures'. 'Reason gave this harmony, as Cassiodorus says, only

---

159 The *quaestiones* (as yet unedited) are at least as early as the *Tract. de configurationibus*.
160 See p. 283 ff. above.
161 Clagett, ed. cit., I.xxvi–xxxi, passim.
162 Ibid., pp. 243–5.
163 The chapter is entitled 'de causis occultis quarundam naturalium actionum'.
164 Ibid., p. 299. See esp. n. 1.

to the mind, nature not producing it for the ears'[164], and 'change of this harmony or difformity can be one of the causes why sometimes heavenly bodies emit below more benign, and at other times less benign, influences'.[165] This can be regarded as the most sophisticated reduction of celestial influence to mathematical harmony of the middle ages. Clagett was surely right to see Oresme's application of the configuration doctrine to what he calls 'remarkable phenomena' as 'the outgrowth of an older doctrine which explained such phenomena on the basis of similar and dissimilar ratios and or dispositions of primary qualities'.[166] Oresme had referred to William of Auvergne's *De universo* (c. 1231-36), a work in which William had been inspired by his knowledge of the astrological doctrine of aspects to extend it to other, intractable, phenomena (medical, herbal, psychological, and so forth). And now we find Oresme simply accepting the extended principle, and turning it back on itself to explain astrological influence!

These ideas are, in a way, just an extension of the principle that celestial and other causes and effects have to be similar. Henry of Langenstein, for instance, applied the configuration doctrine to explain the similarity of the mandrake with the human form, and again there were precedents in plenty. Henry had plenty to say against 'occult' qualities, hoping that influences and powers of an occult sort in things here below and in the stars above need not be posited.[167] When he wrote of 'spiritual qualities which fall beyond the human senses', though, it is not clear that he is using the word 'occult' in the same sense as Oresme, and he seems to regard 'harmonic influences', and the way such powers 'radiate on to a patient' as non-occult.[168] It would be more to the point to say that phrases like 'celestial influence' and 'occult virtue' can be further explicated, in terms of qualities (see below) and partly along lines laid down by Oresme. The 'marvellous effects' produced by harmonious combinations '*we often attribute* to special influences of the heavens and to occult virtues in inferior things because of the fact that such combinations are hidden and unnoticed by us'.[169] It is obvious that care is needed in describing him as a critic of the occult. Furthermore, as Lynn Thorndike observed, although he had no wish to have recourse to the occult and mysterious, 'we are repeatedly told that these intensions, combinations and proportions are for the most part unknowable in detail'.[170] Thorndike went on to say that therefore 'the practical value of

---

[165] Loc. cit. The word used is *influentia*, later much used by Henry, as we shall see.
[166] Ibid., p. 543, commenting on an earlier chapter, I.xxv.
[167] For further references, see Clagett, op. cit., pp. 115-9.
[168] Ibid., 117-8, for texts that justify this statement.
[169] Ibid., pp. 118-9. All these quotations are from his *De reductione effectuum*.
[170] Ibid., p. 121.

the new theory becomes dubious' as a result, and Clagett added that it was 'not surprising, then, that the configuration doctrine as a technique to account for phenomena died with Oresme and Henry of Hesse'. My own concern, of course, is with ideas that never had any clear practical value; but as we see, that did not prevent the survival of at least some of them for seventeen centuries!

From this account of what was essentially Oresme's brave new interpretation of 'celestial influence' we pass to a short review of a more traditional assemblage of ideas in *Le Livre du ciel et du monde*. Like Aristotle, he would not accept that there was heat in the heavens, for there was no violence or friction there, he said, even as between the sphere of fire and that of the Moon.[171] The highest reaches of the air move like the heavens, but more slowly, 'because the causes are feebler and reduced the further one descends'.[172] Furthermore, one 'might reasonably state that one cause is the influence of the heavens or the motive power of the intelligence which inclines the air towards following the motion of the heavens', and another cause is friction between the air and the fire. Later in the same work he turns a principle used by Averroes against the Commentator himself: the spheres do move about different centres and therefore are of different species – differences in their motions arise from differences in their natures, and not mere accidental or chance occurrences. He was evidently rather relieved to be able to add that this was in accordance with what we find in books of astrology, according to which different celestial bodies have different virtues. Here again, it all has a familiar ring.[173] In fact he seems to me to have his eyes very firmly on Buridan's commentary at this point – as elsewhere –, but this is difficult to prove. The manoeuvres he went through are fairly conventional. He noted Averroes' likening of the heavens to a city, the regularity of the one corresponding to the government of the other. The intelligences in the heavens correspond, in this analogy, with the officers of the city; and no doubt he was grateful for an opportunity of making reference to the sovereign intelligence.[174] I hope that Charles V took the point. He noted, too, Averroes' likening of the heavens to a living body.[175] One should mention here that for Oresme, the incorporeal movers of the planets were neither forms nor souls.[176] The heavens were for him the instruments of God; he quoted Scripture to support yet another commonplace. And then, at the very end of the

---

[171] Ed. cit. (Menut and Denomy), pp. 441–3.
[172] Loc. cit., lines 235–6, my translation.
[173] Ibid., pp. 462–3.
[174] Ibid., pp. 464–5.
[175] Cf. p. 87 ff. above, for Buridan's reference to the same text.
[176] See his II.5.

chapter (II. 22) he threw in a passage that only the scholars in his audience are likely to have recognised as alluding to the work of twenty years earlier:

> I say, therefore, that the position and ordering of the heavenly bodies, the number, speed, quality, diversity or difference of their movements – all these things are expediently arranged with, as the end in view, the generation and conservation of terrestrial bodies.[177]

Aristotle, Christian theology, and configuration theory (if we are not imagining too much) are all in the melting pot, and it is hard to make out their outlines any more.

Like Oresme, Henry of Langenstein was an opponent of astrology as it was practised, while believing in the causal influences by which its procedures were supposedly justified. Their strategies, though, differed in many respects. Both naturally made capital out of the iniquities and follies of astrologers, the evident inconsistencies in their procedures, and the theological dangers they courted. Henry's attack, however, had a potentially more trenchant effect, for he set out a rather complicated theory of causation that, had it been accepted by astrologers, would have implied a drastic revision of their procedures. In effect he told them to give their attention to the terrestrial end of the process of celestial causation, to the terrestrial objects from which were elicited effects already present. Of course his theory had in it many of the old ingredients, but the blend was his own. It appears not only in the two works already listed, but in his *De habitudine causarum,* his *De reductione effectuum,* his *Sentences* commentary, and *Lecturae super Genesim,* the last being the subject of a useful recent study by Nicholas H. Steneck.[178]

Henry of Langenstein brought back into prominence a number of Augustinian and neo-Platonic ideas, in particular the doctrine of seminal reasons – invisible potentialities, as it were seeds implanted by God at the creation of the world – but he also gave some attention to theories of perspective as an important part of a theory of influence. In other word, he revived a thirteenth century fashion. His notion of a Golden Chain, *catena aurea,* also has about it a ring of an earlier period. He took it for granted that God, the Creator and First Mover, acted on inferior things through a chain of secondary causes. In the controversy over the nature of what drives round the spheres he opted for the alternative that had the celestial movers as souls, not human or angelic,

---

[177] Ibid., p. 510, my translation.
[178] *Science and Creation in the Middle Ages. Henry of Langenstein on Genesis,* (Notre Dame, 1976). This study does of course also take the other writings into account, and I shall rely on it often.
[179] Ibid., pp. 92–4, quoting from the *Lecturae.*

whether good or bad, but rational creatures of some other type.[179] But what was the next link in the chain, that between the intelligences and the inferior elements? He made four assumptions in the *Tractatus,* and these he says had been deduced in a certain *'Tractatus de natura communi',* which might be his *De habitudine causarum et influxu naturae communis.*[180] The assumptions are: (1) the intelligences can only act through mediating influential qualities of the spheres and stars; (2) the intelligences united to the spheres are essentially free, acting in a natural mode and of necessity on external matter, and not 'by free contradiction', philosophically speaking; (3) no secondary intelligence by intellect and command of will alone produces an effect on the inferior world; (4) the same star and part of the sphere always has the same influential quality. The third supposition was regarded as a consequence of the first two, and simply meant that the intelligences could not interfere with God's purpose. The fourth statement seemed to be directly opposed to the standard astrological doctrine of aspects, according to which the angle separating the planets has a marked effect on their powers. Henry went to some lengths to show that the doctrine was misconceived, and that the angles are not as they are conventionally drawn on a zodiacal aspects diagram. His was the attitude of one trained in 'perspective' (geometrical optics). Of course, if planets were in conjunction or near conjunction their rays (*influencie*) would tend to reinforce; it was the other aspects (thirty, sixty, ninety degrees, and so on) that were misunderstood.

This does not mean that he took the influences to be identical with light. He postulated four, and only four, species of influence -- the qualities of hotness, coldness, wetness, and dryness. We have already seen how, like Oresme, he accepted that they worked in accordance with principles of harmony, expressible with the help of the configuration doctrine.[181] He does not seem to have had much to do with this doctrine in his later works, although it is naturally in the background whenever there is talk of qualities. The influential qualities were supposed to pour down on the elements, which were then either disposed or not disposed to receive them -- the doctrine of seminal reasons being used to explain when co-operation could and could not be expected.[182] The four primary qualities could effect three different sorts of change: accidental change, preserving the substantial form (as when heat makes water warmer); 'generation and corruption' such as the alchemists hoped to achieve, changing the actual substance; changes in secondary qualities, as when

---

[180] Pruckner, ed. cit., pp. 160-1 has the allusion, and a footnote by the editor suggesting a different solution.
[181] See p. 291 above.
[182] Steneck, op. cit. pp. 95-6.

heat, in evaporating water, turns gravity into levity. The primary qualities might give rise to softness, hardness, odour, flavour, rarity, density, and so forth, but also to occult secondary qualities – such as we have already discussed briefly in connection with Henry's name.[183] The mode of travel of the primary qualities was a sort of radiation, so that like light they could be reflected and refracted, and like light they could be weakened and strengthened by processes describable, as I have said, in terms of the doctrine of intension and remission of forms.[184] If, as Steneck suggests, he introduced perspective and the latitude of forms into his work 'to introduce into the sublunar region some of the precision so obviously present in the celestial realm', his ambitions do not seem to have persisted into later life. At the end of the *Tractatus contra astrologos* he tried to apply his own general principles, but he cannot have been privately very content with the results. In a medley of botanical, meteorological and medical material he offered explanations that to us, at least, have always the mark of special pleading. But how could it have been otherwise? His tract might have been against the astrologers, but it was, after all, effectively arguing for the reform of a limited part of astrology!

The one form of astrological prediction that Henry of Langenstein felt to be reasonably secure was meteorological astrology. This subject, in keeping with Aristotle's ideas about their terrestrial origin, included the theory of comets, but here Henry drew the line, and his *Quaestio de cometa* comes down as heavily against predictions based on the appearance of the comet of 1368 as his later tract against the astrologers was to do. It is an earlier work than those I have been discussing, though, and one or two of its characteristics are worth mentioning. In it, Henry dismisses several later accounts of cometary origins, and follows more or less the Aristotle version: comets are hot and dry exhalations, passing upwards through the air until they burn in the sphere of fire.[185] We are not here concerned with its sceptical thesis, but with a few small clues it offers as to the drift of his thought. In this work light (*lumen*) is identified with one of the influences – it is of course the *influencia calefactiva*. The planets are often spoken of as 'attracting' the elements, as the Moon, for instance, quite traditionally, is said to attract and gather up moisture.[186] There was no planetary attraction on the comet, on the other hand, as is shown, he said, by the fact that it moved from the zodiac in the direction of the arctic pole, which no planet does.[187] The

[183] Ibid., pp. 96–7.
[184] Ibid., p. 98, from the *De reductione effectuum*.
[185] Pruckner, ed. cit., esp. pp. 89–99; Steneck, op. cit., p. 86.
[186] Pruckner, pp. 93–4.
[187] Ibid., p. 106.

planets exert a force (*vis*), as Saturn, for instance, a cooling force; but there is also a *virtus cometalis*.[188] All in all, this earlier work can be said to have as its main assumption the idea of influence through 'connaturality'. When he wrote that the magnet has a greater connaturality (*connaturalitas*, the English word is known to the dictionaries) with iron than the heavens with the elements, he places himself in a tradition we have seen often before. To add another reference: according to Albertus Magnus in the *De proprietatibus elementorum,* the Sun draws up moisture, while the Moon affects the sea connaturally.[189] On the score or so of occasions when he speaks of the influences of the heavens, Henry, in sum, gave few signs in the *Quaestio de cometa* that he had learned to think about astrology in the relatively clear-cut way of the later works.

We have seen repeatedly how natural philosophers of the middle ages tended to regard the celestial spheres as God's instruments. In Henry's case this was a view given expression through his Golden Chain analogy. In a very broad sense one can see here the makings of a divide between medieval and protestant theology. Both Luther and Calvin, for example, played down the importance of natural powers, and put more emphasis on immanent divine causation. The very example Luther took to illustrate this point has to be seen against the background of theories of celestial influence: he was discussing procreation within the egg in the hen's body. 'The philosophers', he wrote, 'advance the reason that these events take place through the working of the Sun and her belly. I grant this. But the theologians say, far more reliably, that they take place through the workings of the Word...'.[190] There is the well known disagreement over the nature of miracles and divine intervention in the universe generally, that gave rise to the correspondence between Leibniz and Clarke, and crux of their disagreement is very well illustrated by their use of the clock analogy. For Leibniz, the Newtonian universe was like a clock that had to be constantly interfered with by its maker – it was an inferior machine. By contrast, the Leibnizian world comprised monads with their future programme intrinsic to them, capable of running on for ever without God's interference, after the act of Creation. To Clarke, this threatened atheism, for God's providence was no longer guaranteed at every stage of history.[191]

Now in the light of this later development, Henry of Langenstein's lectures on *Genesis* are of interest for the fact that he too made use of the

---

[188] Ibid., pp. 94 and 106.
[189] II.4.
[190] For this whole problem see Keith Hutchison, 'Supernaturalism and the mechanical philosophy', History of Science, 21 1983, pp. 297–333. The quotation is at p. 315.
[191] H. G. Alexander (ed.), *The Leibniz-Clarke Correspondence* (Manchester, 1956) is the most convenient edition, in which see L.I.4, L.II.6f., L.III.13f., L.IV.38f. and L.99f., and the corresponding Clarke paragraphs.

clock analogy (as did Oresme, in *Le Livre du ciel et du monde* II. 2). Here God, the Creator and First Mover of the universe, is said to be like a clockmaker who, having fashioned his masterpiece bit by bit, set the machine in motion.[192] What is noteworthy here, however, is that Henry cannot be simply put on the Leibnizian, 'conservative', side of the divide, for God, according to him, did not withdraw from nature. Were he to do so, 'we should not live, nor move, nor exist'. This does not mean that God was supposed to involve himself directly in every natural event. As we saw at the outset, he acts only through secondary causes in the Golden Chain, at least in most cases. Parallels with the seventeenth century case are not exact, but in his use of the clock analogy (he applies it also to animals and ecological systems of animals) he fits more naturally with the proponents of mechanical explanation in science than with the organic philosophy of Leibniz. Perhaps the best way of differentiating Henry from the Newtonians, for instance, would be to say that, for both, God stands outside the clock that is the universe, but for Henry this was true in a spatial sense, whereas for the later scholar it was not. Further, for Newton (if we are to accept a language which was, after all, Leibniz's), God needs to attend to the clock from time to time fo reform its movements; in the case of Clarke, God's continuing providence was needed – a point over which Leibniz found it impossible to disagree in a broad sense; and for Henry of Langenstein, much the same could be said, except for the mode by which God interferes in the world. In point of fact I do not think that any of these writers had thought the clock analogy carefully through. To take Henry's use of it: to what do the intelligences correspond? To the source of motive power, the weights? God's plan, which they execute, must be the plan of the clock itself, and its planned motion. How is God needed? To wind the clock? To keep it in being? The first might be seen as a concretization of the second, and both as piously grafted on to the analogy to make it theologically acceptable. We can get a clearer insight, perhaps, into the change that was really taking place in scientific outlook if we take Henry as representing the view of God as essentially a part of the chain of efficient causation with which natural science is concerned, while for the seventeenth century philosopher, while God was written in to the story, so to speak, he could be omitted in most cases without affecting the scientific details. The former alternative, too, left open the way to a scientific investigation of the world, untrammeled by theology, from the stars down, so to speak. Celestial influences were potentially a part of nature, and few of the medieval accounts of their workings can be seen as in any way a threat to natural science. The real threat was from dogmatic

---

[192] Steneck, op.cit., pp. 92 and 149. The passage belongs to day four; but the setting in motion was understood as belonging to the seventh day, of course.

utterances of the sort quoted from Luther – which of course had medieval precedents. The effect of excessive contemplation of God's direct intervention in the world must surely be to throw scientific practice into disarray. God as a sustaining cause of all creation is one thing, but God as intervening miraculously in nature is another. It is there that we use the word 'supernatural', in the sense of something that is not in principle open to study and incorporation in a system of natural philosophy (extended, in Aristotelian contexts to include a study of the heavens). Perhaps one should be more careful here, since there are two broadly different approaches to a doctrine of miracles: acording to one, they are a suspension of the laws of nature, and thus plainly supernatural, or praeternatural. The Augustinian approach, though, was to say that nothing is against nature that happens by God's will; and many scholastic philosophers insisted that while a miracle might not be natural to us, it will be so to God. There were for them two orders of nature, miracles being simply a part of the order unknown to us. They are supernatural only in the sense of 'beyond what can be known as natural to us'. The point is that in neither of the two senses have most of the ideas on celestial causation I have introduced so far any connection with the supernatural.

The main exceptions to this generalisation are where there was an element of *will* in the explanation. Even here, though, the wills of angels, intelligences, demons, or whatever, were usually taken to be entirely subject to the will of God, so they are no less a part of the natural order than all else that depend on God's will. This was all very well until infiltrated by the notion (largely from Gnostic sources) that the planets were associated with a spirit of evil that had to be overcome, for example by exercises in the purification of the soul, and talismanic methods. The cult of planetary angels was something else – for Aquinas it was in the same category as that of saints, for both were God's servants. For philosophical reasons that should by now be obvious, if one wanted to draw planetary influences to oneself it should be by imitation in some sense. The standard methods used stones, metals, plants, or anything conceived as having the same elementary qualities, or music, incantation, prayer, writing, or whatever had the same rational structure, or consonance. Such magical methods using natural substances were acceptabel even for Aquinas, although he drew the line at writing, which involved bad demons. (Ficino frequently cited him.)[193] When Ficino,

---

[193] D. P. Walker, *Spiritual and Demonic Magic from Ficino to Campanella* (London, 1958), p. 43. Cajetano explained that this was because writing required intelligence, and was not a natural physical effect. Ficino insisted, though, that the characters were directed at his own intelligence, and that he, the operator, was thus changing himself, so the act was acceptable.

Pico, and Campanella manipulated, as they thought, astral influences, the justification for their acts, philosophically speaking, was grounded in medieval philosophy. As D. P. Walker writes:

'In the third Book of [*Adversus astrologiam*] Pico states as his own a theory of astral influence which is almost identical with Ficino's. The heavens are the universal cause of all motion and life in the sub-lunar world. They operate by means of a heat which is not elemental, but which contains 'in perfection and virtue' all the elemental qualities. This heat is borne by a 'celestial spirit', which penetrates everywhere, nourishing, tempering, forming, vivifying. It is analogous to the spirit which, in men and animals, unites body and soul, 'a very fine invisible body, most closely akin to the light and heat of the stars'.'[194]

These remarks, which sound more like Aquinas and the thirteenth century than the fourteenth, are a touch more disciplined than Campanella's, according to which the bodies of the planetary angels are the visible planets, these being wholly spirit, which they continually pour forth. For him the *spiritus mundi* was identical with the visible Sun, the body of the *anima mundi,* dominating all the other planets.[195] (In his earlier magic he had used bad demons; later he turned to Ficino's spiritual magic, now, like Ficino, having come to fear demonic magic.)[196] Notions of celestial influence, in moving off in neo-Platonic directions, are losing contact with the natural sciences. (Whether they are to be described as 'natural' is a semantic point of no great importance.) One might ask why this should matter, when celestial influence is an illusion, a concept leading nowhere. As a purely contingent matter it is true that it led nowhere. It would be very special pleading to hold that it contributed much to seventeenth century mathematical theories of gravitation, for instance – although the 'philosophical' embellishment of these theories was replete with medieval notions and terminology. There were attractions, and instincts of bodies to unite with others, and effluvial spirits. The almanac makers, purveyors of astrology to the public, could speak of terrestrial events moved by corpuscles whose first mover was the stars, or of 'balsamic atoms emitted from the stars, as well as corroding ones', or even of stellar operation by 'some invisible wires, some unknown magnetisms'.[197] At the outset of the seventeenth century we find Gilbert treating the magnet as a fragment of the *anima mundi,* and soon Kepler was to make use of numerous animistic analogies in his

---

[194] Ibid., p. 55.
[195] Ibid., p. 227.
[196] For some central distinctions, see ibid., pp. 76–7.
[197] Bernard Capp, *Astrology and the Popular Press. English Almanacs 1500–1800,* (London, 1979), p. 184.

Harmonice mundi. Newton was to make use of subtle spirits to explain vegetation, mineral formation, and fermentation. When another member of the scientific establishment, Robert Hooke, told the Royal Society of a man with a hole in his head, whose brain grew turgid at full moon and flaccid at new, the Society ordered an investigation.[198] Robert Boyle, whom the astrologer John Bishop claimed as a patron, believed that the Sun, Moon, and stars poured down 'subtle but corporeal emanations' distinct from heat and light, and wished to identify them.[199] There still prevailed a widespread belief in celestial influence, mingled though it often was with scepticism over astrological practice. The belief, one might say, led nowhere, just as – when assessed with a mean pragmatic eye – the corresponding beliefs of the middle ages led nowhere; and yet they had an important, even crucial, intellectual binding power within the cosmological systems that incorporated them.

---

[198] Ibid., p. 207. Cf. p. 237 above.
[199] Ibid., p. 189.

19

INTIMATIONS OF COSMIC UNITY? FOURTEENTH-CENTURY VIEWS ON CELESTIAL AND SUB-LUNAR MOTION

One of the most striking things about modern discussion – historical discussion, that is to say – of medieval commentary on Aristotle, is the sheer reverence for the texts, especially for the texts of Aristotle, and for their putative consistency. There seems to be a Principle of Magnanimity at work – inconsistencies must not be multiplied beyond necessity – and this principle is never more dangerous than when it is extended to the detection of precursors for later scientific doctrine. I shall be largely concerned in what follows with a doctrine in which there are glaring inconsistencies from beginning to end, which were to have unfortunate scientific consequences for nearly two millenia. I am speaking of the division of the world into two essentially different regions, the celestial, in which natural motion is circular, and the lower regions where natural motion is straight up or straight down, away from or towards the centre of the world. It was left to the seventeenth century to show that motions in both regions are subject to the same laws. There is a strong temptation here to reconstruct history. I shall note several examples of historians who have succumbed to temptation. I have no intention of arguing, for instance, that universal laws of motion might have been put forward sooner had Aristotle not stood in the way, and had scholastic philosophers been more critical of him. This is obviously true, although it overlooks the sheer stature of the Philosopher, earned for him by the monumental quality of his writings. I am not inclined to examine history entirely retrospectively, but it does seem to me that there is something to be gained by considering fourteenth century natural philosophers such as Buridan, Oresme, Marsilius of Inghen, and others both earlier and later, as critical spirits bringing pressure to bear on an essentially indefensible conceptual distinction which we *know* from later history to have been best removed. It is not a serious case of reading history backwards to regard these people as having been beating against a barrier – even though we happen to know what was to come later, when the barrier finally crumbled.

For Aristotle's case, what is natural to a body, that is to say, the law governing its motion, depends on the *quantitative* nature of the body, and its *place*. (I leave aside forced motion, for the moment.) Now one can excuse Aristotle for drawing a distinction between two sorts of natural motion; after all, even in his early 'Platonic' phase, before he saw the heavenly motions as 'natural', he certainly treated them as essentially *different* from terrestrial motions. But Aristotle did something much more serious: he attempted to give a rational proof of what he took to be the *utterly irreconcilable nature of linear and circular motions, and hence to divorce the two cosmic regions permanently*.

I want now to review Oresme's handling of the argument from the supposedly fundamental differences between the two types of motion. We shall, in the first test, find Oresme to have been somewhat of a conservative.

In *De Caelo* I.2 Aristotle conveniently forgets certain exceptions he has made elsewhere, and insists that a thing can have only one contrary. When he says that the contrary of upwards is downwards (269a9 ff.) we have no difficulty in agreeing, until we see where his argument leads, namely to the conclusion that an element that naturally moves up or down will *un*naturally move down or up, and that it cannot (and he is considering that fire might do this) move in a circular sense even *unnaturally* (for that would imply that its natural motion had *two* contraries). Oresme, however, introduces the idea that there are *three* kinds of motion: not just natural and unnatural, but also motion *outside of nature*, 'hors nature'. Fire, for instance, moved partly transversely 'in its place or in its sphere' (which I assume to mean the same thing), is being given a motion of the third sort.

Oresme now proceeds to make silent amendments to Aristotle's reasoning as to natural motion and contraries. The text offers difficulties, and we can see why: the Aristotelian argument will no longer work. Suppose we try to prove that the substance moving circularly in the heavens is not fire. It must be moving with a motion 'outside of nature'. More than that we cannot say. We cannot prove that it is *not* fire that we see in the heavens, unless by 'outside of nature' Oresme means 'outside the natural world'. It is doubtful whether he did mean that. The alternative, towards which it is easier to be sympathetic, would be something like 'not with nature, not against nature, but such that it would not be appropriate to say either'. The great weakness of Oresme's formulation of the argument, with this interpretation, is that in a sense we can now have two contraries to natural motion. Fire, for example, *could* now be moving in the heavens, albeit 'outside nature'. In trying to help Aristotle, Oresme has effectively pulled the rug from under his feet.

The whole question of natural motion is of course closely connected with that of natural place, and here problems arise as soon as we start asking for movers or *causes* in natural motion. There was considerable medieval discussion of these things. A distinction was drawn – whether true to Aristotle or not was a matter for dispute – between things that move themselves and those requiring a mover, as in the case of violent motion. I will not go into the history of these things – a big subject.

Buridan, incidentally, in his own commentary on the *De caelo*, collected together four explanations of the cause of acceleration in the motions of falling bodies: first, a heating of the medium which progressively reduces its resistance; second, proximity to natural place; third, the progressive reduction in the air below the falling body, entailing progressive reduction in resistance; and an impetus explanation. He opted, of course, for the last one. Note that he did not consider antiperistasis, or the modified version of it offered by a supposed 'Jordanus Nemorarius' (possibly his pupil).

By the mid-fourteenth century, the doctrine of natural place was leading natural philosophers into ever deeper waters. How was it to be reconciled, for instance, with the view (as found in Giles of Rome and Walter Burley, for example) that speed of fall depends on distance fallen, rather than on some function of natural place? Air in the place proper to fire descends, while air in the region of water ascends. Are these two motions natural? Are they not contrary? Oresme offers a thought experiment which shows very clearly how difficult it is to interpret Aristotle's doctrine of natural place coherently. He imagined a closed tube extending from the centre of the Earth to the innermost of the heavens. Filled with water, apart from a small quantity of air introduced at the bottom, the tube would see the air rise to the heavens themselves. Is the upward movement forced or not? Armed with Archimedean ideas on relative density, we should say that it was forced. Oresme says explicity that the motion is *natural* until the air leaves the water, and then the air is thrown upwards in a *forced* manner, 'because the water lifts it up and pushes under it by reason of its heaviness'. (One can imagine echoes of antiperistasis here.) Concentrating only on the natural phase of the motion, we can say that the natural motion of the air depends not only on *place* (in the sense of distance from the Earth's centre) but on the *surrounding elements*. A simple body is evidently by its nature capable of two simple contrary motions. This idea strikes at the heart of the Aristotelian principle that an element that moves naturally in one direction will move unnaturally in the opposite direction, unless, that is, we add a reference to surrounding elements.

We tend to put the surroundings into the 'external forces' side of the equation of forces. The problem when one is trying to preserve a doctrine of natural place is that this solution is not appealing. Oresme is trying to be a good Aristotelian when he explains apologetically how, when Aristotle said 'that there is but one simple movement for each simple body, we are to understand this to mean *within the same medium*'. There is a similar comment in *Le Livre du ciel et du monde* IV.5: 'one should not understand that air and water possess within themselves the two motive qualities of weight and lightness in the absolute manner that, in the opinion of some people, two contraries have in relation to their intermediate state . . .' This might have seemed to him to be all in an

Aristotelian spirit, but in truth it extends the earlier reading of Aristotle (to the effect that there is natural motion, unnatural motion, and motion outside nature) with the principle *that considered in isolation a body has nothing that can be called a natural motion at all!* I consider this to be a small but useful contribution to the slow erosion of the doctrine of natural place, and hence indirectly to that of a distinction between terrestrial and celestial sciences.

It has to be said that Oresme's unorthodoxy does not seem to have meant very much to him. He was perfectly at ease with the old crude principles, such as 'earth rises by violence and falls by nature, etc'. He could accept them, in a loose sense of that word, however, only because he did not really need them. Men knew that fire rises long before Aristotle gave as an explanation that it was moving to its natural place in the universe. The firm knowledge *that* fire rises was always so commonplace that Aristotle's explanation of it was never really needed to decide what would happen in actual cases. When Oresme produced his thought experiments with the tubes, I doubt whether he was relying on Aristotle to decide what would happen – although he wanted as much consistency with Aristotle's philosophy as possible and so made more show of principle than of instinct. That the strain of achieving consistency was too great for him is evident elsewhere in *Du ciel et du monde*.

Oresme provided two examples which he considered to show how position, in the sense of 'proximity or distance from their proper places', might indeed affect the nature or inclination of simple bodies. Once again, he proceeded from 'thought experiments' reminiscent of Aristotle's own:

(1) A portion of fire perfectly centred on the world, if it could be guaranteed to hold together in one piece, would not move. Displaced even slightly, it should move to its proper place. (Oresme likens this to the ideal situation when a sword is imagined to stand on its point.)

(2) A piece of earth equidistant from two worlds, on the line joining their centres, if it could split would do so, half going to one world and half to the other. If it were to stay united, it shold remain unmoved, like a piece of iron between two magnets of equal strength, 'comme un fer entre .ii. aymans egalz et equalement'.

Where Aristotle had denied that things (constituted from elements) might change their nature, and hence their tendencies to move, according to their places, Oresme clearly refused to be convinced. To be sure, he did not consider any other cases than purely symmetrical ones, but he had a clear feel for an equilibrium between *two* gravitational forces. Part of the task of breaking down the barrier between Aristotle's terrestrial and celestial regions was to get rid of the notion that gravitational forces are unique, in the sense of being directed to one centre.

Oresme, in the place indicated, tacitly rejected the requirements of a unique gravitational centre. He did not argue his case, but he provided us in passing with an argument. The 'nature' of a body, that which decides how it will move, was not to depend on the body in isolation. He had earlier held it to depend on its surroundings; and now, implicity, he shows himself willing to take relations with more than one gravitational centre into account. But this having been said, it must be emphasized that he had *no* intention of advocating the concept of a centre of gravitational attraction, and in this he was following in the footsteps of Aristotle and other commentators.

Before considering Oresme's treatment of the problem of other worlds, the theological basis for Buridan's ultimate rejection of Aristotle (on other worlds) ought to be noticed. Like most Parisian theologians of his century, and in conformity with the decrees of 1277, he acknowledged that God could, if he so wished, make more worlds than one. God could, if he wished, 'produce dissimilar effects', by which he meant that God could operate on a single species in *different* ways in different worlds, *or* could operate on different species using the *same* principle in one world. This is his answer to Aristotle's argument concerning the body that would have to move up in one world and thus down in another. The first alternative was more or less that suggested by Averroes, while the pair of alternative was taken over by Oresme, in his *Livre du ciel et du monde*.

On the face of things, Aristotle has universality on his side: *all* heavy things go down to one centre. The Averroes-Buridan argument, loosely expressed (and I will say shortly why it was wrong), was that matter in one world might count as anti-matter in another, there moving naturally up rather than down. Oresme, while seeming to accept this 'higher' sort of universality, in the end drops it in favour of what one might call an argument from conceptual insulation. If God were to make another world, its centre, he says, would be neither higher nor lower than our world. That, surely, represents a move *contrary* to the spirit of universal science.

Before considering more closely the differences between the 'dynamics' of the sublunar region and the heavens, we ought to review in very general terms Aristotle's ways of distinguishing the regions. They contain different sorts of matter, partly as a consequence of their different sorts of motions. The kinematic differences were limited by the possibilities known to Aristotle, who had the misfortune of living before the high-water mark of Greek astronomy. His homocentric kinematics inevitably coloured his view of the dynamical (kinetical) distinctions to be drawn between the two regions. This in turn had implications for medieval views on the nature of the intelligences thought to cause heavenly motions. Angelology, in short, was subject to intellectual links with the geometry of the cosmos, and Ptolemaic astronomy was for some an embarrassment, precisely because this link was held to be significant. Another aspect of the kinematical distinction is that which concerns contrary motions, their possibility in sublunar realms and their impossibility in the heavens; while yet another aspect concerns the seeming eternity of circular, as opposed to linear, motions. As far as possible I shall try to consider these different distinctions separately, beginning with the last.

The Christian scholar was obliged to disagree with the principle that the heavens were never created and will last for ever. The circular motion of the heavens is perfect, and – as Oresme says – the perfect contains the less perfect. Circular heavenly motions are endless; and there must be a perpetual motion *causing* those motions that indeed have a beginning, this perpetual motion being the motion of the heavens. Aristotle had argued that the heavens move eternally, and therefore there must be something else at the centre which is *still*. The Earth (or whatever we choose to call it) must then exist, the argument goes, and therefore so, incidentally, must fire exist, by what is sometimes called Aristotle's principle of plenitude. Oresme will not accept this conclusion, only that it is *possible* that if there are contraries a thing might exist in both forms. He also saw that the something, the Earth that is at rest at the centre of the moving heavens, might be indefinitely small.

Before we go any further we ought to note carefully how the fundamental cosmological division under discussion rests so heavily on the difference between straight lines and circles, the first indefinitely extensible and hence imperfect, the second not so, but prior to all other two- and three-dimensional figures. *All* the heavens were held to be perfect because they nested (without the dreaded void spaces) perfectly within the first sphere; and the evidence for this was held to be close at hand – the surface of the sea being spherical, the air above it following its form, and so on.

What then of Ptolemaic astronomy, with its eccentrics, epicycles, and the rest? Here is an interesting difference of attitude as between Buridan and Oresme. Both are pulled along by the arguments of Aristotle, but Oresme a little more reluctantly, for he was much more in sympathy with the immensely successful astronomical theories of his day than was Buridan. But even Buridan was driven to accept a modicum of 'modern' astronomy at the expense of the homocentric principle, though only as long as he could preserve the integrity of the intelligences that he supposed moved the spheres. He hints at the fact that he might even have been prepared to accept epicycles, had it not been for the evidence from the Moon's markings. It is an interesting subject, but I do not have time to discuss it here.

Oresme's position in *Le Livre du ciel et du monde* was to accept the affinity of the heavens with angels (intelligent creatures). The heavens are moved, he said, 'according to Aristotle and Averroes . . . by immaterial and incorporeal and spiritual things called intelligences or separate substances, so it is reasonable to assume that these are *with* the celestial bodies which they move'. His acceptance of Ptolemy, that is, of epicycles, more or less forced on him the notion of moving intelligences: the intelligence in the main (deferent) circle might perhaps be held to be stationary, but if one gives an intelligence to the epicycle, one is more or less obliged to link it to that moving epicycle. No very precise location could be assigned to the angels, though 'the body moved and the angel are in one place or nearly so', if the angel wills it. Angels may change shape, and move around with resistanceless motion as rapidly as they wish, subject only to God's will. Actually Oresme is rather vague about resistance in the heavens, seeming to suggest that it exists, but is of a special type. 'When God created the heavens, he put into them qualities and motive powers just as he put weight into terrestrial things, and put into them resistance against these motive powers [namely their weights], In the heavens the powers and the resistances are of such a kind and of such matter that they differ from any sensible thing or quality down here below. They are moderated, tempered, and harmonized so that celestial motions are made without violence, as when (violence apart) a clock is made to run by itself. They move without expenditure of effort, because they have no tendency downwards, no weight. (Oresme does not have a very clear idea of what it is that makes for effort on the part of men and animals in their labouring; weight is the only resistance he recognizes.) The heavens were evidently noble. (Both he and Aristotle found it easy to equate nobility with effortlessness.) The understanding and the will were not of a corporeal nature. Like Aristotle he had to reconcile his view of resistanceless heavens with the fact that velocities in the heavens were finite, and movement not instantaneous. That the intelligences acted by *will* was the solution. The intelligence could *choose* the speed of its sphere. As Oresme wrote:

'If finite power moves without resistance, this motion cannot require all its strength; and the same goes for infinite power.' In other words, in the absence of resistance we need another sort of restraint, something not imposed from outside the movement (for that would be a resistance), but something self-imposed. The great virtue of the hypothesis of angelic intelligences moving the spheres is that they may be expected to have restraining wills. The same, hints Oresme, would be true of clods of earth if they were to fall with finite speeds under no resistance. As though to underline the essential *difference* between the two realms, Oresme wrote in a later part of his commentary that although in other kinds of motion there is 'a proportion between mover and moved', the same is not true for celestial movement, simply because things in the heavens are moved by an absolutely voluntary force, under no resistance.

There are several places in Oresme's commentary, serving to reinforce the view that he is not sensitive to any need to reduce the Earth/sky contrast. He did not want the primum mobile to be thought to attract the other spheres to itself by virtue of its 'great size and force of motion', for this moves in a way somewhat unnatural to them, added to which it could not pull the spheres below it round, owing to their concentricity and the fact that the concavity 'is very completely and perfectly polished, planed, and smoothed, so that it could not be more so'. Here we see his instincts telling him to use the same criteria in discussing celestial dynamics as one uses in day to day life; and yet not long afterwards we find him assuring us that the angels are *not* to be regarded as relative forces acting between the Earth and the heavens; and even when he makes a hypothesis as to an intelligence moving the Earth, he expressly refuses to allow that this same intelligence has anything to do with the movement of the heavens. This is hardly surprising, considering the draughts of Aristotle he had imbibed, dealing with the essential differences between matter in the different parts of the universe. Not even all circular motions are of the same species, for these can depend on the *poles* of the motions; there are vital distinctions between the nobilities of the intelligences, let alone between theirs and ours; there are three sorts of music in the heavens which he cannot hear; and even when he seems to be on the point of drawing parallels between terrestrial and celestial heating caused by motion, he pulls back. The heavens cannot, he says, cause heat, because they are not receptive to active and passive qualities, added to which there is no violence in the heavens. And let us not forget that polished form they have. In the end he settles for the heavens causing the fire to rotate, which produces frictional heat with the air below. He is not very keen on the idea that the intelligences might incline the air to follow the circulation of the spheres more immediately. If there is a break in the conceptual barrier, it has to be said that it is barely perceptible.

Before trying to identify the causes of motions in the heavens, Aristotle had given a qualitative description of them, as regular and uniform. Oresme could see that epicycles and eccentrics combine to give *non-uniform* (accelerated or 'difform') motions. (In fact the same could have been said of the Eudoxan model.) Aristotle's aim was to point out the essential difference between uniform heavenly motion and non-uniform terrestrial motion. At first one has the impression that Oresme disagrees; but in the end it emerges that he wishes only to show that Aristotle reaches the right conclusion for the wrong reasons. The Aristotelians thought that the motion of a heavy or light (terrestrial) object could not be regular. No, suggested Oresme, it might be possible to choose forces *and resistances* so that in part motion was so. The heavens are devoid of necessity for change, thought the Aristotelians. But an *angel's will* may be changed, added Oresme, admitting that God's will is another matter.

Uniformity of motion apart, Aristotle attacked the problem of the causes of celestial motions through attention to the qualitative differences between planetary velocities. Planets at greater distances take longer to circle the Earth because they are nearer the primum mobile, which therefore has greater power over them, and slows them down accordingly. The idea of a dragging of the daily motion, said Oresme, was a view held by his contemporaries; but he disagreed with it, since – as we have seen already – heavenly motions are *natural*, a fact hard to square with the idea of a forced retardation. He had his own ideas about the causes concerned: of course we are back to intelligences.

For Oresme, then, causes of the heavens' motions are the intelligences. He used the concept of impetus, however, of terrestrial motions, in both of his commentaries on the *De caelo*, the Latin and the French. Clagett, following Maier, argues that his is a new concept, different from Buridan's, in that it is a function of acceleration as well as of velocity, and no longer considered to be a permanent nature. For these two reasons it would have been thought unsuitable, according to both writers, to explain celestial movement in the manner Buridan tentatively proposed. (Buridan had said, very tentatively, that God *might* have given the planets impetus at their creation, and that they might have kept it thereafter.) As for the first statement that it is a function of acceleration, for Oresme, I cannot see any clear and marked difference between his and Buridan's account. I take the position that Oresme's was a badly managed version of Buridan's doctrine, and that he had no definite view on a relationship between impetus and acceleration as such.

He did not introduce impetus into his account of the causes of celestial motions, but his reasons were not that this would have demanded accelerations in the heavens, as Clagett and Maier imply. Beyond saying this, one can proceed only by conjecture. My own view is that he was conservative, theologically speaking, and preferred to stick with angelic intelligences. He had his theory of celestial causation ready made, and had no wish to change it.

We have now seen something of the Parisians' attitudes to Aristotelian doctrine as to the causes of terrestrial motion, and the causes of celestial motion. We have seen very few signs of a preparedness to apply the same principles on both sides of the boundary between the two regions of the universe, but there is another way in which the boundary might be thought to have been under attack. What of causal (or other) interaction linking the two regions? More is needed here than the standard acknowledgement of celestial influence that as often as not rests on the authority of Aristotle in the first book of *Meteors* or *De generatione et corruptione* II.10.

There are several places where Oresme seems to be pondering the possible reduction of celestial influence to law. This hardly squares with the image of the sceptic and iconoclast, who repeatedly noted flaws in astrological presuppositions – such as the presupposition of commensurable angular velocities. As has been pointed out, however, by G. W. Coopland, editor of Oresme's *Livre de divinations*, Oresme was typical of his time in criticizing attempts to apply astrology to the particular and individual, in offering saving clauses as to free will, and in recognizing the illicit possibilities of obtaining knowledge of the future and the secrets of nature. 'Such differences as distinguish Oresme from earlier advocates of restraint in the use of judicial astrology are those of emphasis, of particular concrete argument,' writes Coopland, 'in short, those native and peculiar to him as an individual.' Again and again Oresme's text shows that he simply thought the astrology books defective in fact, not in principle, at least not all of the time.

Le livre du ciel et du monde might seem a strange vehicle for astrological doctrine, but there are several instances of it. Comets are 'the causes or signs of notable events here below' – but perhaps we should discount this example, since comets were taken to be sub-lunar, a manifestation of burning in the air (he followed with speculations on celestial influence, following *Meteors*). Later in the work Oresme reports on the views of Aristotle and Averroes that the Moon is in nature like the Earth, and the other celestial bodies are in virtue like the elements too. Supporting his argument with reference to astrological tracts, he seems inclined to accept that 'the virtues of celestial bodies are not only different but contraries (benevolent/malevolent, and so on). The general context of all this – and there is more of it – is the discussion of whether the heavens are all of the same species. If the heavenly bodies were all of the same special kind they would move alike, but they do not; added to which their astrological powers are contraries, etc. And just because circular motion has no contrary, it does not follow that all circular motion is of the same species, for species can depend on the pole of the motion; besides which, similar movements in two bodies do not imply identical species, any more than multiple motions in a single body imply that it is of several species. Oresme is admittedly going through a fairly standard Aristotelian exercise, but it is at least interesting to see him siding with Avicenna and Albert against Aristotle, Averroes, and Macrobius, and insisting that the heavens might well be of one very general species, but are certainly not all of the same very special species. This admission of diversity in the universe, however mild the diversity might seem to us, is unlikely to have obstructed improvement of the scientific world picture.

In what must be one of the most surprising asides in another work by Oresme, his *De configurationibus qualitatum et motuum*, we find a tentative hypothesis as to how the influence of the heavens comes about. He begins Part II chapter 12 'De figuratione celestis velocitatis' by noting that although individual circular motions in astronomy are at constant angular velocity, the fact that motions are about different centres and different poles gives rise to difform motions in the end result (accelerations and retardations). Even the fixed stars are said to move in this way, he adds, thus hinting at the theory of trepidation. He obviously thought that it would be unfitting for the difform motion, when 'figured' (graphed, 'figurari' is the word), to yield a figure that was not beautiful.

It is as though Oresme's delight at his appreciation of a new sort of harmony – he was wise to give no examples – overcame all his old scruples. Older forms of harmony, whether of number, traditional geometrical form, or music, had been given their turn. Now it was the turn of 'configurations of motions'. Here is a sign that Oresme was *proud* of what he had found in that connection: one does not offer one's smallest lamb to the gods. It is also a sign that he thought astrology a subject for which rational explanations were demanded. The heavens and our sublunar region were certainly linked, by astrological radiation and harmonic influence; but Gilbert, and Kepler, and those who in the seventeenth century would try to couple this qualitative idea with quantitative theories of planetary movement were more than two centuries in the future, and it would perhaps be forced to suggest debts to the fourteenth century on this score, except of the very remotest sort.

KINEMATICS: MORE ETHEREAL THAN ELEMENTARY

Moreover, rotatory motion is prior to rectilinear, because it is simpler and more complete . . . Aristotle, *Physics*, VIII.9.

The number of [planetary] movements can be determined only by that mathematical science which is most akin to philosophy, that is, astronomy, which alone deals with substance that is perceptible but eternal, whereas the others—arithmetic and geometry—do not treat of substance at all. That the motions are more numerous than the bodies in motion is obvious to anyone who has given even moderate attention to the matter . . . Aristotle, *Metaphysics*, Λ. 8.

THIS paradox—namely that celestial motions are simpler than rectilinear, and yet more complex (numerous)—lies at the very root of Aristotle's division of the universe. Circular motions were for Aristotle essentially simpler, and in that sense prior, because they were thought not to involve us in all the complexities of infinity, as the notion of unlimited rectilinear motion seemed to do.[1] Aristotle's finite universe was suitably divided into an inner spherical region (bounded by the sphere of the Moon) within which all was subject to continual alteration, growth and decay, and an outer spherical shell, the ethereal region, containing the spheres of the planets. Natural motions within the inner region were rectilinear but finite, trapped as they were within a finite boundary. They were directed either towards or away from the center of the universe. Celestial natural motions were circular, and on that account enjoyed a measure of perfection. These motions had an eternal character, according to Aristotle, and "a motion that admits of being eternal is prior to one that does not."[2] Needless to say, Aristotle will accept a temporal infinity where he would not accept the equivalent in space. This equivocation gave Averroes,

---

1. "The impossible does not happen; and it is impossible to traverse an infinite distance." Aristotle. *Physics*: 265a. Aristotle goes on to say that if something travelling along a straight line turns back on its path, the motion is composite, i.e. not simple.
2. See Aristotle,[1] and, showing the persistence of Aristotle's ideas in Copernicus' work, Copernicus, N., *De revolutionibus*. Book I, section 4.

Aquinas, and other loyal commentators certain problems,[3] but these are not my concern. I simply want to direct attention to that hard and fast boundary between two fundamentally different cosmic regions, each with its own type of motion. And I want to do so at the outset so that I may all the more plausibly distinguish between two different sorts of fourteenth-century scholar, each concerned with the analysis of one of those motions. I am of course referring on the one hand to those who—especially in Oxford and Paris—laid the foundations of terrestrial kinematics, and on the other hand to those who advanced the techniques of Ptolemaic astronomy.

Did neither group contribute to an understanding of the problems posed by the other? Was specialization such that the groups were made up of entirely different individuals? These questions have not been asked sufficiently often. If we are to judge by the sharp division between modern historical studies of the two groups, there was precious little interaction. I think this is broadly true, and by drawing attention to the Aristotelian distinction I have tried to explain why this fact should not surprise us. The Aristotelian distinction must surely bear at least part of the blame for any intellectual barrier we happen to find. Even so, we know of many medieval scholars who knew at least the rudiments of both Ptolemaic astronomy and terrestrial kinematics. If you put two sets of related ideas into one man's head, is it not reasonable to expect some sort of interaction? I cannot say that my study of this side of the problem is a very thorough one, but such evidence as I have found tends to suggest that first impressions are more or less correct, and that interaction was negligible across the relevant boundaries in the university curriculum. In other words, as far as original kinematic analysis is concerned, it seems that there were indeed two sorts of medieval scholar. It is therefore hardly surprising that each has produced its own sort of medieval historian. (And *there* is a barrier we have it in our power to abolish.)

In what follows, the most I can hope to do is draw together some of the loose threads, referring to arguments that have already been separately and expertly treated. For the fine detail I must refer to other commentaries. I shall collect together instances where interaction between the concepts of astronomy and terrestrial kinematics seems plausible, and I shall indicate a number of interesting, and sometimes intricate, examples of kinematic analysis that bear comparison with those most widely discussed by historians of kinematics. Above all, I want to show how astronomical kinematics was far from being a stagnant subject in the fourteenth and fifteenth centuries.

I shall begin with a few truisms, for the sake of those who are strangers

---

3. See, for example, *Comm. in Phys.* 250 b.11 to 153 a.21.

to medieval scientific thought. The fourteenth-century discussion of freely falling bodies was not an empirical one. The difficulties of turning it into an empirical study comparable with the study of planetary motions were entirely practical. Even the Moon takes a leisurely month to move once round the sky against the background of fixed stars, while Saturn takes almost 30 years. On the other hand, a stone dropped from the highest point of Merton College in Bradwardine's day would have reached the ground in less than 2 seconds. (So far as I know, no one thought of investigating mathematically that other sub-lunar natural motion, namely the rising of fire.) Secondly, the links between force (and resistance) and local motion—in other words, between dynamics and kinematics—were without a counterpart for celestial motions. If the astronomer pondered the cause for these, he fell back on intelligences and the Aristotelian prime mover.

Those who advanced the study of kinematics at Oxford and Paris—and the names of Thomas Bradwardine, William Heytesbury, Richard Swineshead, John Dumbleton and Nicole Oresme come most readily to mind—may all be reasonably described as philosophers trained in mathematics. The philosophical problem was one of explaining how forms or *qualities* could vary in strength or *intensity*. This was the problem of intension and remission of forms, which is to say, of increase and decrease in their intensity. The Mertonians treated this problem in general (in regard to intensity of color, heat, and so on) by analogy with the treatment of problems of motion in space.[4] The vocabulary used is to a casual observer very reminiscent of that used in astronomy, with such words as *"motus"* ("movement," but often meaning "position," and even at times functioning as "velocity"), *"gradus"* (as in *"gradus motus"*—"degree of movement"), and *"latitudo"* (as applied to measurements on the velocity axis of a velocity-time graph). It is difficult to decide whether these resemblances are of any significance. Graphical methods were of course well established in astronomy, not only for the simple representation of position through the allocation of coordinates, and in transformations from one system to another, but also for computation using geometrical loci with what would now be called "analogue methods."[5] On the other hand, we should not forget that what are to our eyes technical mathematical terms were in the past simply common words—for movement, step, and breadth, in the instances mentioned. Before the recognizably modern coordinate graph,

---

4. For the general problem of intension and remission of forms, see Maier, Annaliese, *Zwei Grundprobleme der Scholastischen Naturphilosophie*, 2nd edit., Rome, 1951.
5. J. D. North, ed., *Richard of Wallingford*. An edition of his writings with translation and commentary. 3 vols. Oxford: Clarendon Press, Oxford University Press, 1976, passim.

as in the work of Giovanni di Casali and Nicole Oresme, there is to be found a two-dimensional *geometrical* analogy to contrast the intensity of a quality and its quantity of extension.[6] There is no mention in any of these contexts of an astronomical association, and it seems to me to be quite mistaken to suppose that in kinematics the words *"latitudo"* and *"longitudo"* have any but their everyday meanings of length and breadth.

Although it should hardly be necessary to point out that astronomy has operated with kinematic concepts since Babylonian times, this is by no means to say that there was any conscious attempt to *define* the concepts so used. In his invaluable collection of materials on medieval mechanics, Marshall Clagett mentions a number of definitions from early astronomical writings.[7] Autolycos of Pitane (ca. 310 B.C.) defined uniform movement in terms of equal distances ("lines") in equal times (with the consequence drawn that distances were in the same ratio as corresponding times), and he applied the definition in some theorems on the uniform rotation of a sphere. Two centuries later, Geminos emphasizes the importance of geometry for astronomy, but by then this was something of a platitude in the Greek world. Even by the time of Autolycos, the extraordinary system of homocentric spheres as devised by Eudoxos had shown the great power of the geometric method. The treatise by Autolycos was translated into Latin by Gerard of·Cremona in the twelfth century. Gerard of Brussels, however, who can be regarded as having launched medieval kinematics on its career in the thirteenth century, makes no obvious use of Autolycos. Gerard's geometric approach stems from Euclid and Archimedes. He had the same main object as Autolycos, namely to treat of the rotation of the parts of a sphere, and of other bodies—not a very pressing astronomical problem, as it happens. Throughout all this work there is lacking the concept of velocity as the ratio of essentially different quantities, viz., as a quotient of a length and a time. To speak of uniform motion in terms of equal distances in equal times is of course the counterpart of the astronomer's tables in which the *medium motus* is given for standard time intervals; for here "mean motion" is simply an angular distance measured from some base direction. The mental block was largely of Aristotle's making: lengths and times were of different species, and even a comparison of movement along a straight line and a curved was according to him out of order, since they too were supposedly of different species.[8] (This last idea Gerard of Brussels rejects, although he is squarely in the Aristotelian tradition.) Before they could readily progress towards

---

6. M. Clagett, *The Science of Mechanics in the Middle Ages*. Madison, Wisc.: University of Wisconsin Press, 1959, p. 336.
7. See Clagett,[6] pp. 164-168; 178-182; 217-218.
8. Aristotle. *Physics*: 248a–249a.

the view of instantaneous velocity as a limit of a ratio of a length and a time, medieval writers on kinematics had to pass through the stage at which different "motions" (different velocities, as we might anachronistically say) were compared through the ratio of distances traversed in equal times. Gerard of Brussels, Bradwardine, and later writers have this idea explicitly; but of course it is implicit in the work of those astronomers who, like Ptolemy, tabulated motions and compared them (notably in eclipse calculations, which one may loosely describe as the problem of deciding whether two ships, of different sizes, on different (but intersecting) courses, and moving at different speeds, will collide). One should not underestimate the achievement of Ptolemy, or of those who later adapted his remarkable algorithm for this problem. Richard of Wallingford, for example, devised an instrument for the solution of this problem, as a component of his so-called "albion" (all by one). Just how ingenious this analogue device is can be seen by referring to his treatise on the albion and my commentary on his treatise.[9] For those who are not familiar with the problem, let me simply point out that its solution involves finding, not perhaps "instantaneous" velocities, but at least the hourly motion of the Moon as a function of a certain angle (the argument, so-called, when the Moon's epicycle is at apogee). Although, had they been challenged, Ptolemy and his followers would probably have taken the standard Aristotelian way of discussing velocity, in their working they came perilously near to a treatment of instantaneous velocities, and of instantaneous velocities in terms of quotients of small differences. It would be too pretentious to talk of differentials.

The sort of compound instrument designed by Richard of Wallingford —abbot of St. Albans and only a few years Machaut's senior—was part of a long tradition stretching back to antiquity. I cannot possibly begin to review this tradition here, but let me point out that it includes the graphical representation of planetary motions generally, and involves the generation of curves (usually circular arcs) on which position is correlated in some way with *time*. It is well known that three great problems of Greek geometry—the squaring of a circle, the duplication of a cube, and the trisection of an angle—led to the invention of curves involving movement, that is, using time as a parameter. There is no reason to suppose that the works of Hippias, Pappus, and Archimedes (on spiral lines) influenced in any way the computing tradition of which I spoke, but there too we find examples of curves generated by kinematic means. One notable example I might mention in particular is in fact to be found both in the computing tradition and in straightforward astronomical doctrine. I refer

---

9. As, for example, Richard of Wallingford. See ref. 5, vol. 2, especially pp. 206-209, and the corresponding sections of vols. 1 and 2.

to the resultant deferent curve for the planet Mercury. Mercury was given a more complicated movement than any of the higher planets. Ptolemy supposed it to move on a circle (epicycle) whose center was to move on another circle (the deferent proper), with a center moving on yet a third circle. For all planets other than the Moon, two circles were enough, namely, a (moving) epicycle and a (fixed) deferent circle. Clearly it would be convenient if Mercury's second and third circular movements could be combined. In fact they were combined into a single (oval) curve on an eleventh-century instrument designed and described by the Hispano-Moorish astronomer known as Arzachel. Richard of Wallingford, possibly by a process of independent invention, makes use of comparable ovals twice over, for one of his scales transferring the time divisions from the oval Mercury deferent to a concentric scale, and for another of his curves producing what can be described as an approximation to the geometrical inverse curve. The latter was so designed as to have its radius vector proportional to a certain correction term ("equation") needed for the calculation of Mercury's position. It would, however, hardly be relevant if I were to pursue further the history of the Mercury deferent. I introduced it simply as an example of the way in which medieval astronomers used geometrical techniques for the "static" solution of problems involving *times*, but times which were no longer, as it were, thrown away when the parametric representation had done its work. The time-value corresponding to each point of an Archimedean spiral was of no great interest to the geometer of the ancient or medieval world. Not so the times corresponding to the points of curves analogous to the Mercury deferent, which therefore more truly belong to the subject of kinematics.

Another subject intimately connected at once with both astronomy and mundane kinematical principles is the analysis of the motions of trains of gear wheels. If I may be allowed to introduce him once more, Richard of Wallingford gives evidence not only of personal ingenuity but of traditions that we can now only dimly perceive. At first the geared planetarium, and later the mechanical and particularly astronomical clock, served as a meeting point of theory and practice. If we search only for anticipations of Galileo, the concept of velocity as a ratio, or the equation of free fall, we shall overlook entirely the calculus of gear ratios. We might take a high moral line over this matter. Certainly most trains were designed merely to preserve a constant ratio between the members of a set of constant mean angular velocities. Bearing in mind the very awkward ratios that obtain between the planetary motions, it is easy to see that the difficulties were rather greater than those of linking the hour-hand and the minute hand of a watch; but awkward as they were, and transported trains of wheels though they might have required, the motions in ratio were usually constant. Consider, however, Richard of Wallingford's solution to the problem of making the Sun's wheel move with a continuously varying velocity. This velocity was not to be equal to the velocity of the real Sun (i.e. in the ecliptic), but of the projection of the Sun on the celestial equator. He gave the solution, but without discussing the way of arriving at it, or even the formulation of the problem. The solution was to engage an *oval* contrate wheel (i.e. with teeth at right angles to its plane) with a pinion lying along a radius vector.[10] (The pinion must be of such a length that it will mesh with the contrate teeth at both greatest and least distance from the center.) How then did he decide on the very precise form of the oval contrate wheel? I have hazarded a guess in my edition of his works, and I suggest that as a part of his plan he effectively plotted the reciprocals of solar velocities (velocities in right ascension) in polar coordinates.

But there's the rub! Do I really mean *velocities*, or simply daily movements taken from an astronomical table? Of course in point of procedure I mean the latter, but there is surely a case to be made out for saying that by making something—namely a radius vector—a *function* of these small daily bits of "movement," Richard of Wallingford was at least groping for the idea of velocity as a richer conceptual entity than that of mean motion. But even granted that we have not caught the will-o'-the-wisp of instantaneous velocity, it must surely be admitted that—blacksmith's work or not—Richard of Wallingford's was a piece of kinematic wizardry comparable with that produced by the philosophers of his own university—philosophers who usually took only the crudest of astronomical examples to illustrate their classifications of motions. John of Holland, for instance, a mid-fourteenth century Oxford man, takes the so-called ninth sphere of the astronomers as an example of a sphere "moving uniformly with respect to time. . ." He makes no mention, however, of the movement with which so many of his astronomer contemporaries were obsessed, namely, the periodic (and therefore accelerative) motion of the eighth sphere. The steady movement of the stars with respect to the equinoxes was discovered by Hipparchos, in the second century B.C., but the kinematically complex explanation later offered for the movement (which was not looked upon as a precession of the equinoxes) seems to have originated with Thābit ibn Qurra of Harrān in the ninth century (Christian era). An admirable survey of this whole subject of precession has been given recently by Raymond Mercier, who incidentally touches upon the difficulties some astronomers have apparently had in trying to interrelate even steady periodic motions.[11] And while on the subject of precession,

---

10. See *Richard of Wallingford*, ref. 5, vol. 2, pp. 342-350.
11. R. Mercier, "Studies in the medieval conception of precession." *Archives Int. d'Hist. des Sciences*, 26 (1976). pp. 197-220; 27 (1977), pp. 33-71.

it should be noted that Hipparchos obviously had a clearer vision than most; one cannot help wondering whether the course of scientific history would have been in any way accelerated had his lost work on falling bodies been handed down to the Middle Ages.

I want to turn now to some of the more abstract parts of astronomy, namely planetary tables, in which kinematic ideas are often only to be inferred. I will try to give the barest outline of the sort of procedure followed by an astronomer who wished to calculate the position of a planet at a particular moment of time. Consider the planet Venus (P, Fig. 20)

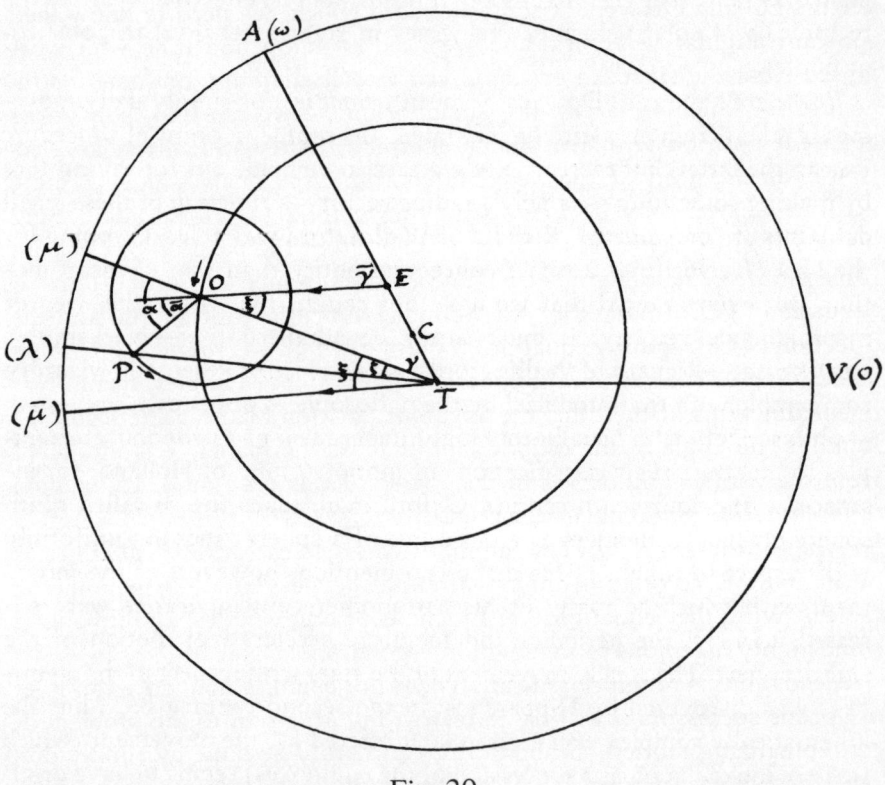

Fig. 20

which, according to Ptolemaic principles, moved steadily round an epicycle with center O, while O moved round the deferent circle, with center C. The Earth (T) was at some distance from C, and a point E, known as the equant, was the same distance to the other side of C. The equant was important because it was the EO, and not CO, which turned at constant angular speed. The resulting locus of P is a curve with a series of loops; but (if we overlook the complication that not all our circles were thought to be in quite the same plane) the planet would mostly be seen from the Earth to proceed in the same general direction as the Sun (in its annual path), but to turn back on itself from time to time, with the so-called "retrograde" motion. Its apparent place on the ecliptic, the path it seemed to follow through the background of stars, is at the point marked (λ). It is this point the astronomer wishes to find.

Now, once the relative dimensions of the circles and the direction of the line of symmetry TCE have been established, there are essentially only five further pieces of information needed to solve the problem. (We overlook the complication of precession.) We need to know where the radius vectors EO and OP were at some standard epoch, we need to know their constant angular velocities, and we need to know the time which has lapsed from the standard epoch. Given a scale diagram, or some instrumental equivalent—namely our "equatorium"—we can find the planet's apparent position with reasonable accuracy. Better still, given an electronic calculator and a knowledge of coordinate geometry, an even more accurate answer may be found. But the method that the Middle Ages learned from Ptolemy was a different one. From the position of EO with respect to the line of symmetry TCE, a small angular correction-term or "equation" EOT (marked ξ) was found. From this information, and the direction of OP, another equation PTO (marked ζ) was found. Combining by addition or subtraction the equations and the given angles, themselves linear functions of the time, the apparent position of the planet could be found. (The angle λ, the ecliptic longitude of the planet, was measured from the vernal equinox V.) In the course of this procedure, 10 or more different tables might have been used.[12]

What of these tables? They fall into two broad categories: (1) tables of mean motions, correlating angles with times; (2) essentially trigonometric tables, giving the "equations" as functions of other angles, and involving some highly ingenious processes of interpolation.

Although this rough classification does not begin to do justice to Ptolemy or his successors, I will try to restrict my attention to the problem of conceptualizing motion, although this can hardly be done by simply disregarding tables of type (2). It is the astronomer's grasp of the whole system that counts. Consider, for example, the simplest "planetary" movement, that of the Sun, which was supposed to travel at constant speed around the circumference of an eccentric circle. Not much cause for excitement here! But when an almanac is compiled—i.e. a list of consecutive solar positions, usually at one-day intervals—and provided, as was often the case, with columns of first differences (i.e. daily motions), and

---

12. For a fuller account, see *Richard of Wallingford*, ref. 5, Appendix 29.

even of corresponding hourly motions, then the astronomer must have begun to sense what it was to look at a table of changing velocities. We know, for example, that Thābit ibn Qurra looked at the subject with a keen analytical eye, and wrote a book on "the deceleration and acceleration of movement along the ecliptic as dependent on distance from the eccentric center".[13] Even allowing for the modern cast given to Thābit's words, on which I am not qualified to comment, I have no doubt that when he discussed maxima and minima in the movement, and when he found the point at which the "velocity" was equal to the mean value, he had a clear understanding of the idea of acceleration. This has been shown true of a yet greater astronomer, al-Bīrūnī, who early in the eleventh century (Christian era) discussed the acceleration of the Sun in a perfectly explicit way.[14] Bīrūnī's discussion rests on his analysis of first and second differences of the true solar motion, this having been conceived of as comprising linear and non-linear components.

I have come across nothing so explicit in any writer in the medieval West, although it is not unusual to find tables of differences; and these, on closer inspection of associated texts, might prove to hide a deeper understanding than we yet appreciate. There are certainly matters of kinematic interest in astronomical tables, especially in those that stemmed from the tables of Alfonso X, King of Leon and Castile, which were assembled at some time between 1263 and 1272.[15] The history of the tables after they were first compiled is uncertain. They seem to have reached Paris a little before 1320, and from there they spread to all quarters of Europe within a decade or two. They were modified and rendered more convenient in Paris and Oxford in particular, in Paris especially by John of Lignères and his pupils John of Murs and John of Saxony, and in Oxford by an unknown astronomer, and later by John Killingworth. The aim of these people was to lighten the task of using astronomical tables. Thus John of Lignères had tables (*tabulae magnae*) which gave, by double-entry, a single combined equation, rather than the two separate equations I mentioned earlier. Some Oxford tables, drawn up so as to be of use from 1348, go much further than the *tabulae magnae*, and allow planetary longitudes to be taken out more or less directly. (Correction for precession was necessary. Although the word "almanac" was sometimes applied to them, following John of Lignères' usage, they were not an almanac in the usual

---

13. O. Schirmer, *Sitzungsber. der Med.-Phys. Soc., Erlangen*, 58 (1926), pp. 33-88.
14. W. Hartner, & M. Schramm, "Al-Bīrūnī and the theory of the solar apogee." In *Scientific Change*, ed., A. C. Crombie. London. pp. 206-218.
15. A detailed account of the matters raised in the last part of this paper will be found in: J. D. North, "The Alfonsine Tables in England." In ΠΡΙΣΜΑΤΑ. *Naturwissenschaftsgeschichtliche Studien. Festschrift für Willy Hartner*, eds. Y. Maeyama & W. G. Saltzer. Wiesbaden: Steiner Verlag, 1977, pp. 269-301; see below, pp. 327-59.

sense, of course, for it was first necessary to use tables for the mean motions with which to enter the final table.) The Oxford tables can be shown to have been grounded in the tables of John of Lignères, but where his ideas came from is uncertain. Earlier examples of double-argument tables are known from the Islamic world, but that there is a connection between them and the European tables at present seems unlikely. The Oxford tables of 1348 seem to have been well received. They are provided in one manuscript with an alternative set of canons, ascribing them to "Battecombe alias Bredon sive Bradwardynes." "Battecom" is alone mentioned at one later point and Bredon at another. I think we must simply be satisfied to accept that the tables were "in Oxonia constituta."

The fact that the 1348 tables were so close to an almanac or ephemeris meant that they could be easily used to carry information of a general kinematic sort. They showed whether motions were direct or retrograde, for instance, and they naturally therefore showed the planet's stationary points. A chapter of the canons for their use is given over to this general problem of planetary acceleration and deceleration. (One might even say that the writer was obsessed with this descriptive problem, for he refers not to the planets but to the *"retrogradi"*!) Another set of tables, meant to be used along lines that would take too long to explain, were drawn up by John of Murs, and effectively yield a picture of the *relative* motion of each of the planets with reference to the Sun, i.e. in terms of the age of the planet reckoned from the time of its conjunction with the Sun. These tables were rather difficult to use, and seem not have found much favor. They date perhaps from a little after 1320.

After the Oxford tables of 1348 I know of little important activity in this connection until the work of the Merton College astronomer John Killingworth, who died in his mid-thirties in 1445. A sumptuous copy of his tables, heavily interlined with gold leaf and done for Duke Humphrey of Gloucester, is to be found in the British Library (MS Arundel 66). The aim was to simplify the production of ephemerides, and not merely to yield isolated planetary longitudes. Without trying to explain, even in outline, John Killingworth's entire procedure, I will discuss one or two of his more interesting achievements. For a particular planet for much of the calculation he reckons mean motions as functions of the number of days from the noon when the planet was nearest the line of symmetry I alluded to earlier. (In my Venus example it was the line *TCE*. This can be called the "line of aux," or the "line of apogee.") The planets Jupiter and Saturn move very slowly, however, and the line of symmetry itself is moving slowly, and therefore he needs tables to allow us to make proper correction for this slow relative motion. This table obviously required clearer kinematic thinking than many medieval writers—Roger Bacon, for example—were capable of; but a third type of correction table was of an altogether higher order of originality.

As I have already hinted, John Killingworth reckoned the values of the constantly changing angles (μ and α in my figure) from the noon nearest the time the planet reached aux. Suppose, now, that we wish to find the planet's position at some later noon. Because we are working from noon to noon, an error is introduced into the calculation. The number of days between the two noons corresponds to a certain mean angular movement (μ_0), and this, added to the aux position, gives a position for the vector EO to which corresponds a certain equation (ξ_2, Fig. 21). But the planet was

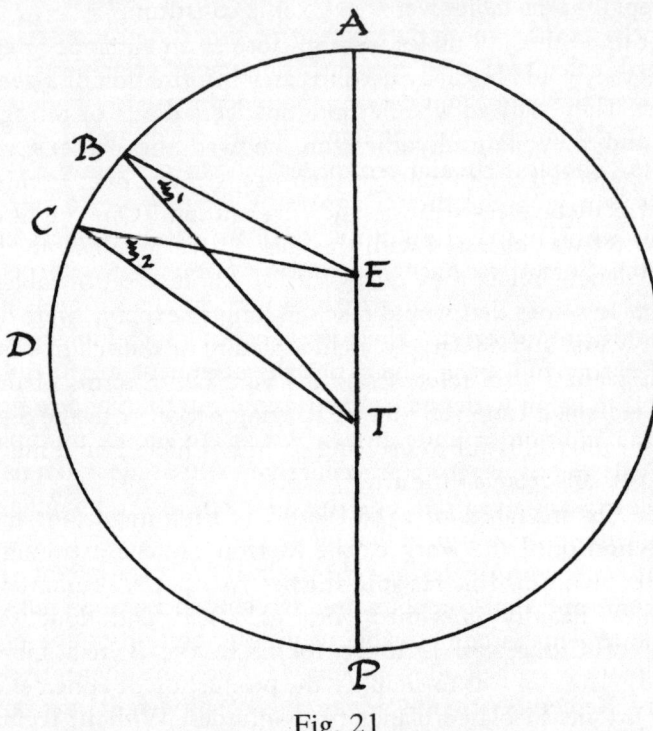

Fig. 21

not at aux, at the first noon. It was at some other point, to which, when μ_0 is duly added, a different equation (ξ_1) corresponds. John Killingworth gives us therefore a table whereby we may find a correction factor proportional to the difference ($\xi_2 - \xi_1$), by which factor it is in due course necessary to correct both mean motions (μ and α). The trigonometric knowledge necessary for the tabulation of the correction factor was far from simple, but John Killingworth clearly understood what was needed. By reasoning on the basis of the differential calculus we ourselves can show that for errors as small as those ever encountered in this system, the error (ξ_2 —

ξ_1) is directly proportional to the initial error, i.e. the displacement of the planet from aux at the first noon. John Killingworth somehow convinced himself of this proportionality, as can be seen from a closer examination of his method than I have been able to give here. He goes on, again in a way of his own, and making use of a new sort of double-entry table, to deduce the planet's longitude. However tedious even my bare outline might seem, the method is accurate and relatively easy to apply.

At the end of all this we have an ephemeris, looking much like any other ephemeris—that is to say, with no more continuity than in the readings on a digital clock. Where does John Killingworth stand, *vis-à-vis* the concepts of kinematics? Are we to say that those who know only digital clocks cannot know the concept of time? Astronomers like John Killingworth calculated with constant component velocities (if I may be allowed to use the word), but they compounded these to simulate the complex accelerated motions of the planets. (Indeed I do not need to remind you that the simplest circular motion entails an acceleration.) This was kinematics, not with the formal apparatus appropriate to freely falling bodies, but with an analytical structure of its own, a subject in its own right. There is no purpose in entering into a contest, as it were, between the two sorts of kinematics, but an extreme argument agains the astronomical sort is not difficult to devise. "There is a sense in which Ptolemaic astronomy is not in any full sense kinematic" the argument might run, "for its technique is to begin with one geometric configuration and to pass to another by the addition of finite angles, which are simply multiples of the time interval; and from such a procedure you will at most extract the concept of velocity and even *that* only by a stretch of the imagination." As against this, it might be argued that *all* kinematics is concerned with separate events, and with devising a theoretical structure potent enough to connect them; and that compounding accelerated motions out of steady circular movements is one possible way to a solution of the problem of planetary motion—albeit a less potent way than the alternative later discovered by Kepler. With the advantage of hindsight we can see the Ptolemaic kinematics of the Middle Ages not as a subject giving conceptual clarity to the notion of linear acceleration, and not as having given rise to the Galilean theory of free-fall, but as the father of Kepler's description of planetary motion. And that was certainly no mean kinematic offspring.

Plate 5. Memorial brass to John Killingworth (*see* pp. 320, 343-6), from the transept of the chapel of Merton College, Oxford, where he was a fellow.

THE ALFONSINE TABLES IN ENGLAND

'Les Tables Alfonsines', wrote Feller in his 'Biographie Universelle', 'lui ont acquis plus de gloire que ses combats'. Whether or not he was right to add that 'il est certain que son attachement à l'astronomie lui fit faire de grandes fautes dans la politique', the fact remains that Alfonso X left his kingdom of Leon and Castille impoverished and divided, and that 'El Sabio' must to many of his subjects have seemed a double-edged epithet. The irony of the historical situation is that the astronomical tables which brought Alfonso such renown were not in the strictest sense Alfonsine. Not only were there extensive borrowings from earlier sets of tables, but the Alfonsine tables which circulated in such large numbers outside Alfonso's kingdom owed almost as much to their editors as the original revisions had apparently owed to Alfonso and his scholars. No copy of the original tables can now be identified, with any certainty, but time might change this state of affairs, since the headings are known. The Castillian canons for the use of the tables have now been available in a printed version for more than a century [1], although it has long been realized that the tables associated with the canons by their editor, Rico y Sinobas, have nothing whatsoever to do with those canons.

The tables are of course only one aspect of Alfonso's scholarly achievement. He pointedly encouraged the translation from Arabic and Castillian of many philosophical and scientific writings, work which had begun under the patronage of his father, San Fernando. This was all in continuance of an older tradition of translating from Arabic into Latin, a tradition well established at Toledo at the courts of the archbishops. After his accession (1252), Alfonso made a determined attempt to incorporate the best cosmological elements from Islamic and Jewish culture into the young civilisation of Castille. He did so, as is well known, by the creation of a school of Christian, Jewish, and Arab savants, over which he himself presided, revising their work and often writing his own introductions to it. The names of fifteen collaborators are known from the introductions to their works, and the canons to the tables begin with the words 'Dixo Yhuda fi de Mose fi de Mosca et Rabiçaq Aben Cayut' - two Jewish writers, and not the vast concourse of astronomers sometimes suggested [2]. Jehuda ben Moses Cohen and Isaac ben Sid compiled the tables, but it was the king himself who 'porque anaba los saberes et los preciaba' had commanded instruments to be constructed and observations to be made at Toledo - and this because of the discrepancies between predictions made in accordance with the old Toledan tables and current observations. The two Jews observed the Sun throughout an entire year, we are told, as well as the conjunctions of the planets both with each other and with the fixed stars, and also solar and lunar eclipses. The two men are, as it happens, called 'of Toledo' in the texts, and are presumed to be natives of the city; but as Miss Procter pointed out, it would be a mistake to suppose that the king and his court resided habitually at the ancient capital of the Visigothic kingdom, or that all the astronomical work was done there, for Seville was where Alfonso spent the longest consecutive periods of his reign [3]. He visited Toledo less often than Burgos, for example, and much of the work of revision of the 'Libros del saber de astronomia' was apparently done there rather than at Toledo. He is known to have been in Toledo between 3 February and 19 May 1254, from 26 January 1259 to 6 February 1260, for short visits in 1268 and 1269, and between 8 December 1278 and 10 April 1279.

Only the two short visits mentioned can be related to what is known of the dates of compilation of 'El libro de las taulas Alfonsies'. Most of the material in the 'Libros del saber' seems to have been assembled in 1276 or 1277, some years after 'El libro de las taulas' [4]. The 'Alfonsine era' adopted in the tables is 1 June 1252, the day of the king's coronation, but the 'Prologo' is moderately explicit, in placing the tables 'en la primera dezena del quarto centenario del segundo millar de la era del Cesar' (that is, between 1301 and 1310 of the Spanish era, namely between A.D. 1263 and 1272). Duhem, unaware of this testimony, criticizes severely a remark by Andaló di Negro, Genoese astronomer and teacher of Boccaccio, a remark which merely confirms the statement in the Castillian [5]. Andaló - writing in 1323 - was defending Profatius from a charge of encouraging inaccuracy through the use of the Toledan tables, when drawing up his almanac (c. 1300). Andaló mistakenly supposes Profatius to have used the Alfonsine tables, which were of course Toledan, but were not the 'Toledan Tables' in the accepted fourteenth century use of that phrase [6]. He set the 'correction' of the tables in the year 1272 which Duhem thought to be a mistake for 1252, the 'era of Alfonso', but which must be more or less correct. If, as is most probable, Andaló took the information from the 'prólogo' itself, then he clearly assumed that the decade named was complete at the time the work was finished. There is no justification for this assumption, and the best we can do is to leave the date of compilation between the broad limits 1263 and 1272. These limits are wholly compatible with another important piece of evidence. The 'Book of the Foundation of the World' by the Toledan Jewish astronomer Isaac Israeli ('the younger') provides us with important circumstantial evidence of observations made by Isaac ben Sid in the king's service. The earliest was an eclipse of the Sun on 5 August 1263, and three others were 'calculated and observed' between then and 13 December 1266 [6].

Although the discovery of a copy of the original tables might at length allow us to answer a number of outstanding questions, what matters most from the point of view of their influence is the way in which they were transmitted to the rest of Europe. This is an extremely involved subject, but broadly speaking the tables may be said to have travelled along two chief routes - one a mediterranean route of little consequence in the long term, and the other a route to Paris, and thence to all Europe in several revised forms.

I am not in a position to say much about the diffusion of the tables through Spain, southern France and Italy, although it is worth noting that among the scholars who collaborated in the revision of 'El libro de las estrellas fijas' in 1276 are 'Maestre Joan de Mesina e Maestre Joan de Cremona', while the name of the Parmesan scholar Egidio de Tebaldis appears in the Preface to the 'Quadripartitum' and elsewhere [7]. Although these men were permanently in the royal service, we are reminded that until 1275 Alfonso regarded himself as a candidate for the imperial crown, and that embassies were continually exchanged between the court of Alfonso and the Italian states [8]. As for the dissemination of the tables from Paris, after about 1320, its history is almost coextensive with the history of European astronomy as a whole, at least until the sixteenth century. If we overlook criticisms made by a handful of expert astronomers, we can say that during these centuries the Alfonsine tables were endowed in some quarters with authority of a sort reserved elsewhere for Aristotle himself. But again, this is a subject which I am not in a position to discuss. What I shall consider is the transmission of the Alfonsine table northwards from Paris to England, and in particular to Oxford. I shall try to establish the chronology of this transmission, and in doing so it will be necessary to revise the thesis put forward in 1920 by J.L.E. Dreyer, to the effect that the original form of the Alfonsine tables is to be found only in an Oxford recension due to William Rede [9]. More interesting by far, however, are the ways in which the astronomers of Paris and Oxford chose to modify the tables, in order to reduce the labour of calculating individual planetary positions or complete ephemerides. The new tables that resulted were clearly looked upon with some favour in their time, and yet they appear to have been completely overlooked by historians. They resemble in some ways tables from both earlier and later periods of Islamic astronomy, but they were very probably devised independently.

The Alfonsine Tables in Paris

The ambiguity inherent in the phrase 'Toledan tables' has encouraged numerous writers, medieval and modern, to suppose that the Alfonsine tables were widely known within a decade or two of their first being compiled, but the evidence is that they were not known outside Castille until roughly half a century had passed. The evidence is in the form of a series of texts composed in Paris by a small group of scholars, of whom John of Lignères appears to have been the central figure, with John of Murs and John of Saxony his most influential pupils. These three astronomers were at the very centre of the Alfonsine explosion. They worked together, judging by their writings no less than by the effusive testimony of John of Saxony to the virtues of his master, John of Lignères. In 1317, when he was writing his critique of the ecclesiastical computus, 'Autores calendarii nostri ...' [10], John of Murs showed himself to be a supporter of the Toulouse tables [1]. The introduction of the Castillian tables to Paris was therefore not complete by 1317. In 1321, however, in his 'Expositio intentionis regis Alfonsii circa tabulas eius', John of Murs refers to an observation made in 1318, underlining retrospectively the excellence of the new tables; while John of Lignères, who had in perhaps 1320 written a work independent of the Alfonsine tables, by about 1322 had prepared a first draft of a new recension of them.

It is hardly possible to confine within closer limit the date of the acceptance of the Alfonsine tables by the Parisians. Leaving aside for the time being the interesting 'tabulae magnae' by John of Lignères [12], tables which must have been compiled within a few years of 1320 (the first radix date they contain), and which will be shown to have incorporated some recognisably Alfonsine material, he composed two chief works of relevance to the problem. The first is dated 1322, and includes a section on spherical astronomy (the appropriate canons beginning 'Cuiuslibet arcus ...'), and another section on planetary movement (with canon 'Priores astrologii'..., sometimes separated from the work as a whole) and eclipses (this part often separated from the rest, and usually then beginning 'Diversitatem aspectus Lune ...') [13]. This first work rests heavily on Arzachel's canons to the Toledan tables. There is in it, however, allusion to the Alfonsine tables, degrees are grouped into signs of 30° (and not 60°), as the 'libro de las taulas' makes plain was the case with the Castillian tables, and the motions of the 'auges' incorporate terms for a steady precession of the eighth sphere as well as for a trepidational term - another Alfonsine characteristic. The star positions are seemingly not Alfonsine, and yet there is a sense in which the same could be said of the star positions in the Alfonsine version of the star catalogue of as-Sūfī. More significant is the fact that the planetary equations for Jupiter and Venus, which distinguish the Alfonsine from the Toledan tables, are typical of the latter. The hybrid character of this work of John of Lignères is evident from the grouping of years by scores, starting at 1320 - a system followed in the Alfonsine as well as the Toulouse tables, although of course in those cases with different radices [14]. This new grouping of years seems to have exerted strong influence on later English tables. To simplify future references to this arrangement, I shall refer to tables of this sort as 'tables with cyclical radix dates'. The John of Saxony tables have multiple radix dates, but one is meant only to work with that one of them which is appropriate to the calendar accepted.

The other significant work by John of Lignères was a set of canons beginning 'Quia ad inveniendum loca planetarum ...', and designed to explain the use of what was to become the definitive version of the 'Alfonsine tables' on the European continent. To all intents and purposes, this is the version of the tables later printed together with the canons of John of Saxony (A.D. 1327), 'Tempus est mensura motus ...' [15].'Quia ad inveniendum ...' was evidently written after 1322 and possibly even earlier than 1327, although John of Saxony in his work of that year abstracted eclipse canons from John of Lignères' 1322 work ('Cuiuslibet arcus ...') and made no allusion to 'Quia ad inveniendum', so far as I can detect.

In speaking of a definitive continental version of the Alfonsine tables, I am of course alluding to one or two mathematically rather superficial characteristics. The most obvious of these concerns the mean motus and argument, which were now to be calculated by determining the difference between the date in question and the beginning of the Christian era (which thus became the 'radix'), converting it to sexagesimal parts and then using only a s i n g l e table for the mean motus (or mean argument) of each planet. Diminishing the unit of time by a factor of 60, one simply shifts the headings of the table of motions, as it were, one place to the left; and so for smaller time divisions. Sexagesimal time division was used by Ptolemy and in some versions of the Toledan tables. Familiarity was a convincing motive for adopting the system, and yet, for ease of understanding, not to say calculation, the original Alfonsine system surely had all the advantages at this point in history. This, the system of cyclical radices (or, more properly, the system of radices 'ad annos Christi collectos'), had been used, as I have said, in the tables accompanying the canons 'Cuiuslibet arcus'. The radices were usually given at 20 year intervals (1320, 1340, and so on, these being the 'anni collecti'), and then mean motus and argument were tabulated for 1, 2, 3 ... 20 year periods ('anni expansi'), for the months of the year (beginning with January), and finally for hours (usually 1, 2, ... 10, 20). It is often said that this system was mathematically less virtuous than a consistent sexagesimal system, but the calculators in question would have been quite properly indifferent to the charge. They lived in a mathematically imperfect society which reckoned time in the same illogical way as we do today, and they were not looking for mathematical virtue, but for an easy life. The Parisians preferred to copy fewer tables of motions, and to add tables for the conversion of time intervals to sexagesimal parts. Alfonso, the Toulouse astronomers, John of Lignères for a time, and most of the English, preferred to copy more and think less [16].

The Parisian tables in England

When J.L.E. Dreyer examined a large number of codices of 'Alfonsine Tables' of various sorts over fifty years ago, he was struck by the fact that he had 'only come across two types of tables: the sexagesimal ones, identical with the printed tables, and the original tables in the form described in the "Libro de las taulas", but reduced to the meridian of Oxford' [17]. I shall not follow his discussion very closely, since it is clear that he failed to distinguish between several essentially different sorts of Oxford table. I shall explain the differences in due course. What Dreyer did, by implication at least, was to give the impression that the original tables arrived in Oxford, where they were preserved in something close to their original form, without passing through the hands of the Parisian group of editors [18]. This is almost certainly not what happened. Dreyer did hazard the guess that John of Lignères was responsible for the sexagesimal arrangement [19], and this was a reasonable enough conjecture. We know that John of Murs was engaged in reshaping the original tables so as to yield the sexagesimal arrangement, and that in 'Si quis per hanc tabulam ...' he composed simple canons to his so-called 'tabula tabularum' of 1321, a sexagesimal multiplication table likely to have been used in the conversion. But this does not answer the question as to when, and through what channels, the Alfonsine Tables reached their Oxford editors.

Surprising as it may seem, in the light of the Parisian activity of c. 1320, the tables do not appear to have been widely known in England before the 1330s. One hears so much of the intellectual mobility of the period. With an appreciable effort the fifty years from Castille to Paris may be explained away in terms of the linguistic barrier between the vernacular and the Latin of the universities. But why did the foremost English astronomer of the century, Richard of Wallingford, not mention the Alfonsine tables in his works of 1327 [20]? (He seems to have learned of them before his death in 1336, but it is impossible to say precisely when he first encountered them.) The problem is a difficult one, but it seems to me that the explanation for the preservation in England of a style closer than John of Saxony's to that of the original tables is simple enough. It is that the two early compilations of John of Lignères were of paramount importance in the transmission of the tables to England, and that the earliest format [21] was to the taste of those who were most influential in disseminating the 'new' material. The fist part of this suggestion can be supported by reference to two or three codices still extant in English libraries, and illustrating, as I believe, the initial reception of the Alfonsine tables in England.

The set of 'tabulae magnae' or 'magnum almanach' of John of Lignères is associated with canons beginning - with the spelling of the Gonville and Caius copy - 'Multiplicis phylozophie variis radiis illustrato Domino Roberto Lombardo de Florentia Glassuensis ecclesie inclito decano ...'. They were addressed, in other words, to Robert the Lombard of Florence, dean of Glasgow. This was no superficial inscription on a single copy, for references to the dean are found in the text itself. The tables are very comprehensive. After a prologue with a number of platitudes concerning the use of geometry to survey the heavens, and one wherein the names of Grosseteste, Ptolemy, Geber and Arzachel (but not Alfonso) are mentioned, the labour of using ordinary tables for planetary longitude is contrasted with the ease of using planetary equatoria. Because equatoria are not to be found everywhere, however, and because they are none too precise, John of Lignères compiles - as he tells us - his labour-saving tables.

The first of the tables after the canons for their use is for the auges of the planets at intervals of 20 years, from 1320 to 1520 inclusive, differences (minutes and seconds of arc) and second differences (seconds and thirds) being tabulated [22]. (The second differences diminish by virtue of the fact that the aux positions for the year 1320 are later given without trepidation, and differ from the first set for that year by 8; $^{\circ}$17, 6, 48.) [23]. After the aux positions, there are tables of radices for mean conjunctions and oppositions of the Sun and Moon from 1321 to 1609, at intervals of 24 years, and tables of the 24-year cycles and the monthly cycle. There are tables of mean motus and mean argument for the planets for single years (up to 20) for periods of 40, 60, 80, 100, 200, 300, ... 1000, 2000 years. The same functions are tabulated for months, days (1 - 31), and hours, and appropriate radices for Paris are listed for the time of Christ and for A.D. 1320. There are tables of equations of the Sun and Moon, and of their hourly motions, but the usual tables of planetary equations are omitted. In their place there are tables which show John of Lignères to have been a man of some originality.

The new tables give, by double-entry, a single combined equation which, when added to the mean motus ($\bar{\mu}$), gives the true position of the planet. In the left-hand margin of the table the mean argument ($\bar{a}$) is given, and along the upper margin the mean centre ($\bar{\gamma}$). The explanation in cap. 5, on the use of this simple table is largely occupied with the business of double (but linear) interpolation. Much of what I have to say concerns double entry tables such as these early examples by John of Lignères, but before discussing them in detail I shall assemble enough manuscript evidence to show that, notwithstanding Dreyer, the several works of John of Lignères are all that we need to explain the common English form of the Alfonsine tables. It is necessary, first, to examine more closely Dreyer's thesis, and the canons and tables of William Rede.

Dreyer was very reticent about the date of Rede's work. He drew attention to the fact that the radices usually ran from 1340 to 1600 with intervals of 20 years, but that in two manuscripts (Digby 48 and Laud Misc. 594) at least one table commences at 1320. The only conclusion he drew from this observation was that 'the tables were probably not much in use at Oxford till towards the middle of the century' [24]. In fact the 1320 exceptions are without significance. One manuscript (Digby 48) has a table of auges commencing at 1320, but this is in addition to a standard table beginning at 1340. The other manuscript (Laud. misc. 594) has an incomplete set of tables, ascribed to Rede, admittedly, but containing so many small (and seemingly deliberate) disagreements with Rede's figures as to rob it of any significance, vis-à-vis the problem of dating Rede's work. Dreyer's reference, furthermore, to a radix date of 1327 in the tables of the fifteenth-century manuscript Laud misc. 674 is a mystery. There is no such year mentioned at the page to which he refers (f. 46^r), and the only suggestion I can make is that he misread a gloss (in the top corner of the page) which runs 'tabule Arzachelis facte sunt anno Christi 732 (sic)', the year being written carelessly in a way which could lend itself to a reading of 1327.

William Rede's canons and tables can hardly have been compiled much before the year 1340, by which time the tables of John of Lignères had been available for roughly two decades. A number of phrases in Rede's work seem to reflect the style of John of Saxony's 'Tempus est mensura motus', even allowing for the differences in the structure of the two sets of tables. Such resemblances are admittedly often no more than a reflection of the cliché-ridden style of writers of canons generally. At all events, William Rede simply cannot be the crucial figure historians have made him out to be in this matter. He cannot have been born much more than a decade before 1320, and might even have been born after that date [25]. The straightforward version of the Alfonsine tables accompanying the canons 'cuiuslibet arcus' by John of Lignères (as opposed to his 'tabulae magnae') are so close to Rede's tables, moreover, that even had they not been circulating in England during Rede's youth, he would have deserved precious little credit for his merely having adjusted them to the longitude of Oxford. That the earlier tables were indeed circulating in England can, however, be seen by reference to a number of extant MSS:

The Alfonsine Tables in England

(i) Bodleian Library, MS. Digby 114. At the beginning of this rather untidy volume there are, among tables excerpted from Campanus, two or three Alfonsine tables and some canons excerpted from John of Lignères (ff. 8^r - 11^v), to go with the 1320, 1340, ... 1520 tables. The date of the copy was probably 1323, since some contemporary notes on f. 1^r reckon 'anni expansi' (in the context of the Campanus tables) for 7, 8, 9, 10, 11 and 12 years after 1316. The manuscript is of little textual value, and has in any case deteriorated to a point where it is now difficult to read, but I note the curious ascription at the head of the first relevant canon:

> Canones super conjunctiones et opposiciones secundum Mag. Johannem de Lineri et Mag. Philippum Aulyn Alunn [or Alioun ?] ...

Does the second name refer to an Englishman in Paris? The only persons I can find whose names resemble that of the colleague of John of Lignères are, first, Philip, Chancellor of Paris University [26], and second, one 'Philip Alanfranc the Englishman', author of a medical work in British Museum MS Sloane 3124 (15c.), ff. 196^v - 220^r. (This work has the rubric 'Pratica Magistri Philipi Alanfrancii Englici, magistri in artibus et medicina super aliquas infirmitates corporis'.) Whatever the answer to this problem, it seems that these early sections of MS. Digby 114, judging from a note on f. 5^v naming the two towns, were from somewhere in the neighbourhood of Leicester and Northampton [27].

(ii) An incomplete set of tables with single radices and 'anni expansi' set out in John of Lignères' decimal manner, but the radices now being adapted to the longitude (taken as common to all) of London, York, and Lincoln, is in the British Museum, MS. Egerton 889, ff. 1^r - 5^v. Signs of $60°$ and $30°$ are curiously mixed. Other tables (including Rede's) are bound together indiscriminately in the volume, until ff. 79^r - 100^r, where another (and now very fine) set of tables in the same John of Lignères style occurs, this time with radices for Toledo, Oxford, London, Colchester and Paris. There is also a full set of tables of (Alfonsine) equations. The heading 'Cambridge' is occasionally added, and data are then filled in by a later hand. (The tables are followed by others, and canons to them, which I ignore.) [28]. As an example of the contents of these English tables, I set out below the radices (for the Incarnation) for the Sun's mean motus, as found in the manuscript (signs are of $60°$):

Toledo	4^s	38°	21^I	0^{II}	30^{III}	28^{IV}	54^V	26^{VI} 38^{VII}
Oxford	4	38	20	21	4	55	49	33 49
London	4	38	20	6	17	50	55	14 0
Colchester	4	38	19	46	35	4	21	47 36
Paris	4	38	19	2	13	49	39	48 10
Cambridge	4	38	20	13	41	23	22	etc.

These figures, to seven sexagesimal places, get their appearance of exceptional accuracy only from the method whereby they were obtained. The Toledo figure comes (presumably) from some basic radix which might have been quoted as far as minutes, or even seconds, but which has been adjusted in part by being combined with some fractional 'eighth sphere' term. The effort of discovering this would hardly be worth while. Once a Toledo figure had been found, however, by the Alfonsine astronomers, all that was needed for its adaptation to the other towns was that the mean motus of the Sun be multiplied by the time difference corresponding to the longitude from Toledo. The quoted radices may be shown to correspond exactly (given the number of sexagesimal places) to time differences of 16^m (Oxford), 22^m (London), 30^m (Colchester), 48^m (Paris), and 19^m (Cambridge), from Toledo (all towns being to the East). The radices are in principle non-terminating sexagesimals, and there was no limit to the spurious accuracy to which an astronomer might lay claim by deriving radices in this way. The Cambridge astronomer John Holbrook, for example, in the very same manuscript (see ff. 134^r - 160^r), quotes mean motus as far as sexagesimal tenths.

I cannot say when the tables at issue were adapted from their Parisian prototype, but I note that William Rede could have used them, since he assumes a 16^m time difference for Oxford. It is more probable, however, that he worked from the tables of 'Cuiuslibet arcus', or some English modification of them; see (iv) below. The tables to be considered next below are certainly earlier than Rede's, and adopt a London-Toledo difference of 32^m (not 22^m); but 22^m is commonly encountered in writings of this period. John of Lignères in one place writes of Paris as being 40°30' from the West. Toledo was traditionally taken as 11° from the West. This is inconsistent with the 48^m derived above, and yet adjustments made in John of Lignères' tables do conform with the figure of 48^m, which I note that John of Murs used [29].

(iii) MS. Bodley 790 is a late fourteenth-century or early fifteenth-century English collection of works. It opens (ff. 1^r - 2^r) with fragments of tables of auges resembling those of John of Lignères in the 'tabulae magnae', but now with radix dates from 1380 to 1520 only, and with radix date 1412 at the foot. These fragments are for the Paris meridian, and might come from an intermediate work of 1330 [30].

More interesting are the London tables (ff. 5^r - 33^r), accompanied by a short canon (ff. 2^v - 4^r), and with an addition at 4^v, the latter beginning 'Ad noticiam tabularum sequencium secundum quod ...' These tables, as already explained, are adapted to London by taking the town to be 32^m from Toledo in time. No reference is made to Alfonso, but, allowing for appreciable miscopying, there is reasonable consistency between these London radices (given both for the time of Christ and for 1336) and the various Parisian lists. The radix dates are not of the cyclical type, and the 'anni expansi' follow the decimal style of John of Lignères' 'tabulae magnae'. Planetary equations are of the traditional sort, however, and are characteristically Alfonsine [31].

(iv) The (incomplete) tables ascribed to Rede in the English MS. Laud misc. 594 (see p. 274 above) could well derive from an earlier, or at least independent, adaptation of the tables with John of Lignères' canons 'Cuiuslibet arcus'. It seems very probable that the latter were those on which Rede based his work, directly or indirectly. I note that in giving essentially the same table of auges as John of Lignères, William Rede rounds off to seconds the column of aux movement per annum, whereas it had earlier been given to seconds and thirds of arc. In some interesting tables in Bodleian MS. Ashmole 393 (where the aux table runs from 1400 x 20 to 1680, but where a table of access and recess starts at 1320), aux positions (and all differences) are given to fourths of arc. See f. 24^r, section I. In the Bodleian MS. Digby 97, among a miscellany of tables copied in the late fourteenth century, a gloss at the table of auges in what is ostensibly a set of Rede's tables explains that the auges are 'non secundum Rede, set verius'. In fact they are modified trivially by not more than a second of arc in any one case.

(v) An English copy of the sexagesimal tables of John of Lignères is in British Museum, MS. Sloane 407, as the first among a rich volume of fourteenth- and fifteenth-century tables from many sources. Here there are the 'Quia ad inveniendum' canons, now barely legible (14th century, ff. 4^r - 6^v), with the tables (ff. 8^r - 33^v, a table of longitudes of towns being included at f. 7^r). These have very densely written canons, now faded, round the margins. Oxford radices have been added by a later hand to the margins. There is a full set of tables of equations. Another English copy of the material, but of no great interest, is in British Museum, MS. Royal 12. D. VI [32]. Radices are here given for Toledo and York. A good English set of the 'Quia ad inveniendum' canons and tables is in MS. Digby 168 (tables ff. 139^r - 144^v and 146^v; canons ff. 145^r - 146^r), but this set is not modified in any way. Many more English codices could be named with wholly sexagesimal Alfonsine tables, and it would be wrong to suppose that Rede's tables had a monopoly of English astronomy.

Although it would not be difficult to add to this list of English adaptations of the early (c. 1320) Parisian Alfonsine material, we now have evidence enough that William Rede's contribution to the Alfonsine element in English astronomy was wholly unoriginal. That he had the good fortune to collect together one version of the tables, and to write a simple set of canons which were to be much copied, can best be explained on the grounds of his Oxford position, rather than of any intrinsic merit in his work. And once and for all, it should be pointed out that the multiplicity of rubrics and incipits for his writings, as reported in the several biographies and bibliographies in which his name appears, has obscured the fact that as far as astronomy goes we know him to have written only (i) the one short set of astronomically trivial canons; (ii) the few calculations needed to adapt the Alfonsine tables to Oxford, which had almost certainly already been done by others; (iii) an almanac of the Sun covering the years 1341 to 1344; and (iv) a few other isolated computations of celestial events [33]. This is a slender basis for an astronomical reputation of the sort Rede has somehow achieved.

It is not possible to say precisely how the Parisian tables came to the different English centres of learning. There is no mystery about the exchanges of knowledge in general between England and France in the fourteenth century. In the eighteenth century, however, Bishop Thomas Tanner, basing his argument on references in a treatise by Geoffrey of Meaux in MS. Digby 176 (ff. 26^v - 27^r) to the times of eclipses of the Moon at Oxford in 1325 and 1345, suggested that he was in Oxford in person in those years. Duhem and Thorndike both enlarge upon the idea [34]. Notwithstanding Geoffrey's reluctance to use the Alfonsine parameters in 1320, might he in fact have had a change of heart, and have actually taught the new ideas in Oxford? Although Royal Physician at the coronation of Charles V, Geoffrey writes as though he customarily lectured, and he is known to have been defended on a matter of astrology by a clerk from the university of Paris[35]. An immediate reaction to Tanner's remark is to look for Oxford tables in which the eclipses are predicted with the properties assigned to them. Geoffrey of Meaux gave the time of the eclipse - now past - to minutes and seconds. In fact all the predictions I have found are several hours wide of the mark, whereas the times given by Geoffrey's for 1345 correspond closely with those in Oppolzer's 'Canon der Finsternisse', which are unlikely to be more than fifteen minutes in error. There are, however, no precise details given for 1325, and it is not improbable that Geoffrey was quoting Oxford data (about astrological houses), merely because he wished to draw conclusions about the English nation. He could easily have taken the data from a written record. Even so, Tanner's suggestion still seems sufficiently plausible for the whole question to be left open.

New styles in planetary tables

Having mentioned some tables for the longitude of Salamanca, with 'anni collecti' running from 1348 (x 20) to 1628, and some similar tables done for the meridian of Prague by Albert of Brudzevo (Copernicus' teacher at Cracow), Dreyer asked himself why the 'anni collecti' should start at 1348. He decided that 'most probably this arrangement dates from Oxford, and was due to William Rede, since several sets of tables for the meridian of Oxford start with radices for 1348 and motion in "anni expansi" up to 10.000' [36]. In a footnote referring to three sets of tables with 1348 radices - but none, as he should have seen, having any connection with William Rede - Dreyer asked whether 1348 was chosen as 'the year in which the Black Death came to England'. 'Is that perhaps the reason', he went on to ask, 'why it was chosen as an epoch, like the Deluge formerly?'

The idea has a certain romantic charm, but is no more than conjecture. What is certain is that the 1348 tables are very different from ordinary astronomical tables, their author having gone one stage further in the direction defined by John of Lignères' 'tabulae magnae'. The tables of 1348 seem to have been little heeded outside England. Apart from the works of John of Lignères, the tables and canons 'Tempus est mensura motus' of 1327 [37] by his pupil John of Saxony soon penetrated to every corner of Europe, including Spain, eventually to become even better known in the printed editions. A rarer work by John of Murs, written ostensibly in 1339, beginning 'Prima tabula docet differentiam unius ere', is along conventional lines. Admittedly John of Speyer (perhaps also in 1348 [38]) wrote a commentary ('Circa canonem de inventione augium...') on John of Lignères' 'tabulae magnae' with their canons 'Multiplicis philosophie', but the commentary is now also rare, as are the canons and tables which gave rise to it [39]. Literally scores of variants on the different canons and ordinary tables began to appear at this time, contributing nothing to the underlying principles of calculation. It was the last-named work of John of Lignères (dedicated to Robert the Lombard of Glasgow, it will be remembered) that seems to have led an English movement towards simplification of the process of calculating planetary longitudes.

As has already been explained, the 'tabulae magnae' included double-entry tables yielding a single combined equation. We may now write the planet's longitude (λ) as a function of mean centre ($\bar{\gamma}$), mean motus ($\bar{\mu}$), and mean argument ($\bar{\alpha}$) as follows, where L denotes John of Lignères' table for the combined equation (taking into account the correct algebraic sign):

$$\lambda = \bar{\mu} + L(\bar{\gamma}, \bar{\alpha}). \qquad (1)$$

The Oxford double-entry tables of 1348 were written out in a superficially similar way, except that $\bar{\gamma}$ was now down the left, while $\bar{\alpha}$ was at the head of the tables. The real difference, however, was that, with some slight qualification, the Oxford tables gave λ directly. If ω_0 denotes the longitude of the planet's aux at some fundamental moment of time, and B the function embodied in the Oxford tables, then we may write

$$B(\bar{\alpha}, \bar{\gamma}) = \omega_0 + \bar{\gamma} + L(\bar{\gamma}, \bar{\alpha}), \qquad (2)$$

assuming - as I shall show is the case - that the same eccentricity and epicycle radius were used in both sets of tables. The 'canons' for the Oxford tables, beginning 'Vera loca omnium planetarum in longitudine', show no sign of any dependence on John of Lignères' canons, which are at the same time longer, in much better Latin, and yet in some ways less thorough than the 1348 canons. There can be little doubt, however, that the earlier set of tables is the source of the later. The 'anni collecti et expansi' are arranged similarly on both occasions (1, 2, 3,... 20, 40, ... 100, 200, ... 1000, 2000, ... 10000 in the Oxford case), for example; but in the end it is the system of equations adopted, and their evaluation, that settles the matter.

Before considering more closely the contents of the longitude tables, and in particular those of true place (or combined equation), we ought to consider pos-

sible Islamic influences on John of Lignères. Claus Jensen has recently described
a set of lunar tables with double-entry for a single combined equation, and using
a parabolic law of interpolation, from the thirteenth century Zīj of al-Baghdādī [40];
and even more recently D.A. King has described a double-argument table for the
lunar equation, in a work attributed to the tenth-century Egyptian astronomer Ibn
Yūnus [41]. From references to the works of the priest Qiryāqūs, the Egyptians
Ibn al-Majdī and Shams al-Dīn al-Sūfī, and the Samarqand astronomer al-Kāshī,
all four of the fifteenth century, it is clear that there was an established tradition of this sort of table for the lunar equation in the Islamic world [42]. The
'Zīj-i Khāqānī' by al-Kāshī has been studied by M. Tichenor, who finds that similar tables are outlined for all the planets - with mean centre ($\bar{\gamma}$) down the left,
and mean argument ($\bar{\alpha}$) along the top, as we find in the 1348 tables - but that in
the two manuscripts considered only the tables for Venus and Jupiter have been
filled in [43]. Here, as with al-Baghdādī, the interpolation procedures go beyond
the linear methods of the European tables: interpolation with respect to $\bar{\alpha}$ is linear, but with respect to $\bar{\gamma}$ it is non-linear, and there are special directions for
interpolation across an extremum. Al-Kāshī's parameters for Venus are different from the Ptolemaic and the Alfonsine. If there was ever any connexion between the European and Islamic occurrences of tables of combined equations, it
remains to be uncovered. As David King has pointed out, al-Kāshī was himself
ignorant of the authorship of the underlying idea [44]. For the time being, it
looks as though the European tradition is quite separate from the Islamic.

The Oxford tables of 1348

Whatever their ancestry, the Oxford tables seem to have had a moderately
vigorous life. There are at least six extant manuscripts with the 'Vera loca omnium planetarum' canons, or the tables, or both; and one of these manuscripts (of
the fifteenth century) provides evidence of considerable concern with the tables,
which are provided with a complete set of worked examples, much supporting
material on interpolation methods, and a unique copy of an alternative 'canon
optimus', which begins as follows [45]:

> Quicunque voluerit quinque planetas retrogrados equare secundum tabulas
> de Battecombe alias Bredon sive Bradwardynes vocatas. Primo queritur
> centrum medium, etc.

The tables are thus here ascribed to Batecombe, Bredon or Bradwardine; but
the last two can be dismissed on the grounds of the poor Latinity of the 'Vera
loca' canons, and indeed Bradwardine's knowledge of astronomy was not, in
my opinion, equal to the relative sophistication of the work [46]. Earlier in the
codex there are the worked examples, beginning:

> Exemplum precedentis capituli quo calculabis in Sole, Luno et Saturno et
> sic in ceteris pluris anno Cristi inperfectis ... 4º die mensis Augusti,
> per tabulas Battecom predictus que dicitur eius almonacke perpetuum.
> Quere medium motum argumenti Solis ... [47]

Batecomb's name is now mentioned alone, and it is worth noting the idiosyncratic use of the word 'almanac', which echoes that in the title given to John of
Lignères' canons ('super magnum almanach') in the Gonville and Caius copy
of 'Multiplicis philosophie'. The copyist of the later work, one Thomas Cory,
monk of Muchelny (who dates various parts of the copy as 1440, 1441, and 1459)
[48], clearly has little idea as to the name of the author, however, for at

another place [49] he seems to settle for Bredon. As for the internal evidence of the 1348 canons, they simply include a statement in the opening paragraph that the tables were 'ad hoc in Oxonia constituta', and the same form of words is often used as a running heading to the tables in MS. Digby 57. In MS. Bodley 432, where the beginning is missing, the work ends with fulsome praise thus:

> Explicit almanak nobilissimum a diversis et sapientissimis elaboratum ingeniis facilimum atque perpetuum ... [50]

In MS. Laud misc. 594, there is at the head of the tables a simple 'anno cristi 1348º ad meridiem Universitatis Oxoniensis' [51]. This last ascription is perhaps as far as we can reasonably go. Virtually nothing more is known of William Batecomb than is to be found in the rubrics under discussion [52]. His concern with the problem of simplifying planetary computation is suggested by the ascription to him of a lost work on the equatorium [53], but this was a concern he shared with many others. (Perhaps he simply passed off John of Lignères' equatorium as his own!)

An analysis of the tables of mean motion is easily given. Mean centres, rather than mean motus are tabulated. There is no difference except in the radices, of course, unless the aux be regarded as moving. Corresponding tables associated with 'Multiplicis philosophie' are headed 'motus' rather than 'centrum', and yet read the same. In fact the 1348 canons and tables make allowance by advocating the inclusion of a (tabulated) quantity which we may write as $\delta\omega$, where ω signifies the longitude of aux. Mean argument is of course usually tabulated only in the case of the Moon, Venus, and Mercury, and evaluated as the difference between the mean motus of the Sun and mean motus of the planet in the case of Saturn, Jupiter, and Mars:

$$\bar{\alpha} = \bar{\mu}_s - \bar{\mu}. \tag{3}$$

In a majority of surviving copies of the 1348 tables, mean argument is actually also tabulated for the superior planets [54]. The mean centres of the Sun and Venus are, on the Ptolemaic theory, identical. (The mean centre of the Sun is referred to as 'argumentum Solis'.) If, now, $\bar{\gamma}$ is the mean centre, as before, then for the superior planets

$$\bar{\alpha} = (\bar{\gamma}_s - \bar{\gamma}) - (\omega - \omega_s), \tag{4}$$

while for Mercury,

$$\bar{\gamma} = \bar{\mu} - \omega = \bar{\mu}_s - \omega = \bar{\gamma}_s - (\omega - \omega_s). \tag{5}$$

The quantity here denoted by $(\omega - \omega_s)$ is, in the treatise 'Vera loca omnium planetarum', given the name 'addicio communis'. It was reckoned to be constant by most astronomers, who thought the planetary auges to move with the fixed stars. The 'common additions' tabulated for the planets are as follows [55]:

Saturn:	161;º	58,	19
Jupiter:	82;º	11,	37
Mars:	43;º	46,	50
Mercury:	119;º	14,	10.

These are interesting data, for (unlike aux positions) they are independent of any precessional term, and therefore allow us to judge of their source. They all differ appreciably from the data of the Toledan tables, but agree precisely with the differences derivable from the leading 'Alfonsine' tables - such as those of John of Lignères' 'Multiplicis philosophie', John of Saxony, and William Rede. As for the 1348 auges, had they been found from Rede's tables they would all have been 1" lower, and one or two marginalia in the copies I have seen show an awareness that Rede's figures differ somewhat [56]. These several comparisons might seem superfluous - until we find, as we do subsequently, that the equations implicit in the tables of true place are not wholly compatible with all 'Alfonsine' prototypes.

The procedure for finding the longitude (λ) of a planet is now very straightforward: mean centre and mean argument are found directly from appropriate tables. (If a full set is not available, then the 'common differences' of equations (4) and (5) will be used). All Alfonsine tables seem to incorporate the secular precessional term in the mean motions. One must nevertheless modify all centres and arguments where necessary by a term for trepidation. Once $\bar{\alpha}$ and $\bar{\gamma}$ are found, the 'table of true place', earlier denoted here by B ($\bar{\alpha}, \bar{\gamma}$), gives λ immediately [57] (see equation (2)) except inasmuch as the position of aux has changed since the fundamental epoch. The difference ($\delta\omega$) between the new and old positions is simply the movement of the eighth sphere, however, and for this there are the usual Alfonsine tables [58]. (The 'table of true place' is therefore not strictly so named.) Thus in the end we have

$$\lambda = B(\bar{\alpha}, \bar{\gamma}) + \delta\omega. \tag{6}$$

For the Sun, the procedure is simpler, since there is only one argument. For the Moon, the elongation ($\bar{\eta}$) (from the mean Sun) is first found in the usual way, after which the procedure is to add the 'true place' found from the table (namely B ($\bar{\alpha}, \bar{\eta}$)) to the position found for the mean Sun. There is now, that is to say, no need to add for the movement of the eighth sphere.

One of the advantages of the 1348 tables is that they can be easily used to carry information about the direct motions, stations, and retrogradations of the planets, and whether they be in an easterly or westerly part of the epicycle. A chapter of the 'Vera loca' canons is devoted to these matters, which had an astrological import, although this was not alluded to by their author. It is perhaps worth pointing out the unusual way he has of referring generally to the planets, namely as the 'retrogradi'.

In a note at the end of these canons of 1348 there is a statement summarizing rather baldly equation (2) above, under the heading 'modum autem componendi tabulas precedentes ostendere'. The note suggests that such tables as these may be compiled by combining mean centres, arguments, equations and auges, on the basis of ordinary tables of Alfonsine equations. That the writer drew up his own tables in this way is unlikely, as I shall show. And it is even less likely that he calculated his tables from first principles.

The function B like the earlier function L, has a form which it is easy for us to write down analytically, even though it was not evaluated in this way when the corresponding tables were drawn up. We can certainly analyse the tables to extract from them the underlying parameters by modern methods applied to the Ptolemaic models; but if we do so, we run the risk of ignoring inconsistencies of a sort that medieval tables are known to contain [59]. Within one or two minutes of arc, all derived values for the epicycle radius in the tables which are of greatest relevance to ours (those of the 'Almagest' (a), al-Battānī (B), the Toledan tables (T), and the Alfonsine (H)) are alike. The same is true of the eccentricities derived from the tables for Saturn, Mars, and Mercury. For the Sun, however, taking always a standard radius of 60 parts, we find an eccentricity of 2;30 in A, 2;05 in B and T, and of 2;16 in H. For Venus, we find 1;15 in A and H, 1;02 in B and T, and, by another method, 1;08 in H. This double standard in H is found again in the Jupiter tables, those of A, B, T, and H all implicitly containing a figure of 2;45, but H also embodying 3;07. The Moon is a yet more difficult case, an eccentricity of 12;28 (as on the Canobic inscription) being found by one method from all tables; 12;46 (all tables) by another; 12;32 in B and H by a third method; and yet further figures of 12;23 (B and T) and 13;39 (H) using other criteria. The reason for this diversity is that different columns (equation of centre, equation of argument, proportional minutes, etc) were calculated by different people, or at least from different parameters. It ought to be emphasized, however, that the parameters abstracted here are not necessarily those on the basis of which the tables were calculated, for the methods used to abstract them assumed initially in some cases a measure of consistency in the model - in the cross-linking of columns, as it were. The parameters are simply offered as a 'reductio ad absurdum', proving inconsistency, and as a rough guide to influence when we make a comparison with other tables. It should also be noted that by 'Alfonsine tables' I mean those which became canonical - namely the later Parisian tables, Rede's version of them and so on.

Turning now to the 1348 'tables of true place', simple methods of abstracting comparable parameters offer themselves, even without assuming total consistency in the tables from which they were compiled. It should be clear from their structure that the entry corresponding to $\bar{\gamma} = 0$, $\bar{\alpha} = 0$ is in all cases the longitude of aux for 1348 (or rather noon, 31 December 1347) [60]. In all cases there is agreement with the separately tabulated aux positions. Keeping the zero value of $\bar{\gamma}$, and considering 'true place' for $\bar{\alpha} = 90°$ and $\bar{\alpha} = 270°$, figures should be obtained at equal angular distances from the position of aux. Let these distances be denoted by z_1, with $t_1 = \tan z_1$ [61]. Taking next the entry for $\bar{\gamma} = 180°$, $\bar{\alpha} = 0$, we have the opposite of aux, that is, perigee of the epicycle. (The tables are always consistent in making the new longitude 180° removed from aux.) Holding to the same value of $\bar{\gamma}$, and considering $\bar{\alpha} = 90°$ and 270°, as before, we derive new separations z_2 from the mean, with $t_2 = \tan z_2$. It is now a simple matter to show, on the basis of the four Ptolemaic models, that in all cases (except of course the Sun),

$$r = (60 - e)t_2, \qquad (7)$$

where r is the epicycle radius and e is the eccentricity. For all planets except Mercury (but including the Moon),

$$e = 60(t_2 - t_1)/(t_1 + t_2), \qquad (8)$$

and for Mercury,

$$e = 60(t_2 - t_1)/(t_1 + 3t_2). \qquad (9)$$

Applying these equations to the data found in the tables of 1348, we find, as was to be expected, that values close to the usual ones are derived for the epicycle radii. Much the greatest error is that in the case of the Moon, and even there the error is barely 1 per cent. It is unlikely that any of the errors is significant. For the eccentricities we similarly find much that was to be expected, with values close to the widely used 3;25 for Saturn, 6;00 for Mars, 3;00 for Mercury, and 12;45 for the Moon. Again in the case of the Moon there is an error of somewhat large dimensions (0;02,39), but one which, as in all other cases, is a consequence of the relative crudity of our method. For the Sun, a characteristically Alfonsine figure is obtained within reasonable limits of error (the figure is 2;16,12 - cf. the Ptolemaic 2;30 and the Toledan 2;05). For Jupiter and Venus, however, there are dangers inherent in our simple method. While for Jupiter we find an eccentricity close to 2;45, a figure to be found in 'Almagest', the Toledan tables, and the Alfonsine tables of the equation of the argument, it should be clear that the equations (7) - (9) are incapable of indicating any different parameter built into the equation of centre. We know that the Alfonsine tables do indeed there make use of an eccentricity of 3;07 for Jupiter. The same caveat applies in the case of Venus, for which planet, although we find an eccentricity of about 1;08 by our simple method, we should expect to find the Ptolemaic 1;15 in the equation of the centre [62].

This is precisely what we do find, on looking more closely at the tables. Before explaining briefly how this is so, we can simplify the discussion by anticipating agreement between the equations incorporated into the 1348 tables and those of the tables which John of Lignères had sent to Robert the Lombard almost thirty years before. Those earlier tables, which for brevity I shall call the 'tabulae magnae', are now extremely rare, and there appears to be no copy with the equations in any British library [63]. Approximate initial values for the underlying parameters can be extracted from them along similar lines, but much more rapidly, since our z_1 and z_2 (for example) can now be taken from them immediately. Proceeding as before, with data taken from the Erfurt copy, we find that for all planets except the Sun (where the equation is now tabulated to seconds) and Moon (unfortunately not represented in the copy in question), the values of z_1 and z_2 agree to the minute with the 1348 values.

This is not a conclusive proof of borrowing. Much more to the point is the fact that the 'tabulae magnae' and the 1348 tables are found to be completely compatible over their entire range, with the exceptions of a few obvious errors of calculation and transcription. Another superficial indication of borrowing is to be seen in the fact that both sets of tables have columns and rows in the double-entry tables at intervals of six degrees, except where the Mars table (between 120° and 180°) and the Venus table (between 150° and 180°) switch to intervals of three degrees for the mean argument. This is not a common practice, but is admittedly reminiscent of Ptolemy, although not quite the same as his procedure in the 'Almagest' [64]. (The Alfonsine and Toledan tables give the ordinary equations at intervals of a single degree.)

Turning now to the Venus and Jupiter tables of the 'tabulae magnae', we can first verify the results already outlined, in a more careful analysis of two complete columns. For every value of the mean argument there will be a useful pair of entries, one under a mean centre of 0° and the other under 180°. From each of these conjugate pairs it is a simple matter to derive values for r and e. With our previous notation, and now denoting by t' the reciprocal of t, i.e. the cotangent of the angle, we first find r from the equation

$$\frac{60}{r} = \frac{(t_1' + t_2')}{2} \sin \bar{\alpha} - \cos \bar{\alpha} \qquad (7')$$

and then e from

$$e = \frac{(t_1' - t_2')}{2} r \sin \bar{\alpha}. \qquad (8')$$

For Jupiter, for example, we obtain highly consistent results for e and r. Discarding four of the twenty-nine pairs of entries on the grounds that they differ from those of the 1348 tables, and hence that one or the other is miscopied, we find for r a mean value of 11;30,24, with a standard deviation of only 0;00,47. For e we find a mean value of 2;41,56, with a standard deviation of 0;07,30. These are the values we expect. The only pair of matched equations which give for e an aberrant result is the last of the twenty-nine — and this gives 3;07,29, the familiar alternative which we can expect to find when we look beyond the two isolated columns of the equations in the 'tabulae magnae'. We do this by holding $\bar{\alpha}$ constant, and considering the pattern of variation of the combined equation as $\bar{\gamma}$ is allowed to vary. The curve is sinusoidal, with turning points close to 90°

and 270° for the mean centre. From the maxima, accepting no a priori value for r, we can derive e. If we take the radius of Jupiter's epicycle to be the ubiquitous 11;30, and take a zero value for the mean argument, we have a particularly straightforward (but quadratic) solution for e. By all methods we find a value very close to 3;07.

Both here, and with the Venus tables, we get no more than we expected - namely 'Alfonsine' results. There is every reason, obviously, to believe that John of Lignères simply calculated his tables in the standard manner from the 'Alfonsine' tables of equations. This would have been more difficult to prove from a simple examination of the tables alone, attempting to repeat his calculations, since it is not at all clear from any individual case that he rounded off his intermediate data in any particular way. In the knowledge that he was indeed using the Alfonsine tables, however, we are able to form an opinion as to how, when finding the equation of centre, he combines it with the mean argument to give true argument (which will thus in general no longer be an integral number of degrees) and then interpolates in the table for the equation of the argument. He seems, in fact, to alternate between taking a correctly interpolated figure and taking the closest entry in the table. Occasionally also he seems merely to round off to the lower entry. In view of this evidence, and in view of the double standards in all these Alfonsine's tables of equations for Jupiter and Venus, the hazards of laying down categorically 'Alfonsine' planetary positions - for example, in attempting very precise solutions of historico-literary problems - are plain to see.

The planetary tables of John of Murs

The planetary tables of John of Lignères and the Oxford tables of 1348 were not the only ones with double-entry to be produced in Europe in the late middle ages with a view to reducing the tedium of calculation. At much the same time as John of Lignères was combining the two Ptolemaic equations into one, John of Murs was designing a very different procedure, clearly inspired by an older method for lunar longitude. In general terms we may say that where John of Lignères and the author of the 1348 tables - and also, as we shall see, John Killingworth - looked for symmetry around the line of aux, that is to say, symmetry in the function $\bar{\gamma}$, John of Murs looked for symmetry with respect to the mean argument, $\bar{\alpha}$. Turning back to equation (3), it should be clear that the period of this function is the interval between successive conjunctions of the mean planet with the mean Sun. John of Murs tabulated these mean conjunctions (for the meridian of Toledo, according to the standard Alfonsine radices and motions), and for dates in the years 1319 or 1320 and in appropriate years later in the century [65]. A typical line in one of the conjunction tables ('tabulae principales') can be transcribed as follows:

$$1320, \text{day } 11| = \text{Jan.} 11^d 18^h 55^m | \bar{\mu}_o = 299°40 | \bar{\gamma}_o = 128°41 | \lambda = 295°36.$$

This is the first line of the table for Jupiter. The suffix 'o' is used here to indicate values at the moment of mean conjunction. The year (1320) is complete (i.e. we are in 1321), but the month not. Noon of 11 January has passed, and $18^h 55^m$ more; i.e. we are in 12 January, civil reckoning. (Almost all of the 'Alfonsine' astronomers use this astronomical system, working from the last noon of the previous year.)

With each planet's 'tabula principalis' there is an extensive 'contratabula', a double-entry table in which $\bar{\gamma}_o$ (at intervals of 12°) is the abscissa and the 'age' of the planet the ordinate, given in weeks and days (to the nearest day) for approximate use, and in days, hours, and minutes. In the body of the table is given the difference between the true place (λ) of the planet at its particular 'age' and its mean motus ($\bar{\mu}_0$) at mean conjunction. The 'contratabula' has 30 lines for equal intervals of the 'age'. (Thus the last line of the Jupiter tables is for $398^d\ 21^h\ 12^m = 30$ x the first line, that for $13^d\ 7^h\ 6^m$. The first figure is only 8.73 seconds below the Alfonsine synodic period of Jupiter, that is, the period to cover 360° in mean argument 12° in excess of the previous line, and also to a value of the mean centre of a different but constant amount in excess of that for the previous line. (For Jupiter, for instance, the tacit increment in $\bar{\gamma}$ will be $1;^o06$.) There is a column for the 'semidifferencia', this being half the difference between successive entries on the same line, namely for a change of 6° in mean centre, and yet there is clearly a precarious piece of double interpolation to be undertaken in using the 'contratabula'.

If conjunction between mean planet and mean Sun always took place at aux, there would be need for only one column in the 'contratabula'. Each entry would correspond to a given mean centre and a mean argument, each starting from zero at aux, and for each pair of values there would be a definite combined planetary equation, and therefore a definite correction to be added to $\bar{\mu}_0$ (which would then be the longitude of aux, in fact). But $\bar{\gamma}_0$ (the value of $\bar{\gamma}$ at mean conjunction) is not in general zero, and a separate column is in principle needed for each initial value $\bar{\gamma}_0$. An example will show the way of calculating a typical entry for the simplest column, namely $\bar{\gamma}_0 = 0$. The tenth line in a Jupiter table corresponds to values $\bar{\alpha} = 120^o$ and $\bar{\gamma} = 0^o + 11;^o03, 10$. The latter figure, used in Alfonsine tables of equations, gives as equation of centre $\xi = 1;^o05$, wherefore $\alpha = 121;^o05$. From the same standard tables: proportional minutes $\Pi = 60$ ('longior'); 'diversitas diametri' $\Delta = 0;29$; and therefore the 'pars proportionalis' $\delta = 0;^o29$ [66]. The basic equation of the argument is $10;^o19$, wherefore the corrected equation of argument $\zeta = 10;^o19 - 0;^o29 = 9;^o50$. Combining the equations ξ and ζ with $\bar{\gamma}$ (subtracting the first and adding the second), we find $19;^o48$ as the entry for the 'contratabula'. This is exactly John of Murs' figure, which therefore embodies the 'Alfonsine' eccentricity for Jupiter.

It is difficult to judge how widely known was the method of John of Murs [67]. Surviving manuscripts are certainly few, although his related works on lunae-solar calculation and on the calculation of eclipses are somewhat more numerous [68]. So far as I am presently aware, the latter half of the fourteenth century produced in northern Europe nothing at all resembling any of the three systems so far described. There was continued concern with the general problem of producing planetary almanacs (ephemerides), but further refinement of the fundamental Ptolemaic methods was seemingly wanting. One might mention such interesting trivialities as the calendar of daily mean motus and mean argument throughout a whole year calculated for the Oxford meridian (radices 1374 -) and surviving in perhaps only manuscript [69]; but this sort of thing is of a much lower intellectual order than the methods we have been discussing. There is one further example to be considered, however, of tables showing a mastery of the mathematical techniques of the time. John Killingworth's work of the mid-fifteenth century, done shortly before his early death (1445), attained to an unusual level of originality and sophistication.

The planetary tables of John Killingworth

Killingworth was perhaps the best of the Merton College astronomers. He was described in the 'Catalogus vetus' of the college as 'nobilis astronomus qui multas tabulas composuit', and most of the other works with which he is credited have some bearing on tables [70]. Thus his 'Algorismus', discussed many years ago in a short article by L.C. Karpinski, includes an example of the continued addition of the mean motus of Saturn for eight-day intervals - precisely as required in his astronomical tables [71]. The date of the example is given in the text itself as 1444. John Killingworth was probably only in his mid-thirties when he died, a year later, and his tables were probably prepared within a year of two of 1442 [72]. Although known copies are not numerous by comparison with those of the standard Alfonsine tables, Killingworth's tables survive in greater numbers than those discussed here of John of Lignères, John of Murs, or the author of 1348. The ascription to Killingworth is secure. It is in four of the manuscripts, even though the associated canons, beginning 'Multum conferre dinoscitur ...' are twice described as 'editi per Magistrum [Thom.] Pray'. (Pray was a fellow of University College, and somewhat younger than Killingworth. It is not improbable that Killingworth died before completing the canons.) The untiring Lewis of Caerleon copied the tables, calculated new tables for the sixteenth century, and wrote the short supplement, 'Restat de compositione tabularum ...', all perhaps half a century after the original work was done [73]. The most sumptuous copy I have seen of any set of astronomical tables, namely British Museum MS. Arundel 66, opens with Killingworth's work, heavily interlined with gold leaf. This volume, associated with Duke Humphrey of Gloucester, has the canons ending with these words:

> Expliciunt canones tabularum facilis composicionis Almank [sic] secundum modum Universitatis Oxoniensis. Et incipiunt tabule que dicuntur Kelyngworth.

As so often before, there is concern not merely with the problem of calculating occasional planetary positions, but with the calculation of complete almanacs, i.e. ephemerides. Killingworth's methods were designed with this general purpose in mind.

Like the tables of John of Lignères and the Oxford tables of 1348, those by John Killingworth make use of symmetry around the line of aux. The first step in their use is to convert the date in question to a sexagesimally expressed number of days from the Incarnation radix. (This is done through a 'tabula reductionis annorum Christi' [74] and a table of months. Thus 1 February 1461 = one day in excess of January (complete) of 1461 (1460 complete) = 02, 28,07,48;00 days plus 31;00 days plus 1;00 day.) [75] Turning to the second set of tables ('tabule radicum') we find, roughly speaking, the days elapsed from 'radix Christi' when the mean planet was last at aux. This difference, that is to say the time from aux, will later be used in the 'tables of revolutions' to give those additional movements which are to be combined with that at the time when aux was reached. But first to the 'tabule radicum'.

For the planets Jupiter and Saturn there are three short tables. For Saturn, for example, they are headed (a) 'Tabula radicum Saturni'; (b) 'Tabula correctionis radicum Saturni'; and (c) 'Tabula reductionis radicum Saturni ad meridiem'. For the Moon there is a table only of type (a), and for the Sun, Mercury, Venus and Mars of types (a) and (c) only. Table (a) gives the times (rounded to minutes of a day) when the planet is at aux. Increments would in principle be constant, were it not for the irregular movement of aux, due to trepidation [76]. (For the Moon, the interval is for 49 'synodic' periods.) The second column of dates (a) lists 'motus ad proximam meridiem' (degrees and minutes), which is the radix of mean motus ($\bar{\mu}$) for the noon nearest the moment of passing aux. The third column gives radices for mean argument ($\bar{a}$) for the same times. The fourth column gives 'motus pro tempore vere radicis' (except for the Moon);

that is to say, in contradistinction to the second column, the fourth gives the motus for the radix actually listed in the first column. A final column, for Jupiter and Saturn only, gives the movement of aux in the interval.

Tables (b) and (c) are for correction of entries in (a). The motion of Saturn and Jupiter is so slow that their auges move appreciably (with the eighth sphere) in any significant fraction of the cycle, and table (b) corrects for this, giving a term to be added to, or subtracted from, $\bar{\mu}$ and $\bar{\alpha}$. Table (c) is of a wholly different sort and order of sophistication. As a device for dealing with planetary anomalies, it must surely be reckoned one of the most ingenious productions of the later middle ages. The constituent tables of (c) are unpretentious, and for some planets are not even calculated with much care - although this impression is partly due to John Killingworth's decision to round off to minutes of arc alone. Their merit is to be seen in the conception rather than the execution of the tables. Referring to fig. 1 (where the separation of points is greatly exaggerated, for clarity), let us suppose that A marks the aux (apogee) of the deferent circle, and P the perigee. T is the Earth, and E the equant point. We recall that John Killingworth's prime concern was with ephemerides, and that he listed $\bar{\mu}$ and $\bar{\alpha}$ at the noon nearest the time the mean planet reached aux. Suppose, now, that one wishes to find a planet's position at some later noon. There is an obvious gain in simplicity if we work only with whole numbers of days. This will introduce an error, however, due to the fact that $\bar{\gamma}$, the mean centre (which is implicit in the time interval used in subsequent tables), will be misjudged by an amount corresponding to the movement of the mean planet between aux and its position at the nearest noon. At some later noon, for example, the centre of the epicycle might be at B, with equation of centre ξ_1. The integral number of days used for the calculation, however, will imply that the centre of the epicycle is rather at C, with equation ξ_2. Table (c), in effect, is a table for the error $\Delta\xi (= \xi_2 - \xi_1)$, by which both the motus and the argument will in due course need to be corrected. More precisely, the table gives a multiplying factor proportional to $\Delta\xi$. It is accepted by John Killingworth, on the basis of reasoning at which we can only guess, that there is strict proportionality between $\Delta\xi$ and the initial error (which I will call $\Delta\bar{\gamma}$) in the mean centre. (The initial error is known from the 'tabula radicum'.) For small angles this is true. Abbreviating the sine and cosine of $\bar{\gamma}$ to s and c respectively, and writing 60/e as k, we can ourselves proceed from this expression for the equation of centre:

$$\tan \xi = s/(k+c) \tag{10}$$

to

$$\Delta\xi = \Delta\bar{\gamma}[(1 + kc)/(1 + 2kc + k^2)]. \tag{11}$$

In fig. 2 I have plotted (as points) the data recorded in John Killingworth's 'tabula reductionis radicum ad meridiem' for Mars, and have drawn against them a continuous curve plotted from equation (11), and confirming both the significance of the table and its author's grasp of his materials. As to the general pattern of the error curve, it should be clear from fig. 1 that the difference between ξ_1 and ξ_2 is least in the neighbourhood of D, and that in ratio to $\Delta\bar{\gamma}$ (an angle at E rather than T), it is somewhat greater when B and C are near P than when they are near A.

The next tables of importance are the 'tabule revolutionum', in which, as a function of the time (in days), quantities are tabulated under the following headings: (col. 1) 'motus correctus'; (col. 2) 'minuta proportionalia'; and (col. 3) 'argumentum verum'. Since time from aux is directly proportional to the mean centre, a definite time corresponds to a unique value of ξ, the equation of centre, and therefore to a definite (corrected) movement ($\delta\mu$) in longitude from the radix date. This is given in col. 1. By the same token, a corrected value for the change in argument ($\delta\alpha$) can be stipulated. This is given in col. 3. These quantities are listed at intervals of 3 days (Sun, Mercury, Venus), 4 days (Mars), 6 days (Jupiter), 8 days (Saturn), or daily (Moon). Simple interpolation is therefore necessary at this stage. The radix values of $\bar{\mu}$ and $\bar{\alpha}$ are now to be taken from the 'tabula radicum' and added - with any of the corrections from the 'tabula correctionis radicum' - to $\delta\mu$ and $\delta\alpha$ respectively. It should be clear from the contents of the previous paragraph that while the correction for aux movement (table (b)) can be made to all intents and pupposes before or after the 'tabula revolutionum' is made use of, the nature of table (c), for 'correction to the nearest noon', is such that it is needed after the 'tabula revolutionum'.

The 'true argument' which results is next used in a final table of equations to give the equation of the argument. This table of equations is a table with double entry, with true argument as abscissa (with signs of 30°) and proportional minutes as the ordinate. Combined with the corrected mean motus, this gives the planet's true place. This double-entry table appears - like so much of John Killingworth's work - to be completely original in conception.

Although the method is somewhat tedious to describe, it is relatively rapid to use. Double interpolation is not called for in the tables of equations, which are calculated for intervals of a minute of arc, and which are in consequence very extensive. The calculation of longitudes for other times than noon could, of course, be done by interpolation at the stage of discovering the time-interval from aux. (In fact 'correction to the nearest noon' need not have been done in the manner described, had we been prepared to interpolate to fractions of a day in the early stages.) But John Killingworth again preferred to leave the odd fraction of a day to the end, when, knowing the daily motions, one can interpolate with the help of tables included for that purpose. The most elaborate of these interpolation tables is for the Moon. (The procedure for the longitude of the Moon is not very different from that for the planets, and I shall not comment further on it, or on the tables of latitudes or of aspects included in the complete set.)

Doubts and certainties about the Alfonsine material

The 'tabule revolutionum' of John Killingworth have entries to minutes of arc only, but they allow us to form an idea of their origins. The table for Jupiter, for example, has a 'corrected motus' of 84° opposite the number of days corresponding to a quarter of the full cycle; which is to say that for $\bar{\gamma}$ = 90°, ξ is rounded to 6°. This clearly corresponds to the 5;°57 of standard Alfonsine tables, rather than to the 1;°59 of the Toledan or the 2;°24 of the Ptolemaic. There is a single table of revolutions for the Sun and Venus. (We are instructed when working with Venus to keep the corrected motus and use it for the true place of the Sun.) The values we deduce for ξ from this table correspond closely with the standard Alfonsine figure (2;°10). It is probable that John Killingworth was working on the basis of the John of Saxony tables, or just possibly William Rede's. There is no reason to suppose that he knew the underlying parameters, and there is abundant evidence that many good astronomers were ignorant of them. In MS. Digby 68, for example, a fourteenth-century volume, there is, in the midst of Alfonsine material, a table of eccentricities and epicycle radii which is - apart from some miscopying - wholly Ptolemaic [77]. Exactly the same can be said of an excellent corpus of Alfonsine material in a Brussels codex of c. 1445 [78]. A similar table in the hand of Lewis of Caerleon, towards the end of the fifteenth century, was probably copied uncritically from a work by the Franciscan John Somer, but later in the same volume [79] Lewis performs a calculation unusual for the period, when he derives the solar eccentricity from the Alfonsine tables. (He finds 2;⁰16,26, the maximum equation being 2;⁰10,19.)

The adjustment of the fundamental parameters of the Ptolemaic models accepted by most medieval astronomers required a combination of theoretical abilities and observational skills very rarely encountered between Ptolemy's time and the late sixteenth century. When criticism of the Alfonsine tables was indulged in, it was destructive rather than constructive in character. 'The Alfonsine astronomers had said that such and such would happen on a particular day and at a particular time', the typical argument went, 'but it happened differently, and three days later'. Star positions seem to have been subject to closer scrutiny than those of the planets, on the whole, and consequently criticism of precessional (or rather trepidational) predictions was sometimes made to reflect, rather unfairly, on the planetary tables - unfairly in the sense that the first can be changed radically while minimal changes are made to the second. The medieval criticism of authoritative tables on empirical grounds is a subject which deserves a study to itself. Without trespassing on this separate preserve, there is a short work of a critical character with a bearing on this subject worth mentioning, a work which is as interesting from the point of view of its later history as for its rather unsystematic criticism. The work in question was at length to be incorporated in the printed 'opera' of Nicholas of Cusa, with the title 'Correctio tabularum Alphonsi' (beginning 'Ad sciendum quid veritatis et quid dubii ...'). Following Duhem, it is customary to doubt the ascription, but without good reason, as it seems to me [80]. Duhem argues that the work is simply an accidental concatenation of writings left among Nicholas of Cusa's papers at his death! As authors, he suggests Geoffrey of Meaux (conjectured) William of St Cloud (supposedly paraphrased), and Henry Bate of Mechelen. In fact the sources of passages by the last two are openly acknowledged by the writer. As for the text (about 2600 words long) tentatively but quite improbably assigned to Geoffrey of Meaux, Duhem is referring to the work 'Bonum mihi quidem videtur ...', dated 20 April 1347 [81]. When we compare texts, it turns out that the 'Correctio' merely paraphrases the work of 1347. There is no reason for suggesting that Nicholas of Cusa was in any way original in compiling the 'Correctio' but neither is there any good reason for doubting that he did so; and if so, his readiness to disparage the Alfonsine tables is very interesting. I note that the 'Correctio' mentions 1424 in a way which suggests that it was assembled during that year. Cusanus was then at Padua, where he had just graduated 'doctor decretalium', and where he attended the lectures of the astrologer Prosdocimo de Beldomandi. This in itself merits comment, since Prosdocimo wrote a canon to James de' Dondi's extract from the Alfonsine tables for the meridian of Padua [82], and thus presumably propagated 'docta ignorantia' in the eyes of the future Cardinal. Or was it merely that the young scholar saw criticism of the Establishment as the surest way of attracting attention?

The Alfonsine tables passed, as we have seen, from relative Iberian obscurity to the rank of a great European institution. The means by which they travelled from Castile, to reach Paris half a century after their composition, is never likely to be known, but there is a curious historical note of relevance to this problem, in the 'Bonum mihi quidem videtur'. After considering the longitude of the star 'Cor Leonis', the writer continues:

> Vidi namque librum stellarum fixarum scriptum in Hispanico continentem radices stellarum fixarum eodem modo, qui liber extractus fuit de armario regis Alfontii. Servus (?) dixit mihi qui extrahit eum procuravit. Vidi eciam stellas fixas situatas isto modo in spera solida facta pro ipsomet Alfonsio[83].

In other words, a Parisian astronomer writing in 1347 had seen 'El libro de las estrellas fijas' (1256), from Alfonso's own shelves, as well as a celestial globe made for the king in accordance with it. The tenour of the passage might be thought to suggest that these things were seen in Paris, rather than in Spain; but this is far from clear. One can only suppose that the rapidly growing interest in astronomy evident in Paris in the late thirteenth and early fourteenth centuries had created a demand for new texts of a sort Castillian scholars were able to supply.

Canonized and criticized though the tables became, they were never taken for granted by those scholars who invested untold energy in copying, verifying, and using them. A full set of conventional tables like Rede's, for example, might contain 40,000 digits or more, while a full set of Killingworth's tables might contain a quarter of a million. Lewis of Caerleon, as a prisoner in the Tower of London, no doubt knew the analgesic value of endless calculation, but this can hardly explain away the feverish computational activity of many scores of astronomers throughout Europe and Islam. Two reasons in particular for this phenomenon are so obvious that they need to be emphasized. First, there is a sense of intellectual satisfaction in the creation of a new system - such as John of Murs' or John Killingworth's - or even in the mastery of an established one. Secondly there was the sheer utility of the tables. The medieval scholar was an intensely practical man. He could make nothing but the simplest of astrological prognostications without a knowledge of planetary positions, and if he could not calculate them himself, then he must needs rely on those who could do so. The compiler of ephemerides was a man in whose debt many considered themselves to stand. In those excessively astrological centuries, the sixteenth and seventeenth, his methods of exacting rewards for his labours were raised to a fine art. It became increasingly difficult for those not adept at calculation to stay in business. In the Merton College of the fourteenth century, John Ashenden retailed baseless astrological dogma by the yard, and yet was seemingly obliged to have William Rede perform any difficult calculations he might need. The fact that an astrologer made demands of a man with more mathematical knowledge does not, needless to say, imply a difference in attitude towards astrological prognostication, and there are few who practised the mathematical arts of astronomy who can be shown to have been true sceptics. John of Lignères, John of Murs, John of Saxony, and John Killingworth are all credited with astrological writings. An astronomer adept at the use of the Alfonsine tables was a man whose services were at a premium, and it is significant that most of those astronomers we know to have been in the royal service - often as physicians - were demonstrably expert in calculation. To take but a few of the names we have mentioned already: John Somer composed an almanac (1387-1462) at the instance of Joan, princess of Wales, while Nicholas of Lynne did likewise at the request of John of Gaunt, duke of Lancaster. John Holbrook was one of those who cast a figure for Henry VI, a fact - and a figure - recorded by Lewis of Caerleon, along with the names of other royal advisers on these matters. The advisers were Thomas Southwell (who died in 1441 on the eve of his planned execution for necromancy with the duchess of Gloucester); John Somerset (physician to Humphrey, duke of Gloucester and then to Henry VI); and John Langton, later bishop of St Davids, and, like Somerset, Lewis's tutor in astronomy) [84]. We recall the magnificent MS. Arundel 66, a symbol, if one were needed, of the place of astronomical tables at the court of duke Humphrey [85]. Lewis himself was successively employed as a physician to Elizabeth, widow of Edward IV, to the mother of Henry duke of Richmond, later Henry VII, and to Henry and his queen [86]. John of Murs was for some years in the employ of Philippe III d'Evreaux, king of Navarre. Geoffrey of Meaux was sometime physician to Charles V. All told, it was clearly no bad thing to be a member of the calculating classes, with such distinguished royal forebears, so to speak, as 'Ptholomeus rex' (a common medieval misconception) and Alfonso the Wise.

NOTES

[1] Manuel Rico y Sinobas, 'Libros del Saber de Astronomia del Rey D. Alfonso X de Castilla', vol.iv (Madrid, 1866), pp. 111-183.
[2] For the names of other scholars and an excellent summary of the external evidence of the introductions to the Alfonsine works, see E.S. Proctor, 'The scientific works of the court of Alfonso X of Castille: the king and his collaborators', 'The Modern Language Review', xl (1945), pp. 12-29. For the sources of the myth of the 'vast concourse' see pp. 26-27.
[3] Ibid., p. 28.
[4] Rico's edition of the 'Libros del Saber' appeared in four volumes, and part I of a fifth, between 1863 and 1867, 'El libro de las taulas' (in vol. IV) being correctly treated as a strictly distinct book.
[5] 'Le Système du Monde', vol. IV (Paris, 1916), p. 271.
[6] That Andalo was indeed mistaken is proved by G.J. Toomer in 'Prophatius Judaeus and the Toledan tables', 'Isis' lxiv (1973), pp. 351-355. The relevant part of Isaac ben Sid's book has been printed only once, and even then only in the form of a synopsis in German, which appeared in part iv of B. Goldberg and L. Rosenkranz, 'Liber Jesod Olam, seu Fundamentum Mundi ... auctore R. Isaac Israeli, Hispano', Berlin, 1848 (sectio prior). The synopsis of this part was prepared by David Kassel. I owe this reference to Professor Bernard Goldstein. Since the book is hard to trace, I give the Bodleian shelfmark: Opp. Add. 4º. 1. 186.
[7] Miss Proctor, op.cit., p. 25 also supposes that his associate 'Petrus de Regio' was an Italian from Reggio in Lombardy, and that the Spanish apellation 'Pedro del Real' involves a misunderstanding.
[8] In 1256 the ambassadors of the Commune of Pisa had offered to recognize Alfonso's imperial claim.
[9] 'On the original form of the Alfonsine tables', 'Monthly Notices of the Royal Astronomical Society', lxxx (1920), pp. 243-262. Dreyer well summarizes the astronomical content of Rede's tables, and as far as possible I shall avoid covering the ground again.
[10] I shall usually refer to texts by their incipits, which are listed in alphabetical order, with further textual details, at the end of this chapter.
[11] Here and elsewhere I have made use of Emmanuel Poulle's excellent articles on the three Johns, in 'Dictionary of Scientific Biography', vol. VII (New York, 1973) pp. 122-128; 128-133 and 139-141. As Poulle points out, the ascription to John of Murs of the text 'Autores calendarii nostri ...' is in a fifteenth-century manuscript, but there is no reason to doubt it. The 'Expositio' by John of Murs begins 'Alfonsius Castello rex illustris florens ...'; the 1322 tables by John of Lignères are accompanied by the canon 'Quia ad inveniendum loca planetarum ...'
[12] See below. The canon begins 'Multiplicis philosophie ...'
[13] The date is usually assumed, following references to it in examples in the text. A writer at the end of the copy in British Museum, MS. Egerton 889, at f. 68^r, working out the movement of the solar apogee between Alfonso and 1428, and between John of Lignères' treatise and 1428, gives data which imply quite precisely that he took the year 1323 for the treatise.

[14] The copies available to me are not accompanied by tables, and for their contents I follow Poulle, op.cit., pp. 123-124, where bibliographical details will be found. Note that John of Saxony's 'Quia plures astrologorum ...' was a commentary to the canons on spherical astronomy.

[15] Editio princeps Venice, 1483 (4 July, Erhard Ratdolt), followed by many others, of which the best known are probably those of 1492 and 1518 (Venice), 1545 and 1553 (Paris).

[16] The evidence that the original Alfonsine tables were arranged by 'anni collecti' at 20 year intervals is in 'capitolo' XV (ed. Rico, p. 133).

[17] Op.cit., p. 251.

[18] Cf. p. 251, ibid.: ' ... there can be no doubt that the Oxford tables represent the original Alfonsine Tables in the form in which they were issued from the Toledo Observatory about the year 1272. Only the lengthy introduction to the tables does not seem to have found its way to England, probably because it was written in Castillian and not in Latin ...'

[19] Ibid., p. 252. I ought to emphasize here that the sexagesimal division of time is at issue, and not simply the question as to whether signs were of 30° (common signs) or 60° (natural signs).

[20] 'Rectangulus' and 'Albion'. My edition of the complete writings of Richard of Wallingford is currently in press.

[21] That is, the truly Alfonsine format, which meant tables of mean motion with cyclical radices, and tables of equations of the ordinary type (both as found in John of Lignères' tables for 'Cuiuslibet arcus'). See below for the alternatives.

[22] In the Gonville and Caius MS., this table is copied twice, with only the copyist's mistakes distinguishing the two versions (at pp. 7 and 15).

[23] Ibid., p. 17.

[24] Op.cit., p. 253.

[25] The earliest known dated event in his career is that he was a fellow of Merton College in 1344. He lived until 1385. See A.B. Emden, 'A Biographical Register of the University of Oxford to 1500', vol. III (Oxford, 1959), p. 1556.

[26] The 'Speculum astronomie' usually ascribed to Albertus Magnus is in MS. Digby 228 (14 C.) ascribed to this Philip. See f. 76^r.

[27] Later sections of the codex, with some Oxford tables, and some for London, seem to have been bound in with the first section in a later century.

[28] In particular I note the John of Saxony canons 'Tempus est mensura motus' (ff. 101^r - 108^v), Rede's canon (ff. 111^r - 112^v) and tables (ff. 113^r - 119^r), and Holbrook's adaption of the Alfonsine tables (ff. 134^r - 160^v), which I mention again briefly below.

[29] Bodleian Library, MS. Canon. misc. 501, f. 55^r.

[30] A table at f. 2^r for the entry of the Sun into the signs instructs us to begin by subtracting 1330 from the year.

[31] The greatest equation of centre for Jupiter and Venus, that is to say, being 5°57' and 2°10' respectively. For a more detailed summary of the differences, see my forthcoming 'Richard of Wallingford', vol. iii, Appendix 29.

[32] A fifteenth century miscellany of Alfonsine tables, which I will not describe, except to point out some radices at yearly intervals, from 1400 onwards. The text of 'Quia ad inveniendum' is at ff. 74^v - 79^v.

[33] A brief guide to these materials is as follows: (i) The incipit I take for the canons is 'Volentibus futuros effectus planetarum'. For the MSS.,

see my index at the end of this chapter. Other incipits which have been used for the same work are (a) 'Medios motus et argumenta' (the heading to the first chapter); (b) 'Anni collecti tempus communis' (Thomas Tanner - probably a table heading); (c) 'Quoniam quidem oportet secundum opus tabularum' (a 15th century copy, supplemented with other material). There is much variation in the way different copies of the canons end, but the original work seems to have had only seven conclusiones. (ii) For the simple character of these changes, see above.
(iii) Usually known by the rubric 'Almanak sive tabula Solis pro iv annis 1341-1344'. (iv) There are computations to be found in MS. Digby 176 (ff. 9-13, 34), where they are accompanied by John Ashenden's prognostications, based on them.

[34] For full references, see Thorndike, op.cit., vol. iii, pp. 283-4.
[35] Thorndike, loc.cit.
[36] Op.cit., p. 254.
[37] There is an autograph manuscript of 1331 at the Bibl. S. Dominici, according to Thorndike and Kibre.
[38] The year of a worked example. The coincidence seems to be quite fortuitous.
[39] For the manuscripts, see the list at the end of this chapter.
[40] 'The lunar theory of al-Bahdādī', 'Archive for History of Exact Sciences', vii i (1972), pp. 321-328. The 'Zīj' was completed in A.D. 1285. Al-Baghdādī gives three methods for lunar longitude (of which this is the third), the other two originating with his predecessors Ḥabash and Yaḥyā.
[41] 'A double-argument table for the lunar equation attributed to Ibn Yūnus', 'Centaurus', xviii (1974), pp. 129-146.
[42] For bibliography, see King, ibid., notes to p. 131.
[43] 'Late medieval two-argument tables for planetary longitudes', 'Journal of Near Eastern Studies', xxvi (1967), p. 126-128.
[44] Op.cit., p. 142.
[45] Magdalen College, Oxford, MS. 182, f. 37^v.
[46] I say this on the strength of some elementary astronomical errors in the 'De Causa Dei'. There seems to me to be further evidence against Bredon's authorship in MS. Digby 178, where Bredon's autograph (?) at f. 13^r leads us to suppose that earlier parts of the MS. are in his hand; and these earlier parts include latitude tables differing from those of the 'Vera loca' work.
[47] Ibid., f. 31^r.
[48] Loc.cit., ff. 37^v, 51^r, 33^r.
[49] Ibid., f. 32^v: 'Et nota quod canon Bredon superius vult quod ...' The reference cannot be to either of the canons 'Si vis scrire ...' or 'Medium motum omnium planetarum ...'
[50] f. 57^r.
[51] f. 54^r. The canons in B.N. Lat. 7281 (15th century) are headed 'Canones tabularum Oxonie anno Cristi 1348 ex tabulis Alfonsii factarum', and they end in an equally simple style.
[52] The entry on 'William Batecumbe' by C.H. Coote in the 'Dictionary of National Biography' is without value. Following Tanner, Coote identifies his subject with a man of the name who died in 1487. Following Leland, who lists - perhaps with good reason - works by Batecombe on the astrolabe, the solid sphere, the hollow sphere, and the saphea, Coote suggests that he was professor in the reign of Henry V. This would hardly be compatible with tables prepared in 1348.

[53] See Emden, op.cit., p. 130.
[54] An exception is MS. Digby 57, where the mean arguments of the superior planets are not tabulated, but the mean centres of the inferior planets (omitted in the other MSS.) are tabulated!
[55] Following MSS. Rawlinson D. 1227 and Digby 57. Although a very fine copy, MS. Magdalen 182 has had many radices carefully altered.
[56] An unjustifiably enthusiastic note is struck by Thomas Cory in MS. Magdalen 182, f. 32^v, where he suggests that greater accuracy can be had by reference to Rede's tables. Rede's 'error' stemmed from his having listed the annual movements of aux to only one second of arc, rather than to seconds and thirds. Note that the radix date of the 1348 tables is the last noon of 1347.
[57] I am putting aside the problem of double interpolation, which appears to have given fourteenth century astronomers some difficulty. Tables of proportional parts are included in most copies of the 1348 work.
[58] In MS. Magdalen 182, at f. 3^r, there is a table for the equation of the eighth sphere ($\cong 9° \sin \theta$), at the foot of which is written 'Requirem meliorem autem tabula Walteri'. This presumably refers to Walter of Evesham. At f. 30^v there is a table of additions to be made to the planetary longitudes found, proceeding by single years from 1349 to 1600. Specimen entries are: 1' for 1350; 32' for 1400; 1°30' for 1500; 2°23' for 1600.
[59] For a more extensive treatment of the contents of this paragraph, see Appendix 29, vol. iii, of my 'Richard of Wallingford', Clarendon Press, forthcoming. A treatment reaching similar conclusions along different lines is reported in Emmanuel Poulle and Owen Gingerich, 'Les positions des planètes au moyen age: applications du calcul électronique aux tables Alfonsines', 'Comptes Rendus de l'Académie des Inscriptions et Belles-Lettres', Paris (Mai, 1968), pp. 531-548. See pp. 541-542.
[60] The Moon's true place', being referred to the mean Sun, will be zero. For the Sun, there is of course no $\bar{a}$.
[61] This is easily visualized without a diagram: roughly speaking, $2z_1$ is the 'apparent' angular size of the epicycle at its greatest distance, for a geocentric observer, while $2z_2$ (see below) is the approximate angular size at the perigee of the epicycle.
[62] See again my 'Richard of Wallingford', Appendix 29, section 9.
[63] I have used a microfilm of the tables in Erfurt MS. Amploniana, Q. 388 ff. 1^r-35^r (radices at 38^r) to supplement the Gonville and Caius copy, which lacks the equations. The Erfurt copy is late - in fact, judging by a list of aux positions at f. 35^r, it dates from ca. 1446.
[64] Ptolemy's intervals ('Almagest', Book XI) are for all planets 6° up to 90°, and 3° thereafter.
[65] I follow the fifteenth century copy in the Bodleian Library, MS. Canon, misc. 501, ff. 55^r - 92^r. The ranges of the conjunction tables are: Saturn from 1320 and for 59 years thereafter; Jupiter from 1320 and for 83 years; Mars from 1320 and for 79 years; Venus for 1319 and for 75 years; Mercury from 1320 and for 46 years; Moon from 1320 and for 76 years. The signs are of 60 degrees, rather than 30. In this particular manuscript there is evidence of the continued use of John of Murs' technique, for between ff. 92^v and 99^v a new set of 'principal' tables has been calculated, starting at 1452 and continuing for 59 years for Saturn, and so on. The short canons 'Si vera loca planetarum per presentes tabulas invenire ...' follow the second tables at f. 103^r-v. After

The Alfonsine Tables in England 353

an explicit, a related work begins 'Sequitur composicio tabularum: sic fuerit tabule ...' (ff. 103^v - 105^r).

[66] For further explanation of these terms, again see my work, 'Richard of Wallingford', vol. iii, Appendix 29.

[67] I am here overlooking a number of rather trivial ancillary tables, and the latitude tables.

[68] See 'Omnis utriusque ...' and 'Patefit ex Ptolemei ...' in the list of incipits below.

[69] MS. Digby 57, ff. 9^r - 23^v, beginning with a short canon 'Medium motum Solis per kalendarium ...' There are no equations. The tables of radices, meant to cover the span 1374-1405, are filled in for all planets in 1374 and with a few scattered entries up to 1380. At f. 24^r there begins a tabulated chronological history of the world, with spaces to 1530. It is difficult to judge where the original script ends in this table, but it seems to date from the end of the fourteenth century or the beginning of the next. A planetary almanac, possibly related to the first tables, follows at ff. 111^r - 118^v and 119^v - 120^v, and includes some eclipse tables. Dates are 0^d and 9^d for each month. The Sun and Moon are given longitudes to minutes, but the other planets only to degrees. The lists are complete between 1375 and 1380, but there are some other rulings to 1390.

[70] For a short list, with what biographical details are available, see Emden, op.cit., pp. 1049-1050. Agnes Clerke, in the 'Dictionary of National Biography', confused two Mertonians of the same name.

[71] See 'The Algorism of John Killingworth', 'English Historical Review', xxix (1914), pp. 707-717. Karpinski did not appreciate the connexions between the two works, and he is perhaps too critical of Killingworth's failure to grasp the possibilities of the Hindu numerals, which he says is 'characteristic of the period'.

[72] This is the earliest date in the MS. Savile 38 copy, in the first table. Killingworth became principal successively of Corner Hall in 1436 and Olifant Hall in 1438, and was later (northern) University Proctor. His memorial tablet, with portrait brass, is still to be seen in Merton College chapel.

[73] Bodleian Library, MS. Savile 38 is Lewis's autograph, devoted entirely to the Killingworth tables. It opens (ff. 2^v - 4^r) with 'Restat de compositione', which text closes with the words 'Expliciunt canones tabularum Mag. J. Kylyngworth ab ipsomet edite preter unum capitulum de correctione radicum Saturni et Jovis quod ego Lodowicus Caerlion superaddidi cum istis tribus tabulis de divisione motus augis.' The short table follows. (Cf. Lewis's 'Circa compositionem tabularum elevationum signorum ...', which - despite its opening words - actually covers the calculation of many sorts of table.) The 'Multum conferre' text and Lewis's recalculated tables follow (ff. 4^v - 5^r and 6^r - 7^r), with a copy of the original tables occupying almost all of the remainder of the volume.

[74] Here MS. Magdalen and MS. Bodley run in single years 1460 to 1560. MS. Savile has 1442 to 1500 (which was probably the original span), supplemented by 1501 - 1612 (Lewis's additions).

[75] The decimal equivalent is 533297. The incarnation radix is J.D. 1721423, and 1 February 1461 has indeed a Julian date 2254720.

[76] Approximate values of the increments are: 2,59,15; 33 (Saturn); 1,12,12;02 (Jupiter); 11,26;58 (Mars); 6,05;15 (Sun, Venus, and Mercury); 24,7;00 (Moon). Lewis of Caerleon's intervals are rather more accurately calculated.

[77] See f. 108^v. I have discussed this volume above.

[78]   Brussels, Bibliothèque nationale, MS. 10117-26, f. 79^V. The suggested date is that of some aux positions on ff. 117^V - 118^r.

[79]   Cambridge University Library, MS. Ee. 3.61. The first table is at f. 81^V, and the calculation is at ff. 189^V - 190^r.

[80]   Duhem, op.cit., vol.iv, pp. 70-72, I am using that version of the 'Correctio' which appears in the Basel 1565 (Henricus Petri) edition of the 'Opera', at pp. 1168-1·73.

[81]   Duhem is following a fifteenth century ascription, which was probably prompted by Geoffrey's known reluctance to use Alfonsine radices for his calendar of 1320. (For his objections, see Duhem, op. cit., p. 70, and Thorndike, op. cit. p. 284, n. 14.) Duhem quotes also the report by the writer of the 1347 treatise that he had read of William of St. Cloud's work in a book in the Sorbonne. Did Geoffrey of Meaux belong to this school? 'Cela semble probable', Duhem decided, on the strength of the statement by Thomas Tanner (1748) that Geoffrey taught at Oxford in 1325 and 1345. See above.

[82]   See Bodleian Library, MS. Canon. misc. 554, ff. 73^V - 93^V. Another Bodleian manuscript with similar tables but no canon is MS. Lat. misc. d. 88.

[83]   MS. Bodley 790, f. 56^r. A very corrupt version of the same passage occurs in the printed Nicolas of Cusa work, op.cit., p. 1169; but the printed 'quemadmodum' is perhaps preferable to 'servus' (or 'Suis'?). The 'procuravit' should perhaps also be omitted, as in the printed version.

[84]   Cambridge University Library, MS. Ee, 3.61,f. 161^V, etc.

[85]   Curiously enough, it was John Killingworth, in his capacity as northern Proctor of Oxford University, who wrote to acknowledge duke Humphrey's gift of books, which were to form the nucleus of a new University library. See H. Anstey, 'Epistolae Academicae Oxon.', Oxford 1898, vol.i, pp. 204-205.

[86]   A.B. Emden, op.cit., pp. 337-338.

List of incipits of Alfonsine canons and related works

The following list is not meant to be exhaustive, in terms either of texts or of manuscripts. It could probably be doubled in length in both respects. I have consulted most of the MSS. mentioned from English libraries, and some of the others but I have rarely made a careful collation of texts, since canons to table are so often varied trivially. The list should help to reduce the number of texts so often mistakenly supposed independent. The following abbreviations have been used: T. & K. - Lynn Thorndike & Pearl Kibre, 'A Catalogue of Incipits of Medieval Scientific Writings in Latin', Mediaeval Academy of America, London, England 1963; B.L. - Bodleian Library, Oxford; B.M. - British Museum Library (now called the British Library); B.N. - Bibliothèque Nationale, Paris; C.U. - Cambridge University Library; Ea - Erfurt, Wiss. Bibliothek, Amploniana; VI - Vienna, Österreichische National-Bibliothek.

1. Ad noticiam tabularum sequencium sciendum quod ...
    Canons for Alfonsine tables for London, a. 1336. B.L. Bodley 790, ff. 2^V - 4^r (canons, addition at 4^r) & 5^r - 33^r (tables), with some repetition between 34^r - 45^V in a different hand. Only copy known.
2. Ad sciendum quid veritatis et quid dubii tabulae Alphonsi contineant ...

Printed as Nicholas of Cusa, 'Correctio Tabularum Alphonsi' in the
'Opera' (Basel, 1565), pp. 1168-1173. Related to item 6. See p. 289,
above.
3. Alfonsius Castelle rex illustris florens ...
John of Murs 'Expositio' of 1321. B.N. Lat. 7281, ff. 156^V - 159^V.
Only copy known.
4. Anni collecti tempus communis ...
Tanner's incipit for Rede's tables. See item 41. Probably merely a
rubric to a table.
5. Autores calendarii nostri duo principaliter ...
John of Murs (on the ecclesiastical calendar) favours the Toulouse
tables, 1317. VI 5292, ff. 199^r - 209^v. Only copy known.
6. Bonum mihi quidem videtur ...
Criticism of the Alfonsine tables, 1347/8. Plagiarised by Nicholas
of Cusa (see item 2). Trinity College, Cambridge, MS. O.IX.6, ff.
55^r - 57^v (wrongly ascribed to John of Murs); B.L. Bodley 790, ff.
53^r - 59^v; B.N. Lat. 7281, ff. 172^v - 175^v.
7. Circa canonem de inventione augium sunt tria ...
Commentary by John of Speyer on John of Lignères, item 21. B.N. Lat.
10263 (ascribed), ff. 78^r - 86^v; Ea Q.366 (anon.), ff. 33^r - 37^v.
Worked example dated 1348.
8. Circa compositionem tabularum elevationum signorum ...
Lewis of Caerleon, method of calculating several kinds of astronomical
table. C.U.Ee. 3.61, ff. 3^r - 7^v. Only copy known.
9. Cuiuslibet arcus propositi sinum rectum invenire ...
John of Lignères, canons on primum mobile, etc., 1322 or 1323.
See also items 30 and 39. Items 14 & 27 are parts of it. Printed in part
by Curtze, Bibl. Math. i (1900), pp. 390-413. B.M. Egerton 889, ff.
55^r - 68^r and C.U. Ee.3.61, ff. 86^v - 96^r can be added to the three MSS.
in T. & K. See also Poulle, op.cit. (D.S.B., vol. 7), p. 127.
10. Cum hoc scire placuerit a Sole ...
Opening words of text item 40, as in Magdalen 182.
11. Cum scientia astrorum sine tablulis minime ...
John of Saxony's canons in the version of Ea F. 389, ff. 1 - 55.
12. Cum volueris invenire tempus medie coniunctionis ...
John of Lignères & Philip Alan (?) on conjunctions and oppositions.
See p. 275 above. B.L. Digby 114, f. 8^r & 8^v. Possibly same as B.N.
Lat. 7316A, f. 187ra & 187va.
13. Cunctis Solis et Lune scire desiderantibus vera loca ...
Geoffrey of Meaux, calendar of a. 1320, rejecting Alfonsine tables.
B.N. Lat. 7281, ff. 160^v - 162^r; B.N. Lat 15118, ff. 74 - 75. Other
MSS. listed by Zinner (1925).
14. Diversitatem aspectus Lune ...
John of Lignères eclipse canons, often quoted as a separate work, but
part of item 9.
15. Exultarem non minori gaudio quam tu Augustine suavissime ...
J.L. Santritter, introduction to the Venice printing of 1492 of John of
Saxony's canons and tables (item 39).
16. In nomine Domini nostri Iesu Cristi ...
A common beginning to item 41 (Rede).
17. Incipit canon almanak predicti ad cognoscendum vera loca omnium planetarum ...
Second rubric to version of item 40 in Magdalen 182.

18. Medios motus et argumenta ...
    Given as incipit to a work by Rede, but merely rubric to conclusio 1, item 41.
19. Medium motum Solis et omnium planetarum ...
    Short canon mentioning radix date 1320 and ascribed to John of Lignères. B.L. Digby 114, f. 9^r. Not in T. & K.
20. Medium motum Solis per kalendarium ...
    Mean motus tabulated daily throughout year. See n. 69 above. B.L. Digby 57, ff. 9^r (canon) and 9^v - 23^v (tables). Anno ca. 1374.
21. Multiplicis philosophie variis radiis illustrato domino Roberto Lombardo de Florentia Glassuensis ...
    John of Lignères to Robert of Florence, ca. 1320, canons and Alfonsine tables with tables of combined equations ('tabule magne'). Gonville and Caius College, Cambridge 110 (179), pp. 1-6 (canons) and 7-18 (tables, no equations); Ea Q.366, ff. 28^r - 32^v (canons); B.N. Lat. 7281, ff. 201^v - 205^v (canons); B.N.Lat. 10263, ff. 70^r - 78^r (canons); Ea F. 388, ff. 1^r - 42^r (tables, isolated canons); Lisbon Ajuda 52-VI-25, ff. 67^r - 92^v (tables of equations only, ex inform. E. Poulle).
22. Multum conferre dinoscitur non solum astronomis ...
    John Killingworth, canons and tables, ca. 1442. Magdalen College Oxford, 182, ff. 69ra - 70ra (canons), 68^v - 101^r (tables); B.L. Bodley 432, ff. 57^v - 63^r (canons), 63^v - 73^r (tables); B.L. Savile 38, ff. 4^v - 5^v (canons), 6^r - 8^r (Lewis of Caerleon's additional radices), 8^r - 30^v (tables); B.M. Arundel 66, ff. 1^v - 30^r (canons and tables); B.M. Sloane 407, ff. 57^r - 196^v (?) (tables); B.M. Royal 12. G.X., ff. 1^r - 32^r (canons and tables).
23. Numerum annorum mensium et dierum a principio alicuius ere ...
    Several MSS. - see T. & K. col. 958 - one of them (Ea F. 386, ff. 38rb - 48ra) ascribed to Hermann Stilus de Norchem. But the second paragraph of the printing of item 39 begins in the same way.
24. Omnis utriusque sexus harmoniam ...
    John of Murs, canons and 'tabule permanentes', a. 1320/1. See Poulle, op.cit. (D.S.B.), vol. 7, p. 133. VI 5268, ff. 45^v - 48^v possibly the only MS. with tables. See items 25 and 38 below.
25. Patefit ex Ptolomei disciplinis in libro suo qui dicitur Almagestis ...
    John of Murs, a. 1321, addition to calendar for rapid determination of syzygy. See Poulle, ref. in item 24, and item 38.
26. Prima tabula docet differentiam unius ere ...
    John of Murs, a. 1339. B.L. Bodley 491, ff. 32^r - 35^r (canons, anon., but Paris). 41^r - 64^v (tables); Oxford, Hertford College, e.4, ff. 140-147 (canons; Ea Q.366, ff. 52^r 52^v (canons); B.N. Lat. 18504, ff. 209^r & 209^v (canons).
27. Priores astrologi motus corporum celestium ...
    John of Lignères excerpt from item 9.
28. Pro continuacione temporum vere radicis ...
    Part of item 22, by John Killingworth, explaining the construction of the tables.
29. Quia ad inveniendum loca planetarum ...
    John of Lignères, undated but thoroughly sexagesimal canons and tables (a. 1322-7 ?). B.L. Digby 168, ff. 139^r - 146^v (canons and tables out of sequence); Oxford, Hertford College E.4, ff. 148^v - 155^r; B.M. Royal 12.D.VI, ff. 74^v - 79^v; B.M. Sloane 407, ff. 4^r - 6^v (canons, barely legible), 7^r - 33^v (tables). See also T. & K. col. 1213 and Poulle, op.cit., p. 127.

30. Quia plures astrologorum diversos libros fecerunt ...
    John of Saxony, commentary on item 9, ca. 1335. Canon 37 used to
    end printed ed. of item 39. C.U. Ee.3.61 ff. 178^r - 181^r. Cf. T. & K.
    col. 1228.
31. Quia secundum philosophum 4º Phisicorum tempus et motus mutuo se
    mensurant ...
    Not John of Murs (as suggested by Cat. of Digby MSS.), but a trivial
    variant version of item 39. B.L. Digby 168, ff. 131^r - 135^v; C.U.Mm.
    III.11.ff. 44^r - 51^r; Florence, Laurentiana 131, ff. 147^r - 166^v.
32. Quicumque voluerit quinque planetas retrogrados equare secundum tabulas
    de Battecombe, alias Bredon sive Bradwardynes vocatas: Primo queritur
    centrum medium et ...
    'Canon optimus' for the Oxford tables of 1348 (see item 40). Oxford,
    Magdalen 182, f. 37^v.
33. Quoniam celestium motuum calculus annorum supputatione ...
    John Holbrook, canons and tables. B.M. Egerton 889, ff. 133^v - 151^r;
    C.U. Ee.3.61 ff. 54^r - 57^r.
34. Quoniam tabularum Alfonsi laboriosi ...
    Salamanca tables of ca. 1348. Jacobus de Bona Diei? See Dreyer, op.
    cit., pp. 253-4. B.L. Canonici misc. 27, 33^r - 118^v (?) (tables) and
    122^v - 153^r (canons). Incomplete.
35. Restat de composicione tabularum revolucionum planetarum ...
    Lewis of Caerleon (?) on the tables of John Killingworth. B.L. Savile
    38, ff. 2^v - 4^r; B.M. Royal 12.G.X, f. 2^r & 2^v.
36. Sequitur composicio tabularum: sic fuerit tabule ...
    Related to item 38 (John of Murs). See p. 284, n. 65, above.
    B.L. Canonici misc. 501, ff. 103^v - 105^r. Not in T. & K.
37. Si quis per hanc tabulam ...
    John of Murs, multiplication table for sexagesimals, a. 1321. See Poulle,
    op.cit., p. 128, and T. & K., col. 1461. The closing fragments are in
    B.L. Digby 190, f. 66.
38. Si vera loca planetarum per presentes tabulas invenire ...
    John of Murs, a. 1321, longitude of planets through tables of conjunc-
    tions with the Sun. Cf. items 24 and 25 above. Not in T. & K. B.L.
    Canonici misc. 501, ff. 54^r - 106^v. Tables also in Lisbon Ajuda 52-VI-
    25, ff. 24^r - 66^r. Cf. item 21, associated in latter MS.
39. Tempus est mensura motus ut vult Aristoteles ...
    John of Saxony, Alfonsine canons and tables, a. 1327, much copied
    and later printed (ed. princ. Erhard Ratdolt, 4 July 1483, Venice).
    On eclipses the pr. edn. duplicates John of Lignères, item 9 above.
    MSS. consulted: Brussels, Bibliothèque royal, 10117, ff. 90^r - 100^r,
    and C.U. Ii.I.1, ff. 112^r - 126^r. Cf. T. & K. cols 1561, 1560, 1712,
    etc. Note autograph MS. of a. 1331 at Bologna, Bibl. S. Dominici.
40. Vera loca omnium planetarum in longitudine ...
    Oxford tables and canons of 1348, attributed to Batecombe, Bredon,
    and Bradwardine, but see p. 279 above. Cf. items 10 and 17. Oxford,
    Magdalen College 182, ff. 1^r - 30^v and 35 (tables), 36ra - 37va
    (canons); B.L. Bodley 432, ff. 56^r - 57^r (canons, beginning missing,
    no tables); B.L. Rawlinson D. 1227, ff. 64^r -87^r (tables) and 87va -
    89rb (canons); B.L. Digby 57, ff100^r - 102^r (canons) and 44^r - 99^v
    (tables); B.L. Laud misc. 594, ff. 51^r - 81^v (tables only); C.U. Ii.I.27,
    ff. 17^v - 38^v (tables). In the Magdalen copy the canons by John Somer
    'Verum motum 5 retrogradorum et Lune ...' are added to the end of
    item 40.

41. Volentibus [pronosticare] futuros effectus planetarum in istis inferioribus ... William Rede, Alfonsine canons and tables, ca. 1340. Dreyer, art. cit. in n. 9 above, lists many copies, despite confusion with item 40 above. Further MSS. include: B.L. Ashmole 1796, ff. 90^v - 94^r (canons and tables), not complete); Oxford, Hertford College E.4, ff. 60^r - 65^v (no tables); B.M. Egerton 889, ff. 111^r - 112^v (tables at 113^r - 119^r) approximate to item 40; B.L. Digby 92, ff. 11^r - 14^r (canons), with final note referring to tables of John Trendelay, monk of Christ Church Canterbury; B.L. Bodley 432, ff. 1^r - 27^v (tables, copy of 1456) and 28^r - 33^r (canons; note death of Rede mentioned at 28^r); (from T. & K.), Cambridge, Peterhouse 250, ff. 59^r - 62^v; Oxford, Jesus College 46, 3^r - 36^v (tables) and 37^r - 42^r (canons), MS. having 43 ff.; B.L. Ashmole 191, ff. 59^r - 61^v (canons) and 62^r - 74^r (tables), copied after a. 1439; B.L. Ashmole 393, ff. 1-41 of sect. IV, tables only; B.L. Digby 72, ff. 9^r - 31^r (tables); B.M. Royal 12.D.6, ff. 85^r - 90^r (canons, ending incomplete).

Note that the copy of item 40 in Magdalen 182 begins with the words of 41 (short preamble).

Addenda

In 1981 Prof. B. Goldstein pointed out to me the existence of a Hebrew translation of the Oxford Tables of 1348 done by the fifteenth-century Mantuan Mordecai Finzi. This, strangely enough, returned to Oxford, where it is now MS Lyell 96 in the Bodleian Library (not in the printed catalogue). Finzi mentions an anonymous Christian who helped him in Mantua. Oxford is called 'Osonia', and is said to be 17 deg. 56 min. from the west. Pico's *Adversus astrologiam* ix.11 also mentions the 1348 tables and one 'Angelus Mantuanus', presumably Finzi. In 1982 Prof. J. Dobryzcki drew my attention to other European manuscripts containing the 1348 tables, now in the Jagellonian Library in Cracow: MSS 553 (Silesia?, 15th cent.) and 551 (Prague, late 14th cent.). I have myself found references to 'tabule anglicanae', perhaps the same 1348 tables, in 15th century writings, for example in the work of Henry Arnaut of Zwolle. All told, it seems that the Oxford astronomers had a European influence that remains to be more fully explored.

A
B
C
D
E
T
P

10
Δξ
(min.)
5
0
-5
-10
120
240
360
420
600 days
γ̄/m
(m = mean motus
per diem)

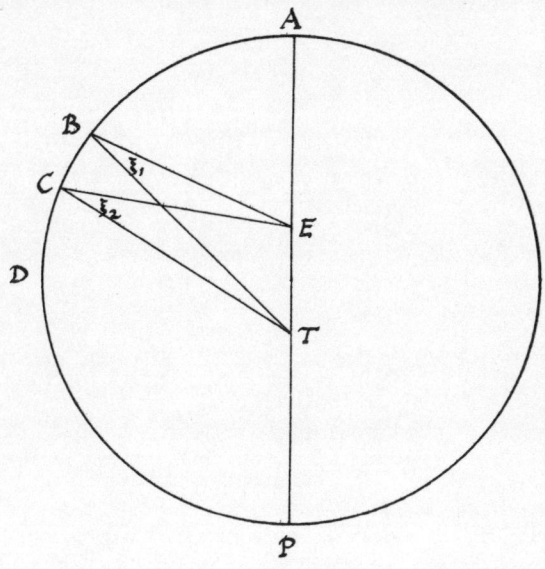

Fig. 22

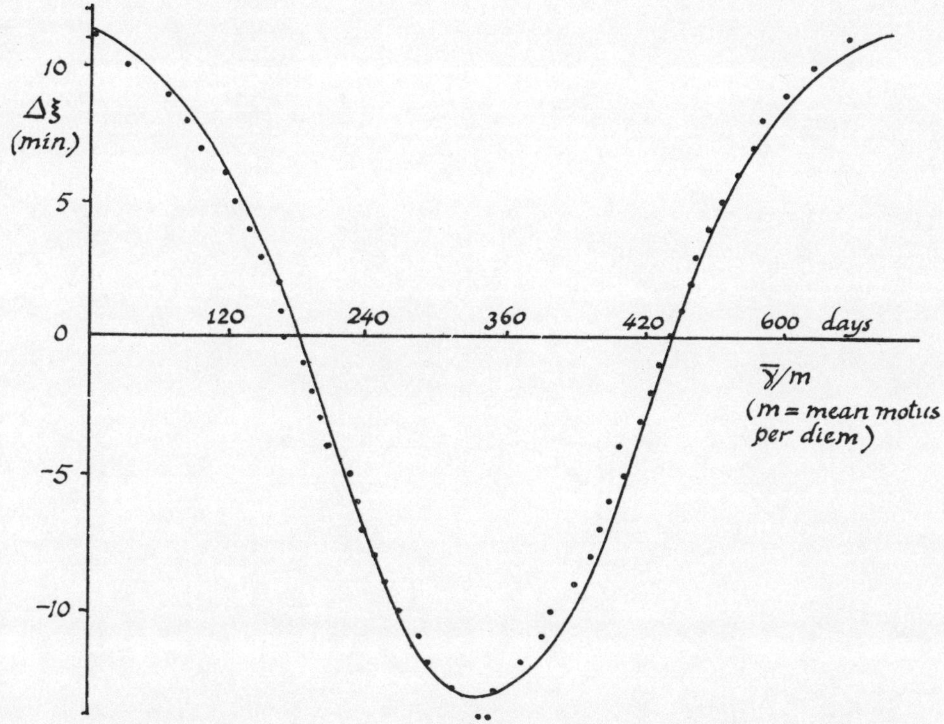

22

1348 AND ALL THAT:

OXFORD SCIENCE AND THE BLACK DEATH

I shall take it as axiomatic that there was something extraordinary about Oxford science in the fourteenth century, and that this is not an illusion created by over-assiduous historians in recent times [1]. Walter Burley, William of Ockham, Richard of Wallingford, Thomas Bradwardine, William Heytesbury, John Dumbleton, William Rede, Richard Swyneshed ... and so one could go on through the century. If one were to do so, however, the list would become less impressive, but is that not perhaps a consequence of the fact that so many recent historians have chosen to concentrate their attentions on kinematics and natural philosophy? Is there no way of assessing the rise and fall of Oxford science without introducing the distortions imposed by modern historical preferences? There is a way, albeit tentative, preliminary, and approximative. We may count heads, books, texts, college affiliations, and the like, all regardless of their quality, and then look for broad trends. The source of our information must be as comprehensive and unbiased as we can make it. Here we are fortunate, for whatever the lacunae in A. B. Emden's biographical register, it is as comprehensive as we have any right to expect, and was compiled without regard for any of the more obvious principles of selection that would stand in our way [2]. Such a demographic method as I propose to apply is certainly lacking in the subtleties of Heytesbury and Dumbleton, but perhaps Asger Aaboe will accept it in memory of a friend we had in common. It is after all possible to draw a very passable circle with a very imperfect piece of string.

The effects of the Black Death on the universities of England and Europe have long been a matter for historical debate, and W. J. Courtenay's cautionary account [1980] is one of the most important modern studies of the problem as it concerns England. Earlier studies tended to put undue emphasis on reports by contemporaries, regretting declining standards in education. (Was it ever otherwise?) Estimates of the decline in university *numbers* cannot be easily related to the mortality rate, either nationally or locally, for the simple reason that masters and students vacated the town when the plague first made its appearance, and returned only when it had abated. The university population at any one time in the fourteenth century is now thought to have been between 1,000 and 1,700 [3]. Emden's first register, although it includes 15,000 biographical sketches of Oxford masters and students, could not possibly encapsulate anything approaching the entire university population. I am concerned not with any absolute measure, however, but with change, and that especially in the upper echelons of scholarship. The numerical distortions consequent on Emden's omissions are likely to be much the same over the second half of the fourteenth century as over the first half. The main qualification here is in the matter of documentation: although this tends to improve with time, the improvement is not a steady one. One might expect also that the very epidemic that concerns us would have a masking effect of some sort – if the university is disrupted, so are its records likely to be – but this does not seem to be a serious problem, and that largely because Emden used so many records from outside the universities.

It appears that there was a much lower mortality rate among university scholars than in the country at large – say five or ten per cent in the university, as against something between twenty and forty over England generally. Dividing the century into five periods of twenty years, Oxford University members known by name from these five periods number 1112, 1192, 1086, 1100, and 1547 respectively, and it should not be necessary to chart these data to see that they seem to represent a serious check in mid-century to a growth that recovers after it [4]. The complicating factor here is the fluctuation in the number and quality of the records from the colleges – note, for instance, the sudden appearance of matriculation lists from New College, founded in 1379 – and the loss of those from the mendicant orders and religious houses generally.

Numbers are not an end in themselves. One also wants to know the effect the plague had on the quality of education by virtue of the removal of key individuals. As Courtenay explains, a number of important intellectuals who died in the period 1348-9 – I might single out John Baconthorpe, William of Ockham, Robert Holcot, Thomas Bradwardine, and probably John Dumbleton – had already left Oxford. The survivors include some impressive names: Richard Fitzralph, Adam Wodeham, Thomas Buckingham, William Heytesbury, John Ashenden, Richard Swyneshed, William Rede, Richard Billingham, Ralph Strode and John Wyclif, for example. "If the Black Death was a factor here, it was not in removing the great minds of that generation but in removing those who might have been the great minds of the next generation" [5]. The question remains: was there a marked decline in scholarly and scientific activity as a consequence of the epidemic? Was there, perhaps, a change in the general character of the activity that we should now describe as scientific, and what were the chief centres of that activity, within Oxford?

The last question is particularly difficult to answer, bearing in mind the great lacunae in our records of activity within the religious houses, contrasted with the surprisingly good Merton records, for example. (These indicate, incidentally, no great alteration in the numbers of scholars there.) If we are interested primarily in the rise and fall of scientific activity in Oxford as a whole we naturally focus attention on writers, that is, on what was most notable – as against the numbers of those who entered the educational system but made no great mark on it –, and then the more public nature of the relevant records will to some extent overcome the imbalance that would have been likely had we relied on the extant administrative records. The latter, however, are needed if we are to assign individual writers to the correct institutions, and this we are unable to do, as we shall see, in about one in six of the cases that interest us.

In character, most natural philosophy remained essentially Aristotelian in cast throughout later medieval Oxford. If anything, it seems to have become plainer, that is, less bound up with the "barbarian subtleties" of those who practised suppositional logic, although this is not to suggest that there was an abrupt severance of those traditions represented by Thomas Bradwardine, William Heytesbury, Richard Swyneshed, or John Dumbleton, traditions that have occupied so many historians of science in recent times. To name but five men who after mid-century continued to write in the same Merton tradition of kinematics, for example, we have Richard Feribrigge, John Wyclif, Ralph Strode, Edward Upton, and John Chilmark. The movement had admittedly lost its edge by the time they were writing, and this is a matter that is in need of an explanation, but the changes that were taking place in the character of Oxford studies do not seem to me to have been apocalyptic.

Even in the face of a colossal decline in the population of the country at large, it seems that there was a general growth in the number of institutions of elementary education, as well as a sharp increase in theological enrolment in the university – perhaps a consequence of a high death rate among parish clergy in the country at large. In paying closest attention to theology, William Courtenay decided that "To the degree that Emden's *Register* is an adequate reflection of what is going on in Oxford in the fourteenth century, the Black Death did not have the effect on higher education often ascribed to it". For other subjects he conceded that it did coincide with a change in interests and possibly in quality, but he did not attempt to say how. "If the Black Death helped to catalyze this change", he concluded, "it was by initially impairing the quality of primary education, and only subsequently higher education" [7]. Judging by the numbers of writers in the various scientific genres, however, these conclusions do not seem to be borne out.

To explain a decline is one thing, to ask whether it is real rather than illusory is another, and a prior matter. Had one access to all the necessary information, quality apart, one might learn much from a chart of pages (or words, even) written each year on scientific themes, or better still, a three-dimensional graph of persons and pages against time. For the middle ages, these are utterly unrealisable ambitions and, for reasons explained earlier, I shall count *names,* in the first instance of men with a title to being called "writer", and at a later stage of men with some sort of scientific connection, such as is implied by ownership of a scientific work, the pledging of an astronomical instrument, or the like.

A writer will be counted once and once only for every one of my eleven categories of subject to which he contributed. In short, we are not so much counting persons as personae. It is often far from easy to decide whether a name shall count at all, for writings on the soul may spill over into the life sciences, and theology into kinematics, arithmetic, logic, or geometry. The Gordian knot will undoubtedly be cut by different analysts in different ways, and I must confess to a strong subjective element in the results that follow.

The main difficulty at the core of any study of temporal change is that of the precise dating of works and people. To arrive at any general picture based on ostensibly precise, numerical, treatment of notoriously inexact data, a measure of insensitivity is called for, but this makes it all the more important that we make the nature of our approximations as explicit as possible. It is of course ultimately the dating of *works* that really interests us, but since this information is rarely accurately known, I have treated all writers alike, and have aimed instead at identifying a point in the typical author's career that might be supposed to approximate to the peak of his activity. This I have taken arbitrarily as the date at which he became B. Th. or attained his thirtieth year. Dates of birth and death are of course sometimes as problematical as dates of authorship, and for this reason I have made my count by five year periods. A typical uncertainty is of the order of five years, but in many cases more. Finally, faced with the statistical problem of small numbers, I grouped these periods of five years into periods of twenty years.

In counting "person-themes" rather than activities, their durations, or their qualities, there is an element of distortion. A man with a hundred books to his name will rank with the author of a few pages, if they both concern a single theme, and a man who wrote briefly on three themes will register his importance as thrice that of a man who wrote extensively on only one; but these inequalities do even out over the period as a whole. (There is no significant change in the average number of themes attempted by authors at different times, for example.) Since I count more than two hundred and forty "writer-themes", or "personae" in the sense explained, this question of weight need not be felt too seriously. There are in any case many instances of dubious authorship: historians are as prone as they ever were to ascribe anonymous writings to anyone famous who happens to be around. The limits of my initial survey are 1170-1510 [see Figure 23].

To avoid undue anachronism in the notion of what is to be deemed "scientific", I have counted writers on these eleven topics:

1. Natural philosophy and related epistemology, theology, etc.
2. Logic
3. Arithmetic and algebra.
4. Geometry.
5. Music.
6. Astronomy, chronology and geography.
7. Astrology.
8. Kalendar, almanac and computus.
9. Chemistry and metallurgy.
10. Medicine and the life sciences.
11. Magic, geomancy, etc.

Music is included by dint of its place in the quadrivium, but is so seldom written on – I count six writers in all in this quadrivial tradition – as to be of no real influence on my broad general findings. Much the same goes for the eleventh category, under which I find only two writers.

As is to be expected, natural philosphy is the theme most strongly repre-

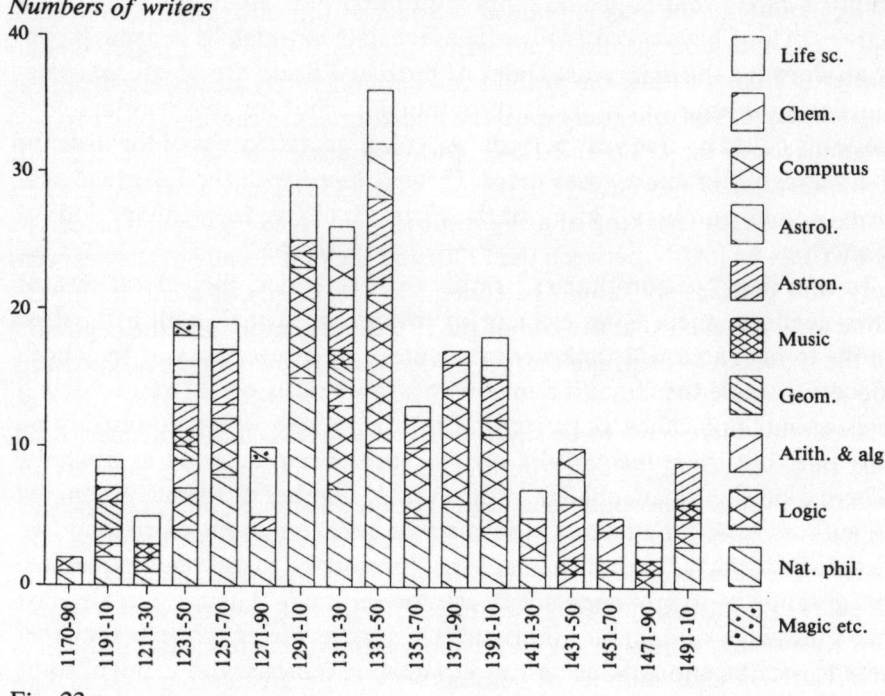

Fig. 23.

sented, both by writings and writers – 73 writers in the period 1180-1510. Logic follows with 61 writers, many of course the same men as under the first heading. Astronomy attracted only about half as many (32), followed by medicine and the life sciences (25). Writers on arithmetic (12) and geometry (7) are surprisingly few, although here one might easily inflate the figures by the use of a more generous interpretation of those words. Seven out of the eleven astrological writers also wrote on astronomy – which is to say that their works on both subjects survive. By counting them separately we emphasize the fact that astrology was by no means a necessary concomitant of academic astronomy, although it might be argued that there is some resulting distortion of the notion of a "writer-theme". There are few works put under the heading of magic or geomancy that are not of dubious provenance, although there were translations of Arabic works in circulation, and magic was evidently occasionally practised. There is a solitary writer on chemistry, Walter of Odington, possibly not an Oxonian. Two or three anonymous works of the period might well by later research be proved to have originated in Oxford.

Reckoning in the way explained, activity as judged from the four main subjects, and from totals of all writer-themes, follows a roughly similar pattern. Taking totals, the graph shows a moderately steady rise from the foundation of the university until the mid-fourteenth century, apart from a slight check in the 1270s and 1280s. More significant by far is the dramatic collapse after the twenty-year period 1331-1350. There does seem to be some reason for speaking of a slight subsequent recovery – much helped by the writers on logic – between the 1370s and the first decade of the next century, but perhaps we should be content to speak only of a continuing decline until its reversal at the end of the fifteeenth century and the beginning of the sixteenth – for which there is evidence not shown on the accompanying chart.

In all this there seem to be slight differences between the fortunes of the best-represented subjects: astronomy and the life sciences were never as popular as natural philosophy and logic, but they attracted fairly constant attention. Attention to logic seems to have reached its peak three or four decades after natural philosophy. It is tempting to see in this material the influences not only of the Black Death but of such edicts as those of Tempier and Kilwardby, although my impression is that the intellectual influences of a relatively small number of scholars were of far greater importance than those prohibitions, for the minor fluctuations. I will name a few who seem to have served as strong natural foci of discussion and debate in the late thirteenth and fourteenth centuries. After Richard of Middleton, Duns Scotus, John of Gaddesdon, Walter Burley, John Baconthorpe and William of Ockham we are brought to roughly the end of the 1320s, and the "science" we are charting is still generally lacking in that strong feeling for mathematics earlier revealed by Grosseteste and Bacon, for example. In the all too brief Golden Age that followed, we have Richard of Wallingford, Thomas Bradwardine, William of Heytesbury, John Dumbleton, Richard Swynneshed, and William Rede – not scholars working in glorious isolation, but men whose works were known, admired, and imitated. After mid-century there is no comparable name until the end of the fifteenth century. (John Wyclif and Paul of Venice can hardly be regarded as such, in the context of Oxford).

From chronology we turn to affiliation. To which institutions did our writers belong? The following list covers men whose writings formed the basis of the above graph, *but only from the period 1285 to 1510*. (The numbers in parentheses should be ignored for the time being).

Benedictine monks	10	(32)	Unknown affiliation	24	(74)
Cistercian monks	1	(2)	Univ. Coll.	4	(8)
Carmelite friars	14	(16)	Balliol Coll.	9	(27)
Dominican friars	3	(5)	Merton Coll.	41	(83)
Franciscan friars	20	(23)	Exeter Coll.	2	(8)
			Oriel Coll.	1	(11)
Magdalen Hall	1	(2)	Queen's Coll.	7	(12)
St Edmund Hall	1	(2)	New Coll.	1	(21)
London Coll.	1	(1)	Lincoln Coll.	0	(4)
St Andrew's Hall	0	(1)	All Souls Coll.	2	(18)
St Mary Hall	0	(1)	Magdalen Coll.	0	(23)
Hart Hall	0	(1)			
St Paul's Hall	0	(1)	Austin friars	0	(2)
Trillock's Inn	0	(1)	Augustinian Canons	0	(3)

Although a writer is in principle included only once, that is under his house, no matter how many themes he wrote upon, there is some slight inflation of the total (ostensibly 142) since a few writers migrated from one institution to another. Our most striking finding is that in this late-medieval period Merton College evidently far outstrips in importance the other colleges and halls taken together. Merton can claim nearly a third of the total number of different writers, even far outnumbering those of unknown affiliation. Of course, having an early foundation date (1264) helps, but is not the whole story, as the cases of University College and Balliol College show.

The importance of the Franciscans was to be expected, perhaps less so the Carmelites, who nevertheless put in a steady appearance throughout our period, beginning before – but no doubt they greatly benefited from it – the acquisition of Beaumont Palace from the king in 1317. The leasing of their old seat to the Black monks of Gloucester is symbolic of the inroads made by the friars into the university at the expense of the monks. (The poor Cistercian showing no doubt reflects the smallness of their establishment at Rewley.) Of the dozen names of important scholars mentioned earlier, only one, Richard of Wallingford, a Benedictine, was not of the three groups which seem to have held the centre of the scientific stage before the end of the fifteenth century, namely the Franciscans, the Carmelites, and the Mertonians. The Dominicans are hardly a force to be reckoned with in our picture; and yet they were apparently in Oxford in the early part of our period, at least, in larger numbers than the Franciscans and Carmelites (say 100 against 75 and 45 respectively, at any one time). The reversal of Dominican fortunes had its origins in the Oxford of Grosseteste, but their history might still have gone very differently, from the point of view of the sciences as here defined, and why it did not remains a matter for explanation.

Although, in the last analysis, it is by their works that the various Oxford institutions should be judged, there are scattered through the pages of Emden's register numerous references suggestive of a loose scientific connection or interest. These are of greatly differing weight, and are often as much a liability as an asset, but a few trends are dimly visible from the indices based on them. These, the numbers in parentheses in my list, are arrived at from a count of persons at the institution in question who have any scientific allusion in Emden's biographical notice of them. Thus 41 Mertonians are counted as writers, and another 42 have scientific associations of other sorts, giving in total the 83 in parentheses.

Merton remains clearly in the lead, followed by the "unknowns", representing in most cases incidents in the lower reaches of learning – say the loan of a copy of Ptolemy, or possession of a magnet. The difference between this new sort of parameter and that which depended on written works is that writings tended to be copied and kept, valued by all and sundry, whereas the preservation of casual references depends to a large extent on the preservation of records that are unique, and thus prone to be lost. The sorts of record I have counted vary very widely. A man might be listed as a surgeon or physician, for instance, quite separately from any record of his having held a medical degree. By either of these criteria, as it happens, Oxford seems to have produced only two or three medical men per decade with academic qualifications of any weight. A man might be listed as a purchaser of astronomical instruments – as when Richard Courtenay is recorded as having patronised the great French maker Jean Fusoris; or he might be listed as an owner of instruments, as was the Austin friar John Ergum. Among other sorts of references are those to the practice of the black arts (John Stacy, Thomas Southwell, and John Stokesley, for instance).

Most of the references counted here are in some way connected with books, however, and in them we see some of the inevitable distortions. Many are records of donation – and here the friars are inevitably ill-represented. Others are names of transcribers or borrowers, or of deposits against loans. The Benedictines come out well in all this, presumably having a stronger sense of real estate than the friars: seven references involve their colleges (Canterbury, Gloucester, and Durham), while the monk William More, alias Peers, apparently owned no fewer than 69 printed books. Not all great donors were from the secular life, of course. Humphry, duke of Gloucester, William Rede of Merton, and William Gray of Balliol are well known donors of works that included many scientific examples; but the Austin friar John Ergum, Chaucer's contemporary and himself a political satirist, was able to bequeath more than 250 volumes to the York

Convent, many of them on prophecy, astrology, astronomy, and "supersticiosa". As mentioned earlier, he also owned astronomical instruments – no fewer than six, in fact.

The broader indices of scientific involvement, the figures in parentheses, do little to alter our earlier conclusions, except inasmuch as the Benedictines, for reasons explained, overtake the Franciscans and Carmelites. These indices are of much slighter significance than the first, although they reveal, of course, something that we knew independently, namely that the surviving records of the colleges are so very much better than those of the religious orders. No fewer than seven Merton teachers of astronomy and mathematics after 1370 are caught by the mere fact that chance references are made to their function; they are not known through any of their writings. There must be many from other institutions whose names are irretrievably lost. If we are to pay heed to the numbers in parentheses, there is rough justice in the fact that John Ergum, for all his gifts to York, should count for only one, as do Thomas Bradwardine and Richard of Wallingford; for where, one might ask, were the Austin friars in Oxford capable of writing on the subjects so well represented by his munificence? In the end it is the first series of figures that provides the better guide to institutional differences in matters of scientific concern.

Rough justice is the most one should ever demand from a demographic approach to intellectual history. It is perfectly possible to refine endlessly the counting of heads, that is, without reference to what is inside them, but I prefer to rest my case with two simple conclusions. The Black Death did have a markedly adverse effect on Oxford science, for whatever reason; and Oxford science was then to a very considerable degree synonymous with Mertonian science.

NOTES

[1] For a cursory survey of Oxford science in the late Middle Ages, see my two chapters in the forthcoming second volume of the official history of Oxford University, edited by R. J. A. I. Catto for the University Press.

[2] I shall refer repeatedly to [Emden 1957-74] by the author's name alone. In making my various counts, I shall occasionally supplement this work from other sources, and always silently. It would be inexpressibly tedious to annotate the many hundreds of pieces of information that go to support my essentially simple conclusion.

[3] Courtenay 1980, 700, especially n. 12.

[4] Courtenay 1980, 702-5. His data come from the Oxford computerized index, based primarily on Emden. This Index, as it stands, is of no use for a study of the present kind.

[5] Courtenay 1980, 706.

[6] Clagett 1959, 629-635.

[7] Courtenay 1980, 711-714, suggests two other explanations for the growth in the numbers

of theologians, namely an increase in chantry endowments and unexpected inheritance bringing a university education within the grasp of some who could not previously afford it. (He does not make the traditional point that catastrophes tend to turn people's thoughts to God.)

BIBLIOGRAPHY

Catto, R. J. A. I. (ed.) 1987 or 1988. *The History of the University of Oxford*, vol. ii: The Late Middle Ages. Oxford: Clarendon Press.

Clagett, M., 1959. *The Science of Mechanics in the Middle Ages*. Madison: The University of Wisconsin Press.

Courtenay, W. J. 1980. The Effect of the Black Death on English Higher Education. *Speculum* 55, 696-714.

Emden, A. B. 1957-74. *A Biographical Register of the University of Oxford, to A.D. 1500*, 3 vols, and *A.D. 1501 to 1540*, in 1 vol. Oxford: Clarendon Press.

Addendum

Although this chapter was not primarily concerned with reasons for imbalance as between the apparent performances of different institutions, some observations by Dr Roger Highfield relating to the end of my first paragraph of p. 368 are worth repeating here. He notes that the date of foundation of a college is comparatively less important than the scale of the foundation and the aims of its founders. Merton was well endowed, whilst the founder of Balliol forced its members to leave when they became masters. These facts taken together explain why, for example, Bradwardine left Balliol for Merton. After 1340 the institution of new Balliol fellowships changed the situation, but prior to this, Merton had some of the magnetism of the All Souls of later history. Dr Highfield notes further that the general picture I have drawn fits well with the fact that Archbishop Islip, who presided over the see of Canterbury in the worst period of pestilence, chose to found Canterbury College at Oxford afterwards in an attempt to remedy an unhappy situation there.

NICOLAUS KRATZER, THE KING'S ASTRONOMER

*Après l'invention du blé ils
voulaient encore vivre du gland.*

English humanism of the early Tudor period is perhaps best interpreted as a series of separate and often isolated phenomena. The lives of Grocin and Linacre, More and Colet, offer analogies with the critical and philological spirit of the Quattrocento, but in their time no English mathematician or astronomer of any stature appears to have been inspired by their example, in a way which might have seemed inevitable had the tradition of humanism been all-pervasive. Linacre might translate Proclus on the sphere, or rather the excerpts from Geminus which he took to be by Proclus; but this proved to be an isolated exercise without significant consequences[1]. Richard Fox's founding of Corpus Christi College could well have provided educational cohesion for such fragments of the new learning as were able to survive in an essentially hostile environment, and yet at least two generations were to pass before the emergence of an Oxford teacher comparable with, for instance, Niccolo Leonico Tomeo. In Padua Tomeo preached actively and systematically against scholasticism. He did translations from Aristotelian and pseudo-Aristotelian treatises, and influenced generations of scholars — occasionally, it is true, with scientifically unfortunate consequences[2]. Tomeo numbered Cuthbert Tunstall, Reginald Pole, and Pietro Bembo among his friends and pupils, but he taught Greek also to a man whom it is natural to compare with Nicolaus Kratzer — namely Nicolaus Copernicus. Kratzer was the younger man. He was more fortunate in the patronage he enjoyed, and was not short of enlightened friends, if that is an apt description of

---

[1] The translation of this work, done for Prince Arthur, was first issued with Firmicus Maternus, *Astronomicorum libri V* (Venice, Aldus Manutius, 1499). Delambre was perhaps the first to discover that the text was Geminus, *verbatim*.

[2] See M. Schramm, "The mechanical problems of the 'Corpus Aristotelicum'...", *Atti del Primo Convegno Internazionale di Ricognizione delle Fonti per la Storia della Scienza Italiana i Secoli XIV—XVI*, Sezione V, Vol. I, pp. 21—33.

Erasmus, More, Tunstall and Colet. That Kratzer achieved much less than Copernicus was obviously not unrelated to a disparity in intellect, but it was due also to the absence of that critical and even pedantic literary spirit which refused to accept medieval versions of classical treatises. The corruption of the scholastic versions was, and still frequently is, much exaggerated. What mattered most was a frame of mind; and there is no evidence for supposing that Kratzer's sense of values was ever more than that of a knowledgeable artificer. It is beside the point to ask whether he was a better or worse astronomer than other Oxford men. Certainly the prospect is bleak, if we take the Merton astronomers of the day for comparison: the learning of Blis, Dens, and Mosgrove for example, would hardly have astonished the Mertonians of two centuries before.

Kratzer never approached the apex of either the astronomy or the mathematics of his day. If he established or confirmed in England any tradition, it related to the craft and rationale of instrument-making. In this practical connection it is natural to compare him with Hans Holbein, the author of Kratzer's portrait and also his friend, and his compatriot at the English court. Holbein came to England in response to a plea from Erasmus, who complained that the arts in Europe were freezing. Holbein faced a religious, rather than a literary, crisis: portraiture was one of the few art forms which the Reformation did not bid to suppress, and Holbein rose brilliantly to meet this changing situation. As far as can be seen, Kratzer was never conscious of any comparable intellectual crisis. In astronomy, he was as conservative as any of the much reviled scholastics. If he introduced anything new into the English astronomical tradition, this was possible only because his source texts were slight variants on the works current in England. There is something to be said for fame as a diagnostic of influence, and even of human value. The signs of Kratzer's influence are few and uncertain. But what could be the fame of a sundial, by comparison with that of the portrait of a king, a duke, or an ambassador?

Kratzer was neither a Holbein nor a Copernicus, but in this, after all, he resembled all but two of his contemporaries. If his life seems to have a certain aimless quality, this is because we know so little of it: from its drift, however, we can at least detect the direction of one or two social currents. It is often said that one of the characteristics of the Renaissance was the seemingly boundless opportunity for self-expression which its rulers had. "Self-expression" is an odd phrase to use of a patron of the arts — as though one could compound the humanity and individualism of Brunelleschi, Ghiberti, Donatello, and Gozzoli to produce a bit of the soul of the early Medici. But just as each of them helped his patrons to fashion the image they treasured of themselves, so did Kratzer, alike at court and at Oxford. Like them, he was drawn by the obvious financial inducements and protective calm of patronage, and in his case it is clear that London had a greater attraction than Oxford. His preference for the metropolis is not really surprising. Many of his compatriots were there. More craftsman, perhaps, than scholar, he might well have been bemused by the intellectual climate of Oxford. His well-developed mercantile instincts would have been more easily indulged and less alien at court than at Oxford. But above all, there was a social gradient running counter to that of the Thames valley. This fact must have been so obvious, at the time, that Kratzer would doubtless have been at a loss to understand why his preference for the court should be thought to need an explanation.

Kratzer seems to have first come to England in 1517, or perhaps 1518. Of his earlier life little is known, and as to the social pressures to which he was subject, only the most obvious guesses can be made. He was born in Munich in 1487[3]. The son of a saw-smith, he no doubt learned something of the art of metal-working from his father, before leaving his native town for the University of Cologne (matriculated 18 November 1506; B. A. 14 June 1509)[4]. There, as it happens, he was an almost exact contemporary of Heinrich Cornelius Agrippa, the occultist natural philosopher and medical writer, who was to visit Colet and the English court for a short time in 1510[5]. From Cologne, Kratzer went to the small University of Wittenberg, a fact known from Oxford records[6]. Needless to say, he must have been there at a time when Luther was still preaching scholastic theology to the university, which had been newly opened in 1502 by Frederick, the Elector of Saxony, and which was as yet blissfully unaware of the storm about to break around it.

Kratzer now disappears from sight for several years. Although he might have practiced as an instrument-maker, we may be reasonably certain that he was resident for some time in the Carthusian monastery of Maurbach, near Vienna, where he copied out a number of astronomical treatises around the year 1515[7]. Peter Giles writes to his erstwhile teacher Erasmus in January 1517 that the skilled mathematician Kratzer is on his way with astrolabes and (armillary) spheres, not to mention a Greek book; and in the following November back comes a mysterious reply, not necessarily the first:

---

[3] The painting of him by Holbein, discussed further below, shows on the table a paper which reads: "Imago ad vivam effigiem expressa Nicolai Kratzeri monacenssis qui Bavarus erat quadragessimum primum annum tempore illo complebat 1528".

[4] Max Maas made a thorough search of the extant registers of Munich and Kratzer's two universities. See *Nikolaus Kratzer, ein Münchener Humanist* (Munich, 1902). Reprinted from the *Allgemeine Zeitung* (München) numbers 64 and 65:18 and 19 March 1902. Maas uncovered details of the trades and addresses in Munich of a number of men by the name of Kratzer, but looked in vain through university lists and histories. Hermann Keussen provided P. S. Allen with the dates of Kratzer's matriculation and degree. See *Opus epistolarum Des. Erasmi Roterodami*, ed. P. S. Allen (Oxford, 1910), II, p. 431.

[5] Henry Morley, *The Life of Henry Cornelius Agrippa von Nettesheim*, 2 vols. (London, 1856). It is known that Kratzer met Heinrich Glareanus at Cologne, and that he renewed the acquaintance years afterward in London. See R. Wolf, *Biographien zur Kulturgeschichte der Schweiz* (Zurich, 1858—61), I, p. 13.

[6] Maas, *op. cit.*, p. 7, notes that the Munich astronomers Wolfgang and Paul Seberger were both at Wittenberg at the time.

[7] See p. 398 below.

Magnifico d[omi]no p[ri]vato
hu[m]ili[m]i[n]g [?] m[i]cola[us] Krarcz[?]

Heri d[omi]no ep[isto]la in [Christ]o amata
horis [?]palentin[us] ad me missi ut v[est]ra[m]
p[rae]stantia[m] Her Io[hannes] halhe[?] dedi ut v[?]
q[ua]e scripta dn[?] in man[?]b[us] gabrieli[?] [?]
antverpie retenta sui ut m[ih]i mitte[?]
Iob rumore a papistis sparsam qu[?]
statu[m] redactu[m] affirmat ut Hig gra[?]
apparet / Scripsit[que] li[tte]ras ad ca[n]cell[arium]
et suo archidiacono quas li[tte]ras illi[?]
missi [et] q[uas] li[tte]ras ep[iscop]o Culmensi / has tu[?]
q[ui] iste mi[?] in curia regis [?]
Wormacie habuit · Nulla nova e[?] [germa]
m[i]si tu[?] p[ro]xima estate venturu[m] esse d[?]
papiste mauie optat ut Lutheran[os] ab[?]
puniret [?] Ep[iscop]i roman[us] p[er] m[?]telg ad h[?]
m[is]si · hi[?] cu[m] [?] p[er]eg[ri]nati fuerut [?]
pauc[?] ab illis habere · h[?] milites g[?]
nescio ut vult quod ali[?] gravi[us] [?]
pl[us] discordiam studet et sem[per] hic o[?]
om[n]is fuit · nulla alia m[?]si magnifi[centia]
pauperem[?] nicolau[m] remem[?]rat ut ali[q]u[?]
felic[?] vivere possim / [?] me [?]
a[?]pan[?] nove / filia fide nove / ita q[?]
est ut sem[per] a[?]pa[?]a fide[?] do[m]i[?]
uxor mea in die assumpt[i]o[n]is mar[ie]
filia[m] maue pep[er]it / Unde pl[us]
q[uo]n[iam] a[?]por[?]on[?]s videat / om[?]

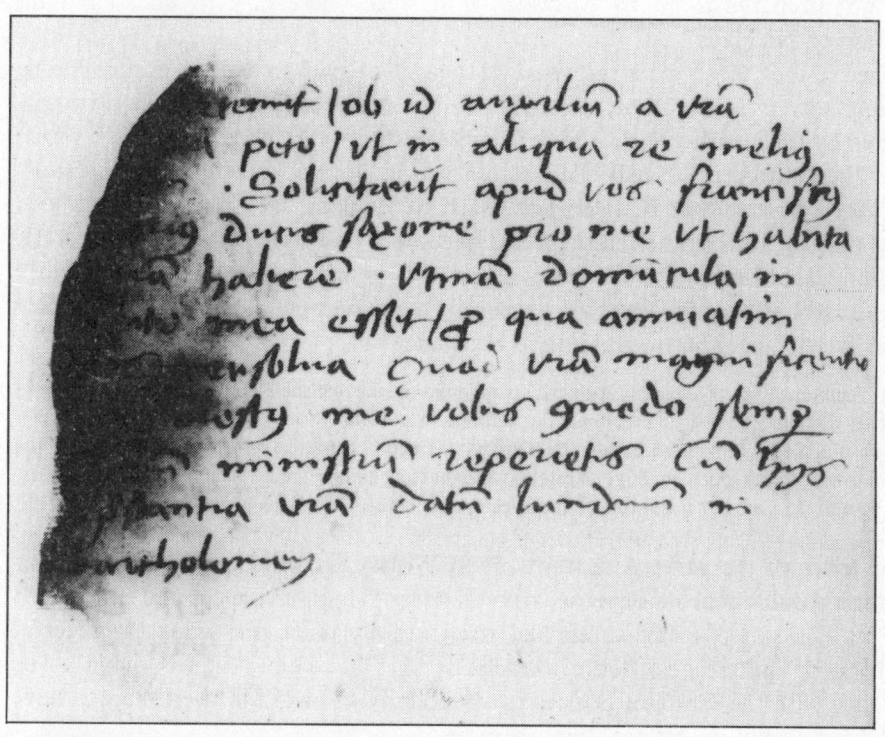

Plate 6. A specimen of Kratzer's hand, in a letter dated 24 August 1538, now British Library MS Add. 21075, fol. 256 (old numbering 276).

Admonish Nicolaus, first of all, to keep the thing secret, and to tell no-one whom he is to visit in England, or who summoned him. He is to invent an excuse as far from the truth as possible. You will understand the problem from the Secretary's letter and from the spoken explanation of my [courier] James[8].

Erasmus was prepared to go to Antwerp, if it were "necessary to do something in the affair of Nicolaus".

Whether or not these were instances of Erasmian paranoia, it is impossible to say. Having secrets is a notable academic foible. It is conceivable that Erasmus was worried by the hostility shown to foreigners in the riots of May Day, 1517, which in July of the same year he had cited as his reason for not returning to England[9]. It seems, nevertheless, that Kratzer first went to England late in 1517 or early in 1518, and that by 1529 at the latest he had become a trusted servant of Henry VIII, in whose household accounts he figures as an astronomer, and who had apparently by 1520 already given him leave of absence. Could he not, Tunstall asked Henry in a letter from Liege, be of diplomatic assistance?

> Met at Antwerp with [Nicolaus Kratzer], an Almayn, deviser of the King's horologes, who said the King had given him leave to be absent for a time. Asked him to stay till he had ascertained if the King would allow him to remain with Tunstal till the coronation and the assem[bly] of the Electors be over. Being born in High Almayn, and having acquaintance of many of the princes, he might be able to find out the mind of the Electors touching the affairs of the empire[10].

A similar letter of the same date went off to Wolsey, in which it was pointed out that the king would incur no expense, since Kratzer "shall have meat and drink with Tunstal"[11]. The son of a saw-smith had risen rapidly under the aegis of powerful English friends, and this, no doubt, largely because he spoke "high Almayn". His acquaintance with the German princes is more likely to have followed than to have preceded his acceptance by the English court.

Kratzer's visit to the Netherlands in 1520 coincided with Dürer's tour, which was described by the artist in a well known journal. There it is recorded that Dürer

---

[8] Allen, *op. cit.*, II, 431–2, Petrus Aegidius from Antwerp, 18 January 1517: "Mei fuit officii ut ad te scriberem, dum isthuc Nicolaus Bavarus, matheseos peritus, profisceretur. Adfert astrolabia et sphaeras aliquot isthuc vendendas". For a reply, see vol. III (1913), p. 128, from Louvain, 3 November 1517: "Mone Nicolaum in primis ut rem dissimulet, neque cuiquam effutiat ad quem adeat in Angliam aut cuius nomine accersitus".

A remark in another letter to Peter Giles (Aegidius) of December 1517 (*ibid.*, p. 166) to the effect that "everything James has said is true, I have complete faith in Nicolaus", almost certainly refers to Nicolaus of Hertogenbosch, rather than to Kratzer, as used to be supposed. See Allen's note, *ibid.*, p. 142. Note that in connection with Erasmus, the name "Cratzerus" in isolation invariably means his Augsburg correspondent, the theologian Matthias Kretz.

[9] *Ibid.*, p. 6: "Et Anglie motus timeo et servitutem horreo".

[10] *Letters and Papers Foreign and Domestic of the Reign of Henry VIII...*, arranged and catalogued by J. S. Brewer, Vol. III, pt. 1 (1867), p. 374, from B. M. Vit. B. xx. 163 (12 October 1520).

[11] *Ibid.*, p. 375, from Ellis, 3 Ser. I. 231.

did a portrait of Kratzer, and also that Kratzer was present on one of the occasions when Dürer portrayed Erasmus. The Kratzer portrait is lost. In 1943, Robert Eisler claimed to have identified it with a painting in Brussels[12], but the original portrait is known to have been in silver point[13], and there are other good reasons for rejecting Eisler's claim[14].

We know almost nothing of Kratzer's duties at the English court, and it would be rash to draw out analogies with the far better documented life of Master William Parron (Gulielmus Parronus Placentinus), the Italian astrologer at the court of Henry VII[15]. Kratzer's surviving manuscripts contain only a few brief and scattered notes on anything remotely resembling judicial astrology. He was himself drawn into the orbit of Sir Thomas More, who frequently professed his distaste for the subject, and who even went as far as to put an avowal of disillusion in the mouth of the dead queen, Elizabeth of York, in his *Ruful lamentacion... of the deth of quene Elizabeth*[16]. He probably influenced Henry, for after he was appointed Chancellor of the Duchy of Lancaster in 1525, he gave the king instruction in matters astronomical[17]. More's stance, however, is not likely to have acted as more than a weak deterrent to an all-pervasive habit of thought. To take but two later facets of court astrology: first, the engraver Thomas Gemini, who had come to England from the Netherlands in or before 1524, and who must have been known to Kratzer, was to make two magnificent astrological astrolabes, one for Edward VI (dated 1552), and one for Elizabeth I (dated 1559)[18]. And, second, at some time shortly after Mary's accession,

---

[12] "Oxford personality of 16th century. Subject of Dürer portrait discovered to be Nicolaus Kratzer". *Oxford Mail*, 5 August 1943.

[13] See Moriz Thausing, *Albert Dürer, His Life and Works*, tr. F. A. Eaton (London, 1882), II, p. 175.

[14] Erwin Panofsky, *Albrecht Dürer*, II, 164, 3rd ed. (Princeton, 1948) did so, pointing out that the Dürer portrait was specifically a *drawing* (which may be the drawing L 380 in the Louvre); and that the features of the sitter of the Brussels painting (Musée des Beaux-Arts) do not remotely resemble those of the Holbein portrait of 1528, in which Kratzer appears younger than the man in the supposed Dürer portrait of 1521. Winkler had actually rejected his attribution to Dürer before Eisler's article appeared.

[15] See C. A. J. Armstrong, "Astrology at the court of Henry VII", [in] *Italian Renaissance Studies*, ed. E. F. Jacob (Faber, 1960), p. 433—54.

[16] Quoted by Armstrong, *ibid.*, p. 453—4.

[17] "A good parte [of More's twenty years and more of service with Henry] used the kinge upon holidaies, when he had done his owne devotions, to send for him into his travers, and there sometyme in matter of Astronomy, Geometry, Devinity, and such other Facultyes, and sometimes of his worldly affaires, to sitt and confferre with him. And other whiles wold he, in the night, have him uppe into his Leades, there for to consider opracions of the starres and planetes. And because he was of a pleasaunt disposition, it pleased the kinge and Quene... commonly to call for him to be merry with them". William Roper, *The Lyfe, Arraignment, and Death of that Mirrour of all true honour and vertue, Syr Thomas More*, 1626 ed., E. V. Hitchcock (E. E. T. S., 1935), p. 11,

[18] See R. T. Gunther, "The astrolabe of Queen Elizabeth", *Archaeologia*, LXXVI (1937). p. 65—72.

John Dee was invited to calculate the nativities of the queen, her husband Philip, and even the Princess Elizabeth[19]. Court astrology was disreputable, but alive.

How long Kratzer stayed in the Low Countries after October 1520 we cannot say, but he might have stayed on the continent until after the Diet of Worms was over, early in the following year. "The Lutheran drama is over", said Erasmus in a letter of 24 May, and no doubt Tunstall would have felt that he could dispense with his intermediary, had he kept Kratzer in Europe so long. Whatever the date of Kratzer's return, it cannot have been long before he found himself in the service of Sir Thomas More, then treasurer of the exchequer. Writing from court on 23 March 1521 to his "school", namely a tutorial group which included his own children, Margaret Gyge, his step-daughter Alice Middleton, and others, More rejoiced that "Mr Drue is returned safe[20]", and admitted that if he loved them less he would envy them their good fortune in having so many great scholars for their masters:

> For I think Mr Nicolas is also with you, so that you have learned astronomy from him. I hear that you have advanced so far in the subject that you now know not only the Pole Star or the Dog Star, or any other of the constellations, but also (which speaks for a skilled and absolute astronomer) from among the chief and principal stars you can distinguish the Sun from the Moon![21]

The school was admonished to remember Lent, and warned against neglecting, by an excessive concern for the mind's elevation, to raise the soul up to heaven. Within a very short time More wrote again, mentioning Nicolaus Kratzer, "amantissimus nostri et astrologicarum rerum peritissimus", and again congratulating the school on its good fortune:

> In the space of a month, and with only a little effort on your part, you will learn so many and such sublime wonders of the Eternal Artificer, which so many men of preeminent and almost superhuman intelligence have discovered only with hot toil and study, not to say with shivering cold and nightly vigils down the ages[22].

---

[19] *Calendar of State Papers (Domestic)*, 1547—80, eds. R. Lemon and M. A. E. Green (London, 1856—1872), p. 67.

[20] Other tutors were John Clement, who later went to Oxford, William Gonell, and Richard Herde, who in 1524 published with More's help a translation of Vives' *De institutione foeminae Christianae*.

[21] "Sed D. Nicolaum nunc superesse vobis puto, ita quicquid habet astronomicarum rerum edicistis. In quibus adeo vos provectos esse audio, ut non iam polarem tantum stellam, aut caninam, aut aliud quidvis e gregariis astris verum etiam (quae res peritum et absolutum Astrologum postulat) in praecipuis illis ac primariis syderibus, solem dinoscatis a luna...," E. F. Rogers, *The Correspondence of Sir Thomas More* (Princeton, 1947), pp. 249—50. The year 1521 is not stated on the letter but there is little doubt as to its correctness.

[22] "[*Dixistis*] quod Nicolaus et amantissimus nostri et astrologicarum rerum peritissimus, iterum vobis caelestium corporum sphaeram auspicatus est, et illi gratias ago et vobis istam gratulor faelicitatem; quibus intra unum mensem continget minimo labore pernoscere tot et tam sublimia Opificis aeterni miracula, quae tam multo, tam praeclara fereque supramortalium sortem ingenia, tot sudoribus et studiis, imo tot algoribus et sub dio pervigilatis noctibus multarum aetatum decursu repererunt". *Ibid.*, p. 254. The year (1521) is fixed by a reference to the marriage of Margaret (to William Roper). She continued to be a member of the school after her marriage at sixteen.

It is difficult to decide whether Kratzer's move to the More household was meant to be for a long or a short term. More's star — if he will forgive the expression — was in the ascendent. He was not yet Speaker, but neither had he yet crossed Wolsey. Kratzer was soon to leave More's household, however, for Wolsey's Oxford foundation; but Kratzer's writing on Holbein's well-known sketch for the portrait of More's family, done around 1527, suggests that the astronomer returned to Chelsea. Payments to him from the royal purse are not all calendared, but were probably made regularly. He was worth £ 5 a quarter, it seems, in 1519, and no more in 1531[23]. It seems, however, that he could supplement his salary by trading[24]. Other adornments to Henry's court of 1519 included spangles for the guards' jackets to the tune of £ 1,419.18s.2d; but Kratzer had no cause to complain. Two physicians could command £ 25 and £ 10 a quarter, admittedly, and William Norrice, the master of the hawks, £ 10; but the painter Vincent Voulp got £ 5, while Anthony Trasillon, a clock maker, received £ 1.10s.5d, and Holbein was never much better off than Kratzer and Voulp. Kratzer's Oxford colleague Vives was apparently considered to be worth the same as he, or appreciably more, according to the interpretation we place on a letter by Erasmus; and therefore it seems unlikely that Kratzer, during his Oxford interlude, benefited financially from the move[25]. More leaves no doubt in our minds that he considered both Oxford fare and Oxford society to be low in his scale of values. After resigning the Great Seal in 1532 he would say:

I have been brought up... at Oxford, at an Inn of Chancery, at Lincoln's Inn, and also in the King's court — and so forth from the lowest degree to the highest; and yet have I in yearly revenues at this present left me little above an hundred pounds by the year[26].

If Kratzer's next patron[27] could offer only Oxford, it must have been an Oxford with considerable promise, and this apparently of a kind transcending financial

---

[23] *Letters and Papers*, ed. Brewer, Vol. III, pt. 2, p. 1535 ("Nich. Kratzar of Baver, 100s." 1519); Vol. III, pt. 1, p. 408 ("Nicholas Craser, an estronomyer, 100 s." 1520); Vol. XIV, pt. 2 (ed. J. Gairdner), p. 311 ("Nich. Crazer, 100s." 1531).

[24] *Ibid.*, Vol. IV, pt. 2 (ed. Beaver), p. 1596. On 28 October 1527 Kratzer was given a license to import from Bordeaux and other parts of France and Brittany 300 tons of Toulouse woad and Gascon wine. At much the same date an usher and a sewer of the Chamber were given similar licenses, which suggests that this was a standard way of obtaining a supplementary income.

[25] Vives' salary was stated by Erasmus (Allen, III, ep. 658, col. 773) to be £ 60 or 60 angels: "Quod si verum est, opinor sexaginta libras propositas esse pro sexaginta Angelatis". (25 Sept. 1523 to Conrad Goclen). If the salary was in angels of 6s. 8d, it tallied with that given to Kratzer from the royal household account.

[26] From Roper's *Life of Sir Thomas More*, ed. R. S. Sylvester and D. P. Harding (New Haven, 1962), p. 226.

[27] There were patrons and patrons, as the astronomer Johannes Schöner found to his cost. At about the time of the Protestant Diet of 1524, held in Nuremberg, Schöner handed over to Cardinal Campeggio books and globes to the value of twenty florins, but was paid nothing in return, on the grounds that he was a "Lutheran" which is to say that he was a married priest. His anxiety to hide his whereabouts are graphically explained in a letter to Willibald Pirckheimer of Easter 1525. See J. H. Jaeck, *Pantheon der Literaten und Künstler* (Bamberg, 1812), p. 1027.

advantage. Wolsey's plans for a new collegiate foundation in Oxford were at first more successful than his machinations both personal and ecumenical, in Europe. In 1523, as in 1522, he failed in his bid for the papal office; but in 1524 he was empowered by bulls from the Pope to dissolve St. Frisewide's Priory and twenty other religious houses to provide finances and a site for his new college. The following year saw the defeat and capture of Francis I at Pavia; but it saw also the issue of a charter for Wolsey's college, and the commencement of building operations[28]. These and subsequent events in the history of the college, which was to be refounded after Wolsey's fall as "King's College", are well documented, but there is much confusion over the nature and timing of Wolsey's initial appointments of lecturers, and over the relationship of the lecturers (or "readers") to the foundation adjacent to Wolsey's both in time and situation, Corpus Christi College.

The accepted accounts of this episode in Oxford history all apparently point back to Anthony Wood, whose version rests largely on an amalgam of circumstantial evidence drawn from the biographies of individual lecturers[29]. Wood's method would have been perfectly sound had not the evidence itself been distorted by that old Oxford collegiate custom of claiming great men as fellows. In this instance Richard Hegge is as much to blame as anyone: he does not make Kratzer a fellow of Corpus Christi in his transcript of the catalogue of admissions to the college (1628 or earlier), but, as we shall see later, he did so on his drawing of the college sundial, which was at least twice copied[30]. In the *Collectanea* of Miles Windsor and Brian Twyne, which was known to both Hegge and Wood, we find this statement[31]:

Quatuor publici lectores Cardinalitii simul in Collegio Corp. Christi: Ludovicus Vives, Tho. Lupsett, Nich. Cratcherus, Tho. Moscroffe.

When, in his survey of the antiquities of the city, Wood discusses the statutes given by Fox to his college in 1517 (approved 20 June), he points out that the twenty fellows were to be qualified in part by their place of birth — five from the Winchester diocese, one from the diocese of Durham, and so forth[32]. No provision was made for any foreigner, despite which there are those who have spoken warmly of Fox's vision in appointing Vives and Kratzer. Wood explained, furthermore, how Fox, from his twenty fellows, appointed three — readers in Humanity, "Greece" and

---

[28] See J. G. Milne and J. H. Harvey, "The Building of Cardinal College", *Oxoniensia*, VIII and IX (1943–4), 137–53; *Statutes of Colleges of Oxford*, Vol. II, sect. 11 (1853), p. 3–184.

[29] His case is set out in J. Gutch's edition of Wood's *History and Antiquities of the Colleges and Halls in the University of Oxford*, Vol. II, pt. 2 (1796), p. 834. The first statutes for Cardinal College set out the conditions of employment of four readers whose duties were to teach only sophistry, dialectic, philosophy, and *litterae humaniores*. There were of course many more academic appointments envisaged — about 180, in fact.

[30] The catalogue is in the college archives. On the sundial, see pp. 387 below.

[31] MS. C. C. C. 280, f. 215a. Cf. MS. Twyne 24, pp. 411–2, where the same names are given.

[32] Andrew Clark, ed., *Survey of the Antiquities of the City of Oxford composed in 1661–6, by Anthony Wood*, I (1889), p. 540–1.

Divinity — to lecture to the society or to ourtsiders; and yet Wood found nothing incongruous in adding a note listing the four readers named by Twyne, of whom Kratzer, and perhaps even Mosgrove, would have been ill-fitted for any of the three positions. Vives, moreover, appears in the *Athenae* as having incorporated LL.D. on 26 October 1523, he "being this Year a Lecturer in *Oxon.* as I have told you among the Writers, under the Year 1544"[33]. At this time Thomas Mosgrove was Commissary (Vice Chancellor), "now or lately, Fellow of *Merton* College". He was almost certainly still a fellow of Merton in 1524, however, since Warden Philips then offered to resign on condition that Mosgrove's name be among three from which the Visitor was to choose a new warden[34]. And if Mosgrove was no fellow of Fox's college, this argument for Vives' and Kratzer's fellowships is void. To continue with this implausible account: Vives, according to Wood, who seems to follow Hegge's Catalogue, was made a fellow in 1517. Wood is constrained to add that Vives was then in Louvian, and that he was invited to England by Wolsey "as it seems" in 1523[35]. Wood is obviously doing his best to reconcile the university records of Vives' incorporation with Corpus mythology. He was right to have qualms, for there is not a shred of evidence to connect Vives with Oxford before 1523. By his own statement, he had spent fourteen years in Bruges, lecturing at the University of Louvain after receiving his licence to teach on 5 May 1520[36]. In July 1522 he wrote to Erasmus from Louvain, and dedicated his commentary on the *De civitate Dei* to Henry VIII from that town, the dedication being acknowledged by Henry in a letter of 24 January 1523. The association of Vives with Oxford obviously began only in 1523. It seems that he came to England under the aegis of More, Mountjoy, and Wolsey, and that he was called on by Queen Catherine, his fellow-countrywoman, to sketch a course of study for Princess Mary, then in her eighth year. With Linacre, Vives was asked to direct her education. He dedicated his work on the education of women to Catherine[37], and for Mary wrote a plan of studies in *De ratione studii puerilis*, which he completed in October 1523[38]. On 16 June 1524 Vives writes to Erasmus from Bruges, explaining that he had left England temporarily, in order to marry. He visited England again in due course, and was even led to an English prison for six weeks, as well as to banishment from court, as consequences of his expressed — and not unnatural — sympathies with Catherine. He was again in Oxford in November 1524,

---

[33] Vives supplicated for incorporation on 10 October.
[34] G. C. Brodrick, *Memorials of Merton College, Oxford* (*O. H. S.*, Vol. IV, 1885), p. 163.
[35] *Op. cit.*, Vol. I, col. 64.
[36] Article by J. B. Mulliner in *D. N. B.* A much more complete biography of Vives, and a general appraisal of his work, is in the introduction to Foster Watson's *Vives: On Education* (Cambridge, 1913). The main part of this book is a translation of the *De tradendis disciplinis* of Vives.
[37] Catherine asked More to translate the Latin edition, but for want of time More did nothing beyond make some corrections to Richard Hird's translation, mentioned previously.
[38] The other plan of studies was for Mountjoy's son, Charles. Watson, *op. cit.*, xxv—xxvi, lxxiv.

but probably not for very long, although between then and 1528 he spent a part of each year in London, returning regularly to Bruges.

Vives had ties with Corpus Christi which were obviously informal but close: he lodged there, like all Wolsey's lecturers before the buildings of Cardinal College were ready, and when he visited Bruges for his marriage, he seems to have been accompanied by two young fellows of Corpus, Richard Pate and Anthony Barker, to both of whom he apparently gave instruction[39]. Despite all this, Vives was always Wolsey's man. In the March following his arrival in Oxford, he reported to Wolsey that the youth of Oxford had grown daily more inclined to solid and true learning, that his *quaestiones* had begun to eject Albertus Magnus from the schools, and that in dialectic and other philosophy he had removed many other *pravas opiniones*[40]. After his marriage Vives promised Wolsey that he would return in September, although of his second stay there is only one trace, in the form of an outspoken letter to the king, urging intervention on the part of Francis I[41]. (Since Vives did in fact return for a time to Oxford, are we to suppose that Wolsey countenanced married lecturers on his foundation?) It has often been supposed that Fox, who gave statutes to his new college in 1517, was then and subsequently on poor terms with Wolsey, but Thomas Fowler gave convincing reasons for rejecting the idea, and supports the belief that Wolsey's lecturers were not only lodged in Fox's buildings, but gave their public lectures there until Wolsey's buildings were ready[42].

What little we know of Kratzer at this time makes much better sense in the light of the movements of Vives than if we accept the more usual account. According to Wood, Kratzer was admitted to Corpus Christi College on 4 July 1517 and

> about that time he reading astronomy in the University by the command of K. Hen. 8 was soon after made by Cardinal Wolsey his Mathematic Reader, when he first settled his Lectures there[43].

The names of Vives and Kratzer nevertheless appear on no contemporary record in relation to Fox's college, nor is any of Wolsey's lecturers mentioned in the *Libri magni* of the appropriate period. In a letter to President Claymond, written in 1519, Erasmus

---

[39] P. S. A[llen], "Ludovicus Vives at Corpus", *The Pelican Record*, VI (Dec. 1902), 157. Vives frequently used Fox's bee-simile. Fox, it will be recalled, had designed his college to be a "bee-garden wherein scholars like ingenious bees are by day and night to make wax to the honour of God, and honey-dropping sweetness to the profit of themselves and all Christians". (Watson, *op. cit.*, xxvii.) To Erasmus, Vives wrote that "Claymund [first president of Corpus] and the best men of the University of Oxford send you their greetings". (*Ibid.*, xxxix.)

[40] *Loc. cit.* At Louvain Vives had lectured chiefly, he tells us, on Cicero, Pliny's *Natural History*, and Virgil.

[41] *Ibid.*, p. 158.

[42] T. Fowler, *The History of Corpus Christi College, Oxford* (O. H. S., Vol. XXV, 1893), p. 15—20.

[43] *Athenae*, 2nd ed., Vol. I (1721), col. 79. Wood lists Kratzer and Vives alike as among those "taken in by the Founder at the entreaty of noble persons, even till the 2nd of July 1524", but without evidence. See Gutch, *ed. cit.*, Vol. IV (1786).

speaks of the great interest which had been taken in Fox's foundation by Wolsey, Campeggio, and Henry VIII himself, and predicted that the college would eventually be numbered "toto terrarum orbe inter praecipua decora Britanniae", and that its "trilingual library" (not a very apt description, in fact) would attract more scholars to Oxford than were formerly drawn to Rome[44]. Kratzer, as we have seen, was known to Erasmus. Had we thought him to have been one of Fox's bees by 1519, it would have been cause for comment that the letter makes no mention of the fact.

Kratzer, in 1521, was in the service of the king, or Sir Thomas More, or both. Between 11 February and 23 March 1522/3 he went through the process, first of supplicating for incorporation as B. A. and proceeding to M. A. He did this all on the strength of his standing as a bachelor of Wittenberg and of Cologne, and when he incepted in arts it was on the recommendation of four masters only[45] — hardly a grace which would have been needed by a man who had held a high university position for five or six years, or even for one. The matter is complicated, however, by a further statement which has all the marks of being in Kratzer's autograph. There he explains that he came to Oxford in 1520 to lecture on astronomical and geographical subjects by command (or permission — *iussu*) of the king, whose servant he was:[46]

Anno 1520 ego Nicolaus[47] Kracerus Bavarus Monacensis natus, servus regis Henrici viii, iussu illius perlegi Oxoniae Astronomiam, super sphaeram materialem Johannis de Sacro busco et compositionem astrolabii et Geographiam Ptholomaei. In illo tempore erexi columnam sive cilindrum ante ecclesiam dive Virginis cum lapicida Wilhelmo Aest, servo regis. Eo tempore Lutherus fuit ab universitate condemnatus, cuius testimonium ego Nicolaus Kracerus in columna manu propria scriptum posui.

It is hard to reconcile the date of 1520 with the supposition that MS. K is Kratzer's autograph. The condemnation (*eo tempore*) of Lutheran doctrine by the uni-

---

[44] Allen, *ed. cit.*, III (1913), 620—1 (27 June 1519).

[45] See university archives, Register H. 7, ff. 100r—101v, 102v and 104r. Cf. MS Twyne 24, p. 409, this being a transcript of relevant entries. There are trivial differences between them and C. W. Boase, *Register of the University of Oxford*, I (*O. H. S.*, Vol I, 1884), 129. Boase overlooks the entries for 11 February (which Twyne mistakenly gives as 19 January) and 20 March. He notes that on 18 March it was agreed that eight masters may suffice "ad deponendum pro eo *in suis inceptionibus*", the reason being, "*quod non est notus apud magistros omnes; apud aliquos tamen est notissimus et probatissimus et in mathematicis et in philosophicis*". But on 20 March it was further agreed that *four* masters would suffice (reg. H. 7, f. 102v).

[46] MS C. C. C. D. 152 (H. O. Coxe's number, the volumes being now deposited in the Bodleian Library), f. 1v. The volume as a whole is discussed more fully below. For convenience I shall henceforth refer to it as MS K. The passage quoted here is copied out in full, with only minor variants, on a scrap of paper now inserted inside the end cover of a beautiful presentation copy of Tycho Brahe's *Astronomiae instauratae mechanica* (1598) with Bodleian shelf mark Arch. B. c. 3. Although this fragment is written in a sixteenth-century hand, which was considered by the compiler of the serial catalogue to be possibly Kratzer's, it bears no resemblance to anything in MS K, which is largely autograph, as will be shown.

[47] Omitted in the Arch. B. c. 3. fragment, where also "Krazerus" is preferred to "Kracerus".

versity, which was the occasion of Kratzer's placing his own testimony to the fact on the sundial he had erected for the university, is well documented. From the university archives it emerges that Wolsey, disturbed by the several ways in which Lutheran doctrines were being manifested in Oxford, called upon the university to nominate delegates to a conference convened by him to examine the problem. A group of senior Oxford theologians — Thomas Brinknell, John Kynton, John Roper, and John de Coloribus — attended the meeting, and brought back their message to convocation[48]. This they did on 21 April 1521, and therefore it would have been perfectly natural for a dial erected in 1520 to have then displayed a notice of the condemnation of Luther by the university.

On balance, however, I do not think that this was so. Kratzer might, of course, have been in the More household for only a few months. It is conceivable that, as the king's servant, he was permitted to read lectures without incorporating, although his stated themes were suspiciously elementary, and were well within the competence of many Oxford masters of the time. There are two much better reasons for doubting the evidence. On the same page of MS K, and in the same hand, there is a verse written for the dial. Although it ends "Anno Domini 1520", it begins with the date 1523 in words:

> Annis mille tribus quingentique adde decem bis
>   Invenies tempus quo hic situatus eram;
> Oxonie rector Thomas Mosgrave, medicinam
>   Qui profitebatur quique peritus erat.
> Me posuit lapicida suis Guilielmus Aestus
>   Perpulchre manibus, hunc dedit atque locum,
> Nicolaus cunctas Kracerus Bavarus horas
>   Dicere me fecit, qui Monacensis erat,
> Quique suis illo prelegerat astronomiam
>   Tempore discipulis multaque tradiderat;
> Et fuit Henrici tum octavi nominis huius
>   Astronomus regis, cui bene charus erat.
> Anglus erat lapicida, fuit Germanus et alter
>   Totius aetatis cum decus ipse fui;
> Ambo viri semper Germano more bibebant,
>   Et poterant potus surgere[49] quicquid erat:
>
> Anno Domini 1520.

---

[48] Brinknell was of Lincoln, Kynton a Minorite, Roper late of Magdalen, and the Walloon, John de Coloribus, a Dominican, who probably became a fellow of Cardinal College (Wood, *Athenae*, 2nd ed., I, col. 21). Edward Powell took the opportunity of the commission's findings to present Wolsey with the first part of a treatise directed against Lutheran teaching. In this he is said to have pleased the king, the university, and Wolsey. Despite this, he later parted ways with Henry, and having been committed to the Tower in 1534, was hanged, after denying the royal supremacy, in 1540.

[49] MS *sugere*.

Thomas Mosgrave did not become "commissary" or "rector" or, as we should say, Vice Chancellor, until 1523[50]. Finally we note a request made of the university on 8 July 1523, for the spending of 12 pence on the making of a dial (*horologium*), if such a dial were to be made[51]. Shortly afterwards the grace was conceded (*ad compositionem novi horologii*; 11 July 1523), and to his transcript of the record, Twyne added the remark that he imagined that he thought the reference to be to Kratzer's dial, still then on its pedestal outside St. Mary's church[52].

That Twyne was correct, and that we in turn are justified in dispensing with the very awkward double occurrence of an autograph "1520" is made more probable from another page of MS K. As we have seen, Nicolaus Kratzer, "who traught his pupils astronomy", who was "greatly loved by his king", and who, sooner or later, had declared against Luther in as public a place as Oxford then knew, did not feel it incongruent to sing the praises of the rector of the university in a verse in which he also boasted of his capacity — and that of the stonecutter William East — for drinking beer. If this was not exactly the High Renaissance, neither did it conform to the sensibilities of Juan Luis Vives, who never lost an opportunity of declaiming against the evils of strong drink[53]. Vives nevertheless composed verses for the university dial, and these are copied out at f. 66r of MS K in a very careless scribble, quite clearly by the man who filled most of ff. 40—80 with rough calculations and notes on dialing — almost certainly Kratzer himself, since the volume was his[54]. Vives scarcely had an opportunity to let himself go. To the east green lines marked the hours for sunrise, to the west were blue lines showing the time from the Sun's last sinking beneath the waves[55]; and so forth. But despite the theme, Vives' neoclassical style is far removed from Kratzer's:

---

[50] In 1520 the position was filled by William Brook, and between his holding of the office and Mosgrave's, Richard Benger was commissary. See Wood's *Fasti*, or University archive, Register H. 7.

[51] *Ibid.*, f. 112r. We are reminded of the story of the Mertonian request for a postern gate put to the king in 1266, not that *ostium fieri*, nor for *ostium factum*, but that *ostium fieri in facto esse*.

[52] *Loc. cit.* Cf. MS Twyne 24, p. 413. The dial may be seen clearly in Plate 11 of David Loggan's *Oxonia illustrata* (Oxford, 1675). It was on the High Street wall of the churchyard, opposite the second buttress of the church from the east end. The wall was removed in 1744. Henry Cotton, *Notes and Queries*, 2nd ser., III (1857), p. 145, thought he could detect marks where the dial stood. The dial had much the same appearance as Turnbull's in the quadrangle of Corpus Christi College but was not quite as high.

[53] See not only the chapter "On the Ordering of the Body in a Virgin" in his *De institutione foeminae Christianae*, but, for example, his dialogues, *Tudor Schoolboy Life: the Dialogues of Juan Luis Vives*, tr. Foster Watson (London, 1908), p. 137—42, and this quotation from the translation of his *An Introduction to Wisdom*, tr. R. Morison (London, 1539), f. Cb. 4r-v: "Thy drynke shalbe that naturall liquour, prepared of god indifferent to all lyvynge creatures, whiche is pure and clene water, or els syngle biere, or wyne alayde with the sayde water".

[54] See p. 388 below.

[55] The hours indicated would then have been "Italian hours".

Carmina inscripta in horologio Universitatis Oxoniensium edita per Ludovicum Vivum
Ad orientem
Per virgas virides notantur horae
Quas monstrant numeri a die renato.
Ad meridiem
Solis meatus lucis alternas vices
Horas diurnas, signa, quae tempus notant,
Umbrae docebunt gnomonum meatis suis.
Ad occidentem
Ceruleae signant ex quo se condidit undis
Temporis interea quot Sol confecerit horas.
Ad septentrionem
Tempora et obliqui Solis Lunaeque meatus
Ostendi mirum possunt mortalibus umbris.
In columna
Tanget quum medii natam diei
Phebus lumine stilus indicabit
A coeli medio polis et horizonte
Ad sudum (?) sycaminum (?) quod esse dices.
OXONIEN. ACADEMI. VETUSTISS.
INCLYT. POSITUM

After this, there is a hastily scribbled note: *Columna..(?).. 1523 a me Nicolao Krazer (?) Oxonie.* The "1523" has here been crudely altered to 1520 on another occasion.

It is not easy for us to enter into a frame of mind which could put an intellectual and even a spiritual value on an object so inert as a sundial, but John Leland was yet another who thought Kratzer's university dial worthy of verse[56], and the assiduous Twyne dutifully copied out his panegyric[57]: If Kratzer left his mark in Corpus Christi College, moreover, it was in the form of yet another dial, this time for the

---

[56] *Columna in Isidis vado, a Nicolao Cratzero Mathematico erecta.*
Marmoreas cantat celeberrima fama Columnas,
Quas claris statuit maxima Roma viris,
Nec minus insignem catabit fama columnam,
Cratzeri artifices quam statuere manus.
Candida Zodiacus circumdat circulus apte
Marmora, quae vario picta colore nitent.
Fulgidus auricomo cum splendet lumine Titan,
Umbriferis spatiis tempora certa docet.
Cum nitidis fulget Phoebe redimita coronis,
Horarum numeros linea tincta notat.
Praeterea variis inscripta columna figuris,
Astrorum motus ingeniosa refert.
Cuius ab exemplo, doctorum turba columnas
Erexit rara sedulitate novas.
From *Principium ac illustrium aliquot et eruditorum in Anglia virorum, Encomia, Tropaea Genethliaca, et Epithalamia* (London, 1589), p. 19.

[57] MS Twyne 21, p. 296.

orchard, and perhaps even finer than that which he designed for the university. We have no original drawing of it, but there is a description of it as detailed as anyone should require in a volume on dialing by Robert Hegge, the fellow of Corpus Christi mentioned earlier. Hegge was much taken by dialing. He has left a volume of theological lectures, for example, which includes one on the "*solarium Achaz*", the dial of Ahaz[58]. He wrote a complete treatise on dialing, in which the second chapter is on Kratzer's dial: "Of the Dial in Pomario C.X.C.OXON". The illustration survives in three versions, all substantially the same, but with small changes in the inscription on the base[59]. The best version reads:

Effigies speciosissimi Horoscopi in pomario collegii Corporis Christi Oxon. Hunc Horoscopum Nico. insignis olim mathematicus et coll. Corp. Christi Socius delineavit[60].

The dial was still in the garden in or around 1668, according to a statement by Hannibal Baskerville[61]. As Kratzer's work had moved Leland, so it moved Hegge, of whose literary conceits this is a fair specimen:

A dial is the visible Map of Time, till whose invention, the Sun seem'd to committ follie to play with a shadow... Heaven itself is but a general dial, and a Dial heaven in a lesser volume[62].

Another instance of "heaven in a lesser volume", a much lesser volume, in fact, was made by Kratzer for Wolsey. There can be little doubt about the ascription, although the only evidence is that of the instrument itself, a portable polyhedral dial of gilt brass, resembling in shape and function the stone orchard dial. The portable dial, which is now in the Museum of the History of Science in Oxford, has a cardinal's hat engraved on each of the two sloping sides of the base, while on the other two sides are the arms of Wolsey and of the cathedral church of York[63]. That Kratzer

---

[58] C. C. C. MS Φ. D. 2. 5, pp. 32—38. See II Kings 20. 10 and 11.

[59] Two copies in Hegge's hand are C. C. C. MS Φ. D. 2. 4. (Coxe no. 40) and Φ. D. 3. 8 (Coxe no. 430). The latter was owned by "Stephen Hegg C. S. I. 1630", and so was probably done as a gift. Some of the drawings in it are executed more carefully than those of the former, but the text is not as complete. It was bought by the college in the last century. The other copy has apparently never left the college. Judging by eclipse predictions at pp. 138—39, it was copied c. 1625. A third version of the drawing of the dial only is included in a manuscript presented to the Museum of the History of Science, Oxford by D. I. Duveen, apparently coming from "Somerton", and done in the mid-seventeenth century.

[60] All versions make the author a fellow. The two other give Nicolaus the surname Kratcher.

[61] MS Rawlinson D. 810. f. 62r.

[62] MS Φ. D. 2. 4, p. 5. The power of Hegge's imagination may be judged from his remarks (pp. 135—36) in Turnbull's dial (1581): "...such is to be seen in that Colossus of Art in Area quadrata C. X. C. whose varietie of invention is such, that if the Authors name had been conceald, I should have thought it one of the Columns that Adam and Sheth had erected to read a lecture to posteritie".

[63] The instrument had nine component dials, and room for a small magnetic compass (now lost) on the uppermost face. It is a good example of the work of the time and is expertly made, if not in the very highest class. (It does not, for example, compare with the work of Kratzer's late contemporaries, Thomas Gemini and Humphrey Cole, although this might be said of almost any other instrument maker). The dial came to the Oxford museum with the collection of Lewis Evans,

not only designed but made instruments is strongly suggested by the Holbein portrait. Wolsey was elevated to the see of York and the dignity of cardinal two or three years before Kratzer came to England[64], and died in 1530. It seems reasonable to suppose that the gilt dial was made for him while an indebted Kratzer was at Oxford.

To what extent Kratzer affected intellectual attitudes in his time it is impossible to say with any confidence, but there is certainly no surviving evidence that he was ever more to astronomy than a competent craftsman and teacher. His two surviving astronomical manuscripts (MS K, and MS Bodley 504, which will be discussed later as "MS H") do nothing to alter this view. The second of these is a slender volume on the use of a dial. It was copied by a professional scribe as a gift for Henry VIII. Although the contents of MS K have occasionally been spoken of as though they were largely original, this is controverted by a note at the foot of the table of contents (f. 1r):

Complura ex veterato libro monisteriae [sic] Charthusae Maurbach 2 miliaria a Vienna Austriae ego Nicolaus Kracerus exstripsi [sic].

That the contents were copied from the old manuscript, rather than abstracted in a physical sense, is assured not only by the manuscript style but by the watermarks. The paper is mostly of two sorts, one known from works printed in Innsbruck in 1515 and Augsburg in 1516, and the other found in Vicenza in 1522, but quite possibly available to Kratzer a few years before[65]. It is clear that he copied the volume, perhaps with paid assistance, shortly before his journey to England of 1517 or 1518. It may be verified easily enough that there is nothing, or almost nothing, original about its contents. Like so many manuscripts copied by a man capable of interpolating his own comment, and of coalescing works, it is difficult to give a precise total for the number of items the volume contains, but of fifteen readily distinguished pieces, I have found ten in older volumes elsewhere, and in each of the remaining five there is clear internal evidence of a medieval source. The works, which are generally well-illustrated, deal with portable sundials, especially polyhedral dials, as we might have expected, but also with ring dials. It deals with the arithmetic of fractions, simple trigonometry, and stereometry; with such instruments as the Jacob's staff, the "new" quadrant of Profatius, the *sphaera solida*, and the astrolabe; and with simple

---

who gave convincing reasons for supposing Kratzer to have been the maker, in "On a Portable Sundial of Gilt Brass Made for Cardinal Wolsey", *Archaeologia*, LVII (1901), 331—334. Especially significant is the shape given to the figure 1/14, which is unusual and yet identical with the shape of that in MS K.

[64] Bull of translation to the see, 15 September 1514; created cardinal-priest 10 September 1515.

[65] Charles-Moïse Briquet, *Les filigranes* (2nd. ed.; Amsterdam, 1968). The first watermark, a bovine head with cross and flower, is Briquet, no. 14556. The second, a balance and star, is Briquet, no. 2596. A third watermark, occurring less often in MS K, is in the form of a capital "A" with the bar collapsed to a "V". This watermark I have not identified in Briquet, although it resembles one found in a Venice book of 1503.

equatoria for calculating planetary positions. There is little here in any way out of the ordinary, unless it be the almost total absence of theoretical astronomy. Apart from a short piece in German, at the end, all is of course in Latin.

The identification of Kratzer's script has already been seen to have some slight bearing on the date of his coming to Oxford, and hence on his very purpose and standing here. Between ff. 40 and 80, MS K contains a series of rough notes and diagrams, clearly meant for personal use. This section includes the Vives verse (f. 66v) where the signature is so carelessly done as to suggest an autograph. The first item in the codex (on portable dials, ff. 2r—80v) is written in an easy but more formal hand, identifiable with that of the more casual sections. The formality of the hand, with the exception already noted, increases steadily as the volume — of over three hundred leaves — is completed. The pen changes, and also the ink, and even occasionally the style of some of the letters. Some of the work might be by an auxiliary hand, although the same batch of paper is found with formal and informal script alike. There remains the problem of the first numbered leaf, where Kratzer's own verse and the statement about the Carthusian monastery at Maurbach are to be found. The pen is there so broad as to detract from what nevertheless seems to be once more a Kratzer autograph, even though the signature is somewhat different[66]. There survives a solitary autograph letter from Kratzer, which is now in the British Museum, and this closely resembles the Maurbach note in general style, even though the signature is written with yet different letter forms.[67] Perhaps Kratzer wrote the note at a similarly late stage of his life — a supposition which would explain the erroneous "1520". At all events, we may not unreasonably take the first treatise in MS K as a specimen of a good Kratzer hand of c. 1516.

Following on this conclusion are two intrinsically more interesting consequences first noted by Otto Pächt, who accepted that MS K was a Kratzer autograph. Following a suggestion by a Paul Ganz, made in 1936, that the handwriting on the famous sketch by Holbein for the portrait[68] of More's family was neither More's nor that of his secretary, John Harris, Pächt had no hesitation in asserting that the hand was Kratzer's[69]. Pächt also maintained that Kratzer was responsible for annotations on another drawing by Holbein, done during his last years.[70] The drawing is of an astronomical clock designed for Sir Anthony Denny, and, as Pächt remarks, quite apart from the style of the script, Kratzer was "the obvious person to expect as their author[71]".

---

[66] There are untidy notes on two (unnumbered) preliminary leaves in the volume, in roughly the same style, even more problematical. The notes embody one or two simple astrological ideas.

[67] MS Cotton Vit. B, f. 256 (f. 276, old foliation). For the contents of the letter see p. 232 below.

[68] This was, as is well known, sent by More as a present for Erasmus.

[69] O. Pächt, "Holbein and Kratzer as Collaborators", *Burlington Magazine*, 84 (1944), pp. 134—139.

[70] Holbein died in 1543.

[71] For further information on the drawings see C. Dodgon, "Holbein's Designs for Sir Anthony Denny's Clock", *Burlington Magazine*, 58 (1931), p. 226.

Pächt's main purpose was to establish that Holbein supplied a decorative initial for what is now MS. Bodley 504 (MS H), which apart from MS K is the only other scientific book associated with Kratzer[72]. This is a beautiful volume, which was copied in a humanist hand in 1528 by Colet's scribe, Peter Meghen, in London[73]. It was dedicated to the king as a new year's present for 1 January 1528/9, and Kratzer explains that the work was originally written for William Tyler (Guilielmus Tylar) the king's chamberlain (*cubicularius*). We are reminded of a much more renowned new year's gift to Henry, namely John Leland's *Laboryeuse...Journey*.

The contents of Kratzer's gift to Henry were a set of instructions for the use of an instrument which he called his *horoptrum*[74]. The table of contents shows that the final blank leaves were intended for associated astronomical tables, and it is doubtful whether Henry can have felt their absence very keenly[75]. As for the instrument itself, original it might have been in its execution, but in its conception it was merely a simplified form of the conventional planispheric astrolabe, with the addition of some easy calendrical aids.

If it is difficult to decide when Kratzer came to Oxford, it is doubly so to discover when he left, although it seems likely that he was here for only a year — a period even less, perhaps, than that spent in Oxford by Vives. On 24 October 1524 we find him writing to Dürer from the English court a letter from which we can learn a great deal about his plans and his personal attitudes. The letter makes no mention of Oxford, but suggests that Kratzer had traveled somewhat in England:

...Know that Hans Pemair [Pömer] has been with me in England. I sent for him. I must write to you because you are all followers of the Gospel in Nuremberg. May God send you grace, that you may persevere to the end, for the adversaries are strong, but God is still stronger ... Dear Master Albert, I pray you to draw for me a model of the instrument that you saw at Herr Pirckheimer's, by which distances can be measured[76], and of which you spoke to me at Andarf [Antwerp], or that

---

[72] Pächt leaves little doubt that the decorative initial *E* of MS Bodley 504 is by Holbein. He confirms that it was copied by Peter Meghen, a fact first discovered by P. S. Allen. See Allen's note in *Erasmi Epistolae*, I, p. 471 and *The Age of Erasmus* (Oxford, 1914), p. 141. Pächt list fifteen MSS in the Bodleian Library by Meghen, including work done for Colet, Wolsey, and Urswick. Finally, Pächt repeats an observation made by W. H. Smyth in 1849 to the effect that a polyhedral dial then in the possession of Lieut. Col. R. Batty, a dial with no fewer than thirty faces, and the frequently repeated date of 1544, might well have been the result of collaboration between Kratzer and Holbein. Pächt was unable to trace the dial.

[73] MS H is bound in green silk velvet, 24 × 17 cm, i + 17 leaves. See F. Madan, *Summary Catalogue of Western MSS in the Bodleian Library*, Vol. III, pt. i (1922), p. 240. The incipit proper, after the dedication, is "*Argumentum huius instrumenti quod horoptrum vocamus...*"

[74] Madan was mistaken to say that the *horoptrum* was the treatise itself, and not an instrument

[75] Madan exaggerates somewhat the volume's omissions. Note that the final calendar, with the daily position of the Sun, is for the use of Sarum. There is a final note to the text which is worth recording: "Nam curiosiores homines, quique omnia certius tenere capiunt, ad tabulas astronomicas de his rebus calculatas remitto".

[76] "...womit man misst in die Fern und in die Weit..."

you will ask Herr Pirckheimer to send me a description of the said instrument... [77]Also I desire to know what you ask for copies of all your prints, and if there is anything new at Nuremberg in my craft. I hear that our Hans, the astronomer, is dead. I wish you to write and tell me what he has left behind him, and about Stabius, what has become of his instruments and his blocks. Greet in my name Herr Pirckheimer. I hope shortly to make a map of England which is a great country, and was not known to Ptolemy; Herr Pirckheimer will be glad to see it. All who have written of it hitherto have only seen a small part of England, no more ... I beg of you to send me the likeness of Stabius, fashioned to represent St. Kolman, and cut in wood...[78]

Dürer's reply of 5 December 1524 explained how Pirckheimer was having the required instrument made for Kratzer. This Pirckheimer had promised to send, together with a letter. The property of "Herr Hans" had been dispersed, and the same was true of the belongings of Stabius. Dürer preferred not to write of new rumors, "but there are many evil things afoot". "You told me once", writes Dürer, "that you were going to translate Euclid into German; I should greatly like to know if you have done any of it".[79] Does this mean that Kratzer knew Greek? This seems unlikely, and if he wished to translate from a printed Latin edition there were at least five different versions in existence by the time he set out for England. There is no known translation by him[80]; but, as Walter Pater was fond of saying of the Renaissance of the fifteenth century, it was great often by what it designed to do, rather than by what it achieved.

Dürer's interest in geometry and perspective scarcely needs remarking upon. As for astronomy, to consider Dürer's connections with the subject is to emphasize how tightly knit was the community of which he and Kratzer were members. Dürer's very house had belonged to the astronomer Bernhard Walther, a pupil of Regiomontanus.[81] Dürer associated with Celtis, Stabius, and Heinvogel. He declared that they found Kratzer, who had helped, incidentally, to carry his fame to England, to be of the greatest assistance. In 1515 Dürer drew for Stabius, from notes by Heinvogel, what was probably the first reasonable perspective terrestrial map of the north-

---

[77] The range-finder was no doubt of the kind sometimes known as a "trigometre", following Philippe Danfrie, who described it in his *Declaration de l'usage du graphometre* (Paris, 1597). Theretofore it had been made by such instrumentmakers as the Swiss Jobst Bürgi and the German Erasmus Habermel, but there appears to be no known example of a date as early as 1524.

[78] See Moriz Thausing, *Albert Dürer, His Life and Work*, (1882), II, p. 241—42. The German original is transcribed at p. 323. Thausing could not identify "Hans". Kratzer's letter was in the possession of a Herr Lampertz of Cologne, when Thausing was writing, but I have not been able to locate its present whereabouts.

[79] Quoted from Thausing, 242—43. See p. 324 for the original German, the letter being in the Guildhall Library, London.

[80] The first German versions to be printed were by Scheubel (Books VII—IX; 1558) and Holtzmann (Books I—VI; 1562). Holtzmann ("Xylander") often omits the proofs, on the grounds that he is writing for artists, goldsmiths, and other craftsmen.

[81] Five years after Walther died in 1504, his house in Nuremberg was sold by his heirs to Dürer (*Albrecht Dürer's Wohnhaus und seine Geschichte*, Nuremberg, 1896, pp. 4—6).

ern and southern hemispheres[82]. He contributed the folio with the armillary sphere in Pirckheimer's edition of Ptolemy, which appeared in the year following Dürer's letter to Kratzer, thus explaining why Pirckheimer would have been more than pleased to have seen the map of England. The vigor with which the practical arts were pursued had ramifications extending far outside their established province, and it is certainly not unreasonable to imagine a connection between Lutheranism and the simultaneous and quite phenomenal rise of the German trade in scientific instruments.

Nicolaus Kratzer's role as a maker of astronomical instruments is brought out nowhere more clearly than in the portrait of him painted by Holbein in 1528[83]. Kratzer holds a pair of dividers and one of his favorite polyhedral dials (P), apparently as yet unfinished, the gnomons (G) lying as they do on the table. Also on the table is a pivoting rule, a ruling knife, a burin, scissors, and another dialing instrument (I). On the wall hang larger dividers, another ruler, and what appears to be a combined parallel rule and square. On a shelf behind is a cylinder dial (C) and an unusual form of adjustable vertical dial (A) with a semicircular scale.

In her very detailed study of Holbein's "The Ambassadors", Mary Hervey suggested that Kratzer might have introduced Holbein to his subjects, identified as Jean de Dinteville, Seigneur Polisy and Bailly of Troyes, and George de Selve, Bishop of Lavaur[84]. The portrait of these two men, which now hangs in the National Gallery, and which was painted in 1533, is not least remarkable for the extraordinary collection of artifacts on the stand between the two men. The flutes, the hymnal, the lute with its broken string, reflect the theme of mortality; it seems unlikely that these have anything to do with Kratzer. The hymnal is opened at Luther's German rendering of "Veni Creator Spiritus" and his shortened version of the Ten Commandments[85]. There is every reason to think that Kratzer remained to his death as orthodox a Catholic as was expedient, but perhaps these things were indeed his, for there is little doubt that he was responsible for the remaining objects. On the upper table we find C, A, I, and P (now fitted with gnomons G and a small magnetic

---

[82] For a good, accessible and well-illustrated biography of Dürer, see W. Waetzoldt, *Dürer and his Times* (1955). The perspective map is at pp. 212–3. Some of the astronomical blocks made for Stabius are still in the National Library in Vienna.

[83] On the year, and the inscription, see p. 377 above. Now in the Louvre, the portrait is in oil and tempera on wood, 83 × 62 cm. This version, formerly in the Arundel collection, came into the possession of Louis XIV after the auction of that collection. It is not the portrait seen by Carel van Mander in the collection of Andries de Loo in 1604, as was previously supposed. That other version was bought by Sir Walter Cope, the builder of Holland House, and belonged to his heirs for three centuries. There is also a miniature copy in the Pierpont Morgan Collection. See Paul Ganz, *The Paintings of Hans Holbein* (London, 1950), p. 233. Van Mander described the de Loo copy as "een feer goedt Conterfeytsel en meesterlijck ghedan". See A. Woltmann, *Holbein and his Time* (1872).

[84] M. F. S. Hervey, *Holbein's "Ambassadors": the Picture and the Men* (London, 1900).

[85] *Ibid.*, p. 219–23.

compass). There is also a celestial globe, a meteoroscope and a torquetum (treatises on which were later published by Apian)[86], and a closed book. The lower shelf contains a terrestrial globe, identifiable as Schöner's (1523)[87], a book, held half-open by a set-square, and identifiable as Apian's *Kauffmanns Rechnung*[88], and a new pair of dividers. It is of interest to note that the celestial globe shows to the fore the constellation Cygnus marked as "GALACIA", looking distinctly cock-like, and symbolizing France. The addition of a score of unusual names to the terrestrial globe (such as "Polisy", the name of Dinteville's Burgundy château) is another move with political significance. These touches of unreality should not, I am sure, persuade us against the idea that these things were largely the real enough property of Nicolaus Kratzer. The ensemble of objects most probably symbolized the quadrivium — music, astronomy, geometry, and arithmetic.

There is no clear indication in Holbein's painting of the ambassadors that he used astrological imagery[89]. He and Kratzer lived, needless to say, among men and women who believed implicitly in the potency of astrology. For a topical example consider the case of Thomas Cranmer, no less, who wrote to Henry VIII in 1532 of a comet (Fracastoro's):

> What strange things these tokens do signify to come hereafter, God knoweth, for they do not lightly appear, but against some great mutation; and it hath not been see (as I suppose) that so many comets have appeared in so short time[90].

It was Cardan who boasted — probably falsely — of having refused a sum of money offered him by Henry on condition that he would restore those titles of which the pope had deprived him. Long before this, in 1510 in fact, Cornelius Agrippa von Nettesheim had been at Henry's court as the servant of Louis of France. He was for long protected by Cardinal Campeggio, who was no stranger to Henry's court, or for that matter, to Wolsey's. Agrippa in 1527 made a notorious prognostication,

---

[86] See my "Werner, Apian, Blagrave and the Meteoroscope", *British Journal for the History of Science*, III (1966), p. 57—65.

[87] Hervey, *op. cit.*, p. 210—18. The identification was first made by C. H. Coote. See H. Stevens and C. H. Coote, eds., *Johann Schöner... A Reproduction of his Globe of 1523 Long Lost* (1888).

[88] Hervey, *op. cit.*, pp. 223—224.

[89] As the dials are set, they all show different times, and some have supposed Holbein to be practicing some sort of numerological game, although I know of no one who has claimed to find a solution. Since some of the hours shown are fractional, the hypothesis seems unlikely. Astrology, too, seems improbable. I note, however, that the cylinder dial (C) is set for the equinox, and the celestial globe is very roughly set for the autumnal equinox, with the sign of Scorpio rising. The future Queen Elizabeth was born on 7 September 1533, a few days from the equinox. Dinteville had promised to act as godfather had the child been a boy. Within a short time he received permission to return to France. See Hervey, *op. cit.*, p. 94.

[90] *The Remains of Thomas Cranmer D. D.*, ed. H. Jenkyns, I (Oxford, 1833), p. 13. We are reminded of the apocryphal story of Pope Calixtus III's modification of the Litany "from the Turk and the Comet, good Lord, deliver us". See, however, J. Stein, "Calixte III et la comète de Halley", *Pubblicazioni della Specola astronomica vaticana*, II (Rome, 1909), 5—39.

advising the Duke of Bourbon to storm Rome, which the Duke did successfully[91]. (The only detail which had been overlooked was that the Duke would be killed in the assault). We simply do not know how Kratzer behaved, faced by this norm of behavior, but at least it is possible to point to one close acquaintance who was patently sincere in his rejection of astrology, namely Vives[92]. Vives had a clear idea of the place of astronomy in the hierarchy of knowledge, as he shows in *On Education*[93]:

> If Latin or Greek authors contain discussions on obscure questions of the higher learning, such as "first philosophy" or the investigation of natural causes, medicine, civil law, or theology, they should be left entirely for the experts in these subjects. If they discuss easier subjects, such as astronomy, cosmography, moral philosophy, practical wisdom, description of nature, or subjects formerly named, so long as they treat of them simply and clearly, I see no reason why these matters should not concern a philologist.

Whatever our view of the Renaissance of letters, it is noteworthy that Vives does not put the philologist at the apex of the academic pyramid. Later, however, *pace* Kratzer, he admits to a preference for the contemplative over the practical arts, although he had no wish to allow mathematical abstractions to direct the mind away from the practical concerns of life, and render it less fit to face concrete and mundane realities. But astronomy, above all, should not be applied to divination, whether of the future or of "hidden things", for both lead through vanity to impiety. Calendrical work, and cosmographical, are legitimate, and the astronomer may assist the navigator. Vives mentions a number of astronomical works, such as those by Ptolemy, Proclus, Sacrobosco, Peurbach, but gives the distinct impression that he is unfamiliar with their contents, although adding that "this is the curriculum for a youth up to the twenty-fifth year, or thereabouts[94]". Unfamiliar as he might have been with astronomy, it is not inconceivable that either he influenced Kratzer or his views reflect Kratzer's own. Of the two, Kratzer was perhaps the more pliable.

To understand the place occupied by a man like Kratzer in the microcosm of New Learning in which he found himself, it is not apposite merely to ask how he viewed his intellectual circumstances — a question which, for want of information, it is difficult to answer. One must at the same time consider what use his friends and pupils could make of his learning, however much it savored of scholasticism and the drabness of medieval Latinity. One sort of answer to this question is easy to divine, by merely turning to writers who followed the style of Pico della Mirandola. Such writers drew their imagery from almost every conceivable source of mystical learning, biblical, mythological, cabbalistic, pseudo-Dionysian and astrological[95]. The astron-

---

[91] H. Morley, *The Life of Cornelius Agrippa von Nettesheim*, II, 1–56, p. 228–29.
[92] See, for example, *De veritate fidei Christianae* in *Opera omnia*, VIII (1790), p. 78.
[93] Foster Watson, *op. cit.*, p. 153.
[94] *Ibid.*, p. 202–7.
[95] And, astrology notwithstanding, Pico's *Disputationes adversus astrologiam divinatricem*, ed. E. Garin, 2 vols. (Florence, 1943).

omy at Kratzer's command was a key not only to an astrology of horoscope and prognostication, but to the astrology of much neo-Platonic literature; and Kratzer's own literary sensibilities had little relevance to the usefulness of the key.

Kratzer's connection with work in metal, easily enough explained in part by his father's profession, and confirmed, perhaps, by the gilt brass dial made for Wolsey, is seen more clearly from a document of 1529. He was then sent with Hugh Boyvell and Hans Bour to search the king's woods and mines in Cornwall to try to melt the ore[96]. Two years later he was paid a large sum — 40 shillings — for mending a clock[97], and it is widely assumed, although without any firm evidence, that he was in some way connected with the clock at Hampton Court palace. The initials stamped on the frame are "N.O. 1540", and not "N.C. 1540", as was at one time suggested[98]. Far better evidence of his involvement in clock design comes from the annotations to Holbein's designs for a clock case of 1544. As was explained earlier, these annotations were shown by Otto Pächt to be in Kratzer's hand[99]. The clock was to be presented to Henry VIII as a new year's gift from his favorite and chamberlain, Anthony Denny (1501-49), who was knighted in the previous September[100]. Reminiscent of the supposed Hampton Court association is a hint dropped by B.L. Vulliamy, a London clockmaker of the early part of the last century, who claimed to find a resemblance between the Hampton Court clock and the clock of Christ Church Cathedral, Oxford — a clock which disappeared during Gilbert Scott's restorations of 1870. It is known, however, that the first cathedral clock was in 1545 transferred from Osney abbey, by an unnamed smith, to the priory church of St.

---

[96] *Letters and Papers*, ed. J. Gairdner, V (1880), p. 314 (Treasurer of the Chamber's Accounts). By a warrant of 12 August they were paid £ 15.

[97] *Ibid.*, p. 754 (Privy Purse Expenses, April 1531): "To Nicholas the astronomer for mending a clock 40s".

[98] Judging by the date, the clock was installed while Henry and Catherine Howard were at the palace. The stamp is on an iron bar on the inside of the dial. See E. Law, *The History of Hampton Court Palace*, especially Vol. I (1885), p. 217—20. Law searched in vain for a designer. "N. O." is now usually taken to be Nicholas Oursian. That the dial was much grander in the sixteenth century than in Law's day is evident from his transcription of instructions to the painter George Gower, who repainted it in 1575 (*op. cit.* p. 220, n. 1). On the subsequent fortunes of the clock see Vol. III (1891), p. 386—90, the many official guides to the palace, beginning with those by Law himself, and the paper by Smyth. Mr. Helier of the Science Museum, South Kensington, has assembled a great deal of information about the mechanism. The present movement was fitted in 1879. A replica (2/3 full size), made by Messrs. Thwaites and Reed, is in the Science Museum.

[99] See p. 389 above, and also the article by Campbell Dodgson, pp. 226—31. Dodgson discusses some pen and wash drawings made not much earlier than 1580 (judging by watermarks) from Holbein's originals.

[100] There is no record of the clock in later inventories. The original (Binyon no. 17, Ganz 220 [Lief. V. 5]) bears, in addition to other notes on the clock, the inscription "*Strena facta pro Anthony Deny... 1544...*"

Frideswide[101]. Since the clock had been repaired in 1544, before it was moved, it is doubtful whether there was much scope, let alone opportunity for Kratzer to exercise his genius on it.

Kratzer has left few traces of his work, and still fewer of his character, but there are reasons for thinking that others found his company amusing. One instance of the astronomer's repartee has come down to us[102]. After mentioning the portrait of "Mr. Niclaes, een Duytsch oft Nederlander wesende", who is quite clearly Kratzer, Van Mander tells how the king asked him how it came about that after thirty years in England he could not speak better English. "Forgive me, your Majesty, but how can one possibly learn English in only thirty years?"[103]

If this story is broadly true, then it shows that Kratzer was still in England in the late 1540s. (Henry died in January 1547, and therefore literal truth is too much to ask, unless Kratzer came to England before his first known visit of late 1517 or early 1518.) One reason for wondering whether he returned to Germany at the end of his life is an isolated record of two parchment instruments dated 1546, both of which were to be found in 1559 in the library of Otto Henry, prince of the Palatinate[104]. They were: (i) *Organum aestivum maris*, and (ii) *Organum motuum humorum humani*. Their existence in a German library probably signifies only that Kratzer continued to travel, perhaps combining his journeying with commerce in wine, or with diplomacy. As for the nature of the disks, there is no reason to think that they were anything but conventional. Kratzer's familiarity with the idea of a parallelism between the tides and the humors of the body merely bespeaks his familiarity with medieval astronomical thought.

Prince Otto Henry was a great patron of astronomy, and his astronomical adviser was at one time Nicolaus Pruckner (*"Pontinus"*). This fact is of interest because Pruckner, a theologian and mathematician who died in 1557 (two years before his patron) while a professor in Tübingen, wrote a dedicatory preface to an edition of the astronomical work of Guido Bonatti, in which preface Kratzer is mentioned[105]. The edition was published in Basel in 1550, and the dedication proper was to William

---

[101] After the See of Oxford was established in 1546 the west end of the priory church was closed to form part of the new cathedral. For further details and references see C. F. C. Beeson, *Clockmaking in Oxfordshire, 1400—1850* (Museum of the History of Science, Oxford, 1967), p. 49—50.

[102] See *Das Leben der niederländischen und deutschen Maler von Carel Van Mander*, ed. and tr. into German by Hanns Floerke (Munich & Leipzig, 1906), I, p. 186. The translator curiously omits the name of the king's astronomer.

[103] "Heer Coningh vergheeft het my, wat Enghels Kanmen leeren in den tijt van dertich Jaren".

[104] See Hans Rott, "Ottoheinrich und die Kunst", *Mitteil z. Gesch. des Heidelberger Schlosses*, V (Heidelberg, 1905).

[105] Guido Bonatti (Guido of Forli, or Forlivio) was a thirteenth century astrologer whose distinction earned him a place in Dante's eighth circle of Hell (*Inferno*, xx. 118). William Lilly's widely known English work, *Anima Astrologiae* (1676) was drawn mainly from Bonatti's writings, and from Cardan's.

Paget, comptroller of the king's household[106]. The book was sent to him and to the king, wrote Pruckner, in order that they might make the same good astrological use of it as Guido's patron, Guido di Montefeltro, the count of Urbino, had done. Pruckner continued:

> Quas vero hic Liber utilitates adferat Mathematicis, quum per te ipse potes existimare, tum etiam ex Mathematico vestro Nicolao Kratzero cognoscere: qui ita bonus et probus est, ut maiore quam Mathematicorum fortuna sit dignus, ut solus isthic artifex haberi debeat.

Kratzer might have agreed that he was worthy of a better position than that of mathematician, without having yearned for the risks which elevation incurred.

The occasion was not the first on which Kratzer had been praised in this way, for in 1536 the French poet Nicolas Bourbon de Vandoevre[107] prefaced his Παιδαγωγειον (Lyons, 1536) with greetings to friends, who included Cranmer, his host in England, Cornelius Heyss (the king's goldsmith), Holbein, and *"D.Nic. Cratzero regio astronomo viro honestis salibus, facetiisque ac leporibus concreto"*. There was no doubt a fellow feeling here, for Nicholas Bourbon was the son of a rich *maître de forges*[108].

The year of Kratzer's death is usually given as 1550, presumably on the grounds that there is no known reference to him after Pruckner's preface of that year. He evidently married one Christiana, when he was about fifty, and had at least one child, a daughter, as emerges from a London letter to Thomas Cromwell[109]. Kratzer no doubt had many dealings with Cromwell of a nature illustrated by an entry for 1533 in Cromwell's *Remembrances*:

To send to Nic. Cracher for the conveyance of Chr. Mount's letters[110].

Christopher Mount (or Mont) was a Lutheran, a native of Cologne, reporting regularly after 1531 to Henry on German affairs, and representing the English point of view to the German princes. It is not absolutely clear whether Kratzer was himself carrying the diplomatic bag, as it were, or acting as a translator. (It is well known

---

[106] *Guidonis Bonati Foroliviensis Mathematici. De astronomia tractatus X universum quod iudiciariam rationem nativitatum. aeris tempestatum, attinet, comprehendentes* (Basel, 1550). Paget was Chancellor of the Duchy of Lancaster and newly (1549) created Baron Paget of Beaudesert. In 1551 his career suffered a temporary reverse when, sharing in Somerset's disgrace, he was sent to the Tower.

[107] The elder of that name (1503 — post 1550). He was for some time given the care of the education of the daughter of Marguerite de Valois.

[108] An interesting and detailed poem he wrote on the subject, at the age of fifteen, (*"Ferraria"*) did not appear in print until 1533 (in his collection *Nugae*, published at Lyons). Scaliger, who made no concession to the writer's youth, deemed it nugatory.

[109] British Museum, MS Cotton Vit. B. xiv, f. 256 (f. 276, old foliation). Dated 24 August 1538. The relevant part of the letter, which was badly damaged in the Cotton fire, reads: "Nulla alia nisi magnificenti [a vestra] pauperrimum Nicolaum rememorat ut aliqu[anto] felicius vivere possum. Quo nunc uxorem h[abeo] Cristiana nomine, filiam Fide nomine, itaque [sperandum] est ut semper Christianam fidem domi h[abeam] ... uxor mea in die Assumptionis Mari[ae alteram] filiam Mariae peperit, unde plus th... quam astronomus videat, omnia ista... [...] tertenuit.

[110] British Museum, MS Cotton Titus B. i. f. 461, printed in *Letters and Papers*, VI (1882).

Hanseatic merchants of the Steelyard were often called upon to undertake both sorts of activity). The former alternative is most probably indicated, since the letter opens by saying that Georg Spalatin[111] was sending two books through Kratzer which he was in turn forwarding to Cromwell by Holbein[112]. Cromwell, furthermore, records in a letter to Henry of the following February that

> Your Highnes servaunt, Nicolas Cratzer, Astronomer, hath brought unto me this mornying a boke. herin enclosed, of *The Solace and Consolation of Princes*, the whiche oon Georgius Spalantinus, somtyme scolemaister to the Duke of Saxon, and nowe oon of the chief prechours, desyred hym to deliver unto Your Majestie[113].

Whether or not Cromwell here refers to yet another book, it seems that Kratzer conveyed it on this occasion in person. We may conjecture that he had known Spalatin from his Wittenberg days, or that they met at the Diet of Augsburg of 1518, or of Worms, both of which Spalatin certainly attended with his prince[114]. His interest in scientific instruments is attested by a letter to the humanist Veit Bild[115]; and here, perhaps, was another point of contact with Kratzer.

Max Maas saw in Kratzer a man who brought German science and technical art to England[116]. For the court, there is some truth in this. English kings have always shown a marked preference for imported fashions, in astronomy as in everything else. As Wolsey's lecturer in astronomy and mathematics, Kratzer doubtless had nothing new to offer of a fundamental kind. Many of his dials were unusual, but his favorite polyhedral dial was perhaps more useful as a repository of verses by Vives, or anti-Lutheran proclamations, than for actually announcing the time with any accuracy[117]. Like that other *Diall of Princes*[118] which Kratzer's contemporary, Anthony of Guevara, presented to his patron, the Emperor Charles V, dials

---

[111] Georgius Spalatinus, or Georg Burkhardt (1484—1545), the great Lutheran and Humanist scholar, was employed by the Elector Frederick III of Saxony, and was an adviser to his successors.

[112] The letter included news (now in part obliterated) of Worms, of the Turks, and of the Pope's mission to Switzerland, where he was seeking mercenaries.

[113] Cromwell to Henry VIII, 5 February 1539. This is an excerpt from a much longer letter in British Museum MS Cotton Titus B. i. f. 257. Printed in *State Papers published under the Authority of His Majesty's Commissioners*, I (1830), p. 594.

[114] For a biographical account of Spalatin by T. Kolde, see Herzog-Hauck, *Realenzyklopädie*, XVIII (1906).

[115] Alfred Schröder, "Der Humanist Veit Bild, Mönch bei St. Ulrich", *Zeitschrift des historischen Vereins für Schwaben und Neuburg*, XX (1893), 173—227. Bild's correspondence is in Augsburg Diözesenbibliothek, MS 81.

[116] *Op. cit.*, p. 20: "Deutsche Wissenschaft und deutsches Kunsthandwerk hat Kratzer nach England gebracht".

[117] One theoretical advantage of a polyhedral dial is that it may be correctly orientated by making the times shown on different faces agree. The advantage hardly applies to a monumental dial; and Wolsey's polyhedral dial included a magnetic compass, like the others depicted in connection with Kratzer.

[118] *Libro de Emperador Marco Aurelio co relox de principes*, 1529; tr. by Thomas North for Mary, 1557. The author might have met Kratzer, for he came to the English court in 1522.

that the have always been valued for the way in which they offered an occasion for moral soliloquy and melancholia. Maas, however, clearly has something else in mind — the artistry and precision with which they and similar instruments were made in Germany, especially in Nuremberg and Augsburg. There was a long established tradition of instrument making in England, not least in Oxford, a tradition which was rich but perhaps in Kratzer's time static and none the worse for an infusion of new ideas from the Low Countries and Germany. But Kratzer neither could nor did bring to England the sort of industry which developed in the great mercantile centers of Germany. He would probably, indeed, have been rather surprised that anyone could ever have seen him in this light. A sawsmith his father might have been, but he himself was a scholar, who made his own instruments as astronomers had almost always been obliged to do. This was certainly nothing new in Oxford. Did the phrase "deviser of the king's horologes" have reference to anything beyond sundials? It would be surprising to find that Kratzer brought with him any new skills relating to the techniques of the clockmaker except, perhaps, those which were concerned with small timepieces. Englishmen might still have been forgiven for thinking that theirs was the home of the craft of the large clock[119]. Did Kratzer, in assessing the potential of the Cornish ore, see his prince as a potential English Fugger? From opposition to Luther's doctrines Kratzer came to a point where he was prepared to traffic with Lutherans. Did he ever abandon his orthodoxy completely? There are too many imponderables in what little we know of this astronomer and diplomatic intermediary. The Attic wit seen by Nicholas Bourbon and others was surely that of a likeable emigré opportunist, whose thoughts were raised above considerations of vulgar nationalism.

Addenda

Dr Helen Wallis has drawn my attention to another piece of evidence as to Kratzer's work, and his collaboration with Holbein. Sydney Anglo, in his *Spectacle, Pageantry, and Early Tudor Policy* (Oxford, Clarendon Press, 1969), pp. 217—9, has given ample evidence that the two men together designed and painted an astronomical allegory on the ceiling of a splendid but temporary building, erected by the king for the entertainment of the French ambassadors, at Greenwich in 1527.

More tangible is a specimen of Kratzer's writing which I found by chance at the very end of the final text of British Library MS Royal 12. G. 1, a codex written c. 1485 by Lewis of Caerleon, and discussed by me in general terms in *Richard of Wallingford* (Oxford, Clarendon Press, 1976), vol. 2, pp. 381, 386, and vol. 3, pp. 219—20. The note in question states that Kratzer bought the volume at St. Paul's on 25 July (?), 1535. I cannot read the vendor's name on microfilm, but it might be Paston, and certainly has the style of doctor.

---

[119] Two centuries earlier Richard of Wallingford, an Oxford astronomer with a smith for a father, had designed for his abbey of St. Albans an enormously complex astronomical clock, which was still to be seen working in Kratzer's day. By comparison, the clock at Hampton Court was a toy. See my *Richard of Wallingford* (Oxford, Clarendon Press, 1976).

Addenda (1988)

In February 1978 the Bodleian Library acquired in a sale at Christie's a Kratzer manuscript entitled *Canones horoptri baculi* (ex info. Dr. A. de la Mare: the shelfmark of the MS is now Lat. misc. f.51). Dr. P. Pattenden, who has written on the Corpus dials (*Sundials at an Oxford College*, Oxford, 1979) and has examined the work, describes it as an elaboration of the work on dialling in MS Bodley 504. The manuscript contains arithmetical tables copied by Peter Meghen in 1537, and it is therefore of special interest inasmuch as it shows him still at work at this late date.

In 1985 Dr. A. R. Somerville sent me a description of a stone polyhedral sundial from Iron Acton Court, near Bristol, bearing the date 1520 and the initials 'N.K.'. He offers arguments for an association with Kratzer. The south-facing polar dial contains errors, however. It is not necessary, of course, that the dial was made for the house where it was found, for the latitude is close to that of Oxford. It might even have been brought from Oxford. Another possibility is that it was copied from a Kratzer dial, perhaps long after 1520.

24

THE MEDIEVAL BACKGROUND TO COPERNICUS

ACCORDING to Aristotle, happiness was not to be found in the search for truth, but in the contemplation of truth previously attained.[1] An all too common view of the astronomer of the late Middle Ages is that he was complacent in his search for this sort of happiness—content, in other words, merely to retail the methods and truths of the ancients. Some writers will go even beyond this, and claim that as late as the time of Copernicus a proficiency in the mathematical methods of Ptolemaic planetary astronomy was something alien to European astronomy.[2] But then, at the other extreme, there are those who regard the medieval astronomer as having lived in a state of intellectual anxiety, constantly worrying about the imperfections of the Ptolemaic system.[3] The truth of the matter lies elsewhere, but it is neither pure nor simple, and in the frame of a short lecture I can scarcely begin to do justice to it. I can say nothing

---

[1] 1177ᵃ 26.
[2] See, for example, T. S. Kuhn, *The Copernican Revolution: Planetary Astronomy in the Development of Western Thought*, 2nd ed., 1969, pp. 123-4: "In fact there was almost no [developed planetary astronomy] in Europe during the Middle Ages, partly because of the intrinsic difficulty of the mathematical texts and partly because the problem of the planets seemed so esoteric. ... Until two decades before Copernicus' birth in 1473 there was little concrete evidence of technically proficient planetary astronomy. Then it appeared in works like those of the German Georg Peuerbach (1423-1461) and his pupil Johannes Müller (1436-1476)."
[3] The imperfections, as Robert Small wrote, were however, "not immediately perceived, especially during the confusions which attended the decline and destruction of the Roman Empire"! *An Account of the Astronomical Discoveries of Kepler*, 1804, p. 81.

of the transmission of Hellenistic astronomy immediately after the time of Ptolemy (say A.D. 141), first to Persia and India, later to Islam, and finally to Europe, which it reached especially through contacts in Sicily and Spain. I shall speak mainly of the *Weltanschauung* of the Christian Europe to which Copernicus belonged. I shall make no attempt to break down his compound thoughts into atoms; much less shall I try to prove that each of the atoms was a constituent of the medieval cosmos, and that therefore Copernicus need astonish you no longer. I am simply here to set the scene, and, therefore, I shall speak of his intellectual precursors rather more than I shall speak of his intellectual debts. It is now fashionable to say that Copernicus gave us old wine in new bottles. However true, or untrue, this may have been of Copernicus, it is an apt description of my own contribution. I have nothing to say that is new, and the most I can hope is that my audience has not heard it all before.

Almost all of the astronomy known to the Christian scholar of the early Middle Ages came from one of seven authors, whose work, judged by the standards of Hellenistic astronomy, was extremely dilute. There were the encyclopedic works of Pliny (first century A.D.), Martianus Capella (fifth century), and Isidore of Seville (seventh century); there was a translation and commentary by Chalcidius (fourth century) of Plato's *Timaeus*; there was another commentary (fifth century) by Macrobius on a work called *Scipio's Dream*, by Cicero; there were the very influential writings of Boethius (sixth century); and there were the many works of Bede (eighth century). Even these seven writers were far from being mutually independent, but their apparently independent repetitions of one and the same idea served to reinforce confidence in that idea, however preposterous it might have been. Worse still, there was in all these writers a sorry absence of underlying explanation.

A more potent influence on the cosmology of early Christian Europe was that extensive body of writings left by the early Fathers of the Church. Their astronomical knowledge, even in aggregate, was pitifully small, and the strong desire they had to reconcile their cosmology with the Scriptures meant that even those of them who were familiar with Greek sources—men like Clement of Alexandria, Origen, Basil, and Augustine—were incapable of transmitting even the spirit of inquiry which had guided Eudoxos, Hipparchos, and Ptolemy. When we consider the allegorical tradition whereby the Earth was symbolized by the Tabernacle built by Moses in the wilderness, whereby the inner Tabernacle was made to correspond to the Kingdom of Heaven, and so forth, and when we consider the harassment of those who taught the sphericity of the Earth and the existence of the antipodes, then we can only marvel at the moderately enlightened attitudes of such Christians as Isidore and Bede, however trifling we ourselves happen to find their writings.

I have no wish to give the impression that there were all-pervading and official Church dogmas restricting cosmological thought throughout the Middle Ages, but we should never lose sight of the Church's monopoly of academic instruction, and of the ways in which all intellectual attitudes were to some extent coloured with religious belief. Man was, above all, a fallen creature, whose instincts were base, and whose reason was often at variance with those things in which he put his faith. With the Last Judgement, thought Augustine, the world would be newly organized, so why trouble to rationalize its present organization? For Isidore, the Fall led to a diminution in the intensity of the light from the Sun and Moon, but Isaiah assured us that after the redemption all would return to its former glory. Such instances of intellectual pessimism do not betoken a complete abandonment of the Greek idea that nature is orderly, intelligible, and accessible to human reasons, but they may be taken, for the purposes of my all too hasty lecture, to characterize the period before the eleventh century. Then, generally speaking, Christian man saw himself as a degraded and miserable creature whose only hope was in prayer and penitence, and as one to whom a rational understanding of the motions of the planets was supremely irrelevant.

This gloomy state of affairs brightened very rapidly with improvements in the organization of cathedral and monastic education. Secular schools were founded, which in turn led to the establishment of the first modern universities. At Bologna, Paris, and Oxford, religion was harmonized with other learning, especially that new learning which, from the twelfth century on, was flooding into Europe in the form of translations from Greek and Arabic. And then Aristotle arrived on the scene, not in a diluted Boethian form, but in respectable renderings of his own writings. Before long, Aristotle's power to stimulate thought became evident, and perhaps the greatest achievement of his greatest commentator, Thomas Aquinas, was to convince men that the universe, like God, could be understood through the operation of reason, and that the best of ancient learning was reconcilable with the Scriptures. It is no accident that in the most majestic of all medieval allegories, the *Divine Comedy*, which was written about a quarter of a century after the death of Aquinas, Dante should have provided a moral theme[4] within a framework of Aristotelian cosmology. Dante's *Paradise* is more than an unbroken succession of descriptions of heavenly bliss. It has an astronomical structure, and when the souls, for instance, are placed rank upon rank, the ranks are simply the planetary spheres.[5] In the *Convivio*, Dante refers to Aristotle as "that glorious philosopher to whom Nature most fully revealed her secrets",[6] and the poet goes on to make a remark which Copernicans will note with interest. Dante will not, he says, give the reasons which Aristotle alleges contradict those who claimed that the Earth is not fixed in its place. "It is enough for those whom I am addressing to know, on his weighty authority, that this Earth is fixed

---

[4] According to Dante (*Epistolae*, xi. 8), "the subject of the work as a whole, literally accepted, is the state of souls after death. . . . But if the work be taken allegorically, the subject is MAN, who makes himself liable, by the good or bad exercise of his free will, to the rewards or punishments of justice."

[5] *Par.* iii. 52. I am also convinced that Purgatory had a structure which was meant to reflect that of the heavens; see above, chapter 13, pp. 187-209.

[6] *Conv.* III. v. 55.

and does not revolve, and that with the sea it is the centre of heaven." This passage should remind us how all those scholars who made commentaries on the *De Caelo* were clearly obliged to come to terms with possible motions of the Earth.

Dante's words also serve to illustrate a typical attitude to Aristotelian authority. On the other hand, the force of Aristotle's arguments was not such that Dante accepted them blindly, and he does contrive to disagree with Aristotle on at least two occasions, first as to the number of spheres, and second as to the number of Intelligences causing the circular movements of the spheres.[7] In the Renaissance, medieval scholastics were often accused of a slavish adherence to Aristotle; but coming from an age obsessed with the imitation of literary ornament no less than with philosophical argument, the challenge need not be taken too seriously. And there is no doubt about Aristotle's influence on the one man who is often said to have banished Aristotle from cosmological speculation—namely Copernicus.

Aristotelian cosmology has two essential ingredients, one a theory of natural motion, the other an explanation of the movements of the planets, an explanation developed from that first given by Eudoxos. This quite extraordinary planetary theory was so debased by the time of the Middle Ages that its explanatory power was almost totally lost. In fact as a purely predictive theory it was no longer needed, the Ptolemaic scheme having superseded it. All that remained was the crude metaphysical idea that the universe was divided into two concentric regions, each with a spherically symmetric structure. The Earth was at the centre, in the elementary region, which contained water above the element earth, air above the water, and fire above the air. The fire reached to the sphere of the Moon, where began the ethereal region, the heavens. Between the thirteenth and seventeenth centuries, by far the best known treatise of astronomy was *the Sphere* by Sacrobosco, and we may quote him on the nature of the heavens:

> Around the elementary region there is the ethereal, which is lucid and immune from all variation in its unchanging essence, and which turns in a circular sense with a continuous motion. It is called the "fifth essence" by philosophers. Of this there are nine spheres ... namely of the Moon, Mercury, Venus, the Sun, Mars, Jupiter, Saturn, the fixed stars, and the final heaven.[8] Each of these spheres encloses the one below spherically.[9]

Sacrobosco's treatise was an elementary work on the astronomy of very simple phenomena, rising and setting, the zodiac, the spherical Earth, the climates, and the Sun's path. (It contained next to nothing on Ptolemy's planetary theory.) And yet

---

[7] See *Conv.* II. v. 20. Aristotle argued that if one Intelligence took care of each of the circular movements, that was enough. Any other intelligence would have been eternally useless, without any function. For Dante the mere *existence* of an Intelligence is its function.

[8] By a simple extension of the idea of concentric spheres, the region beyond the eighth, the empyrean, was itself often given a complex structure. This was done in several ways, that of Boethius being one of the best known. See Gregor Maurach, *Coelum Empyreum*, Steiner, Wiesbaden, 1968.

[9] Translated from Lynn Thorndike, *The Sphere of Sacrobosco and its Commentators*, 1949, p. 79.

even then, I strongly suspect that the passage I have quoted was as much as most educated men tended to remember of it, shall we say a year or two after they had left their university.

The second central doctrine of Aristotelian cosmology concerned natural motion. The quintessential, or fifth, element, out of which the celestial regions were made, was said to be not homogenous, being least pure where it borders on the air around the sublunary region. Its natural motion, circular and continuous, was to be contrasted with the rectilinear motions of the other elements; for earthly bodies fall naturally to the centre of the Earth, while fire rises naturally in the opposite direction, again radially outwards from the centre of the Earth, the centre of the Universe.[10] Forced motions were another matter, which we need not consider here.

Copernicus summarized this Aristotelian cosmology before giving his reasons, in the first book of the *De revolutionibus*, for thinking it inadequate.[11] He, after all, wished to place the Earth in the celestial category. On the other hand, he was enough of an Aristotelian to preserve the doctrine that each body has a natural place in the universe. His categories were Aristotle's reapplied: the Earth had a constant circular motion of planetary type and the same was true of its constituent parts, although a detached portion of Earth moved now with a compound motion,[12] having one naturally circular component and one naturally linear component directed to the Earth's centre. And now that the Earth was displaced from the centre of the Universe, Copernicus felt at liberty to do Aristotle the honour of generalizing his ideas:

> For my own part, I think that gravity is nothing but a certain natural striving implanted in the parts by the divine providence of the maker of all things, so that they may come together in the form of a globe, in unity and wholeness. It is believable that this striving is present also in the Sun, the Moon, and the other bright planets, so that they keep the resultant round shape which they show, even though they make their circular movements in many different ways.[13]

Those who suppose Copernicus to have cut astronomy free from the bonds of the Aristotelian doctrine that the universe is separated into two essentially distinct regions, in one of which matter circulates, and in the other of which it gravitates, must at least reconcile themselves to this passage.

Another occasion on which Copernicus found the spirit of the Peripatetic too strong to resist was in his rejection of the Ptolemaic principle of the equant. Whereas according to Ptolemy, most of the planets move uniformly around deferent circles, the uniformity of the latter motion is not with respect to arc length; that is to say, the

---

[10] Water and air were intermediate elements, whose heaviness and lightness, respectively, were less than those of earth and fire.

[11] See *De rev.* I. 7–8.

[12] Even in the idea of compounding motions he was following Aristotle and the medieval Aristotelians, and he is surely quoting Aristotle with approval when he avers that the division of motion into three sorts is merely an act of reason (*rationis solummodo actus*). *De rev.* I. 8.

[13] *De rev.* I. 9.

radius vector does not move round at a uniform rate. Instead, what rotates uniformly is the vector to the epicycle centre from a point some distance from the centre of the deferent circle—and both are, in general, distinct from the centre of the Earth.[14] In the eyes of Copernicus, the movements along Ptolemy's deferent circles were not uniform (*oportet inaequalem esse*).[15] I will not insist here that his mathematical imagination fell short of Ptolemy's, but merely note that he accepted as axiomatic what he called the "rule of absolute motion", according to which "everything would move uniformly about its proper centre",[16] and that this rule was in perfect conformity with the views put forward by Aristotle.[17]

Cosmology, in the Middle Ages, did not end with Aristotle and the extension of his scheme to an angelic hierarchy, nor did it end with the twin themes of Sun and Salvation as found in Dante. Such writings as Dante's, to be sure, had a reactionary effect, to the extent that they created in the mind of their audience an image of an all too simple universe, an image far too vivid to be easily put aside. Fortunately for astronomy, however, there were men for whom physical and metaphysical speculations were not enough, even in combination with theological and mystical allegory. The marked tendency towards a secularization of learning to be seen in the half century or so before *De revolutionibus*, was accompanied by a number of sound exegeses of the astronomy of Ptolemy's *Almagest*, but it would be a mistake to imagine that there was anything in them which would have surprised the best astronomers of the Middle Ages. Putting sources like Sacrobosco's elementary textbook aside, there are at least five categories of evidence for a competent level of astronomical activity in the Middle Ages.

First, the *Almagest* was far from being unknown *in extenso*. It was twice translated into Latin, in one case directly from the Greek, by an anonymous Sicilian writer in the middle of the twelfth century, and in the other case by Gerard of Cremona, in 1175. Another direct translation from the Greek was done by George of Trebizond in 1451, and this was to be printed in 1528, thirteen years after the printing of Gerard of Cremona's version. Needless to say, there is no fundamental difference between the three translations. In addition to these full translations there were several good epitomes of the *Almagest*, in some cases themselves translations out of the Arabic, but some of them made, so far as one can see, in Western Europe.[18]

---

[14] In his lunar theory, Copernicus obtained a more or less equivalent result by placing the Moon on a second epicycle, the centre of which was carried on the first; see below.

[15] *De rev.* IV. 2.

[16] *Commentariolus*, introduction; translation in E. Rosen, *Three Copernican Treatises*, 2nd edn., 1959, pp. 57–58.

[17] *De Caelo*, II. 6.

[18] One twelfth- or thirteenth-century epitome which was widely known, under one of a variety of titles (*Almagestum abbreviatum*, *Almagestum parvum*, and so forth), was transcribed and annotated by so late a writer as Regiomontanus. The work is currently being studied by Dr R. P. Lorch.

Second, there were the translations of a number of works showing some originality, and first written in Arabic. Such were the works of al-Battānī and Thābit ibn Qurra (both of the ninth century), al-Zarqālī (eleventh century) and Jābir ibn Aflaḥ (twelfth century). The last two astronomers mentioned here (Arzachel and Geber, as they were known in Latin) lived and worked in the Iberian peninsula, from which place also came a number of works in an Aristotelian tradition critical of Ptolemy, and especially of the epicycle mechanism. The alternatives offered involved a hypothesis broadly reminiscent of that of Eudoxos, requiring nested concentric spheres with non-coincident poles. The best known, but astronomically unsuccessful, work in this critical vein was by al-Biṭrūjī (Alpetragius), who died in 1235, and whose principal work was well known in the translation made in 1217 by Michael Scot.

Third, there were the tables used in astronomical calculation. The individual tables fall into many different categories, and a full set of perhaps a hundred tables usually derives from several different sources, although three extremely important and reasonably distinct collections were especially pervasive in Western Europe, namely that of al-Khwārizmī, the Toledan tables, and the Alfonsine. These seem to have circulated in Europe generally from, respectively, the early twelfth, the late twelfth, and the end of the thirteenth centuries. Although they were usually accompanied by canons, that is, treatises explaining in detail their use, they speak for a rather more intense intellectual training than is to be had from undergraduate courses in certain present-day academic subjects.

The fourth respect in which, I suggest, we have evidence of intelligent astronomical activity in the centuries before Copernicus is in the ancillary mathematics. That part of mathematics which we now call spherical trigonometry, came in a well-developed state into the repertoire of European learning through two principal channels, first, in the mathematical chapters of such works as Ptolemy's and Geber's, and second, in the canons to astronomical tables, and especially in Arzachel's canons. Such sources were actively enlarged upon by a number of European writers, of whom Richard of Wallingford, an early fourteenth-century English abbot, was one notable example, and Regiomontanus, more than a century later, another.

My fifth, and last, category of evidence concerns the design and fabrication of instruments, which were used both in observation and as an aid to calculation. (See Figs. 26–31.) There was the ubiquitous astrolabe, of course, with its continuous history from the Hellenistic and Roman world. This instrument was not only often modified in superficial and scientifically trivial stylistic ways, but was redesigned by two eleventh-century Toledan writers, 'Alī b. Khalaf and Arzachel, whose astrolabes were of universal application, in the sense that one plate sufficed for all geographical latitudes.[19] The *saphea arzachelis*, as the universal plate became known in the West,

---

[19] In principle, the conventional instrument requires a different local-coordinate plate for each latitude at which it is to be used.

was to be found, for example, on Richard of Wallingford's albion, an instrument by means of which very many different sorts of astronomical calculation could be made. I mention this last instrument partly because I am familiar with it, having recently edited the treatise devoted to it by its designer, but also because its complexity mirrors that of the astronomical learning of the time. At first sight resembling a rather large astrolabe, it contained upward of sixty different scales, four of which, for instance, comprised a spiral of thirty-one turns, rather as on many modern logarithmic slide rules. It was an observing instrument, and also permitted the calculation of eclipses and of planetary positions. These things it did in ways very different from instruments before it. There were broadly similar devices before this (all known under the generic name *equatorium*), and two European authors whose originality should not pass unnoticed here are Campanus of Novara and Peter of Dacia. The equatorium tradition was never broken, but rather began to fade away towards the end of the seventeenth century. There were, needless to say, several Copernican equatoria.

I cannot do more than hint at the very considerable medieval activity in instrument making, which extended to larger instruments for fundamental observation, as well as to such domestic items as pocket cylinder dials, ring dials, and so on. The art of gnomonics, which Copernicus himself was to essay, was certainly not ignored by medieval writers. Certainly a simple ring dial is very far from being a "world system", such as Copernicus is represented as having overthrown,[20] but its design implies a frame of mind which is altogether congenial to changing the frame of the world.

Medieval speculation on the motion of the Earth does not seem to me to have been significant in itself or to have been a very significant influence on Renaissance thought, but since, for the time being, we have all agreed to act as satellites to Copernicus, I will try at least to put in perspective what speculation there was. I have no wish to belittle medieval physics, but as an alternative to Aristotle (rather than as a more rigorously developed version of Aristotle) it seems to me to have failed in the important sense that the basic theory was not made to yield conclusions with the intention of putting them to empirical test. Speaking generally, the medieval scholastic had a penchant for offering all the potentially acceptable alternative solutions he could devise for a whole range of imaginary problems, and finally choosing one of these alternatives on metaphysical grounds. Now if there was ever any real prospect of loosening the Earth in its fixed central position, this was in regard to a possible rotation of its axis. Ptolemy had seen that, as far as appearances of stellar movements were concerned, the Earth might as well rotate on its axis as

---

[20] Cf. Alfons Kauffeldt's *Nikolaus Kopernikus: der Umsturz des mittelalterlichen Weltbildes*, Berlin, 1954.

be stationary.²¹ He argued, however, that such terrestrial movements as had been proposed by the Pythagoreans Philolaos and Hicetas (both fifth century B.C.), and by Aristarchus (third century B.C.),²² was ruled out by physical considerations—objects thrown up in the air would tend to be left behind, for example. Whether it was from the frequent discussions of Aristotle's *De Caelo*, or from reading Ptolemy's *Almagest*, one does not know, but at all events Jean Buridan in the fourteenth century stated that many people of the time thought it probable that the Earth moves on its axis with a daily rotation, the sphere of the stars being supposed stationary.²³ Like Ptolemy, he held that observed phenomena would be the same, and like Ptolemy he came down—and for the same reason—on the side of a fixed Earth; but he saw merit in the hypothesis of terrestrial rotation, for it seems the simpler of the two, the Earth being smaller than the sphere of the stars.²⁴ It has been suggested that Copernicus might have encountered Buridan's writings at Krakow, as a student, when Buridan's physical writings were required reading.

A younger contemporary of Buridan, Nicole Oresme, translated the *De Caelo* into French,²⁵ and wrote a commentary on it. It is unlikely that Copernicus was directly familiar with the commentary, although it might well have been the subject of contemporaneous debate. Oresme shared a disability with perhaps most of the scholastics: he could usually see both sides of an argument without being able to decide which side deserved his support. Like so many before him, he saw that from a kinematic standpoint there was no inconsistency in the idea of the Earth's rotation. He argued convincingly against Aristotle that movement of the heavens does not require the Earth to be at rest.²⁶ Experience, he says furthermore, cannot decide the

---

²¹ *Almagest*, I. 7. Aristotle (*De Caelo*, 296ᵇ) had used the same argument, but gave priority to two arguments of a different sort. He held that if the Earth were to move at all, it would move with both sorts of motion, which we may call revolution and rotation, since this is true of everything that moves with a circular movement, except the first sphere (*loc. cit.*). As a consequence we should find (i) that the stars would rise and set in different places, as do the planets, and (ii) the shapes of the constellations would change. These things are not observed, and so the assumption is false. Aristotle clearly assumed what Ptolemy did not, namely that the stars were near enough for his conclusions to be tested by experience.

²² Aristarchus, as all who are familiar with the history of Copernicus will know, also believed in the revolution of the Earth in orbit round the central Sun, while the Pythagoreans had said that the Earth moved round a central fire, which provided the motive power for the whole Universe. Ptolemy, however, was discussing only axial rotation.

²³ *Quaestiones super libris quattuor de caelo et mundo*, ed. E. A. Moody, Cambridge, Mass., 1942, p. 227.

²⁴ Buridan's presentation of the case is not wholly consistent with this theory of impetus, but this is not of concern here.

²⁵ Completed 1377. The work, *Le livre du ciel et du monde*, has been edited by A. D. Menut and A. J. Denomy, Madison, Wisconsin, 1968.

²⁶ He shows that no movement—celestial or otherwise—requires in and of itself either the immobility or the movement of another body (*op. cit.*, pp. 372–3). Nicholas of Cusa later (unpublished treatise of 1444) assigned a rotation to both the Earth (which turned once a day) and the heavens (which turned twice, in the opposite sense, in the same period). See J. L. E. Dreyer, *History of the Planetary Systems from Thales to Kepler*, Cambridge, 1905, pp. 285–7.

matter.[27] And yet, after a long disquisition, in which it appears that he must come down on the side of Herakleides, who held that the Earth turned on its axis, he concludes with these words:

> However, everyone maintains, and I think myself, that the heavens do move and not the Earth: for God established the world which shall not be moved, in spite of contrary reasons. ... What I have said by way of diversion or intellectual exercise can in this manner serve as a valuable means of refuting and checking those who would like to impugn our faith by argument.

I have been obliged to omit Oresme's eight pages of detailed argument, but his point has no doubt been taken: the Christian's faith was not a thing to be overturned by mental gymnastics. In ontological terms we may say that the world of the Faith was more real than that of physics, while for many—including Buridan and later Osiander—the world of astronomy was the least real of all. In Buridan's words, "For astronomers, it is enough to assume a way of saving the phenomena, whether it is really so or not."[28] When, in the Middle Ages, the reality of the celestial spheres was discussed—and there is no reason to suppose that Copernicus ever questioned their reality[29]—the discussion came within the province of physics rather than astronomy.

Another question on which the astronomer, as such, was not obliged to form an opinion concerned the bounds of the Universe, and its scale. Some genuinely Ptolemaic ideas on the scale of the Universe—concerning not only the *relative* sizes, but the *actual* dimensions of the orbs, quoted in miles—became widely diffused in Europe through the writings of al-Farghānī (or Alfraganus as he was known in the Latin West).[30] It is frequently said that Copernicus had a truly integrated planetary system, wherein the mean distances of the planets were fixed by his hypothesis that all motions were centred on the Sun. The fact is that the Middle Ages, too, had a system. It was arrived at by an *a priori* argument directed towards a harmonious overall scheme, and this it had in common with the arguments of the Copernican system. Of the two, the older harmony was the more deliberately aimed at; but it was wrong. The distance to the fixed stars was held to be approximately 130 million miles—

[27] *Ibid.*, pp. 520–37.
[28] *Op. cit.*, p. 229.
[29] Here it is noteworthy that there is no reason to suppose that Copernicus had first-hand knowledge of the works of al-Biṭrūjī, even in translation. See E. Rosen, "Copernicus and al-Biṭrūjī", *Centaurus*, vii, pp. 152–6 (1961). Rosen reinforces Birkenmajer's thesis that Copernicus used the *Epitome of Ptolemy's Syntaxis* by Regiomontanus (ed. Princeps, Venice, 1496) as the principal source of his information about Muslim astronomy in general and Biṭrūjī in particular.
[30] When the relative sizes of the deferent and epicycle are known for a given planet, so too is the ratio of the maximum to the minimum distance from the Earth attained by the planet in its orbit. The scale of "Ptolemaic" distances is fixed by supposing that the planets are so nested that the maximum distance of one is the minimum distance of the planet next above. The unit of distance was determined in terms of the lunar distance. Apart from the entirely arbitrary nature of the hypothesis that the Universe is exactly filled by the planetary spheres, there is a weakness in the arguments leading to the Ptolemaic *order* of the planets. Cf. *De rev.* I. 10.

large, but certainly not infinite. Copernicus, however, was himself conservative in regard to the potentially infinite universe, as in so many other respects:

> First it is to be remarked that the universe (*mundus*) is spherical (*globus*), whether because this is of all shapes the most perfect, being an integral whole in need of no joins, or whether because this is the most capacious of figures, and therefore well suited to contain everything there is; or even because the separate parts of the world—the Sun, Moon, stars—are seen to be of such a shape; or because everything in the universe tends to be limited by this form—as is apparent with drops of water and other liquid bodies when they decide their own limits. . . .[31]

This was a conventional enough expression of a conventional view.[32] The decision as between a finite and an infinite universe was yet again one for the natural philosopher, rather than for the astronomer, as Copernicus pointed out with some resignation,[33] although in the same place he registered some surprise that a void beyond a finite universe should by some people have been conceived capable of holding together, as it were, a star sphere turning round with a violent motion.[34]

One of the besetting intellectual sins of the scholastics, that of forcing knowledge into neat categories, might not always have produced very profound results, but it is useful for us to examine occasionally the trees of knowledge delineated in the process. As always, Aristotle is lurking in the shadows, even where Boethius is ostensibly the centre of diffusion of his ideas. The three sorts of science, according to Aristotle, were theoretical (dealing with knowledge of mathematics, physics, and theology), practical (dealing with conduct) and productive (making beautiful or useful things). Mathematics was to be subdivided into the quadrivium of arithmetic, music, geometry, and astronomy—with which we are here concerned. Physics, or natural philosophy, was thus both on a different branch and at a different level, while a subject like calendar reform, to which Roger Bacon applied such astronomical knowledge as he had, would have been yet further removed, along with such "practical" sciences as astrology.[35] Suitably separated thus, astronomy became a subject for university study by most arts men, whereas astrology was rarely studied explicitly as part of the university curriculum, although, one suspects, it was studied assiduously on the side. It was never quite respectable to say that the *purpose* of

---

[31] *De rev.* I. 1.

[32] It was not always clear, however, whether the spherical world was embedded in an infinite void or whether space somehow ended with the spheres. Cf. Oresme, ed. Menut and Denomy, pp. 368–9, and note Oresme's remark that to contradict the proposition that the world actually moves with right motion through an imaginary infinite and motionless space was to maintain an article condemned in Paris. (The proposition would have God's power be limited. Copernicus here left God out of his arguments.)

[33] *De rev.* I. 8.

[34] But this he says in the course of demolishing an earlier argument, and he gives no indication of his own views as to what is beyond the sphere of stars.

[35] On the requirements of the Church and the Lateran Council in relation to calendar reform, see *De rev.*, dedication to Pope Paul III.

studying astronomy was that one might apply it to astrology, but there is no doubt that this attitude was commonplace, especially in the fourteenth century and after, and no less after Copernicus than before him. Copernican astrology was slow to emerge, perhaps because most astrologers were feeble practitioners of mathematical astronomy, or perhaps because the very essence of astrology was to be ancient and mysterious.

Copernicus was no astrologer, but he uses astrological imagery, perhaps unwittingly, when he adapts the common renaissance theme of an equivalence of the Sun, God, the king, and a throne. He quotes Hermes on the Sun as the visible God, and shows a reverence for that Pythagorean tradition which would have the Sun in the centre because it has perfection and is the source of our light and life. This would have been out of place in the Middle Ages. A medieval writer, following Aristotle, might well have referred to Pythagoras, or to his pupil Herakleides, as we have seen, but he would not have done so with the abandon that became so common after the revival of interest in Platonism in the later fifteenth century.[36] Copernicus followed no author's example so sincerely as he followed Ptolemy's,[37] but this was not high fashion in 1539, when Rheticus in the *Narratio Prima* averred that his teacher had followed "Plato and the Pythagoreans, the greatest mathematicians of that divine age".[38] In 1542, when Copernicus wrote the dedication to his greatest work, he quoted Pliny, because Pliny referred to Philolaos (in connection with the Earth's movement round a central fire), and to Herakleides and Ekphantos (in connection with the Earth's rotation). This is purely and simply adulation regardless of rational argument, and adulation of a kind not easily found in the supposedly authoritarian Middle Ages.

Having said all this, I have to admit that, in his Pythagoreanism, Copernicus was doing little more than indulge in a tiresome literary conceit of a significance easily overrated. It is not my brief to ask how solid was the achievement beneath the literary veneer, but merely to contrast the medieval approach to astronomy with his. It is easy to argue for continuity, and to maintain that the Copernican revolution is an illusion since precious few new observations were involved in it, and especially since the theoretical structure of the new scheme was of the same kind as that of the old. Ptolemy was known to the Middle Ages, in the sense that his principal works were available, but he was generally unheeded in one important particular, that is to say, in his techniques for applying to observational data a well-controlled hypothetico-deductive method, finally leading to a planetary model and its parameters. Admittedly

---

[36] Copernicus was an assistant to the Platonist Domenico Maria Novara (*ca.* 1500: see the *Narratio Prima* of Rheticus, trans. Rosen, p. 111) and pupil of a leading humanist, Albert Brudzewski.

[37] Ptolemy comes in for some qualified but genuine praise in *De rev.* I. Prol.

[38] Rosen, *op. cit.*, p. 147.

Copernicus' creative powers ranged less freely than Ptolemy's.[39] The second order epicycle was a far more conservative construct than was Ptolemy's equant;[40] and yet each was the result of thought processes which were almost unknown in medieval Europe, where epitomes of the *Almagest* tended to give only Ptolemy's conclusions. Medieval astronomy was a vigorous activity, but the vigour of its practitioners was directed at making it more comprehensible, more easily applied to practical ends, more concisely expressed. Truth, once established, was not to be tampered with. In the words of a question asked by Dante:

> For what fruit would he bear who should demonstrate once more some theorem of Euclid; who should strive to expound a new felicity, which Aristotle has already expounded; who should undertake again the apology of old age, which Cicero has pleaded? Naught at all, but rather would such wearisome superfluity provoke disgust.[41]

Copernicus managed to look back beyond the Middle Ages, but he could not shake himself completely free from its influence. Occasionally he made use of a consistent sexagesimal system, as opposed to a mixed system (such as we use when taking zodiacal signs of 30°). But one way of representing the Alfonsine tables had been similarly consistent long before. He occasionally changed to a decimal notation; but again there were several medieval precedents. Copernicus is often, and not unreasonably, given credit for his explanation of the slow apparent drift of the equinoxes as a consequence of the movement of the Earth's axis, rather than as a movement of the sphere of fixed stars. But he was not able to appreciate the erroneous character of the prevailing medieval belief in what was known as trepidation, an irregularity in the movement of the equinoxes which was accounted for by a complex of circular movements in three dimensions. Copernicus adopted a model for which he clearly borrowed from the traditional models usually associated with the names of Thābit and Arzachel. There are occasions when it seems to be impossible to decide whether Copernicus was aware of his precursors or not. His secondary epicycle for

---

[39] Copernicus, of course, had the inestimable advantage of having Ptolemy's example to follow. He deliberately kept as close as possible to the conventional Ptolemaic data for the solar distance, and—at least for much of the time—to the Ptolemaic ratio 13/5 for the ratio of the diameters of the Earth's shadow (at the maximum lunar distance) and the Moon. See the masterly study of Copernicus' dependence on Ptolemy by Otto Neugebauer, "On the planetary theory of Copernicus", *Vistas in Astronomy*, Ed. A. Beer, vol. 10, pp. 89–103 (1968), especially p. 101. Even in the theory of planetary latitudes, where there are distinct advantages in the heliocentric approach, Copernicus followed the Ptolemaic method, in which the deferent planes of the planets are taken to pass through the Earth, thus wrongly introducing into the reckoning of latitudes the two eccentricity vectors (of the Earth and of the planet). Cf. Neugebauer, *op. cit.*, p. 103.

[40] There is a sense in which Copernicus preserved the equant. Both models are virtually equivalent, geometrically speaking, and Copernicus clearly aimed at making the discrepancy as small as possible. The planetary longitudes predicted in the two systems should all differ by considerably less than a minute of arc. See Neugebauer, *op. cit.*, pp. 92–95. Kepler explicitly reintroduced the equant into a new heliocentric model, and in doing so earned important dividends, for he was thereby led first to an oval orbit for Mars, and eventually to his elliptical orbits.

[41] *De monarchia*, I. i; trans. by P. H. Wicksteed, pp. 127–8, London, 1904.

the lunar motion had been previously used by Ibn ash-Shāṭir of Damascus, a century before Copernicus was born.[42] Nāsir ad-Dīn aṭ-Ṭūsī, who worked at Marāgha and who died almost exactly two centuries before the birth of Copernicus, devised a geometrical construction for obtaining a linear motion from two circular motions, a construction which was also to be adopted by Copernicus.[43] Even if a day comes when, at last, it proves possible to trace a connection between the two Muslim astronomers and Copernicus, his debt to the Middle Ages is never likely to seem greater than his debt to Ptolemy; the very structure of *De revolutionibus*, let alone its detail, reflects that of *Almagest*. This perhaps explains in part why for so many historians, astronomy was a sort of Rip Van Winkle, sleeping soundly from late antiquity, until it was roused in 1543. (It is perhaps only a matter of time before someone writes a story in which Copernicus was a reincarnation of Ptolemy.) This sort of history is all very well, as long as it is recognized for what it is, namely the pure propaganda of humanism. The plain truth is that astronomy in the Middle Ages, like the ordinary people of the Kaatskill Mountains, was for most of the time very much awake.

---

[42] See Victor Roberts, "The planetary theory of Ibn al-Shāṭir: latitudes of the planets", *Isis*, vol. lxvii, pp. 208–19 (1966), and earlier papers in the series there listed.

[43] See Neugebauer, *The Exact Sciences of Antiquity*, pp. 203–4, 207, Providence, 1957.

INDEX

Names are included in forms familiar in the English-speaking world. As a rule Christian names from before the Seventeenth Century are included: before the end of the Middle Ages they are generally in first place. Notes are indexed when they contain a significant addition to the main text.

Aaboe, A. 32, 34, 361
Aarhus 223, 228
Abelard 126
Abraham Judaeus 108
Abū Ma'shar, see Albumasar
accelerated motion 288-9, 308-9, 320-3
accuracy in choice of gear ratios 184
action, transmission of natural 252
active powers and the elements 277
Adam 112
Adam Marsh 119, 131, 261
Adam Woodham 240-1, 363
Adelard of Bath 63, 100, 251
Aegidius, see Giles, Peter
Aeneid 205-7
Agathodemon 253
Agrippa, Cornelius 375, 393-4
Ahaz, the stairway or dial of 38, 387
Ailly, see Pierre
air, motions of 301
al-Ashraf, Sultan of Damascus 162, 181
al-Baghdādī 337
al-Battānī 108, 407
al-Bīrūnī 57, 136, 66, 153-4
al-Bitrūjī 261, 271-2, 407
al-Farghānī 410
al-Hāshımī 100
al-Kāshī 337
al-Khwārizmī 100, 148, 407
al-Kindī 63, 103, 122, 234, 253
al-Zarqāli see Ibn az-Zarqellu
Albert of Brudzevo 335
Albertus Magnus 69, 272, 276, 308, 350, 382
albion 408
Albumasar 59, 62, 62-73, 97-8, 99-100, 103-5, 112, 250, 251-8, 263, 269
alchemy 262-3

Alexander Neckham 105
Alexander of Aphrodisias 284, 285
Alexander the Great, era of 107
Alexander VI, pope 81
Alexandrian astronomy 54
Alfonsine tables 407, 413
Alfonsine tables in England 327-59
Alfonso X 71, 100, 320, 327, 347-8
Alfraganus 195
Algorismus (John Killingworth) 344
Alhazen, see Ibn al-Haitham
'Alī b. Khalaf 407
alignments of megaliths, astronomical, 11-20 *passim*
Allaeus, F. 113
Allen, P. S. 375-6, 382
Almagest 34, 57-8
Almagest, epitomes of 413
Almagest, translations of 406
almanac (various senses) 320-1, 331, 337, 344
almanac-makers and astrology 297
almucantars 213-16
Altino 207
Ambrose 120
Amen-Ra, Temple of (at Karnak) 26, 31
America and the Deluge 103
Ammisaduqa, Venus tablets of 40, 42
Analemma (Ptolemy's) 56-7
anaphoric clock, see Vitruvius; Salzburg; Vosges
Ancona, public clock in 161
Andalo di Negro 148, 328
angelic hierarchies 406
angelic intelligences 296, 304-5
angelology 303-4
angels and causation 274
angels and light 121

Anglo, S. 399
animism (Kepler's) 297-8
Annianus 112
Anselm 127
Antichrist 68-9, 76, 79, 80, 82, 205
Antikythera, geared device from 136, 153
Apian, Peter 393
apogee of planet (astrology) 77-8
apogee, solar, as indicating era of Creation 109-10
Apollo 50
Apollo (asteroid) 47
Apollonius 56, 122
Apollonius (pseudo-) 258
Aquinas, Thomas 192, 234-6, 272-80, 279, 287, 296-7, 312, 403
Aratus 179
archaeo-psychogenetics 44
Archer, T. A. 241
Archimedes 56, 136, 301, 315
Argonauts 115
Aristarchus 56, 96, 409
Aristophanes 246
Aristotelian cosmology 57, 404 (Copernicus)
Aristotle 5, 119-33 passim, 188, 192, 233-41, 243-98, 299-309, 311-23, 401-14
armourers as clockmakers 183
Armstrong, C. A. J. 377
Arnold of Villanova 80, 82
Arsenal at Venice 209
Artemis 50
Ashenden, see John
aspects (astrological) 283, 292
Assyrian astronomy 32
astrarium, see Dondi, Giovanni de'
astrolabe 120, 138-9, 211-31, 375, 388-90, 407-8
astrolabe dial 157, 171, 178-82
astrolabe, early writers on 178-9
astrolabe, geared 154, 167-9
astrolabe, history of 216-18
astrolabe, uses of 218-20
astrolabes, universal 407-8
astrologers, Frederick II's 163
astrological prognostication 393-4
astrology 33, 34, 51, 56, 58-89, 348, 393, 394, 412
astrology, see also influence
astrology and chronology 114
astrology and innovation 278
astrology, attack on 293
astrology, Indian 200
astrology, meteorological 293
astrology, Oresme on 307-9

astronaut gods 27
astronomy, prehistoric 1-35
astronomy, Ptolemaic 304-5
Atkinson, R. J. C. 29-31
atomism and celestial influence 297
attraction (of planets for elements) 293-4
Aubrey holes at Stonehenge 11-13
Augustine of Hippo 43, 61, 93, 100, 101, 105, 191, 236, 266-7, 274-6, 402-3
Aulus Gellius 250
Autolycos of Pitane 314
automata 160-1, 171
aux 143, 261-2, 266, 321, 338
aux, see also tables, astronomical
aux positions (planets) 149-50
Avebury, prehistoric monument at 15
Aveni, A. F. 25
Averroes 234, 278, 286-7, 290, 303-5, 311-12
Avicenna 125-6, 133, 234, 308
axis of the Earth, shift in direction of 37-8
Azarchel, see Ibn az-Zargellu
azimuth lines (astrolabe) 214
Aztec calendar 52

Babylonian astronomy 32, 34, 49, 53-8 passim
Babylonian Talmud 110
Bacon, see also Roger
Bacon, Francis 132-3
barbarian subtleties 363
Bardesanes 105
Barguet, P. 31
Barker, Anthony 382
Bartholomaeus Anglicus 105
Bartholomew of Messina 246
Barzon, A. 136
Basil 95, 402
Baskerville, Hannibal 387
Bastulos 217
Batecombe 337-42
Batteux, C. 2
Batty, R. 390
Baur, L. 263
Beast, Number of 68, 73
Beatrice 188-92, 195
Beaumont and Fletcher 114
Beckmann, J. 163, 174
Bede, Venerable 68, 82, 100, 108, 402
Bedini, S. A. 166
Beeson, C. F. C. 173, 176
Belgrano, L. T. 177
Bellamy, H. S. 50
Beltane fires 52
Belyaev, E. A. 173
Bembo, Pietro 373

Index

Benedictines at Oxford 368-70
benefactors in science, *see* patronage
Benger, Richard 385
Berkeley, G. 21
Bernard Sylvester 259
Berosus 96-7, 98-9
Bild, Veit 398
Bishop, J. 298
Black Death 71, 336, 361-71
Bladud the Druid 27
Boeckh, A. 9
Boethius 248, 402, 411
Bolingbroke, H. St. J. Viscount 6
Boll, F. 248
bombardment of Earth 47
Bonaventure 125, 234, 237
Bosco, U. 201
Bouché-Leclercq, A. 105, 110, 254
Boulainvilliers, H. de. 117
Bourbon, Nicholas 397, 399
Boyle, R. 298
Bradwardine, Thomas 237, 238-9, 337-42, 363, 241, 313
Brahe, Tycho 16
Bridges, J. H. 130, 271
Brillat-Savarin, J.-A. 5
Brinknell, Thomas 384
Briquet, C.-M. 388
Brook, William 385
Bruckner, J. J. 2
Brudzewski, Albert 412
Bruno Latini 190
Bruno, Giordano 40
Bullough, D. 179
Bunting, H. 114
Burckhardt *see* Spalatin, Georg
Burckhardt, J. 8
Buridan, John 283-8, 290, 301
Burley, Walter 238
Burnett, C. 257, 258
Burtt, E. A. 7
Bury St Edmund's abbey, water clock at 174

Caithness, fan-like rows of megaliths 17, 23
calendar differences (Julian, Gregorian) 211
calendar scale (astrolabe) 212
calendar, Arabic and Syrian 191
calendar, Egyptian 32-3
calendar, Julian 112-13
calendar, prehistoric 15, 24
Callippic cycle 94
Callippus 93
Calvisius, Seth 113-14
Campanella 297

Campanus of Novara 165, 408
Campeggio, Cardinal 379
Can Grande della Scala 202
Canobic inscription 340
Canterbury, clock at 176
Cantor, G. 233
Capelle, W. 246
Cardano, Girolamo 59, 80, 393
Cardinal College, Oxford 380, 382, 384
Carmelites at Oxford 368-70
Casartwlli, L. C. 193
Cassiodorus 288-9
Catherine, of Aragon 381
Catto, J. 233
casual influences from above 273-4
causation, *see also* angels
causes of celestial motions 306-7
causes, astrological 284
causes, efficient 272-3
Cecco d'Ascoli 62, 70, 79
celestial and sub-lunar (Aristotle) 124
celestial regions (versus sub-lunar) 299-309
Censorinus 93, 96
centre (of Universe) 302-3
Chalcidius 141, 260, 402
Chaldaean astrology 247
Chaldaean religion 68, 80
Charlemagne 79, 173
Charles V. emperor 175, 398
Charles V. king of France 290, 335, 348
Chaucer, Geoffrey 173, 214-15
Chauntecleer 173
cherubim 195
Chinese astronomy 33
Chiron 47
Christ, *see also* Jesus
Christ Church clock, Oxford 395
Christ, horoscope of 61, 69, 78-9, 82
Christian theology 56
chronology of the world 91-117
chronology, astrological 72
chronology, Velikovskian 42
Chrysippus 245
churches and sects, fortunes of 59-89
Cicero 97, 135, 179, 192, 245, 247, 249-50, 252, 382, 413
Cistercian Rule, machinery, and clocks 172-4
Cistercians at Oxford 368-70
Clagett, M. 287-90, 314
Clarke, S. 294-5
Claymond, Thomas 382
Cleanthes 245
Clement of Alexandria 402
Clement VI, pope 73, 75

Clement, John 378
Cleomedes 110
clepsydra 171-5, 179
Clerke, A. 353
climate in the ancient world 31
climates *see* plates (astrolabe)
clock (analogy with Universe) 294-5
clockmaking in England 399
clocks 395
clocks, digital 323
clocks, the first mechanical 171-86
Clube, V. 45-52
Cluniac rule and the clock 174
Cole, Humphrey 387
Collingwood, R. G. 1, 3, 7
Cologne, University of 375
Coloribus, John de. 384
Columbus, Christopher 114
Combach, I. 130
comets 37-52 *passim*, 262-3, 293-4, 308, 393
commensurability of motions 307-8
Commentator, *see* Averroes
computation and planetary geometry 56
computus 111
configuration doctrine, *see* graphs
conjunction horoscopes 67
conjunctions (major, minor, etc.) 63-5, 67, 103, 204
conjunctions of Jupiter, Saturn, Mars 59-89
Conrad of Fabaria 162
Conrad, son of Frederick II 162
Constance, Council of 61
conventionalism 124
Coopland, G. W. 308
Coote, C. H. 351, 393
Cope, Sir Walter 392
Copernicus, Nicolaus 107, 239, 311, 335, 373-4, 401-14 *passim*
Corpus Christi College, Oxford 373, 380-99 *passim*
Courtenay, W. J. 362, 364
Cranmer, Thomas 393, 397
craters in the solar system generally 47-8
Creation of the World 91-117 *passim*, 304
creation versus conservation 237
Creation, season of 197-201
Creation, six days of 131-3
Crecy, battle of 73
Critias 99
Crombie, A. C. 121, 124
Cromwell, Thomas 397, 398
Crowther, J. G. 8-9
crystalline heaven 196
Cudworth, R. 6

cup and ring markings 15
cup and ring marks 51
curriculum, medieval 132
Cyprian 111
Cyrus 92-3

Dales, R. C. 131-3, 235, 261, 263-6
Danfrie, Philippe 391
Daniel of Morley 260
Däniken, E. von 27
Dante 127, 187-209, 280-3, 288, 396, 403, 406, 413
De mundo (ps. Aristotle) 246-7
death rates during the plague 364
decimal arrangement of tables 413
Delambre, J.-B. J. 35, 56
Delhaye, P. 236
Deluge, *see* Flood
demons 296-7
demons and the Moon 277
Demosthenes 171
Denny, Sir Anthony 389, 395
density, relative 301
determinism, human 250
dial of Ahaz 387
dial. verses written for 384, 386
dials 395, 398, 408
dials made at Oxford (Kratzer) 385-91
dials, treatise on 388-9
Diaries, Astronomical (Babylonian) 32
Dicks, H. A. 93
Dinteville, Jean de. 392-3
Diodorus 97
Diodorus of Sicily 49
Dionysius (pseudo-) 120-1, 188-97 *passim*
Dionysius the Areopagite 188
Dis 192, 194
distances of the Ptolemaic spheres 195
distances, planetary 410
divination 249-50
Dodgson, C. 389, 395
doge of Venice 208
Dogon tribe's astronomy 27
Dominicans at Oxford 368-70
Dondi, Giacomo de' 135, 149, 165
Dondi, Giovanni de' 135, 136, 152-4, 165-6, 175
Drachmann, A. G. 158-60, 178
Dresden Codex 33
Dreyer, J. L. E. 148, 329, 330-2, 335-6, 409
Dronke, P. 259
Drover, C. B. 173-4, 177
Droysen, J. G. 9
druids 21

Duhem, P. 7, 87, 124, 328
Duke, E. 13-14
Dumbleton, *see* John
Duns Scotus, John 235, 236, 276, 367
Dunstable Priory, clock at 176
Dürer Albrecht 376-7, 390-2
Dux and *Il Veltro* 203

Earth, core of 47-8
Earth, motion of 408-9
East, William 385
Easter, computation of 94-5
Easter, determination of 111
eclipse cycles 52
eclipse observations 328
eclipse prediction, prehistoric 12-14, 17-18
eclipse predictions 335, 387
eclipses and chronology 114
ecliptic ring (astrolabe) 213
economy, mechanical 153
Ecphantus 412
Eddy, J. A. 24
educational reform at Oxford 382
Edward Upton 363
Egyptian astronomy 26, 32, 55
Egyptian chronology 38, 50-1
Egyptian religion 68
Egyptian years 97
eighth sphere, *see* also trepidation
eighth sphere, motion of 98, 198, 317-18
Einstein, A. 39
Eisler, R. 377
electrical influence 286
elements, Aristotelian 236, 246-8
elements, motions of 300
ellipses in prehistoric monuments 22-3
Ely Abbey, clock at 176
emanationism 253
Emden, A. B. 361, 371
Empyrean, 187-97 *passim*
Empyreum 280
Encke's comet 48, 50, 52
Engelbert 246
Epicurus 92
Epigenes of Byzantium 97
Epping, J. 54
equatoria 331, 389, 407-8
equinoxes, precession of 57
eras, various 91-117 *passim* 328
Erasmus 121, 374-6, 381-3, 383
escapement (clock) 185-6
escapement, mechanical 161
Eschuid, *see* John Ashenden
Establishment, the scientific 37-45 *passim*

eternity 92, 233-41, 304
eternity of God 126
eternity of the world 102-3, 132
ethical analogy with physical force 281-2
Etienne Tempier 234, 367
Euclid 122, 413
Euctemon 93
Eudoxus 55-6, 179, 273, 314, 402, 404, 407
Eudoxus, sphere of 115
Eusebius 71, 100, 111
Evans, L. 387
Evesham, *see* Walter of Odington
exaltation (astrology) 199-200
experiment 313
experiment, medieval 119-33
experiments, Grosseteste's 129
experimentum, notion of 124, 130-1
explanation, scientific 127, 128
extension, God's 238

Faber, Joh. Rodolphus 113
Fall, Mankind's 128
fardar 104
Farrington, B. 8
Favorinus 250
Feltro 203-4
Ferdinand of Aragon 79
Finico 296-7
final cause (Aristotle) 189
fire, motion of, 300 ff 313
Firmicus Maternus 66, 72, 113, 199, 373
Firmin of Bellavalle 75
First Cause (God) 279
Fitzralph, *see* Richard
Fitzsimon, H. 91
Flatten, H. 254
Flood 52, 66-9, 72, 75, 91-117 *passim*
force, laws of 249, 268
Forman, Simon 105, 108
forms, intension and remission of 313-14
Fowler, T. 382
Fox, Richard 373, 382
Francesco Piccolomini 107
Franciscans at Oxford 368-73
Frederick II, emperor 162, 181
Frege, G. 241
friction in the heavens 284, 306

Gaius Julius Hyginus, *see* Hyginus
Galaxy, expansion of our 46
Galvano Fiamma, Chronicle of 176
Garin, E. 8
Gatty, A. M. 171
Gayomart, the first man 105

gear trains, calculus of 316-17
gearing, planetary 135-70
gears, various forms of 184
Geber, *see* Jābir
Gemini, Thomas 387
Geminos 314, 373
genealogies, biblical 92-3, 101-2, 108
generation (Aristotle on) 276
Geoffrey of Meaux 75, 335, 347
Geography (Ptolemy) 57
geomancy and ancient trackways 27
George of Trebizond 406
Georgius Syncellus 95, 97-8
Gerard of Brussels 314-15
Gerard of Cremona 406
Gebert 164-5, 179
Gerbert's clock (anaphoric?) 164-5
Gerland 71
Geryon 205
Gianello Torriano 175
Gieben, S 131-3
Gilbert, William 270, 309
Giles of Rome 235, 272
Giles, Peter 375, 376
Gingerich, O. 352
Giovanni di Casali 314
Glareanus, Heinrich 375
Glass, B. 5
Glastonbury 27
globes 379, 393
Goclen, Conrad 379
God as clockmaker 294-5
Golden Chain 291-2, 294, 295
Gonell, William 378
Good Friday (Dante) 201
Gould's Belt 46
Graiff, C. A. 279
Grand Menhir Brisé 24, 30
Grant, E. 269
graphs 288, 313-14
great conjunctions 59-89
great conjunctions, *see also* conjunction
Great Pyramid 26
Great Year 64-5, 69, 72, 93, 96-8, 107-8
Greek language 373-5, 380-1
Greek philosphy 53, 55
Greek rationality and comets 49-50
Greenwich Palace 399
Gregory of Rimini 240
Gregory, pope 190
Greswell, E. 95
Greswell, R. 113-14
Grosseteste, *see* Robert
Guerrucius 148-9

Guevara, Anthony of 398
Guido Bonatti 396
Guido da Montefeltro 195, 204, 317
Guillaume, *see* William
Guillaume de Machaut 311, 315
Gunther, R. T. 167
Gyge, Margaret 378

Haber, F. 92
Hahm, D. E. 245
Hales, W. 98, 115
Halley's comet 48, 52
Haly 105
Hamberger, W. 164-5
Hampton Court clock 395-6, 399
hand, of a clock 141
Handy Tables (Ptolemy) 58
Harmonics (Ptolemy) 58
harmonies, musical 248
harmonies, Pythagorean 247
harmony 288-91, 309
Harris, John 389
Hartner, W. 18, 320
Hārun al-Rashīd 173
Hauser, F. 160, 178
Hawkins, G. S. 11-13, 26, 31, 39-40
heat, *see also* fire
heat in the heavens 284, 306
heat, terrestrial 255
Heath, T. 56
heavens as a cause 282-3
heavens, arrangement of 187-97
Hegel, G. F. W. 2, 58
Hegge, Robert 387
hell, *see* Inferno
Hellenistic astronomy 53-8 *passim*
Henry Bate 347
Henry of Ghent 235-6
Henry of Harclay 80, 240
Henry of Hesse, *see* Henry of Langenstein
Henry of Langenstein 77, 81, 286, 287, 289, 291, 293-5
Henry VI, king of England 348
Henry VII, king of England 348
Henry VIII, king of England 59, 376-99 *passim*
Heracleides 56, 410, 412
Heraclitus 245
Herde, Richard 378
Hereford, school at 119
heretics 194
Hermann of Carinthia 63, 250, 253, 255, 257-8
Hermann of Reichenau 214
Hermes Trismegistos 72, 106, 253, 258, 412
Hermetic philosophy 250-1

Index

421

Hero 174
Hervey, M. 392
Hesiod, 42, 50
Hessen, B. 2
Hezekiah, king 174
Hicetas 409
hierarchies in the heavens 187-97
Hildegard of Bingen 192
Hipparchus 35, 56-7, 98, 115, 178, 216, 402
Hipparchus (pseudo-) 70
Hippias 315
Hippocrates 255, 263
Hippolytus, bishop and martyr 111
Hird, Richard 381
Hissette, R. 238
historicism 3-4
historicism 4
Hobbes, T. 6
Holbein, Hans. jr. 374, 379, 389-90, 395
Holdsworth, R. 112
Homer 50
homocentric spheres 122, 303-4
Hooke, R. 298
horologium, *see* clock, planetarium
horoptrum (an instrument) 390
horoscopes 211
horse (astrolabe) 213
hour-lines (astrolabe) 221-2
hour-striking 186
Hourani, G. F. 200
hours, equal and unequal 174, 219-22
hours, Italian 385
houses, astrological 335
Howard, Catherine, queen of Henry VIII 395
Hoyle, F. 11-12
Hugh of St. Victor 191
Hughes, R. 192, 204
humanism 373-5
humanist propaganda 414
Humboldt, W. 9
Hume, David 1, 6
humours, bodily 396
Humphrey, duke of Gloucester 344, 348
Hussites 79
Hutchinson, K. 294
hydraulic clock drives 160-3, 171-7
Hyginus 179
Hyperboreans, temple of the 24

Ibn al-Haitham 122, 268
Ibn al-Majdī 337
Ibn ash-Shāṭ 414
Ibn az-Zarqellu 148, 150, 272, 316, 329-30, 407, 413

Ibn Tibbon, *see* Profatius
Ibn Yūnus 337
idea (Lovejoy) 4
ideas, history of 1-9
Iliad 38
impetus (Oresme versus Buridan) 307
incipits of Alfonsine works 354-8
Indian chronology 98-100
Indian trigonometry 56
Indiction cycle (15 years) 96
Indo-Persian eras 104
inferiority complexes, historians' 124-5
Inferno (Dante) 187-209 *passim*
infinity 233-41, 311
infinity, Grosseteste on 122-4
influence, celestial 243-98
influence, harmonic 288-90
influence, laws of celestial 307-8
influence, physical (in astrology) 110
Innocent VIII, pope 81
instrument making 373-99 *passim*
instruments, astronomical 369-70
intellect, agent 120
intelligence, divine 253
intelligences 290-1
iron, use in clocks 162, 182
Isaac ben Sid 327, 328
Isidore of Seville 53, 100, 108, 191, 402
Islam 68, 73

Jābir ibn Afflaḥ 407
Jacob's staff 388
Jacobus Aneglus 75
Jean, *see also* John
Jean de Murs 75, 80
Jean Fusoris 369
Jean of Roquetaillade 76
Jehan of Bruges 76
Jehuda ben Moses Cohen 327
Jensen, C. 337
Jerome 71
Jerome Torrella 79
Jesus, name of 60
Jews, and great conjunctions 79
Joachimism 76
Joachimist oracles 74
Joan, princess of Wales 348
Johannes Lichtenberger 79
John, *see also* Buridan, Duns, Wyclif
John Argyropulos 246
John Ashenden 70-5, 97, 101-7, 105-6, 112, 113, 199, 348
John Baconthorpe 239-41, 362-3, 367
John Buridan 303-4, 409

422 *Stars, Minds and Fate*

John Chilmark 363
John Damascene, St. 250
John Dumbleton 313, 363
John Ergum 369-70
John Holbrook 334, 348
John Killingworth 320, 343-6
John Langton 348
John of Cremona 328
John of Gaddesdon 367
John of Gaunt 348
John of Hertford 186
John of Holland 317
John of Jandun 272
John of Lignres 320-1, 329-35
John of Maryns 186
John of Messina 328
John of Murs 320, 329, 331, 342-3, 348
John of Reading 87
John of Roquetaillade 106
John of Sacrobosco, *see* Sacrobosco
John of Saxony 330, 332, 336
John of Seville 63, 250, 260
John of Speyer 336
John Pecham 270-1
John Philoponos 216, 234
John Scotus Erigena 188-9
John Somer 347, 348
John Stacy 369
John Stokesley 369
Jones, C. W. 108, 111
Jonson, Ben. 114
Jordanus Nemorarius 301
Josephus 92, 100, 114
Journal for the History of Ideas 4
jubilees 97
Judaism 68
Julian Period 95-6
Julius Firmicus Maternus, *see* Firmicus
Juno 190
Jupiter (planet and god) 37-8, 50
Juipiter and religion 67-8

kalām 234
kalpa 98
Kant, I. 125, 127
Karpinski, L. C. 344
Kauffeldt, A. 408
Keith, A. B. 193
Kendall, D. G. 32
Kennedy, E. S. 66-7, 103-4
Kent, A. 173
Kepler, Johannes 57, 84, 109, 111, 114, 309
kinematics, astronomical 311-23
kinematics, celestial 311-23

kinematics, Mertonian 236, 363
King's College, Oxford 380
King, D. A. 337
Kintraw 23, 31
Klibansky, R. 141
Knock (Wigtownshire) 22
Knuuttila, S. 240
Kolde, T. 398
Kopenick, der Hauptmann von 9
Koyré, A. 7, 238-9
Krakatoa 43
Kratzer, Nicolaus 373-99
Kratzer's instruments depicted 392
Krupp, E. C. 21-7
Krusch, B. 94
Kugler, F. X. 54
Kuhn, T. S. 401
Kunitzsch, P. 223
Kynder, P. 114
Kynton, John 384

Lamb, H. H. 31
Lansberg, Philip 110
Laplace, P. S., marquis de 41
Lateran Council 411
latitudes of forms 313-14
Lattin, H. P. 136, 164-5, 179
Laud, William 60
Laurence Stoke 184-5
law courts, clepsydras in 171
Law, E. 395
law of motion, universality of 299-309 *passim*
Le Gentil, G. J. H. J. 57
Leff, G. 241
Lehtinen, A. I. 241
Leibniz, G. W. 237-8, 294
Leland, J. 184
Leland, John 175, 386, 387
Lemay, R. 62, 250-4
Leonardo da Vinci 185-6
Leopold of Austria 75-6
Leto 50
Levi ben Gerson 75
Leviathan 205
Lewis of Caerleon 344, 346-7, 348, 399
Lewis, D. H. 30
libraries 369-70
light 119-33 *passim*, 262-3, 265-6, 271-2, 280, 284
light and the heavens 196
light metaphysics 121, 262
Lilly, William 396
Linacre, Thomas 373
Lindberg, D. 122, 129-30

Index

Litt, T. 272, 274–7, 287
Little, A. G. 239, 269
Locmariaquer, menhir at 31
Loggan, David 385
logic 363, 367
Longomontanus 109
Loo, Andries de 392
Lorch, R. P. 406
Lorenzo de' Medici 81
Loth, O. 63
Louis XIV, king of France 392
Louvain, University of 381
Lovejoy, A. 4–5
Lucian 171
Lucifer 192, 194–6
Lucretius 49, 92
lumen and lux 260
lunar alignments, megalithic 15–20
lunar notations, prehistoric 18–19
lunar theory, Babylonian 54
Luther, Martin 296, 375, 383
Lutheran doctrines condemned 383–4
Lutheranism and the trade in instruments 392
Lydyat, T. 113–14
lynx 271

Maas, M. 375, 398–9
Maass, E. 180
Mach, E. 6, 7
MacKie, E. W. 23, 30–1
MacKinney, L. 136
Macnaughton, D. 55
Macrobius, 66, 72, 97, 113, 179, 199, 402
Macrobius, planetarium by 136
Madan, F. 390
Maddison, F. R. 135, 160, 166, 173, 174–5
Maggi, H. 171
magic 269, 296
magnetic compass 387, 398
magnetic influence 269
magnetism 288, 297, 302
magnetism and the heavens 88, 252, 275
magnetization of terrestrial rocks 47–8
Mahābhārata 98
Mahayūga 98
Mahomet 74, 76, 82
Maier, A. 237, 240–1, 276, 313
Maistov, L. E. 32
Maltwood, K. 27
Mander, C. van 396
Mandonnet, P. 279
mandrake 289
Manilius 55
Manitius, K. 34

manna from Venus 37
Manuel, F. 3
manuscripts, see MS
maps 392
Marguerite de Valois 397
mariner's astrolabe 211
married lecturers 382
Marrone, S. 119–33 passim
Mars, planet 38, 43, 52
Marsh, see Adam
Marshak, A. 11–20 passim
Martianus Capella 191, 260, 402
Mary, princess 381
Māshāllāh 63, 66, 69, 103, 200, 214
mathematics and the sciences 122–4
mathematics fostered by astronomy 407
Matteo Moreti 76
Matthew Paris 120
Maurach, G. 404
Maurbach, Carthusian monastery of 375, 388–9
Maya, fire god of the 52
Mayan astronomy 27, 33
Mazdean cosmogony 105
McEvoy, J. 119–33 passim 263
Mecca, Flight from 74
mediation of planets betw. heavens, earth 255–6
megalithic yard 15, 32
megaliths as portrayals of comets 51
Meghen, Peter 390
Melanchthon 80, 111
Menelaus 57
Mercier, R. 317
mercury clock, Alfonsine 174–5
Mercury, librarian of the gods 83
Mercury, motions of 316
Merlin 68, 70
Merton College, Oxford 131, 343, 348, 363–70 passim, 374, 381
Merton College, Oxford, clock at 176
Mesa Verde Sun temple 25
Mesoamerican astronomy 25
metals, formation of 271
meteoroids 46–52 passim
meteorology 246–8
meteorology, astrological 69, 72
meteoroscope 393
Methodius 71
Meton 93
Metonic cycle 93
Metonic cycle (19 years) 96
Metzger, H. 7
Meyerson, E. 7

Michael Scot 407
Michaud-Quantin, P. 267
Middleton, Alice 378
Milan, iron clock in 177
Milan, public clock in 161
millennarianism 74-5, 91-117 *passim*
Miller, W. 24
Minos 195
Møller Pedersen, K. 223
moment (fortieth part of an hour) 142
Mommsen, T. 7
monasticism and the clock 171-86
Moncetti, P. 280
Montezuma 25
Moody, E. A. 283, 284, 409
Moon, action of 243-98 *passim*
Moon, compared with Earth 308
Moon, new 101
Moon, phase of at creation 110, 112-13
Moore, E. 280
Moore, P. 45
moral analogy (Dante) 188, 197
Morbihan, Sea of 30
More, Sir Thomas 377-9, 383, 389
Morley, H. 375
mosaic at Torcello 204-5
Moses 402
Mosgrove, Thomas 380-1, 385
motions 'outside nature' 300
motions as causes 271-2, 280
motions, circular and linear 300, 304, 405-6
motions, compounded 405
motions, natural and forced 299-309
MS, *see also the list of incipits on pp. 354-8*
MS Berlin, Phillipicus 1830 179
MS Cambridge, University Library
  e.3.61 354
MS Cracow, Jagell 551 186
MS Erfurt, Amploniana, Q. 388 352
MS London, British Library, Arundel
  66 344
—, —, Cotton Titus B. i 397-8
—, —, Cotton Vit. B. xiv 397, 399
—, —, Egerton 889 333
—, —, Royal 12. D. 6 335
—, —, Royal 12. G. 1 399
—, —, Sloane 3124 333
—, —, Sloane 407 334
MS Milan, Ambrosiana, H. 75 35, 137
MS Munich, CLM 210 180
MS Oxford, Bodleian Library,
  Arch. B. c. 3 383
—, —, Ashmole 393 334
—, —, —, 802 108
—, —, —, 1522 175
—, —, Bodley 270b 174
—, —, —, 504 (MS H.) 390
—, —, —, 662 117
—, —, —, 790 334, 354
—, —, Can. Misc. 61 137
—, —, —, 190 117
—, —, —, 436 149
—, —, —, 501 350, 352
—, —, Digby 48 332
—, —, —, 57 352, 353
—, —, —, 68 346
—, —, —, 97 334
—, —, —, 114 333
—, —, —, 168 335
—, —, —, 176 335, 351
—, —, —, 178 351
—, —, —, 228 350
—, —, F. 1. 9 179
—, —, Laud Misc. 594 332, 334, 338
—, —, —, 644 150
—, —, —, 674 332
—, —, Rawlinson D 810 387
—, —, —, D. 1227 352
—, —, Sancroft 129 117
—, —, Savile 38 353
—, —, Twyne 21 386
—, —, —, 24 380, 383
MS Oxford, Corpus Christi College 152, 280, 380, 383
—, —, 152 (MS K.) 383-9
—, —, Phi. D. 2. 4 387
—, —, Phi. D. 2. 5 387
—, —, Phi. D. 3. 8 387
MS Oxford, Magdalen College, 182 351-3
MS Oxford, Oriel College, 23 101-2, 104, 106
MS Oxford, University Archives,
  Reg. H. 7 383
MS Paris, Bibl. Nat., Lat. 10117-26 354
—, —, —, 11560 174-5
MS Vatican, Lat. 5367 160
Mullen, W. 44
Müller, Johannes, *see* Regiomontanus
Müller, M. 42
Münster clock 158
Murano and glass 208
Museum of the History of Science, Oxford
  221, 387-8
music 58

Napier, Bill 45-52
Nāṣir ad-Dīn aṭ-Ṭūsī 414

Nastulos 217
nativities 247-8
nativities, *see also* horoscopes
natural philosophy at Oxford 366-7
natural philosophy influenced by Ptolemy 123
Nazca (Peru) 27
necromancy 62
Needham, J. 33, 269
Nemesius 65
Nenet 50
Netherlands visited by Kratzer 376-7
Neugebauer, O. 53-8, 100, 413-14
Newall, N. D. 47
Newark, Ohio, earthen mounds at 25
Newham, C. A. 14, 24
Newton, I. 57, 91-2, 115, 298
Newton, R. R. 33
Nicholas of Cusa 347, 409
Nicholas of Lynn 348
Nicholas of Sicily 246
Nicole Oresme 70, 81, 105, 107, 239
Niero, A. 204
Nixon, R. 31
Norrice, William 379
North American Indian astronomy 24-5
North, Thomas 398
North American Indian astronomy 24-5
Norwich, clock-building at 177-8
Novara, Domenico Maria 412
Nun 50

Obermann, H. A. 237
occult virtues 275-6, 286-7
Ockham, William of 237, 239, 276, 283, 363, 367
Oinopides 93, 96
Opus quarundam rotarum mirabilium 135-70
order and proximity to God 247
ore prospecting 395
Oresme, Nicole 61, 287, 288-92, 301-9, 313-14, 409-10
orientations, shared 26
Origen 275, 402
Orosius 71
Osiander, Andreas 410
Osma Cathedral, MS from 181-2
Østerby, O. 223
Otto Henry, prince of the Palatinate 396
Oursian, Nicholas 395
ovals in astronomy 316-17
Ovid 107
Oxford institutions compared 368-70
Oxford science in the late middle ages 361-71
Oxford tables of 1348 320-2, 337-42

Oxford University 234-5, 361-71
Oxford, astrolabe plate for 229
Oxford, astronomical tables for 331-59

Pächt, O. 389-90
Pacificus, Archdeacon of Verona 181
Paget, William 396-7
Paine, T. 92
Panofsky, E. 377
Pappus, 315
Paradiso (Dante) 187-209 *passim*
parameters of planetary theories 146-70 *passim*
Paris, University of 234-5, 283
Parisian tables 327-59 *passim*
Parker, R. A. 32
Parma, humiliated Frederick II 163
Parron, William 377
Paston(?), Dr. 399
Pate, Richard 382
Pater, W. 391
Paterson, A. M. 40
patronage in science 369-70, 374-5
Paul III, pope 411
Paul of Venice 367
Paul, bishop of Altino 207
Pedersen, O. 34
Pedretti, C. 185
Pelster, F. 239, 261
periodization of World History 98
perspective, *see* light
Petavius, D. 91, 113
Peter of Auvergne 286-7
Peter of Dacia 408
Petit Ménec 30
Petro d'Abano 272
Petrus Peregrinus 269-70
Peuerbach, Georg 394, 401
Phaeton 49
Philip Aulyn(?) 333
Philippe le Bel, owner of silver clock 182
Philo Judaeus 92, 248
Philolaus 93, 96, 409, 412
Philosopher, *see* Aristotle
philosophy and the history of science 6-7
philosophy, history of 2
Pico della Mirandola 59, 78, 82-3, 107, 223, 297, 394
Pierre d'Ailly 61, 78, 82
Pietro d'Abano 69-70, 107, 108
Piggot, S. 29
Pinches, T. G. 54
Pingree, D. 66, 97, 99, 100, 104, 199-200, 251, 253
Pirckheimer, Willibald 379, 390-2

place 299
places, proper 302
Placet, F. 103
Plague 71
planetaria 135-70
planetarium, given to Frederick II 161-4
planetary influence 243-98 *passim*
planetesimals 46-52
planets 138-70 *passim*
planets, like the Earth 308
planets, mechanical representation of 135-70
planets, ordering by distance 147
planispheric astrolabe, *see* astrolabe
plants, astrology and the diversity of 271
plate (astrolabe) 211-14, 225-7
plate tectonics 48
Plato 52, 65, 96, 126, 141, 253, 402, 412
Pleiades 243
plenitude, principle of (Aristotle) 304
Plimpton 232 (Babylonian tablet) 15
Pliny 56, 382, 402, 412
Plotinus 249
Poitiers, battle of 73
Pole, Reginald 373
Policronicon 71
Polynesian navigation 30
Pömer, Hans 390
Pomponazzi, Pietro 60
Posidonius 246, 250
possibility and impossibility 238
Poulle, E. 352
Powell, Edward 384
Power, H. 109
power of God 233-41 *passim*
Prague tables 335
precession 122
precession, *see also* movement (8th sphere), trepidation
precession and chronology (Newton) 115
precession of the equinoxes 107
predestination 279-80
predestination debate 61
prediction, astrological 276-7
Price, D. J. 135, 159, 178, 183
Primum Mobile 189, 251-2, 272-3, 280, 306, 307
Priscian 250
Proclus 57, 171, 394
Procter, E. S. 327
Profatius 148
prophets 73
prophets, advent of 67
propitious times, doctrine of 60, 69
Prosdocimo de Beldomandi 347

Pruckner, H. 287, 292
Pruckner, Nicolaus 396-7
Ptah 50
Ptolemaic models for planetary motion 318-19
Ptolemy 34, 53-8, 72, 82, 122-3, 191, 216, 244, 248, 261, 278, 283, 303-5, 394, 401-14 *passim*
Purgatory (Dante) 193
putrefaction 277
Pythagoras 412
Pythagorean symmetries 190-1

quadrant, Profatius' new 388
quadrivium 411
qualities 307, 313
qualities, primary and secondary 284-5, 292-3
quantitative nature of bodies 299-300
Quiberon Bay, megalithic complex 23-4

Rabanus Maurus 82, 191
radiation, physical theories of 255
radices (in astronomical tables) 332-46 *passim*
radio-carbon dating 50-1
radix dates, cyclical 330
Ralph Strode 363
rationality of the planets 253-4
Ratisbon, library of St Emmerammus 180
Raymond of Marseilles 253, 259
Red Sea, parting of 37-8, 43
Rede, Reed, *see* William
Reeves, M. 77-8
Regiomontanus 62, 401, 407
rete (astrolabe) 213-16, 225-7, 231
retes (of stars and of hours) 157-8
Rheticus, Georg Joachim 109, 412
Rhetorius 200
Rhineland Beaker People 22
Riccioli, J.-B. 6
Richard Billingham 363
Richard Courtenay 369
Richard Feribrigge 363
Richard Fitzralph 240-1, 363
Richard Kilvington 240
Richard Kilwardby 367
Richard of Middleton 238, 367
Richard of St. Germain 162
Richard of Wallingford 136, 139, 153, 158, 160-1, 172, 175, 260, 315-16, 399, 407, 408
Richard of Wallingford, life 183
Richard Swineshead 313
Rinuccio Aretino 246
Robbins, F. E. 111
Robert de Turri, craftsman 177
Robert Grosseteste 102, 119-33, 235, 246, 260-5, 270-1, 367

Index

Robert Halifax 241
Robert Holcot 240, 241, 363
Robert the Lombard 331, 336
Roberts, V. 414
Robertson, J. D. 177
Robertus Anglicus 62, 176
Roger Bacon 67-9, 82-3, 112, 119-33 *passim*, 130, 267-9, 321, 367, 411
Roger of Hereford 95-6
Roger of Stoke, master clockmaker 177
Roper, John 384
Roper, William 377
Rose, L. E. 40
Rosen, E. 401, 406
Ross, H. M. 171
royal patronage of astrology 348-9
Royal Society of London 298
Rupescissa, *see* Jean of Roquetaillade
Ryle, G. 125

Saarinen, E. 240
Sacco, Bernardo 175
Sachs, A. 32, 42, 54
Sacrobosco 62, 70, 105, 120, 176, 394, 404, 406
Salamanca tables 335
Salisbury market and the cathedral clock 173
Salzburg anaphoric clock fragments 178
Salzburg clock fragments 158
Santa Maria de Ripoll, clock of 173
saphea Arzachelis 407-8
Saros 96
Sarton, G. 53
Scaliger, J. J. 95-6, 108, 111, 113
Schiaparelli, G. 93
Schirmer, O. 320
schizophrenic history 312
Schramm, M. 141, 320, 373
Schröder, A. 398
science and philosophy of science 129
science, history of 1-9
scientific disciplines, eleven categories in the curriculum 365-6
Scotland 74
Scott, B. 173
Scott, G. 395
Scriptures, discrepancies between versions 92
sea, *see* tides
Seberger, Wolfgang and Paul 375
seismic stations (lunar) 48
Selve, Georges de 392
Seneca 250
Sennacherib 38
Sentences, Book of (Peter Lombard) 239, 272
Severus Sebokht 214

Seville, *see* Isidore, John
sexagesimal notation and arithmetic 330, 413
Shakespeare, W. 114
Shams al-Dīn al-Sūfī 337
Sharp, D. E. 125
Siger of Brabant 236, 279
Silbury Hill 31
Simon Bredon 337-42
Simplicius 97, 278, 287
Sindhind 100
Sirius, companion of 27
Small, R. 401
Smalley, B. 121
Smiley, C. H. 33
solar cycle (28 years) 96
sollertia, quickness of mind 131
solstice as time of Creation 112-13
Sothis period 97
Spalatin, Georg 398
Speier, Jakob von 62-3
Spengler, O. 9
St. Albans, *see* Richard of Wallingford
St. Albans, clock at 178
St. Albans, installation of abbots 186
St. Mary's church, Oxford 385
St. Pierre 30
Stabius, J. 391, 392
Steneck, N. H. 291-2, 295
Stock, B. 259
Stoics 66, 244-7, 249-50
Stoke, *see* Roger, Laurence (clockmakers)
Stonehenge 11-20 *passim*, 39-40
Stove, D. 39
Strabo 250
Strassmaier, J. N. 54
Strauchius, Aegidius 110
Straus, W. L. 5
Stukeley, W. 21
Sun, action of 243-98 *passim*
Suter, H. 100
Suttor, T. 276
Swineshead, *see* Richard
Sylvester, *see also* Gerbert
Sylvester II, pope 164
Sylvester's clock a sundial? 164
Synesius 57, 102

Tamil eclipse calculation 57
Tanner, T. 335
Tannery, P. 96
Tartar conquests 69
Teeple, J. 33
Telesphorus 77
Temkin, O. 5

Tempier, Etienne 105
Temple, R. K. G. 27
Teotihuacan 26
Thābit ibn Qurra 106, 317, 320, 407, 413
Thackeray, W. M. 2
Thausing, M. 377, 391
Theodoric of Freiberg 130
Theon 55
Theophilus, bishop of Antioch 92
Thiele, G. 179
Thierry of Chartres 259
Thom, A. S. 11-20, 22, 40
Thomas, see also Aquinas; Bradwardine
Thomas Buckingham 363
Thomas Cory 337
Thomas de la Mare 184
Thomas Pray 344
Thomas Southwell 369
Thomas, E. J. 193
Thompson, J. E. S. 33
Thomson, S. H. 263
Thorndike, L. 5, 75, 81, 105-6, 176, 289-90
Thureau-Dangin, F. 54
Thwaites and Reed, clockmakers 395
Tichenor, M. 337
Tideus 122
Titans 49, 50
Tizio of Siena 81
Toledan tables 327-8, 341, 407
Toltec calendar 52
Tomeo, Niccolo Leonico 373
Tommaso Buoninsegni 81
Torcello 204-9
Toulouse tables 329, 330
Trasillon, Anthony 379
Trinity, doctrine of 74-5
Trithemius 162
Troglodytes. Feast of the Repelling of 52
Trüdinger, K. 248
Tunstall, Cuthbert 373-4, 376
Turks and comets 393
Turrel, Pierre 106, 107
Tuscan cosmogony 105
Twyne, Brian 380, 380-1, 385, 386
Tycho Brahe 383
Tyler, William 390
Typhon 50

Universe, centre of 311
Universe, expansion of 46
Ussher, J. 91-2, 96

Velikovsky, I. 7, 27, 37-45, 50-1
Veltro, il 201-209

Venice and il Veltro 206-9
Venus Tables (Maya) 33
Venus, the planet 37-8, 43
Vettius Valens 200
Victoria (Frederick II's ill-fated camp) 163
Victorius of Aquitaine 95
Villard de Honnecourt 174
Vincent of Beauvais 71
Virgil 192, 197, 202, 206
Vitruvian anaphoric clock 178
Vitruvius 160, 171
Vives, Juan Luis (Ludovicus) 378-99 passim
Vogt, H. 56
void 304
Voltaire 3
Vosges anaphoric clock fragments 158, 178
Voulp, Vincent 379

Waetzoldt, W. 392
Walker, D. P. 81, 296-7
Wallingford, see Richard
Wallis, H. 399
Walter, see also Burley
Walter Burley 367
Walter of Odington 101, 107, 112-13, 366
Walther, Bernard 391
Warren, J. 57
water-clock makers in Cologne 173
Watkins, A. 27
Webering, D. 237
week of weeks 97
wheels, 'a device of remarkable wheels' 138-45
 (text & translation)
Whig history 8, 119, 297-9
White, L. Jr. 174, 160
Whitehead, A. N. 7
Wicksteed, P. H. 281, 413
Wiedemann, E. 160, 178
William, see also Ockham
William Gray 369
William Heytesbury 313, 363
William of Auvergne 105, 119-33 passim,
 125-7, 128, 289
William of Coches 235
William of Conches 254
William of Hirsau 176
William of Moerbeke 287
William of St. Cloud 347
William of St. Thierry 254
William Rede 73, 329, 332, 334-6, 348, 363, 369
winds and the tides 266
Windsor, Miles 380
Wittenberg, University of 375

Wolf, R. 375
Wolsey, Cardinal 376, 379-83
Wolter, A. 235
Woltmann, A. 392
women, education of 381
Wood, Anthony 380, 381-2
Woodhenge 22
world periods, Indian 104
world soul 245-6
World Soul and Holy Ghost 254
World Tree (Norse mythology) 50
World Year 66
World, age of 91-117

World, Aristotle's view of 311-12
World, horoscope of the 82
Wyclif, John 363, 367, 241

York, Austin friars 369-70
yuga 98

Zanetti, V. 208
Zeno 245
Zeno of Citium 245
Zeus 50
Zoroaster 70